Student Solutions Manual and Study Guide to Accompany

Understanding Intermediate Algebra
Third Edition

and

Understanding Algebra for College Students

Lewis Hirsch
Rutgers University

Arthur Goodman
The Queens College of the City University of New York

Prepared by
Steven Kahan
The Queens College of the City University of New York

WEST PUBLISHING COMPANY
Minneapolis/St.Paul   New York   Los Angeles   San Francisco

Production, Prepress, Printing and Binding by West Publishing Company.

 TEXT IS PRINTED ON 10% POST CONSUMER RECYCLED PAPER    PRINTED WITH **SOY INK**

COPYRIGHT © 1994 by WEST PUBLISHING CO.
610 Opperman Driv
P.O. Box 64526
St. Paul, MN 55164–0526

ISBN 0–314–03996–1

Contents

CHAPTER 1
THE FUNDAMENTAL CONCEPTS

Exercises 1.1

1. True   3. False   5. True   7. True

9. True   11. True   15. {7,8,9,10,11,12}   19. ∅

21. {41,43,47}   25. {0,6,12,18,24,...}

29. {1,2,3,4,6,9,12,18,36}   33. {0,1,2,3,4,5,6,9,12,15,18,21,24,27,30,33}

37. ∅   39. 2·3·11   41. 2·2·2·2·2·2·2   45. 7·13

Exercises 1.2

1. True   5. True   9. True   11. False

17. True   21. >, ≥, ≠

25.   27.

29.   33.

35.   37.

43.   47.

49. $\{x \mid -4 \le x < 9,\ x \in Z\}$

53. $\{x \mid 1 \le x \le 6,\ x \in Z\}$

57. $\{x \mid 1 \le x \le 6\}$

1

61.

$$\varnothing$$

63.

$$x = 0.674674\overline{674}$$
$$1000x = 674.674\overline{674}$$
$$1000x = 674.674\overline{674}$$
$$-\quad x = \quad 0.674\overline{674}$$
$$999x = 674$$
$$x = \frac{674}{999}$$

$$x = 0.9292\overline{92}$$
$$100x = 92.9292\overline{92}$$
$$100x = 92.9292\overline{92}$$
$$-x = 0.9292\overline{92}$$
$$99x = 92$$
$$x = \frac{92}{99}$$

Exercises 1.3

1. Associative property of addition

5. Distributive property

7. False

11. Multiplicative identity

13. False

15. Associative property of multiplication

19. False

21. Additive inverse property

25. Commutative property of addition

29. Distributive property

31. Closure property for multiplication of integers

33. False

35. We can derive one form of the distributive property from the other by using the commutative property of multiplication three times. For instance, suppose we know that $a \cdot (b + c) = a \cdot b + a \cdot c$. Since

$$a \cdot (b + c) = (b + c) \cdot a$$
$$a \cdot b = b \cdot a, \text{ and}$$
$$a \cdot c = c \cdot a$$

we find that $(b + c) \cdot a = b \cdot a + c \cdot a$. This is really sufficient, but we can make it look exactly like the second version given in the text by replacing b with a, c with b, and a with c.

36. There are no commutative or associative properties of subtraction. Examples illustrate that these "properties" are not always true:

$3 - 5 \neq 5 - 3$, so in general, $a - b \neq b - a$

$5 - (2 - 1) \neq (5 - 2) - 1$, so in general, $a - (b - c) \neq (a - b) - c$.

37. There is no commutative property of division, since $4 \div 2 \neq 2 \div 4$. Thus, in general, $a \div b \neq b \div a$. Similarly, $8 \div (4 \div 2) \neq (8 \div 4) \div 2$, so there is no associative property of division. That is, in general, $a \div (b \div c) \neq (a \div b) \div c$.

38. (1) additive identity

(2) distributive property

(4) additive inverse property

(5) associative property of addition

(6) additive inverse property

2

(7)   additive identity

Exercises 1.4

1. $+5$

3. -11

5. $+24$

7. $-3 - 4 - 5 = -7 - 5 = -12$

9. $-3 - 4(5) = -3 - 20 = -23$

11. $-3(-4)(-5) = 12(-5) = -60$

13. $3(-4 - 5) = 3(-9) = -27$

15. $8 - 4 \cdot 3 - 7 = 8 - 12 - 7 = -4 - 7 = -11$

17. $8 - (4 \cdot 3 - 7) = 8 - (12 - 7) = 8 - 5 = 3$

19. $(8 - 4)(3 - 7) = 4(-4) = -16$

23. $\dfrac{-5 - 11}{-9 + 4} = \dfrac{-16}{-5} = +\dfrac{16}{5}$

25. $\dfrac{-10 - 2 - 4}{-2} = \dfrac{-12 - 4}{-2} = \dfrac{-16}{-2} = +8$

27. $\dfrac{-10 - (2 - 4)}{-2} = \dfrac{-10 - (-2)}{-2} = \dfrac{-10 + 2}{-2} = \dfrac{-8}{-2} = +4$

29. $\dfrac{-10 - 2(-4)}{-2} = \dfrac{-10 + 8}{-2} = \dfrac{-2}{-2} = +1$

33. $8 - 3(5 - 1) = 8 - 3 \cdot 4 = 8 - 12 = -4$

37. $7 + 2[4 + 3(4 + 1)] = 7 + 2[4 + 3 \cdot 5]$
$$= 7 + 2[4 + 15]$$
$$= 7 + 2 \cdot 19$$
$$= 7 + 38 = 45$$

41. $8 - \dfrac{10}{-5} = 8 - (-2) = 8 + 2 = 10$

43. $8\left(\dfrac{10}{-5}\right) = 8(-2) = -16$

47. $\dfrac{-8 + 2}{4 - 6} - \dfrac{6 - 11}{-3 - 2} = \dfrac{-6}{-2} - \dfrac{-5}{-5} = (+3) - (+1) = +2$

51. $(-6)^2 = (-6)(-6) = +36$

55. $2(5)^2 = 2((5)(5)) = 2(25) = 50$

59. $-8 - 2(-4)^2 = -8 - 2(16) = -8 - 32 = -40$

61. $2(-5)(-6)^2 = 2(-5)(36) = -10(36) = -360$

63. $2(-5) - 6^2 = 2(-5) - 36 = -10 - 36 = -46$

67. $-3 - 2[-4 - 3(-2 - 1)] = -3 - 2[-4 - 3(-3)]$
$$= -3 - 2[-4 + 9]$$
$$= -3 - 2(5)$$
$$= -3 - 10$$
$$= -13$$

71. $|3 - 8| - |3| - |-8| = |-5| - |3| - |-8| = 5 - 3 - 8 = 2 - 8 = -6$

75. $(-3 - 2)^2 - (-3 + 2)^2 = (-5)^2 - (-1)^2 = 25 - 1 = 24$

77a. $7\left(\dfrac{3}{7}\right)^2 + 4\left(\dfrac{3}{7}\right) - 3$

$= 7\left(\dfrac{9}{49}\right) + 4\left(\dfrac{3}{7}\right) - 3$

$= \dfrac{63}{49} + \dfrac{12}{7} - 3$

$= \dfrac{9}{7} + \dfrac{12}{7} - 3$

$= \dfrac{21}{7} - 3 = 3 - 3 = 0$

81. $xyz = (-2)(-3)(5) = 6(5) = 30$

85. $|xy - z| = |(-2)(-3) - 5| = |6 - 5| = |1| = 1$

87. $\dfrac{3x^2 - x^3 y^2}{3x - 2y} = \dfrac{3(-2)^2(-3) - (-2)^3(-3)^2}{3(-2) - 2(-3)}$

Since $3(-2) - 2(-3) = -6 + 6 = 0$, our result is undefined (we can never divide by 0).

89. $1.692 - 3.965 + 8.754 = -2.273 + 8.754 = 6.481$

93. $\dfrac{42.4}{1.63 - (2.1)(5.8)} = \dfrac{42.2}{1.63 - 12.18} = \dfrac{42.2}{-10.55} = -4$

99. Step 1: $(-1)x = -x$

Step 2: associative property of multiplication

Step 3: multiplication of whole numbers

Step 4: $(-1)x = -x$

Exercises 1.5

1. $6x + 2 = (6 + 2)x = 8x$

3. $6x(2x) = 6(2)x \cdot x = 12x^2$

9. $3m - 4m - 5m = (3 - 4 - 5)m = -6m$

13. $-2t^2 - 3t^2 - 4t^2 = (-2 - 3 - 4)t^2 = -9t^2$

15. $-2t^2(-3t^2)(-4t^2) = -2(-3)(-4)t^2t^2t^2 = -24t^6$

19. $2x(3y)(5z) = 2(3)(5)xyz = 30xyz$

21. cannot be simplified

23. $x^3(x^2)(2x) = 2x^3x^2x = 2x^6$

25. $-5x(3xy) - 2x^2y = -5(3)xxy - 2x^2y = -15x^2y - 2x^2y = (-15 - 2)x^2y = -17x^2y$

27. $-5x(3xy)(-2x^2y) = -5(3)(-2)xxx^2yy = 30x^4y^2$

31. $10x^2y - 6xy^2 + x^2y - xy^2 = 10x^2y + x^2y - 6xy^2 - xy^2$

$= (10 + 1)x^2y + (-6 - 1)xy^2$

$= 11x^2y - 7xy^2$

35. $6(a - 2b) - 4(a + b) = 6a + 6(-2b) - 4a - 4b$

$= 6a - 12b - 4a - 4b$

$= 6a - 4a - 12b - 4b$

$= (6 - 4)a + (-12 - 4)b$

$= 2a - 16b$

39. $x(x - y) + y(y - x) = xx + x(-y) + yy + y(-x)$
$$= x^2 - xy + y^2 - yx$$
$$= x^2 - xy - yx + y^2$$
$$= x^2 - xy - xy + y^2$$
$$= x^2 + (-1 - 1)xy + y^2$$
$$= x^2 - 2xy + y^2$$

41. $a^2(a + 3b) - a(a^2 + 3ab) = a^2a + a^2(3b) - aa^2 - a(3ab)$
$$= a^3 + 3a^2b - a^3 - 3a^2b$$
$$= a^3 - a^3 + 3a^2b - 3a^2b$$
$$= 0 + 0 = 0$$

45. $(2x)^3(3x)^2 = (2x)(2x)(2x)(3x)(3x) = (2)(2)(2)(3)(3)xxxxx = 72x^5$

47. $2x^3(3x)^2 = 2x^3(3x)(3x) = (2)(3)(3)x^3xx = 18x^5$

51. $(-2x)^4 - (2x)^4 = (-2x)(-2x)(-2x)(-2x) - (2x)(2x)(2x)(2x)$
$$= (-2)(-2)(-2)(-2)xxxx - (2)(2)(2)(2)xxxx$$
$$= 16x^4 - 16x^4 = 0$$

55. $4b - 5(b - 2) = 4b - 5b - 5(-2)$
$$= -b + 10$$

57. $8t - 3[t - 4(t + 1)] = 8t - 3[t - 4t - 4]$
$$= 8t - 3[-3t - 4]$$
$$= 8t - 3(-3t) - 3(-4)$$
$$= 8t + 9t + 12$$
$$= 17t + 12$$

63. $x - \left\{y - 3[x - 2(y - x)]\right\} = x - \left\{y - 3[x - 2y + 2x]\right\}$
$$= x - \{y - 3[3x - 2y]\}$$
$$= x - \{y - 9x + 6y\}$$
$$= x - \{7y - 9x\}$$
$$= x - 7y + 9x$$
$$= 10x - 7y$$

67. $6s^2 - [st - s(t + 5s) - s^2] = 6s^2 - [st - st - 5s^2 - s^2]$
$$= 6s^2 - [-6s^2]$$
$$= 6s^2 + 6s^2 = 12s^2$$

69. (a) $3x - 15y - 11x = 3(2.82) - 15(7.25) - 11(2.82)$
$$= 8.46 - 108.75 - 31.02$$
$$= -100.29 - 31.02 = -131.31$$

(b) $3x - 15y - 11x = -8x - 15y$
$$= -8(2.82) - 15(7.25)$$
$$= -22.56 - 108.75$$
$$= -131.31$$

Exercises 1.6

1. $n + 8$, where n = the number

5. $3n + 4 = n - 7$, where n = the number

7. $x + y = xy + 1$, where x and y are the numbers

9.   Let n = smaller number. Then $2n + 5$ = larger number.

11.   Let n = smaller number. Then $3n$ = middle number and $3n + 12$ = largest number.

15.   Let n = first of the three consecutive even numbers. Then $n + 2$ = second of the three consecutive even integers and $n + 4$ = third of the three consecutive even integers.

19.   Let n = one of the numbers. Then $40 - n$ = the other number.

21.   Let n = one of the numbers. Then $2n$ = another of the numbers and $100 - 3n$ = the third number.

25.   Let n = length of the second side of the triangle. Then $2n$ = length of the first side of the triangle and $n + 4$ = length of the third side of the triangle. So the perimeter of the triangle is $n + 2n + (n + 4) = 4n + 4$.

29.   (a) $n + d + q$ coins   (b) $5n + 10d + 25q$ cents

33.   Let n = number of nickels in the collection. Then $20 - n$ = number of dimes in the collection.

Value of the nickels = $5n$

Value of the dimes = $10(20 - n)$

Value of all the coins = $5n + 10(20 - n)$
$$= 5n + 200 - 10n$$
$$= 200 - 5n$$

Chapter 1 Review Exercises

1. $\{1,2,3,4\}$

3. $\{b\}$

5. $\{x | x \in N, x \neq 5\} = \{1,2,3,4,6,7,8,9...\}$

7. $\{1,2,3,4,6,12\}$

9. $\{12\}$

11. $\{0,12,24,36,...\} = B$

13. $\{x | 3 \leq x \leq 4, x \in Z\}$ or $\{3,4\}$

15.

17.

19.

21.   False

23.   True

25. False

27. True

29. Commutative property of addition

31. Distributive property

33. Multiplicative inverse property

35. False

37. $(-2) + (-3) - (-4) + (-5) = (-5) - (-4) + (-5)$
$$= -5 + 4 + (-5)$$
$$= -1 + (-5)$$
$$= -6$$

39. $6 - 2 + 5 - 8 - 9 = 4 + 5 - 8 - 9 = 9 - 8 - 9 = 1 - 9 = -8$

41. $(-2)(-3)(-5) = (+6)(-5) = -30$

43. $(-2)^6 = (-2)(-2)(-2)(-2)(-2)(-2) = +64$

45. $-2 - (-3)^2 = -2 - 9 = -11$

47. $(-6 - 3)(-2 - 5) = (-9)(-7) = +63$

49. $5 - 3[2 - (4 - 8) + 7] = 5 - 3[2 - (-4) + 7]$
$$= 5 - 3[2 + 4 + 7]$$
$$= 5 - 3(13)$$
$$= 5 - 39 = -34$$

51. $5 - \{2 + 3[6 - 4(5 - 9)] - 2\} = 5 - \{2 + 3[6 - 4(-4)] - 2\}$
$$= 5 - \{2 + 3[6 + 16] - 2\}$$
$$= 5 - \{2 + 3(22) - 2\}$$
$$= 5 - \{2 + 66 - 2\}$$
$$= 5 - 66 = -61$$

53. $\dfrac{4[5 - 3(8 - 12)]}{-6 - 2(5 - 6)} = \dfrac{4[5 - 3(-4)]}{-6 - 2(-1)} = \dfrac{4[5 + 12]}{-6 + 2} = \dfrac{4(17)}{-4} = \dfrac{68}{-4} = -17$

55. $x^2 - 2xy + y^2 = (-2)^2 - 2(-2)(-1) + (-1)^2 = 4 - 4 + 1 = 1$

57. $|x - y| - (|x| - |y|) = |(-2) - (-1)| - (|-2| - |-1|)$
$$= |-2 + 1| - (2 - 1) = |-1| - 1$$
$$= 1 - 1 = 0$$

59. $\dfrac{2x^2y^3 + 3y^2}{zx^2y} = \dfrac{2(-2)^2(-1)^3 + 3(-1)^2}{(0)(-2)^2(-1)}$

Since $(0)(-2)^2(-1) = 0$, this is undefined because we cannot divide by 0.

61. 3.86

63. $(2x + y)(-3x^2y) = 2x(-3x^2y) + y(-3x^2y) = -6x^3y - 3x^2y^2$

65. $\left(2xy^2\right)^2(-3x)^2 = \left(2xy^2\right)\left(2xy^2\right)(-3x)(-3x) = (2)(2)(-3)(-3)xxxxy^2y^2 = 36x^4y^4$

67. $3x - 2y - 4x + 5y - 3x = 3x - 4x - 3x - 2y + 5y$
$$= (3 - 4 - 3)x + (-2 + 5)y$$
$$= -4x + 3y$$

69. $-2r^2s + 5rs^2 - 3sr^2 - 4s^2r = -2r^2s - 3r^2s + 5rs^2 - 4rs^2$
$$= (-2 - 3)r^2s + (5 - 4)rs^2$$
$$= -5r^2s + rs^2$$

71. $-2x - (3 - x) = -2x - 3 - (-x)$
$$= -2x - 3 + x$$
$$= -2x + x - 3$$
$$= (-2 + 1)x - 3$$
$$= -x - 3$$

73. $3a(a - b + c) = 3a(a) + 3a(-b) + 3a(c) = 3a^2 - 3ab + 3ac$

75. $2x - 3(x - 4) = 2x - 3x - 3(-4)$
$$= 2x - 3x + 12$$
$$= (2 - 3)x + 12$$
$$= -x + 12$$

77. $3a - \left[5 - (a - 4)\right] = 3a - \left[5 - a - (-4)\right]$
$$= 3a - \left[5 - a + 4\right]$$
$$= 3a - \left[9 - a\right]$$
$$= 3a - 9 - (-a)$$
$$= 3a - 9 + a$$
$$= 3a + a - 9$$
$$= (3 + 1)a - 9$$
$$= 4a - 9$$

79. $5x - \left\{3x + 2\left[x - 3(5 - x)\right]\right\} = 5x - \left\{3x + 2\left[x - 3(5) - 3(-x)\right]\right\}$
$$= 5x - \left\{3x + 2\left[x - 15 + 3x\right]\right\}$$
$$= 5x - \left\{3x + 2\left[x + 3x - 15\right]\right\}$$
$$= 5x - \left\{3x + 2\left[4x - 15\right]\right\}$$
$$= 5x - \left\{3x + 2(4x) + 2(-15)\right\}$$
$$= 5x - \left\{3x + 8x - 30\right\}$$
$$= 5x - \left\{11x - 30\right\}$$
$$= 5x - 11x - (-30)$$
$$= 5x - 11x + 30$$
$$= (5 - 11)x + 30$$
$$= -6x + 30$$

81. $xy - 5 = x + y + 3$, where x and y are the numbers.

83. Let n, n + 2, and n + 4 be the three consecutive odd integers. Then $n + (n + 2) = (n + 4) - 5$.

85. Let W = width of the rectangle. Then $4W - 5$ = length of the rectangle. So

$$\text{area} = (4W - 5)W = 4W^2 - 5W \text{ and}$$

$$\text{perimeter} = 2(4W - 5) + 2W = 8W - 10 + 2W = 10W - 10$$

87. $x^2 + y^2 = xy + 8$, where x and y are the numbers.

89. Let d = number of dimes. Then $40 - d$ = number of nickels.

Value of the dimes = 10d

Value of the nickels = $5(40 - d)$

Total value of coins $= 10d + 5(40 - d)$
$$= 10d + 200 - 5d$$
$$= 5d + 200$$

Chapter 1 Practice Test

1. (a) $\{2,3,5,7\}$ (b) $\{2,3,5,7,11,13,17,19,23\}$

3. (a)  (b)

5. (a) $-3 - (-6) + (-4) - (-9) = -3 + 6 - 4 + 9 = 3 - 4 + 9 = -1 + 9 = 8$

   (b) $(-7)^2 - (-6)(-2)(-3) = 49 - (+12)(-3) = 49 - (-36) = 49 + 36 = 85$

   (c) $|3 - 8| - |5 - 9| = |-5| - |-4| = 5 - 4 = 1$

   (d) $6 - 5[-2 - (7 - 9)] = 6 - 5[-2 - (-2)] = 6 - 5[-2 + 2] = 6 - 5 \cdot 0 = 6 - 0 = 6$

7. (a) $(5x^3y^2)(-2x^2y)(-xy^2) = 5(-2)(-1)x^3x^2xy^2yy^2 = 10x^6y^5$

   (b) $3rs^2 - 5r^2s - 4rs^2 - 7rs = 3rs^2 - 4rs^2 - 5r^2s - 7rs$
   $$= (3 - 4)rs^2 - 5r^2s - 7rs$$
   $$= -rs^2 - 5r^2s - 7rs$$

   (c) $2a - 3(a - 2) - (6 - a) = 2a - 3a - 3(-2) - 6 - (-a)$
   $$= 2a - 3a + 6 - 6 + a$$
   $$= 2a - 3a + a + 6 - 6$$
   $$= (2 - 3 + 1)a + 6 - 6$$
   $$= 0a + 0 = 0$$

(d) $\quad 7r - \{3 + 2[s - (r - 2s)]\} = 7r - \{3 + 2[s - r - (-2s)]\}$
$$= 7r - \{3 + 2[s - r + 2s]\}$$
$$= 7r - \{3 + 2[s + 2s - r]\}$$
$$= 7r - \{3 + 2[3s - r]\}$$
$$= 7r - \{3 + 2(3s) - 2r\}$$
$$= 7r - \{3 + 6s - 2r\}$$
$$= 7r - 3 - 6s - (-2r)$$
$$= 7r - 3 - 6s + 2r$$
$$= 7r + 2r - 3 - 6s$$
$$= 9r - 3 - 6s$$

9. Let x = number of dimes. Then $34 - x$ = number of nickels. So the total value of Wallace's coins is $10x + 5(34 - x) = 10x + 170 - 5x = 5x + 170$.

CHAPTER 2
EQUATIONS AND INEQUALITIES

Exercises 2.1

1. $x - (x - 3) = 4$
    $x - x + 3 = 4$
    $3 = 4$ contradiction

3. $4(w - 2) = 4w - 8$
    $4w - 8 = 4w - 8$ identity

7. $6(2y + 1) - 4(3y - 1) = y - (y + 10)$
    $12y + 6 - 12y + 4 = y - y - 10$
    $10 = -10$ contradiction

11. $3(a - 5) + 2(1 - a) = a - 8 - (a - 5)$

    CHECK $a = -3$:
    $3(-3 - 5) + 2(1 - (-3)) \overset{?}{=} -3 - 8 - (-3 - 5)$
    $3(-8) + 2(4) \overset{?}{=} -11 - (-8)$
    $-24 + 8 \overset{?}{=} -11 + 8$
    $-16 \neq -3$
    Thus, $a = -3$ is not a solution.

    CHECK $a = 0$:
    $3(0 - 5) + 2(1 - 0) \overset{?}{=} 0 - 8 - (0 - 5)$
    $3(-5) + 2(1) \overset{?}{=} -8 - (-5)$
    $-15 + 2 \overset{?}{=} -8 + 5$
    $-13 \neq -3$
    Thus, $a = 0$ is not a solution.

15. $a(a + 8) = (a + 2)^2$

    CHECK $a = -1$:
    $-1(-1 + 8) \overset{?}{=} (-1 + 2)^2$
    $-1(7) \overset{?}{=} 1^2$
    $-7 \neq 1$

    Thus, $a = -1$ is not a solution.

    CHECK $a = 1$:
    $1(1 + 8) \overset{?}{=} (1 + 2)^2$
    $1(9) \overset{?}{=} 3^2$
    $9 = 9$
    Thus, $a = 1$ is a solution.

17. $2x - 7 = 5$

$2x - 7 + 7 = 5 + 7$

$2x = 12$

$$\frac{2x}{2} = \frac{12}{2}$$

$x = 6$

CHECK x = 6:

$2x - 7 = 5$

$2(6) - 7 \overset{?}{=} 5$

$12 - 7 \overset{?}{=} 5$

$5 \overset{\vee}{=} 5$

21. $m + 3 = 3 - m$

$m + 3 + m = 3 - m + m$

$2m + 3 = 3$

$2m + 3 - 3 = 3 - 3$

$2m = 0$

$$\frac{2m}{2} = \frac{0}{2}$$

$m = 0$

CHECK m = 0:

$m + 3 = 3 - m$

$0 + 3 \overset{?}{=} 3 - 0$

$3 \overset{\vee}{=} 3$

23. $m + 3 = 3 + m$

$m + 3 - m = 3 + m - m$

$3 = 3$

identity

27. $11 - 3y = 38$

$11 - 3y - 11 = 38 - 11$

$-3y = 27$

$$\frac{-3y}{-3} = \frac{27}{-3}$$

$y = -9$

CHECK y = −9:

$11 - 3y = 38$

$11 - 3(-9) \overset{?}{=} 38$

$11 + 27 \overset{?}{=} 38$

$38 \overset{\vee}{=} 38$

29. $5s + 2 = 3s - 7$

$5s + 2 - 3s = 3s - 7 - 3s$

$2s + 2 = -7$

$2s + 2 - 2 = -7 - 2$

$2s = -9$

$$\frac{2s}{2} = \frac{-9}{2}$$

$s = -\frac{9}{2}$

CHECK $s = -\frac{9}{2}$:

$5s + 2 = 3s - 7$

$5\left(-\frac{9}{2}\right) + 2 \overset{?}{=} 3\left(-\frac{9}{2}\right) - 7$

$\frac{-45}{2} + 2 \overset{?}{=} \frac{-27}{2} - 7$

$\frac{-45}{2} + \frac{4}{2} \overset{?}{=} \frac{-27}{2} - \frac{14}{2}$

$\frac{-41}{2} \overset{\vee}{=} \frac{-41}{2}$

33. $3x - 12 = 12 - 3x$

$3x - 12 + 3x = 12 - 3x + 3x$

$6x - 12 = 12$

$6x - 12 + 12 = 12 + 12$

$6x = 24$

CHECK x = 4:

$3x - 12 = 12 - 3x$

$3(4) - 12 \overset{?}{=} 12 - 3(4)$

$12 - 12 \overset{?}{=} 12 - 12$

$0 \overset{\vee}{=} 0$

12

$$\frac{6x}{6} = \frac{24}{6}$$
$$x = 4$$

37.　　　$2t + 1 = 7 - t$

　　　　$2t + 1 + t = 7 - t + t$

　　　　　　$3t + 1 = 7$

　　　　$3t + 1 - 1 = 7 - 1$

　　　　　　　　$3t = 6$

　　　　　　　$\frac{3t}{3} = \frac{6}{3}$

　　　　　　　　　$t = 2$

CHECK $t = 2$:

　　$2t + 1 = 7 - t$

　$2(2) + 1 \overset{?}{=} 7 - 2$

　　$4 + 1 \overset{?}{=} 5$

　　　$5 \overset{\checkmark}{=} 5$

39.　　　$2(x - 1) = 2(x + 1)$

　　　　　$2x - 2 = 2x + 2$

　　$2x - 2 - 2x = 2x + 2 - 2x$

　　　　　$-2 = 2$

contradiction (no solution)

45.　$2(x - 1) + 3(x + 1) = x - 3$

　　　$2x - 2 + 3x + 3 = x - 3$

　　　　　　$5x + 1 = x - 3$

　　　　$5x + 1 - x = x - 3 - x$

　　　　　　　$4x + 1 = -3$

　　　　$4x + 1 - 1 = -3 - 1$

　　　　　　　　$4x = -4$

　　　　　　　$\frac{4x}{4} = \frac{-4}{4}$

　　　　　　　　　$x = -1$

CHECK $x = -1$:

　$2(x - 1) + 3(x + 1) = x - 3$

　$2(-1 - 1) + 3(-1 + 1) \overset{?}{=} -1 - 3$

　　　$2(-2) + 3(0) \overset{?}{=} -4$

　　　　　$-4 + 0 \overset{?}{=} -4$

　　　　　　　$-4 \overset{\checkmark}{=} -4$

47.　$6(3 - z) + 2(4z - 5) = z - (3 - z)$

　　　$18 - 6z + 8z - 10 = z - 3 + z$

　　　　　　$2z + 8 = 2z - 3$

　　　　$2z + 8 - 2z = 2z - 3 - 2z$

　　　　　　　　$8 = -3$

contradiction (no solution)

51.　$x(x - 2) - 15 = x(x + 5) - 3(x + 5)$

　　　$x^2 - 2x - 15 = x^2 + 5x - 3x - 15$

　　　$x^2 - 2x - 15 = x^2 + 2x - 15$

　$x^2 - 2x - 15 - x^2 = x^2 + 2x - 15 - x^2$

　　　　$-2x - 15 = 2x - 15$

CHECK $x = 0$:

　$x(x - 2) - 15 = x(x + 5) - 3(x + 5)$

　$0(0 - 2) - 15 \overset{?}{=} 0(0 + 5) - 3(0 + 5)$

　　$0(-2) - 15 \overset{?}{=} 0(5) - 3(5)$

　　　$0 - 15 \overset{?}{=} 0 - 15$

$$-2x - 15 + 2x = 2x - 15 + 2x \qquad\qquad -15 \overset{\checkmark}{=} -15$$
$$-15 = 4x - 15$$
$$-15 + 15 = 4x - 15 + 15$$
$$0 = 4x$$
$$\frac{0}{4} = \frac{4x}{4}$$
$$0 = x$$

55.
$$5x - 2[x - 3(7 - x)] = 3 - 2(x - 8)$$
$$5x - 2[x - 21 + 3x] = 3 - 2x + 16$$
$$5x - 2[4x - 21] = -2x + 19$$
$$5x - 8x + 42 = -2x + 19$$
$$-3x + 42 = -2x + 19$$
$$-3x + 42 + 3x = -2x + 19 + 3x$$
$$42 = x + 19$$
$$42 - 19 = x + 19 - 19$$
$$23 = x$$

CHECK x = 23:
$$5x - 2[x - 3(7 - x)] = 3 - 2(x - 8)$$
$$5(23) - 2[(23) - 3(7 - 23)] \overset{?}{=} 3 - 2(23 - 8)$$
$$5(23) - 2[23 - 3(-16)] \overset{?}{=} 3 - 2(15)$$
$$5(23) - 2[23 + 48] \overset{?}{=} 3 - 2(15)$$
$$115 - 2(71) \overset{?}{=} 3 - 30$$
$$115 - 142 \overset{?}{=} 3 - 30$$
$$-27 \overset{\checkmark}{=} -27$$

61.
$$3.24x - 5.2 = 7.74x + 0.3$$
$$3.24x - 5.2 - 3.24x = 7.74x + 0.3 - 3.24x$$
$$-5.2 = 4.5x + 0.3$$
$$-5.2 - 0.3 = 4.5x + 0.3 - 0.3$$
$$-5.5 = 4.5x$$
$$\frac{-5.5}{4.5} = \frac{4.5x}{4.5}$$
$$-\frac{55}{45} = x$$
$$-\frac{11}{9} = x$$

CHECK $x = -\dfrac{11}{9}$:
$$3.24x - 5.2 = 7.74x + 0.3$$
$$3.24\left(-\frac{11}{9}\right) - 5.2 \overset{?}{=} 7.74\left(-\frac{11}{9}\right) + 0.3$$
$$-3.96 - 5.2 \overset{?}{=} -9.46 + 0.3$$
$$-9.16 \overset{\checkmark}{=} -9.16$$

67. An identity is an equation that is true no matter what value of the variable is chosen. A conditional equation is one whose truth depends on the value that is assigned to the variable. Put another way, an identity is always true, while a conditional equation is sometimes true.

68. A contradiction is an equation that is false no matter what value of the variable is chosen. A contradiction is always false, while an identity is always true.

69. Two equations are called equivalent if they have exactly the same set of solutions.

Exercises 2.2

1. `Let n = smaller number. Then 4n + 3 = larger number.

$$n + 4n + 3 = 43$$
$$5n + 3 = 43$$
$$5n = 40$$
$$n = 8$$

Then 4n + 3 = 4(8) + 3 = 32 + 3 = 35. Thus, the numbers are 8 and 35.

CHECK: 35 is 3 more than 4(8) = 32, and $35 + 8 \overset{\checkmark}{=} 43$.

5. Let n = first of the three consecutive integers. Then n + 1 = second of the three consecutive integers and n + 2 = third of the three consecutive integers.

$$n + n + 1 + n + 2 = 66$$
$$3n + 3 = 66$$
$$3n = 63$$
$$n = 21$$

Then n + 1 = 21 + 1 = 22, and n + 2 = 21 + 2 + 23. Thus, the consecutive integers are 21, 22, and 23.

CHECK: $21 + 22 + 23 \overset{\checkmark}{=} 66$.

7. Let n = first of the three consecutive even integers. Then n + 2 = second of the three consecutive even integers and n + 4 = third of the three consecutive integers.

$$(n + 2) + (n + 4) = 2n + 6$$
$$2n + 6 = 2n + 6 \qquad \text{identity}$$

Thus, any three consecutive even integers will have this property. For example, if we consider 52, 54, and 56, then 54 + 56 = 110 = 2(52) + 6.

11. Let W = width of the rectangle (in meters). Then 2W = length of the rectangle (in meters).

$$2W + 2(2W) = 42$$
$$2W + 4W = 42$$
$$6W = 42$$
$$W = 7$$

Then 2W = 2(7) = 14. Thus, the rectangle has a width of 7 meters and a length of 14 meters.

CHECK: 14 is twice as large as 7, and $2(7) + 2(14) = 14 + 28 \overset{\checkmark}{=} 42$.

15. Let x = length of the second side of the triangle (in cm). Then x − 5 = length of the first side of the triangle (in cm) and 2(x − 5) = length of the third side of triangle (in cm).

$$x + (x - 5) + 2(x - 5) = 33$$
$$x + x - 5 + 2x - 10 = 33$$
$$4x - 15 = 33$$
$$4x = 48$$
$$x = 12$$

Then $x - 5 = 12 - 5 = 7$ and $2(x - 5) = 2(12 - 5) = 2 \cdot 7 \overset{\checkmark}{=} 14$. Thus, the length of the sides of the triangle are 7 cm, 12 cm, and 14 cm.

CHECK: 7 is 5 less than 12, and 14 is twice as large as 7. $7 + 12 + 14 \overset{\checkmark}{=} 33$.

19. Let n = number of nickels in the collection. Then n + 2 = number of dimes in the collection and n + 6 = number of quarters in the collection.

$$5n + 10(n + 2) + 25(n + 6) = 530$$
$$5n + 10n + 20 + 25n + 150 = 530$$
$$40n + 170 = 530$$
$$40n = 360$$
$$n = 9$$

Then, $n + 2 = 9 + 2 = 11$ and $n + 6 = 9 + 6 = 15$. Thus, there are 9 nickels, 11 dimes, and 15 quarters in the collection.

CHECK: 11 is 2 more than 9 and 15 is 4 more than 11.

9 nickels = $.45

11 dimes = $1.10

15 quarters = $3.75

total $\overset{\checkmark}{=}$ $5.30.

23. Let x = # of 10-cent stamps that Barbie buys. Then 2x = # of 25-cent stamps that Barbie buys and $97 - 3x$ = # of 29-cent stamps that Barbie buys.

$$10x + 25(2x) + 29(97 - 3x) = 2300$$
$$10x + 50x + 2813 - 87x = 2300$$
$$-27x + 2813 = 2300$$
$$-27x = -513$$
$$x = 19$$

Then $2x = 2(19) = 38$ and $97 - 3x = 97 - 3(19) = 97 - 57 = 40$. Thus, Barbie buys 19 10-cent stamps, 38 25-cent stamps, and 40 29-cent stamps.

CHECK: 38 is twice as large as 19. $19 + 38 + 40 = 97$.

19 10-cent stamps = $ 1.90

38 25-cent stamps = $ 9.50

40 29-cent stamps = $11.60

total $\overset{\checkmark}{=}$ $23.00

27. Let x = # of pairs of shoes that Marge sells. Then $30 - x$ = # of pairs of boots that Marge sells.

$$140 + 2(30 - x) + 1x = 187$$
$$140 + 60 - 2x + x = 187$$
$$200 - x = 187$$
$$-x = -13$$
$$x = 13$$

Thus, Marge sells 13 pairs of shoes.

CHECK: salary = $140; commission on shoes = 13(1) = $13; commission on boots = (30 − 13)(2) = 17(2) = $34. Total income $\overset{\checkmark}{=}$ $187.

29. Let x = # of pounds of $2/pound coffee beans in the mixture.

$$2x + 3(30) = 2.60(x + 30)$$
$$2x + 90 = 2.6x + 78$$
$$90 = 0.6x + 78$$
$$12 = 0.6x$$
$$20 = x$$

Thus, 20 pounds of $2/pound coffee beans are required.

CHECK: 20 pounds @ $2/pound = $40

30 pounds @ $3/pound = $90

50 pounds = $130

Price per pound = $\dfrac{130}{50} \overset{\checkmark}{=} $2.60

33. Let x = # of hours that the plumber works.

$$13(x + 2) + 22x = 236$$
$$13x + 26 + 22x = 236$$
$$35x + 26 = 236$$
$$35x = 210$$
$$x = 6$$

Thus, the plumber works for 6 hours.

CHECK: Plumber earns 6($22) = $132

Assistant earns 8($13) = $104

Total earnings $\overset{\checkmark}{=}$ $236

37. Let t = # of hours needed.

$$35t + 50t = 595$$
$$85t = 595$$
$$t = 7$$

Thus, 7 hours are needed.

CHECK: In 7 hours, the slower car travels 7(35) = 245 km, while the faster car travels 7(50) = 350 km.

$$245 + 350 \overset{\checkmark}{=} 595$$

41. Let t = # of hours spent driving to the convention. Then 17 − t = # of hours spent driving home.

$$48t = 54(17 - t)$$
$$48t = 918 - 54t$$
$$102t = 918$$
$$t = 9$$

Then the distance between home and the convention is 48(9) = 432 km.

CHECK: Time going: $\dfrac{432}{48} = 9$ hours

Time returning: $\dfrac{232}{54} = 8$ hours

Total driving time $\overset{\checkmark}{=} 17$ hours

45. Let t = # of hours needed to complete job.

$$60t + 30t = 6750$$
$$90t = 6750$$
$$t = 75$$

Thus, the two workers need 75 hours to complete the job if they work together.

CHECK: In 75 hours, the experienced worker processes 75(60) = 4500 items. At the same time, the new worker processes 75(30) = 2250 items. $4500 + 2250 \overset{\checkmark}{=} 6750$.

49. Let t = # of hours that the trainee works. Then t − 3 = # of hours that the experienced worker works.

$$48t + 80(t - 3) = 656$$
$$48t + 80t - 240 = 656$$
$$128t - 240 = 656$$
$$128t = 896$$
$$t = 7$$

Since the trainee began at 10:00 a.m. and worked for 7 hours, the workers finish packaging the stack at 5:00 p.m.

CHECK: Trainee packages 7(48) = 336 items in 7 hours. Experienced worker packages 4(80) = 320 items in 4 hours.

$$336 + 320 \overset{\checkmark}{=} 656.$$

Exercises 2.3

1. $x + 1 \quad > x - 1$
$x + 1 - x > x - 1 - x$
$\qquad 1 > -1$ identity

5. $-6 \leq r - (r + 6) < 3$
$-6 \leq r - r - 6 < 3$
$-6 \leq -6 < 3$ identity

18

9.   CHECK u = 2: $4 + 3u > 10$

$$4 + 3(2) \overset{?}{>} 10$$
$$4 + 6 \overset{?}{>} 10$$
$$10 \not> 10$$

Therefore, u = 2 is not a solution.

CHECK u = 3: $4 + 3u > 10$

$$4 + 3(3) \overset{?}{>} 10$$
$$4 + 9 \overset{?}{>} 10$$
$$13 \overset{\checkmark}{>} 10$$

Therefore, u = 3 is a solution.

13.   CHECK z = $-$2: $-3 \le 3 - (4 - z) < 3$

$$-3 \overset{?}{\le} 3 - \left(4 - (-2)\right) \overset{?}{<} 3$$
$$-3 \overset{?}{\le} 3 - 6 \overset{?}{<} 3$$
$$-3 \overset{\checkmark}{\le} -3 \overset{\checkmark}{<} 3$$

Therefore, z = $-$2 is a solution.

CHECK z = 4: $-3 \le 3 - (4 - z) < 3$

$$-3 \overset{?}{\le} 3 - (4 - 4) \overset{?}{<} 3$$
$$-3 \overset{?}{\le} 3 - 0 \overset{?}{<} 3$$
$$-3 \overset{?}{\le} 3 \not< 3$$

Therefore, z = 4 is not a solution.

15. $5 < x < 8$ is satisfied by x = 6 and thus makes sense.

19. $-3 > x > 2$ does not make sense, since there is no number that is both less than -3 and greater than 2.

23. $2 > x > -3$ is satisfied by x = 0 and thus makes sense. Its standard form is $-3 < x < 2$.

25.
$$5x - 2 < 13$$
$$5x - 2 + 2 < 13 + 2$$
$$5x < 15$$
$$\frac{5x}{5} < \frac{15}{5}$$
$$x < 3$$

29.
$$3x + 4 \le 2x - 5$$
$$3x + 4 - 2x \le 2x - 5 - 2x$$
$$x + 4 \le -5$$
$$x + 4 - 4 \le -5 - 4$$
$$x \le -9$$

33.
$$11 - 3y < 38$$
$$11 - 3y - 11 < 38 - 11$$
$$-3y < 27$$
$$\frac{-3y}{-3} > \frac{27}{-3}$$
$$y > -9$$

37.
$$3a - 8 < -8 - 3a$$
$$3a - 8 + 3a < -8 - 3a + 3a$$
$$6a - 8 < -8$$
$$6a - 8 + 8 < -8 + 8$$
$$6a < 0$$
$$\frac{6a}{6} < \frac{0}{6}$$
$$a < 0$$

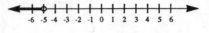

39.
$$3t - 5 > 5t - 13$$
$$3t - 5 - 3t > 5t - 13 - 3t$$
$$-5 > 2t - 13$$
$$-5 + 13 > 2t - 13 + 13$$
$$8 > 2t$$
$$\frac{8}{2} > \frac{2t}{2}$$
$$4 > t \text{ or } t < 4$$

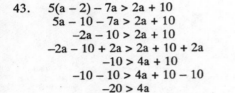

43.
$$5(a - 2) - 7a > 2a + 10$$
$$5a - 10 - 7a > 2a + 10$$
$$-2a - 10 > 2a + 10$$
$$-2a - 10 + 2a > 2a + 10 + 2a$$
$$-10 > 4a + 10$$
$$-10 - 10 > 4a + 10 - 10$$
$$-20 > 4a$$
$$\frac{-20}{4} > \frac{4a}{4}$$
$$-5 > a \text{ or } a < -5$$

47.
$$7 - 5(t - 2) \leq 2$$
$$7 - 5t + 10 \leq 2$$
$$-5t + 17 \leq 2$$
$$-5t + 17 - 17 \leq 2 - 17$$
$$-5t \leq -15$$
$$\frac{-5t}{-5} \geq \frac{-15}{-5}$$
$$t \geq 3$$

49.
$$2(y + 3) + 3(y - 4) < 3(y + 2) + 2(y + 1)$$
$$2y + 6 + 3y - 12 < 3y + 6 + 2y + 2$$
$$5y - 6 < 5y + 8$$
$$5y - 6 - 5y < 5y + 8 - 5y$$
$$-6 < 8$$

identity

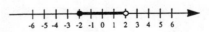

53.
$$1 \leq c + 3 < 5$$
$$1 - 3 \leq c + 3 - 3 < 5 - 3$$
$$-2 \leq c < 2$$

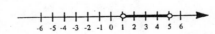

57.
$$-1 < 4 - t < 3$$
$$-1 - 4 < 4 - t - 4 < 3 - 4$$
$$-5 < -t < -1$$

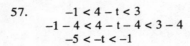

$$\frac{-5}{-1} > \frac{-t}{-1} > \frac{-1}{-1}$$

$$5 > t > 1 \text{ or } 1 < t < 5$$

61.　$2 < 5 - 3(x + 1) < 17$

$2 < 5 - 3x - 3 < 17$

$2 < 2 - 3x < 17$

$2 - 2 < 2 - 3x - 2 < 17 - 2$

$0 < -3x < 15$

$$\frac{0}{-3} > \frac{-3x}{-3} > \frac{15}{-3}$$

$0 > x > -5 \text{ or } -5 < x < 0$

65.　$-2 \le 5 - x < 1$

$-2 - 5 \le 5 - x - 5 < 1 - 5$

$-7 \le -x < -4$

$$\frac{-7}{-1} \ge \frac{-x}{-1} > \frac{-4}{-1}$$

$7 \ge x > 4 \text{ or } 4 < x \le 7$

69.　$0.39 \le 0.72x - 1.5 < 8.1$

$0.39 + 1.5 \le 0.72x - 1.5 + 1.5 < 8.1 + 1.5$

$1.89 \le 0.72x < 9.6$

$$\frac{1.89}{0.72} \le \frac{0.72x}{0.72} < \frac{9.6}{0.72}$$

$$\frac{21}{8} \le x < \frac{40}{3}$$

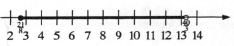

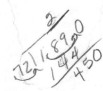

71.　(a)　The inequality sign should not reverse in the final step, since we are dividing by a positive number.

(b)　When we divide 4 by –2, the resulting quotient should be –2, not 2.

(c)　The inequality sign should reverse in the final step, since we are dividing by a negative number.

(d)　When we subtract the same quantity from each side of an inequality, the inequality sign should not reverse.

72.　An identity is an inequality that is always true; a conditional inequality is one that is sometimes but not always true.

73.　A contradiction is an inequality that is always false; an identity is an inequality that is always true.

74.　The properties of equations and the properties of inequalities are the same, with one exception. When we multiply or divide both signs of an inequality by the same <u>negative</u> quantity, we must reverse the inequality sign in the resulting inequality.

75.　Two inequalities are called equivalent if they have exactly the same set of solutions.

76.　It is impossible to check every number in the set of solutions of an inequality, since there are infinitely many such numbers. We check an inequality by checking to see if an endpoint is the solution to the corresponding equation. Then we choose one additional value from the set of solutions to see if the inequality sign is facing the right direction.

Exercises 2.4

1. Let x = a number satisfying the given condition.

$$3x - 4 < 17$$
$$3x - 4 + 4 < 17 + 4$$
$$3x < 21$$
$$\frac{3x}{3} < \frac{21}{3}$$
$$x < 7$$

Thus, any number less than 7 satisfies the given condition.

5. Let x = a number satisfying the given condition.

$$4x - 3 \geq 7x - 3$$
$$4x - 3 - 4x \geq 7x - 3 - 4x$$
$$-3 \geq 3x - 3$$
$$0 \geq 3x$$
$$\frac{0}{3} \geq \frac{3x}{3}$$
$$0 \geq x \text{ or } x \leq 0$$

Thus, the largest number satisfying the given condition is 0.

9. Let L = length of the rectangle (in cm).

$$2L + 2(8) \geq 80$$
$$2L + 16 \geq 80$$
$$2L + 16 - 16 \geq 80 - 16$$
$$2L \geq 64$$
$$\frac{2L}{2} \geq \frac{64}{2}$$
$$L \geq 32$$

Thus, the length must be at least 32 cm.

13. Let W = the width of the rectangle (in feet). Then 3W = length of the rectangle (in feet).

$$100 < 2W + 2(3W) < 200$$
$$100 < 2W + 6W < 200$$
$$100 < 8W < 200$$
$$\frac{100}{8} < \frac{8W}{8} < \frac{200}{8}$$
$$12.5 < W < 25$$

Thus, the width is between 12.5 feet and 25 feet.

17. Let x = length of the shortest side (in cm). Then x + 2 = length of the medium side (in cm) and 2x = length of the longest side (in cm).

$$30 \leq x + x + 2 + 2x \leq 50$$
$$30 \leq 4x + 2 \leq 50$$
$$30 - 2 \leq 4x + 2 - 2 \leq 50 - 2$$
$$28 \leq 4x \leq 48$$
$$\frac{28}{4} \leq \frac{4x}{4} \leq \frac{48}{4}$$
$$7 \leq x \leq 12$$

Thus, the shortest side of the triangle is at least 7 cm and no more than 12 cm.

19. Let x = price of a ticket at the door (in dollars). Then x + 2 = price of a reserved seat ticket (in dollars).
$$150x + 300(x + 2) \geq 3750$$
$$150x + 300x + 600 \geq 3750$$
$$450x + 600 \geq 3750$$
$$450x + 600 - 600 \geq 3750 - 600$$
$$450x \geq 3150$$
$$\frac{450x}{450} \geq \frac{3150}{450}$$
$$x \geq 7$$

Thus, the minimum price that the organization can charge for a reserved seat ticket is 7 + 2 = 9 dollars.

23. Let d = # of dimes that Xerxes has. Then 40 − d = # of nickels that Xerxes has.
$$5(40 - d) + 10d \leq 285$$
$$200 - 5d + 10d \leq 285$$
$$200 + 5d \leq 285$$
$$200 + 5d - 200 \leq 285 - 200$$
$$5d \leq 85$$
$$\frac{5d}{5} \leq \frac{85}{5}$$
$$d \leq 17$$

Thus, the maximum number of dimes Xerxes has is 17.

27. Let x = # of hours Ian tutors. Then 30 − x = # of hours Ian works at a restaurant.
$$8x + 6(30 - x) \geq 250$$
$$8x + 180 - 6x \geq 250$$
$$2x + 180 \geq 250$$
$$2x + 180 - 180 \geq 250 - 180$$
$$2x \geq 70$$
$$\frac{2x}{2} \geq \frac{70}{2}$$
$$x \geq 35$$

Since the problem implies that x is at most 30, it is impossible for Ian to make at least $250.

Exercises 2.5

1. $|x| = 4$

   $x = 4$ or $x = -4$

   CHECK x = 4:   CHECK x = −4:

   $|x| = 4$ $|x| = 4$

23

$$|4| \overset{?}{=} 4 \qquad\qquad\qquad |-4| \overset{?}{=} 4$$
$$4 \overset{\checkmark}{=} 4 \qquad\qquad\qquad 4 \overset{\checkmark}{=} 4$$

5. $|x| > 4$

   $x < -4$ or $x > 4$

7. $|x| \le 4$

   $-4 \le x \le 4$

11. $|x| = -4$ has no solution, since the absolute value of an expression cannot be negative.

13. $|x| > -4$ is an identity, since the absolute value of x is at least 0 and is therefore greater than -4.

17. $|t - 3| = 2$

   $t - 3 = 2$ or $t - 3 = -2$

   $t = 5$ or $t = 1$

   CHECK $t = 5$:
   $|t - 3| = 2$
   $|5 - 3| \overset{?}{=} 2$
   $|2| \overset{?}{=} 2$
   $2 \overset{\checkmark}{=} 2$

   CHECK $t = 1$:
   $|t - 3| = 2$
   $|1 - 3| \overset{?}{=} 2$
   $|-2| \overset{?}{=} 2$
   $2 \overset{\checkmark}{=} 2$

19. $|t| - 3 = 2$

   $|t| = 5$

   $t = 5$ or $t = -5$

   CHECK $t = 5$:
   $|t| - 3 = 2$
   $|5| - 3 \overset{?}{=} 2$
   $5 - 3 \overset{?}{=} 2$
   $2 \overset{\checkmark}{=} 2$

   CHECK $t = -5$:
   $|t| - 3 = 2$
   $|-5| - 3 \overset{?}{=} 2$
   $5 - 3 \overset{?}{=} 2$
   $2 \overset{\checkmark}{=} 2$

23. $|a - 5| < 3$

   $-3 < a - 5 < 3$

   $-3 + 5 < a - 5 + 5 < 3 + 5$

   $2 < a < 8$

25. $|a - 1| \ge 2$

   $a - 1 \le -2$ or $a - 1 \ge 2$

   $a - 1 + 1 \le -2 + 1$ or $a - 1 + 1 \ge -2 + 1$

   $a \le -1$ or $a \ge 3$

27. $|2a - 5| < -1$ is a contradiction, since the absolute value of a quantity cannot be negative.

31. $|3x - 2| - 3 = 1$

   $|3x - 2| = 4$

   $3x - 2 = 4$ or $3x - 2 = -4$

   $3x = 6$ or $3x = -2$

   $x = 2$ or $x = -\dfrac{2}{3}$

$$|3x - 2| - 3 = 1$$

$$|3(2) - 2| - 3 \overset{?}{=} 1$$

$$|6 - 2| - 3 \overset{?}{=} 1$$

$$|4| - 3 \overset{?}{=} 1$$

$$4 - 3 \overset{?}{=} 1$$

$$1 \overset{\checkmark}{=} 1$$

$$|3x - 2| - 3 = 1$$

$$\left|3\left(-\frac{2}{3}\right) - 2\right| - 3 \overset{?}{=} 1$$

$$|-2 - 2| - 3 \overset{?}{=} 1$$

$$|-4| - 3 \overset{?}{=} 1$$

$$4 - 3 \overset{?}{=} 1$$

$$1 \overset{\checkmark}{=} 1$$

35. $|2(x - 1) + 7| = 5$

$$|2x - 2 + 7| = 5$$

$$|2x + 5| = 5$$

$$2x + 5 = 5 \text{ or } 2x + 5 = -5$$

$$2x = 0 \text{ or } 2x = -10$$

$$x = 0 \text{ or } x = -5$$

CHECK x = 0:

$$|2(x - 1) + 7| = 5$$

$$|2(0 - 1) + 7| \overset{?}{=} 5$$

$$|2(-1) + 7| \overset{?}{=} 5$$

$$|-2 + 7| \overset{?}{=} 5$$

$$|5| \overset{?}{=} 5$$

$$5 \overset{\checkmark}{=} 5$$

CHECK x = -5:

$$|2(x - 1) + 7| = 5$$

$$|2(-5 - 1) + 7| \overset{?}{=} 5$$

$$|2(-6) + 7| \overset{?}{=} 5$$

$$|-12 + 7| \overset{?}{=} 5$$

$$|-5| \overset{?}{=} 5$$

$$5 \overset{\checkmark}{=} 5$$

39. $|5 - 2a| > 1$

$5 - 2a < -1$	or	$5 - 2a > 1$
$5 - 2a - 5 < -1 - 5$	or	$5 - 2a - 5 > 1 - 5$
$-2a < -6$	or	$-2a > -4$
$\dfrac{-2a}{-2} > \dfrac{-6}{-2}$	or	$\dfrac{-2a}{-2} < \dfrac{-4}{-2}$
$a > 3$	or	$a < 2$

43. $|x - 1| = 5$

$$x - 1 = 5 \text{ or } x - 1 = -5$$

$$x = 6 \text{ or } x = -4$$

47. $|2x + 7| > 1$

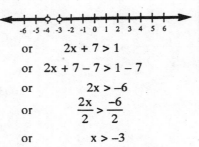

$$2x + 7 < -1 \qquad\qquad \text{or} \qquad 2x + 7 > 1$$
$$2x + 7 - 7 < -1 - 7 \qquad \text{or} \quad 2x + 7 - 7 > 1 - 7$$
$$2x < -8 \qquad\qquad \text{or} \qquad 2x > -6$$
$$\frac{2x}{2} < \frac{-8}{2} \qquad\qquad \text{or} \qquad \frac{2x}{2} > \frac{-6}{2}$$
$$x < -4 \qquad\qquad \text{or} \qquad x > -3$$

51. $|3 - 4x| < 1$

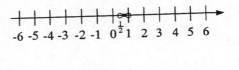

$$-1 < 3 - 4x < 1$$
$$-1 - 3 < 3 - 4x - 3 < 1 - 3$$
$$-4 < -4x < -2$$
$$\frac{-4}{-4} > \frac{-4x}{-4} > \frac{-2}{-4}$$
$$1 > x > \frac{1}{2} \text{ or } \frac{1}{2} < x < 1$$

53. $|5t - 1| = |4t + 3|$

$$5t - 1 = 4t + 3 \qquad\qquad \text{or} \quad 5t - 1 = -(4t + 3)$$
$$t - 1 = 3 \qquad\qquad\qquad \text{or} \quad 5t - 1 = -4t - 3$$
$$t = 4 \qquad\qquad\qquad\quad \text{or} \quad 9t - 1 = -3$$
$$\qquad\qquad\qquad\qquad\qquad 9t = -2$$
$$\qquad\qquad\qquad\qquad\qquad t = -\frac{2}{9}$$

CHECK $t = 4$:   CHECK $t = -\frac{2}{9}$:

$$|5t - 1| = |4t + 3| \qquad\qquad\qquad |5t - 1| = |4t + 3|$$

$$|5(4) - 1| \overset{?}{=} |4(4) + 3| \qquad \left|5\left(-\frac{2}{9}\right) - 1\right| \overset{?}{=} \left|4\left(-\frac{2}{9}\right) + 3\right|$$

$$|20 - 1| \overset{?}{=} |16 + 3| \qquad\qquad \left|-\frac{10}{9} - 1\right| \overset{?}{=} \left|-\frac{8}{9} + 3\right|$$

$$|19| \overset{\checkmark}{=} |19| \qquad\qquad\qquad \left|-\frac{19}{9}\right| \overset{?}{=} \left|\frac{19}{9}\right|$$

$$\qquad\qquad\qquad\qquad\qquad \frac{19}{9} \overset{\checkmark}{=} \frac{19}{9}$$

57. $|a - 5| = |2 - a|$

$$a - 5 = 2 - a \qquad\qquad \text{or} \quad a - 5 = -(2 - a)$$
$$2a - 5 = 2 \qquad\qquad\qquad \text{or} \quad a - 5 = -2 + a$$
$$2a = 7 \qquad\qquad\qquad\quad \text{or} \quad -5 = 2$$
$$a = \frac{7}{2} \qquad\qquad\qquad\qquad \text{a contradiction}$$

CHECK $a = \dfrac{7}{2}$:

$$|a - 5| = |2 - a|$$

$$\left|\dfrac{7}{2} - 5\right| \overset{?}{=} \left|2 - \dfrac{7}{2}\right|$$

$$\left|-\dfrac{3}{2}\right| \overset{?}{=} \left|\dfrac{3}{2}\right|$$

$$\dfrac{3}{2} \overset{\checkmark}{=} \dfrac{3}{2}$$

59.　$|3x - 4| = |4x - 3|$

$3x - 4 = 4x - 3$　　　　　　　or　$3x - 4 = -(4x - 3)$

$-4 = x - 3$　　　　　　　　　or　$3x - 4 = -4x + 3$

$-1 = x$　　　　　　　　　　　or　$7x - 4 = 3$

$7x = 7$

$x = 1$

CHECK $x = -1$:

$$|3x - 4| = |4x - 3|$$

$$|3(-1) - 4| \overset{?}{=} |4(-1) - 3|$$

$$|-3 - 4| \overset{?}{=} |-4 - 3|$$

$$|-7| \overset{\checkmark}{=} |-7|$$

CHECK $x = 1$:

$$|3x - 4| = |4x - 3|$$

$$|3(1) - 4| \overset{?}{=} |4(1) - 3|$$

$$|3 - 4| \overset{?}{=} |4 - 3|$$

$$|-1| \overset{?}{=} |1|$$

$$1 \overset{\checkmark}{=} 1$$

63.　The inequality $|x| < a$ tells us that x is a number that is less than a units away from 0 on the number line. To solve such an inequality, we replace it with the double inequality $-a < x < a$. The inequality $|x| > a$ tells us that x is a number that is more than a units away from 0 on the number line. To solve such an inequality, we replace it with two separate inequalities: $x < -a$ or $x > a$.

Chapter 2 Review Exercises

1.　$5x - 2 = -2$

$5x = 0$

$x = 0$

CHECK $x = 0$:

$5x - 2 = -2$

$5(0) - 2 \overset{?}{=} -2$

$0 - 2 \overset{?}{=} -2$

$-2 \overset{\checkmark}{=} -2$

3.　$3x - 5 = 2x + 6$

$x - 5 = 6$

$x = 11$

CHECK x = 11:
$$3x - 5 = 2x + 6$$
$$3(11) - 5 \stackrel{?}{=} 2(11) + 6$$
$$33 - 5 \stackrel{?}{=} 22 + 6$$
$$28 \stackrel{\checkmark}{=} 28$$

5. $11x + 2 = 6x - 3$
$$5x + 2 = -3$$
$$5x = -5$$
$$x = -1$$

CHECK x = -1:
$$11x + 2 = 6x - 3$$
$$11(-1) + 2 \stackrel{?}{=} 6(-1) - 3$$
$$-11 + 2 \stackrel{?}{=} -6 - 3$$
$$-9 \stackrel{\checkmark}{=} -9$$

7. $5(a - 3) = 2(a - 4)$
$$5a - 15 = 2a - 8$$
$$3a - 15 = -8$$
$$3a = 7$$
$$a = \frac{7}{3}$$

CHECK $a = \frac{7}{3}$:
$$5(a - 3) = 2(a - 4)$$
$$5\left(\frac{7}{3} - 3\right) \stackrel{?}{=} 2\left(\frac{7}{3} - 4\right)$$
$$5\left(-\frac{2}{3}\right) \stackrel{?}{=} 2\left(-\frac{5}{3}\right)$$
$$\frac{-10}{3} \stackrel{\checkmark}{=} \frac{-10}{3}$$

9. $6(q - 4) + 2(q + 5) = 8q - 19$
$$6q - 24 + 2q + 10 = 8q - 19$$
$$8q - 14 = 8q - 19$$
$$-14 = -19 \qquad\qquad \text{contradiction (no solution)}$$

11. $3x - 5x = 7x - 4x$
$$-2x = 3x$$
$$0 = 5x$$
$$0 = x$$

CHECK x = 0:
$$3x - 5x = 7x - 4x$$
$$3(0) - 5(0) \stackrel{?}{=} 7(0) - 4(0)$$
$$0 - 0 \stackrel{?}{=} 0 - 0$$
$$0 \stackrel{\checkmark}{=} 0$$

13. $6(8-a) = 3(a-4) + 2(a-1)$

$48 - 6a = 3a - 12 + 2a - 2$

$48 - 6a = 5a - 14$

$48 = 11a - 14$

$62 = 11a$

$\dfrac{62}{11} = a$

CHECK $a = \dfrac{62}{11}$:

$6(8-a) = 3(a-4) + 2(a-1)$

$6\left(8 - \dfrac{62}{11}\right) \overset{?}{=} 3\left(\dfrac{62}{11} - 4\right) + 2\left(\dfrac{62}{11} - 1\right)$

$6\left(\dfrac{26}{11}\right) \overset{?}{=} 3\left(\dfrac{18}{11}\right) + 2\left(\dfrac{51}{11}\right)$

$\dfrac{156}{11} \overset{?}{=} \dfrac{54}{11} + \dfrac{102}{11}$

$\dfrac{156}{11} \overset{\checkmark}{=} \dfrac{156}{11}$

15. $7\big[x - 3(x-3)\big] = 4x - 2$

$7\big[x - 3x + 9\big] = 4x - 2$

$7\big[-2x + 9\big] = 4x - 2$

$-14x + 63 = 4x - 2$

$63 = 18x - 2$

$65 = 18x$

$\dfrac{65}{18} = x$

CHECK $x = \dfrac{65}{18}$:

$7\big[x - 3(x-3)\big] = 4x - 2$

$7\left[\dfrac{65}{18} - 3\left(\dfrac{65}{18} - 3\right)\right] \overset{?}{=} 4\left(\dfrac{65}{18}\right) - 2$

$7\left[\dfrac{65}{18} - 3\left(\dfrac{11}{18}\right)\right] \overset{?}{=} \dfrac{260}{18} - 2$

$7\left[\dfrac{65}{18} - \dfrac{33}{18}\right] \overset{?}{=} \dfrac{224}{18}$

$7\left[\dfrac{32}{18}\right] \overset{?}{=} \dfrac{224}{18}$

$\dfrac{224}{18} \overset{\checkmark}{=} \dfrac{224}{18}$

17. $5\{x-[2-(x-3)]\} = x - 2$

$5\{x-[2-x+3]\} = x-2$

$5\{x-[5-x]\} = x-2$

$5\{x-5+x\} = x-2$

$5\{2x-5\} = x-2$

$10x-25 = x-2$

$9x - 25 = -2$

$9x = 23$

$x = \dfrac{23}{9}$

CHECK $x = \dfrac{23}{9}$:

$5\{x-[2-(x-3)]\} = x-2$

$5\left\{\dfrac{23}{9} - \left[2-\left(\dfrac{23}{9}-3\right)\right]\right\} \overset{?}{=} \dfrac{23}{9} - 2$

$5\left\{\dfrac{23}{9} - \left[2-\left(-\dfrac{4}{9}\right)\right]\right\} \overset{?}{=} \dfrac{5}{9}$

$5\left\{\dfrac{23}{9} - \left[2+\left(\dfrac{4}{9}\right)\right]\right\} \overset{?}{=} \dfrac{5}{9}$

$5\left\{\dfrac{23}{9} - \dfrac{22}{9}\right\} \overset{?}{=} \dfrac{5}{9}$

$5\left(\dfrac{1}{9}\right) \overset{?}{=} \dfrac{5}{9}$

$\dfrac{5}{9} \overset{\checkmark}{=} \dfrac{5}{9}$

19. $|x| = 3$

$x = 3$ or $x = -3$

CHECK $x = 3$:

$|x| = 3$

$|3| \overset{?}{=} 3$

$3 \overset{\checkmark}{=} 3$

CHECK $x = -3$:

$|x| = 3$

$|-3| \overset{?}{=} 3$

$3 \overset{\checkmark}{=} 3$

21. $|x| = -3$ is a contradiction, since the absolute value of a quantity cannot be negative.

23. $|2x| = 8$

$2x = 8$ or $2x = -8$

$x = 4$ or $x = -4$

CHECK x = 4:

$$|2x| = 8$$

$$|2(4)| \stackrel{?}{=} 8$$

$$|8| \stackrel{?}{=} 8$$

$$8 \stackrel{\checkmark}{=} 8$$

CHECK x = –4:

$$|2x| = 8$$

$$|2(-4)| \stackrel{?}{=} 8$$

$$|-8| \stackrel{?}{=} 8$$

$$8 \stackrel{\checkmark}{=} 8$$

25. $|a + 1| = 4$

$$a + 1 = 4 \text{ or } a + 1 = -4$$

$$a = 3 \text{ or } a = -5$$

CHECK a = 3:

$$|a + 1| = 4$$

$$|3 + 1| \stackrel{?}{=} 4$$

$$|4| \stackrel{?}{=} 4$$

$$4 \stackrel{\checkmark}{=} 4$$

CHECK a = –5:

$$|a + 1| = 4$$

$$|-5 + 1| \stackrel{?}{=} 4$$

$$|-4| \stackrel{?}{=} 4$$

$$4 \stackrel{\checkmark}{=} 4$$

27. $|4z + 5| = 0$

$$4z + 5 = 0$$

$$4z = -5$$

$$z = -\frac{5}{4}$$

CHECK $z = -\frac{5}{4}$:

$$|4z + 5| = 0$$

$$\left|4\left(-\frac{5}{4}\right) + 5\right| \stackrel{?}{=} 0$$

$$|-5 + 5| \stackrel{?}{=} 0$$

$$|0| \stackrel{?}{=} 0$$

$$0 \stackrel{\checkmark}{=} 0$$

29. $|2y - 5| - 8 = -3$

$$|2y - 5| = 5$$

$$2y - 5 = 5 \text{ or } 2y - 5 = -5$$

$$2y = 10 \text{ or } 2y = 0$$

$$y = 5 \text{ or } y = 0$$

CHECK y = 5:

$$|2y - 5| - 8 = -3$$

$$|2(5) - 5| - 8 \stackrel{?}{=} -3$$

$$|10 - 5| - 8 \stackrel{?}{=} -3$$

$$|5| - 8 \stackrel{?}{=} -3$$

$$5 - 8 \stackrel{?}{=} -3$$

$$-3 \stackrel{\checkmark}{=} -3$$

CHECK y = 0:

$$|2y - 5| - 8 = -3$$

$$|2(0) - 5| - 8 \stackrel{?}{=} -3$$

$$|0 - 5| - 8 \stackrel{?}{=} -3$$

$$|-5| - 8 \stackrel{?}{=} -3$$

$$5 - 8 \stackrel{?}{=} -3$$

$$-3 \stackrel{\checkmark}{=} -3$$

31. $|x - 5| = |x - 1|$

$$x - 5 = x - 1$$

$$-5 = -1$$

or $x - 5 = -(x - 1)$

or $x - 5 = -x + 1$

$$2x - 5 = 1$$

a contradiction

$$2x = 6$$
$$x = 3$$

CHECK $x = 3$:
$$|x - 5| = |x - 1|$$
$$|3 - 5| \overset{?}{=} |3 - 1|$$
$$|-2| \overset{?}{=} |2|$$
$$2 \overset{\checkmark}{=} 2$$

33. $|2t - 4| = |t - 2|$

$$2t - 4 = t - 2 \qquad \text{or} \quad 2t - 4 = -(t - 2)$$
$$t - 4 = -2 \qquad \text{or} \quad 2t - 4 = -t + 2$$
$$t = 2 \qquad \text{or} \quad 3t - 4 = 2$$
$$3t = 6$$
$$t = 2$$

CHECK $t = 2$:
$$|2t - 4| = |t - 2|$$
$$|2(2) - 4| \overset{?}{=} |2 - 2|$$
$$|4 - 4| \overset{?}{=} |2 - 2|$$
$$|0| \overset{\checkmark}{=} |0|$$

35. $3x - 4 \le 5$
$$3x \le 9$$
$$x \le 3$$

37. $5x - 4 \le 2x$
$$3x - 4 \le 0$$
$$3x \le 4$$
$$x \le \frac{4}{3}$$

39. $5z + 4 > 2z - 1$
$$3z + 4 > -1$$
$$3z > -5$$
$$z > -\frac{5}{3}$$

41. $5(s - 4) < 3(s - 4)$
$$5s - 20 < 3s - 12$$
$$2s - 20 < -12$$
$$2s < 8$$
$$s < 4$$

43. $3(r - 2) - 5(r - 1) \ge -2r - 1$
$$3r - 6 - 5r + 5 \ge -2r - 1$$
$$-2r - 1 \ge -2r - 1$$

identity

45. $-3 \le 2x - 1 \le 5$

47. $-5 \le 6 - 2x \le 7$

$$-2 \le 2x \le 6$$

$$-1 \le x \le 3$$

$$-11 \le -2x \le 1$$

$$\frac{11}{2} \ge x \ge -\frac{1}{2} \text{ or } -\frac{1}{2} \le x \le \frac{11}{2}$$

49. $-3 \le 3(x-4) \le 6$

$$-3 \le 3x - 12 \le 6$$

$$9 \le 3x \le 18$$

$$3 \le x \le 6$$

51. $3\big[a - 3(a+2)\big] > 0$

$$3\big[a - 3a - 6\big] > 0$$

$$3\big[-2a - 6\big] > 0$$

$$-6a - 18 > 0$$

$$-6a > 18$$

$$a < -3$$

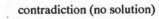

53. $5 - \big[2 - 3(2-x)\big] < -3x + 1$

$$5 - \big[2 - 6 + 3x\big] < -3x + 1$$

$$5 - \big[-4 + 3x\big] < -3x + 1$$

$$5 + 4 - 3x < -3x + 1$$

$$9 - 3x < -3x + 1$$

$$9 < 1$$

contradiction (no solution)

55. $3 - \Big\{2 - 5\big[q - (2-q)\big]\Big\} \le 3\big[q - (q-5)\big]$

$$3 - \Big\{2 - 5\big[q - 2 + q\big]\Big\} \le 3\big[q - q + 5\big]$$

$$3 - \Big\{2 - 5\big[2q - 2\big]\Big\} \le 3\big[5\big]$$

$$3 - \big\{2 - 10q + 10\big\} \le 15$$

$$3 - \big\{12 - 10q\big\} \le 15$$

$$3 - 12 + 10q \le 15$$

$$-9 + 10q \le 15$$

$$10q \le 24$$

$$q \le 2.4$$

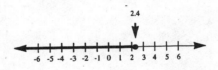

57. $|x| < 4$

$$-4 < x < 4$$

59. $|s| \ge 5$

$$s \le -5 \text{ or } s \ge 5$$

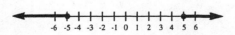

61. $|t| < 0$

contradiction

63. $|t - 1| < 2$

$-2 < t - 1 < 2$

$-1 < t < 3$

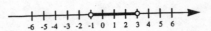

65. $|a - 6| \geq 3$

$a - 6 \leq -3$ or $a - 6 \geq 3$

$a \leq 3$ or $a \geq 9$

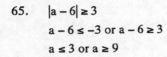

67. $|r + 9| \leq 4$

$-4 \leq r + 9 \leq 4$

$-13 \leq r \leq -5$

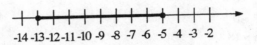

69. $|2x - 1| \geq 2$

$2x - 1 \leq -2$ or $2x - 1 \geq 2$

$2x \leq -1$ or $2x \geq 3$

$x \leq -\dfrac{1}{2}$ or $x \geq \dfrac{3}{2}$

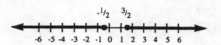

71. $|3x - 2| < 4$

$-4 < 3x - 2 < 4$

$-2 < 3x < 6$

$-\dfrac{2}{3} < x < 2$

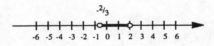

73. $|2x - 3| \leq 5$

$-5 \leq 2x - 3 \leq 5$

$-2 \leq 2x \leq 8$

$-1 \leq x \leq 4$

75. $|3x + 5| + 2 > 7$

$|3x + 5| > 5$

$3x + 5 < -5$ or $3x + 5 > 5$

$3x < -10$ or $3x > 0$

$x < -\dfrac{10}{3}$ or $x > 0$

77. $|3 - 2x| + 8 \leq 4$

$|3 - 2x| \leq -4$

contradiction (no solution)

79. Let n = the number

$3n = 4n - 4$

$-n = -4$

$n = 4$

Thus, the number in question is 4.

CHECK: 3 times 4 is 12, which is 4 less than 4 times 4 or 16.

81. Let n = the number
$$5(n + 6) = n - 2$$
$$5n + 30 = n - 2$$
$$4n + 30 = -2$$
$$4n = -32$$
$$n = -8$$

Thus, the number in question is –8.

CHECK: 5 times the sum of –8 and 6 = 5(–2) = –10, which is 2 less than –8.

83. Let n = # of nickels in Harold's pockets. Then 35 – n = # of dimes in Harold's pockets.
$$5n + 10(35 - n) = 345$$
$$5n + 350 - 10n = 345$$
$$-5n + 350 = 345$$
$$-5n = -5$$
$$n = 1$$

Then 35 – n = 35 – 1 = 34. Thus, Harold has 1 nickel and 34 dimes in his pockets.

CHECK:

$$1 \text{ nickel} = 1(5¢) = 5¢$$
$$\underline{34 \text{ dimes} = 34 \,(10¢) = 340¢}$$
$$\text{total value } = \ 345¢ \overset{\checkmark}{=} \$3.45$$

85. Let n = # of nickels that Alex has. Then 42 – n = # of quarters that Alex has.
$$5n + 25(42 - n) = 750$$
$$5n + 1050 - 25n = 750$$
$$-20n + 1050 = 750$$
$$-20n = -300$$
$$n = 15$$

Then 42 – n = 42 – 15 = 27. Thus, Alex has 15 nickels and 27 quarters in his pockets.

CHECK:

$$15 \text{ nickels} = 15(5¢) = 75¢$$
$$\underline{27 \text{ quarters} = 27(25¢) = 675¢}$$
$$\text{total value } = 750¢ = \$7.50$$

87. Let q = # of quarters that Cindy has. Then q + 3 = # of nickels that Cindy has and 70 – q – (q + 3) = 67 – 2q = # of dimes that Cindy has.
$$25q + 5(q + 3) + 10(67 - 2q) = 715$$

$$(x+1)(x-10) = x^2 - 9x - 10$$
$$(x-1)(x+10) = x^2 + 9x - 10$$
$$(x+2)(x-5) = x^2 - 3x - 10$$
$$(x-2)(x+5) = x^2 + 3x - 10$$

Thus, the values of k such that $x^2 + kx - 10$ factors are ± 9 and ± 3.

(c) Possible factors of 3 are $3 \cdot 1$. Possible factors of 2 are $2 \cdot 1$.
$$(2x-1)(x+3) = 2x^2 + 5x - 3$$
$$(2x+1)(x-3) = 2x^2 - 5x - 3$$
$$(2x-3)(x+1) = 2x^2 - x - 3$$
$$(2x+3)(x-1) = 2x^2 + x - 3$$

Thus, the values of k such that $2x^2 + kx - 3$ factors are ± 5 and ± 1.

69.   Suppose that $x^2 + 3x + p$ factors into $(x + a)(x + b)$. The a and b are two integers with the property that their sum is 3 and their product is p. Since $-20 < p < 20$, there are only five possibilities to consider for the pair of numbers a and b:

$$-3 \text{ and } 6 \rightarrow p = -18$$
$$-2 \text{ and } 5 \rightarrow p = -10$$
$$-1 \text{ and } 4 \rightarrow p = -4$$
$$0 \text{ and } 3 \rightarrow p = 0$$
$$1 \text{ and } 2 \rightarrow p = 2$$

70.   (a) Let $u = x + y$. Then
$$3(x+y)^2 - 4(x+y) - 4 = 3u^2 - 4u - 4 = (3u + 2)(u - 2)$$
$$= \big(3(x+y) + 2\big)\big((x+y) - 2\big) = (3x + 3y + 2)(x + y - 2)$$

(b) $2a^2 + 4ab + 2b^2 - 7a - 7b + 3 = \left(2a^2 + 4ab + 2b^2\right) - 7(a + b) + 3$
$$= 2\left(a^2 + 2ab + b^2\right) - 7(a + b) + 3$$
$$= 2(a + b)^2 - 7(a + b) + 3$$

Let $u = a + b$. Then this becomes
$$2u^2 - 7u + 3 = (2u - 1)(u - 3) = \big(2(a + b) - 1\big)\big((a + b) - 3\big)$$
$$= (2a + 2b - 1)(a + b - 3)$$

Exercises 3.5

3. $x^2 + 4xy + 4y^2 = x^2 + 2(x)(2y) + (2y)^2 = (x + 2y)^2$

5. $x^2 - 4xy + 4y^2 = x^2 - 2(x)(2y) + (2y)^2 = (x - 2y)^2$

7. $x^2 - 4xy - 4y^2$ is not factorable

11. $81r^2s^2 - 16 = (9rs)^2 - 4^2 = (9rs + 4)(9rs - 4)$

13. $49x^2y^2 - 42xy + 9 = (7xy)^2 - 2(7xy)(3) + 3^2 = (7xy - 3)^2$

17. $25a^2 - 4a^2b^2 = a^2(25 - 4b^2) = a^2(5^2 - (2b)^2) = a^2(5 + 2b)(5 - 2b)$

19. $64x^4 - 16x^2y + y^2 = (8x^2)^2 - 2(8x^2)(y) + y^2 = (8x^2 - y)^2$

23. $8a^3 - b^3 = (2a)^3 - b^3 = (2a - b)((2a)^2 + (2a)b + b^2) = (2a - b)(4a^2 + 2ab + b^2)$

25. $x^3 + 125y^3 = x^3 + (5y)^3 = (x + 5y)(x^2 - x(5y) + (5y)^2) = (x + 5y)(x^2 - 5xy + 25y^2)$

27. $4x^2 + 25y^2$ is not factorable

31. $18a^3 + 24a^2b + 8ab^2 = 2a(9a^2 + 12ab + 4b^2)$
$$= 2a((3a)^2 + 2(3a)(2b) + (2b)^2) = 2a(3a + 2b)^2$$

35. $9a^6 - b^6 = (3a^3)^2 - (b^3)^2 = (3a^3 + b^3)(3a^3 - b^3)$

39. $4x^6 + 28x^3y^3 + 49y^6 = (2x^3)^2 + 2(2x^3)(7y^3) + (7y^3)^2 = (2x^3 + 7y^3)^2$

43. $25a^6 + 10a^3b + b^2 = (5a^3)^2 + 2(5a^3)(b) + b^2 = (5a^3 + b)^2$

45. $24x^4y - 54x^2y^3 = 6x^2y(4x^2 - 9y^2) = 6x^2y((2x)^2 - (3y)^2)$
$$= 6x^2y(2x + 3y)(2x - 3y)$$

49. $(a + b)^2 - 4 = (a + b)^2 - 2^2 = (a + b + 2)(a + b - 2)$

53. $8x^4 - 14x^2 + 3 = (2x^2 - 3)(4x^2 - 1) = (2x^2 - 3)((2x)^2 - 1^2)$
$$= (2x^2 - 3)(2x + 1)(2x - 1)$$

57. $9a^5b - 12a^3b^3 + 3ab^5 = 3ab(3a^4 - 4a^2b^2 + b^4) = 3ab(3a^2 - b^2)(a^2 - b^2)$

$$= 3ab(3a^2 - b^2)(a+b)(a-b)$$

61. $(x+y)^3 - (a+b)^3 = \left[(x+y) - (a+b)\right]\left[(x+y)^2 + (x+y)(a+b) + (a+b)^2\right]$

$$= (x+y-a-b)(x^2 + 2xy + y^2 + xa + xb + ya + yb + a^2 + 2ab + b^2)$$

65. $x^2 + 6x + 9 - r^2 = (x+3)^2 - r^2 = (x+3+r)(x+3-r)$

67. $a^3 + a^2 - 4a - 4 = a^2(a+1) - 4(a+1) = (a+1)(a^2 - 4)$

$$= (a+1)(a^2 - 2^2) = (a+1)(a+2)(a-2)$$

71. $a^5 - a^3 + a^2 - 1 = a^3(a^2 - 1) + (a^2 - 1) = (a^2 - 1)(a^3 + 1)$

$$= (a^2 - 1)\left((a+1)(a^2 - a + 1)\right) = (a+1)(a-1)(a+1)(a^2 - a + 1)$$

73. The first and last terms of a perfect square trinomial are perfect squares. The middle term is either plus or minus twice the product of the quantites whose squares appear as the first and last terms.

74. The difference of the cubes of two quantities factors into a binomial and a trinomial. The binomial is the difference of the two quantities; the trinomial is the sum of the square of the first quantity, the product of the two quantities, and the square of the second quantity.

The sum of the cubes of two quantities also factors into a binomial and a trinomial. The binomial is the sum of the two quantities; the trinomial is the sum of the square of the first quantity, minus the product of the two quantities, and the square of the second quantity.

75. $a^3 - b^3$ is the difference of two cubes. We first cube the quantities, then take the difference of their values.

$(a - b)^3$ is the cube of a difference. Here, we begin by finding the difference of the two quantities, then computing the cube of the result.

In general, $a^3 - b^3$ and $(a - b)^3$ are not equal. For example, $2^3 - 1^3 = 8 - 1 = 7$, but $(2-1)^3 = 1^3 = 1$.

76. (a) $x^6 - y^6 = (x^3)^2 - (y^3)^2 = (x^3 + y^3)(x^3 - y^3)$

$$= (x + y)(x^2 - xy + y^2)(x - y)(x^2 + xy + y^2)$$

$$= (x + y)(x - y)(x^2 - xy + y^2)(x^2 + xy + y^2)$$

(b) $x^6 - y^6 = (x^2)^3 - (y^2)^3 = (x^2 - y^2)((x^2)^2 + (x^2)(y^2) + (y^2)^2)$

$$= (x^2 - y^2)(x^4 + x^2y^2 + y^4) = (x + y)(x - y)(x^4 + x^2y^2 + y^4)$$

(c) To verify that the two factorizations are the same, we must show that

$$(x^2 - xy + y^2)(x^2 + xy + y^2) = x^4 + x^2y^2 + y^4$$

But we can rewrite the left side as
$((x^2 + y^2) - xy)((x^2 + y^2) + xy)$, which then equals

$(x^2 + y^2)^2 - (xy)^2$. Continuing, we get

$$(x^2)^2 + 2x^2y^2 + (y^2)^2 - x^2y^2 = x^4 + 2x^2y^2 + y^4 - x^2y^2 = x^4 + x^2y^2 + y^4,$$

as required.

77. The factor $a^2b - b$ can be factored further as follows:

$$a^2b - b = b(a^2 - 1) = b(a + 1)(a - 1).$$

Thus, the complete factorization of $a^3b + a^2b - ab - b$ is

$$b(a + 1)(a - 1)(a + 1) \text{ or } b(a + 1)^2(a - 1).$$

Exercises 3.6

1.
$$
\begin{array}{r}
x + 4 \\
x - 5 \overline{\smash{)}\, x^2 - x - 20} \\
\underline{-(x^2 - 5x)} \\
4x - 20 \\
\underline{-(4x - 20)} \\
0
\end{array}
$$

Answer: $x + 4$

5.
$$
\begin{array}{r}
7z - 2 \\
3z + 1 \overline{\smash{)}\, 21z^2 + z - 16} \\
\underline{-(21z^2 + 7z)} \\
-6z - 16 \\
\underline{-(-6z - 2)} \\
-14
\end{array}
$$

Answer: $7z - 2$, R-14 or

$$7z - 2 - \frac{14}{3z + 1}$$

9.
$$\begin{array}{r}
3x^2 + x - 1 \\
x - 4 \overline{\smash{\big)}\ 3x^3 - 11x^2 - 5x + 12} \\
\underline{-(3x^3 - 12x^2)} \\
x^2 - 5x + 12 \\
\underline{-(x^2 - 4x)} \\
-x + 12 \\
\underline{-(-x + 4)} \\
8
\end{array}$$

Answer: $3x^2 + x - 1$, R8 or

$$3x^2 + x - 1 + \frac{8}{x - 4}$$

11.
$$\begin{array}{r}
2z - 3 \\
2z + 3 \overline{\smash{\big)}\ 4z^2 + 0z - 15} \\
\underline{-\left(4z^2 + 6z\right)} \\
-6z - 15 \\
\underline{-(-6z - 9)} \\
-6
\end{array}$$

Answer: $2z - 3$, R $- 6$ or

$$2z - 3 - \frac{6}{2z + 3}$$

15.
$$\begin{array}{r}
3y^3 - 2y^2 + 5 \\
y - 1 \overline{\smash{\big)}\ 3y^4 - 5y^3 + 2y^2 + 5y - 10} \\
\underline{-(3y^4 - 3y^3)} \\
-2y^3 + 2y^2 + 5y - 10 \\
\underline{-(-2y^3 + 2y^2)} \\
5y - 10 \\
-(5y - 5) \\
-5
\end{array}$$

Answer: $3y^3 - 2y^2 + 5$, R $- 5$ or

$$3y^3 - 2y^2 + 5 - \frac{5}{y - 1}$$

17.

$$\begin{array}{r} 2a^3 - 5 \\ a + 2 \overline{\smash{\big)}\ 2a^4 + 4a^3 + 0a^2 - 5a + 6} \\ \underline{-(2a^4 + 4a^3)} \\ 5a + 6 \\ \underline{-(-5a - 10)} \\ 16 \end{array}$$

Answer: $2a^3 - 5$, R16 or
$$2a^3 - 5 + \frac{16}{a + 2}$$

21.

$$\begin{array}{r} 4a^2 + 2a + 1 \\ 2a - 1 \overline{\smash{\big)}\ 8a^3 + 0a^2 + 0a + 1} \\ \underline{-(8a^3 - 4a^2)} \\ 4a^2 + 0a + 1 \\ \underline{-(4a^2 - 2a)} \\ 2a + 1 \\ \underline{-(2a - 1)} \\ 2 \end{array}$$

Answer: $4a^2 + 2a + 1$, R2 or
$$4a^2 + 2a + 1 + \frac{2}{2a - 1}$$

25.

$$\begin{array}{r} 3z^3 - 6z^2 + 11z - 25 \\ z^2 + 2z - 1 \overline{\smash{\big)}\ 3z^5 + 0z^4 - 4z^3 + 3z^2 + 12z - 10} \\ \underline{-(3z^5 + 6z^4 - 3z^3)} \\ -6z^4 - z^3 + 3z^2 + 12z - 10 \\ \underline{-(-6z^4 - 12z^3 + 6z^2)} \\ 11z^3 - 3z^2 + 12z - 10 \\ \underline{-(11z^3 + 22z^3 - 11z)} \\ -25z^2 + 23z - 10 \\ \underline{-(-25z^2 - 50z + 25)} \\ 73z - 35 \end{array}$$

Answer: $3z^3 - 6z^2 + 11z - 25$, R $(73z - 35)$ or
$$3z^3 - 6z^2 + 11z - 25 + \frac{73z - 35}{z^2 + 2z - 1}$$

27.

$$
x^2 - 2 \overline{\smash{\big)}\ 7x^6 + 0x^5 - 14x^4 + 0x^3 - 3x^2 - 3x^2 + x + 1} \quad \overset{\textstyle 7x^4 - 3}{}
$$

$$
\underline{-(7x^6 - 14x^4)}
$$

$$
-3x^2 + x + 1
$$

$$
\underline{-(-3x^2 \qquad + 6)}
$$

$$
x - 5
$$

Answer: $7x^4 - 3,\ R(x-5)$ or

$$
7x^4 - 3 + \frac{x-5}{x^2 - 2}
$$

Chapter 3 Review Exercises

1. $\left(2x^3 - 4\right) - \left[\left(2x^3 - 3x + 4\right) + \left(5x^2 - 3\right)\right] = \left(2x^3 - 4\right) - \left(7x^2 - 3x + 1\right)$

    $\qquad = 2x^3 - 4 - 7x^2 + 3x - 1 = 2x^3 - 7x^2 + 3x - 5$

3. $3x^2\left(2xy - 3y + 1\right) = 3x^2\left(2xy\right) - 3x^2\left(3y\right) + 3x^2\left(1\right) = 6x^3 y - 9x^2 y + 3x^2$

5. $3ab(2a - 3b) - 2a\left(3ab - 4b^2\right) = 3ab(2a) - 3ab(3b) - 2a(3ab) + 2a\left(4b^2\right)$

    $\qquad = 6a^2 b - 9ab^2 - 6a^2 b + 8ab^2 = -ab^2$

7. $3x - 2(x - 3) - \left[x - 2(5 - x)\right] = 3x - 2(x - 3) - \left[x - 10 + 2x\right]$

    $\qquad = 3x - 2(x - 3) - (3x - 10)$

    $\qquad = 3x - 2x + 6 - 3x + 10 = -2x + 16$

9. $(x - 3)(x + 2) = x^2 + 2x - 3x - 6 = x^2 - x - 6$

11. $(3y - 2)(2y - 1) = (3y)(2y) + (3y)(-1) + (-2)(2y) + (-2)(-1)$

    $\qquad = 6y^2 - 3y - 4y + 2 = 6y^2 - 7y + 2$

13. $(3x - 4y)^2 = (3x)^2 - 2(3x)(4y) + (4y)^2 = 9x^2 - 24xy + 16y^2$

15. $\left(4x^2 - 5y\ \right)^2 = \left(4x^2\right)^2 - 2\left(4x^2\right)(5y) + (5y)^2 = 16x^4 - 40x^2 y + 25y^2$

17. $\left(7x^2 - 5y^3\right)\left(7x^2 + 5y^3\right) = \left(7x^2\right)^2 - \left(5y^3\right)^2 = 49x^4 - 25y^6$

19. $\left(5y^2 - 3y + 7\right)(2y - 3) = 5y^2(2y - 3) - 3y(2y - 3) + 7(2y - 3)$

$$= 5y^2(2y) - 5y^2(3) - 3y(2y) + 3y(3) + 7(2y) - 7(3)$$

$$= 10y^3 - 15y^2 - 6y^2 + 9y + 14y - 21 = 10y^3 - 21y^2 + 23y - 21$$

21. $(2x + 3y + 4)(2x + 3y - 5) = ((2x + 3y) + 4)((2x + 3y) - 5)$

$$= (2x + 3y)^2 - 5(2x + 3y) + 4(2x + 3y) - 20 = (2x + 3y)^2 - (2x + 3y) - 20$$

$$= (2x)^2 + 2(2x)(3y) + (3y)^2 - (2x + 3y) - 20$$

$$= 4x^2 + 12xy + 9y^2 - 2x - 3y - 20$$

23. $\left[(x - y) + 5\right]^2 = (x - y)^2 + 2(x - y)(5) + 5^2 = x^2 - 2xy + y^2 + 10(x - y) + 25$

$$= x^2 - 2xy + y^2 + 10x - 10y + 25$$

25. $(a + b - 4)^2 = ((a + b) - 4)^2 = (a + b)^2 - 2(a + b)(4) + 4^2$

$$= a^2 + 2ab + b^2 - 8(a + b) + 16 = a^2 + 2ab + b^2 - 8a - 8b + 16$$

27. $6x^2y - 12xy^2 + 9xy = 3xy(2x - 4y + 3)$

29. $3x(a + b) - 2(a + b) = (a + b)(3x - 2)$

31. $2(a - b)^2 + 3(a - b) = (a - b)\left[2(a - b) + 3\right] = (a - b)(2a - 2b + 3)$

33. $5ax - 5a + 3bx - 3b = 5a(x - 1) + 3b(x - 1) = (x - 1)(5a + 3b)$

35. $5y^2 - 5y + 3y - 3 = 5y(y - 1) + 3(y - 1) = (y - 1)(5y + 3)$

37. $a^2 + a(a - 10) = a\left[a + (a - 10)\right] = a(2a - 10) = 2a(a - 5)$

39. $2t(t + 2) - (t - 2)(t + 4) = 2t(t) + 2t(2) - \left[t^2 + 4t - 2t - 8\right]$

$$= 2t^2 + 4t - \left[t^2 + 2t - 8\right]$$

$$= 2t^2 + 4t - t^2 - 2t + 8$$

$$= t^2 + 2t + 8, \text{ which is not factorable}$$

41. $x^2 - 2x - 35 = (x - 7)(x + 5)$

43. $a^2 + 5ab - 14b^2 = (a + 7b)(a - 2b)$

45. $35a^2 + 17ab + 2b^2 = (7a + 2b)(5a + 2b)$

47. $a^2 - 6ab^2 + 9b^4 = (a)^2 - 2(a)(3b^2) + (3b^2)^2 = (a - 3b^2)^2$

49. $3a^3 - 21a^2 + 30a = 3a(a^2 - 7a + 10) = 3a(a - 2)(a - 5)$

51. $2x^3 - 50xy^2 = 2x(x^2 - 25y^2) = 2x((x)^2 - (5y)^2) = 2x(x + 5y)(x - 5y)$

53. $6x^2 + 5x - 6 = (3x - 2)(2x + 3)$

55. $8x^3 + 125y^3 = (2x)^3 + (5y)^3 = (2x + 5y)((2x)^2 - (2x)(5y) + (5y)^2)$
$$= (2x + 5y)(4x^2 - 10xy + 25y^2)$$

57. $6a^2 - 17ab - 3b^2 = (6a + b)(a - 3b)$

59. $21a^4 + 41a^2b^2 + 10b^4 = (7a^2 + 2b^2)(3a^2 + 5b^2)$

61. $25x^4 - 40x^2y^2 + 16y^4 = (5x^2)^2 - 2(5x^2)(4y^2) + (4y^2)^2 = (5x^2 - 4y^2)^2$

63. $20x^3y - 60x^2y^2 + 45xy^3 = 5xy(4x^2 - 12xy + 9y^2)$
$$= 5xy((2x)^2 - 2(2x)(3y) + (3y)^2) = 5xy(2x - 3y)^2$$

65. $6a^4b - 8a^3b^2 - 8a^2b^3 = 2a^2b(3a^2 - 4ab - 4b^2) = 2a^2b(3a + 2b)(a - 2b)$

67. $6x^5 - 10x^3 - 4x = 2x(3x^4 - 5x^2 - 2) = 2x(3x^2 + 1)(x^2 - 2)$

69. $(a - b)^2 - 4 = (a - b)^2 - 2^2 = (a - b + 2)(a - b - 2)$

71. $9y^2 + 30y + 25 - 9x^2 = (3y)^2 + 2(3y)(5) + 5^2 - 9x^2 = (3y + 5)^2 - (3x)^2$
$$= (3y + 5 + 3x)(3y + 5 - 3x)$$

73.
$$x - 2 \overline{)3x^3 - 4x^2 + 7x - 5} \qquad \begin{array}{c} 3x^2 + 2x + 11 \end{array}$$

$$\underline{-(3x^3 - 6x^2)}$$

$$2x^2 + 7x - 5$$

$$\underline{-(2x^2 - 4x)}$$

$$11x - 5$$

$$\underline{-(11x - 22)}$$

$$17$$

Answer: $3x^2 + 2x + 11$, R17 or $3x^2 + 2x + 11 + \dfrac{17}{x-2}$

75.
$$2a - 3 \overline{)8a^3 + 0a^2 + 0a - 27} \qquad \begin{array}{c} 4a^2 + 6a - 9 \end{array}$$

$$\underline{-(8a^3 - 12a^2)}$$

$$12a^2 + 0a - 27$$

$$\underline{-(12a^2 - 18a)}$$

$$18a - 27$$

$$\underline{-(18a - 27)}$$

$$0$$

Answer: $4a^2 + 6a + 9$

Chapter 3 Practice Test

1. $(9a^2 + 6ab + 4b^2)(3a - 2b) = 9a^2(3a - 2b) + 6ab(3a - 2b) + 4b^2(3a - 2b)$
    $$= 9a^2(3a) - 9a^2(2b) + 6ab(3a) - 6ab(2b) + 4b^2(3a) - 4b^2(2b)$$
    $$= 27a^3 - 18a^2b + 18a^2b - 12ab^2 + 12ab^2 - 8b^3$$
    $$= 27a^3 - 8b^3$$

3. $(x - y - 2)(x - y + 2) = ((x - y) - 2)((x - y) + 2)$
    $$= (x - y)^2 - 2^2 = x^2 - 2xy + y^2 - 4$$

5. $10x^3y^2 - 6x^2y^3 + 2xy = 2xy(5x^2y - 3xy^2 + 1)$

7. $x^2 + 11xy - 60y^2 = (x + 15y)(x - 4y)$

9. $2x^2 - 3x + 7$ is not factorable

11. $r^3s - 10r^2s^2 + 25rs^3 = rs(r^2 - 10rs + 25s^2) = rs((r^2) - 2(r)(5s) + (5s)^2) = rs(r - 5s)^2$

13. $(2x + y)^2 - 64 = (2x + y)^2 - 8^2 = (2x + y + 8)(2x + y - 8)$

15.

$$\require{enclose}\begin{array}{r}x^2 + 3x + 16 \\ x - 3 \enclose{longdiv}{x^3 + 0x^2 + 7x - 8} \end{array}$$

$$\underline{-(x^3 - 3x^2)}$$
$$3x^2 + 7x - 8$$
$$\underline{-(3x^2 - 9x)}$$
$$16x - 8$$
$$\underline{-(16x - 48)}$$
$$40$$

Answer: $x^2 + 3x + 16$, R40 or $x^2 + 3x + 16 + \dfrac{40}{x - 3}$

Cumulative Review: Chapters 1–3

1. $\{11, 13, 17, 19, 23, 29, 31, 37\}$

3. $\varnothing$

5.   True

7.

9.   Closure property of addition of real numbers

11. $-2 + 3 - 4 - 8 + 5 = 1 - 4 - 8 + 5 = -3 - 8 + 5 = -11 + 5 = -6$

13. $(-9 + 3)(-4 - 2) = (-6)(-6) = +36$

15. $|y - x| - (y - x) = |2 - (-4)| - (2 - (-4)) = |2 + 4| - (2 + 4)$
$$= |6| - 6 = 6 - 6 = 0$$

17. $-3x(x^2 - 2x + 3) = -3x(x^2) + 3x(2x) - 3x(3) = -3x^3 + 6x^2 - 9x$

19. $8n + 3$, where n is the number

21.   Let W = width of the rectangle. Then $4W - 3$ = length of the rectangle.

      perimeter = $2(4W - 3) + 2W = 8W - 6 + 2W = 10W - 6$

      area = $(4W - 3)W = 4W^2 - 3W$

23. $4x - 3 = 5x + 8$

        $-3 = x + 8$

     $-11 = x$

CHECK: $4x - 3 = 5x + 8$

    $4(-11) - 3 \overset{?}{=} 5(-11) + 8$

       $-44 - 3 \overset{?}{=} -55 + 8$

          $-47 \overset{\checkmark}{=} -47$

25. $2x + 3 - (x - 2) = 5 - (x - 4)$

     $2x + 3 - x + 2 = 5 - x + 4$

           $x + 5 = 9 - x$

         $2x + 5 = 9$

            $2x = 4$

             $x = 2$

CHECK: $2x + 3 - (x - 2) = 5 - (x - 4)$

    $2(2) + 3 - (2 - 2) \overset{?}{=} 5 - (2 - 4)$

        $4 + 3 - 0 \overset{?}{=} 5 - (-2)$

          $7 - 0 \overset{?}{=} 5 + 2$

             $7 \overset{\checkmark}{=} 7$

27. $3(x - 4) + 2(3 - x) = 5[x - (2 - 3x)]$

     $3(x - 4) + 2(3 - x) = 5[x - 2 + 3x]$

     $3(x - 4) + 2(3 - x) = 5[4x - 2]$

       $3x - 12 + 6 - 2x = 20x - 10$

              $x - 6 = 20x - 10$

                $-6 = 19x - 10$

                  $4 = 19x$

               $\dfrac{4}{19} = x$

CHECK: $3(x - 4) + 2(3 - x) = 5[x - (2 - 3x)]$

$$3\left(\frac{4}{19} - 4\right) + 2\left(3 - \frac{4}{19}\right) \overset{?}{=} 5\left[\frac{4}{19} - \left(2 - 3\left(\frac{4}{19}\right)\right)\right]$$

$$3\left(-\frac{72}{19}\right) + 2\left(\frac{53}{19}\right) \overset{?}{=} 5\left[\frac{4}{19} - \left(2 - \frac{12}{19}\right)\right]$$

$$-\frac{216}{19} + \frac{106}{19} \overset{?}{=} 5\left[\frac{4}{19} - \frac{26}{19}\right]$$

$$-\frac{110}{19} \overset{?}{=} 5\left[-\frac{22}{19}\right]$$

$$-\frac{110}{19} \overset{\checkmark}{=} -\frac{110}{19}$$

29. $|2x + 1| = 9$

    $2x + 1 = 9$ or $2x + 1 = -9$

        $2x = 8$ or $2x = -10$

          $x = 4$ or $x = -5$

    CHECK $x = 4$:   CHECK $x = -5$:

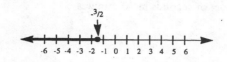

$$|2x + 1| = 9$$
$$|2(4) + 1| \stackrel{?}{=} 9$$
$$|8 + 1| \stackrel{?}{=} 9$$
$$|9| \stackrel{?}{=} 9$$
$$9 \stackrel{\checkmark}{=} 9$$

$$|2x + 1| = 9$$
$$|2(-5) + 1| \stackrel{?}{=} 9$$
$$|-10 + 1| \stackrel{?}{=} 9$$
$$|-9| \stackrel{?}{=} 9$$
$$9 \stackrel{\checkmark}{=} 9$$

31.
$$3x - 5 \leq x - 8$$
$$2x - 5 \leq -8$$
$$2x \leq -3$$
$$x \leq -\frac{3}{2}$$

33.
$$6(x - 2) + (2x - 3) < 2(4x + 1)$$
$$6x - 12 + 2x - 3 < 8x + 2$$
$$8x - 15 < 8x + 2$$
$$-15 < 2$$

identity

35.
$$|x - 5| < 4$$
$$-4 < x - 5 < 4$$
$$1 < x < 9$$

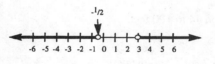

37.
$$|5 - 4x| > 7$$
$$5 - 4x < -7 \text{ or } 5 - 4x > 7$$
$$-4x < -12 \text{ or } -4x > 2$$
$$x > 3 \text{ or } x < -\frac{1}{2}$$

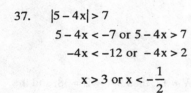

39. Let n = the number.
$$4n = 2n - 3$$
$$2n = -3$$
$$n = -\frac{3}{2}$$

Thus, the number is $-\frac{3}{2}$.

CHECK: $4\left(-\frac{3}{2}\right) = -6$ is less than $2\left(-\frac{3}{2}\right) = -3$, as required.

41. Let x = # of 300-plate packages purchased. Then 28 − x = # of 100-plate packages purchased.
$$2x + 1(28 - x) = 33$$
$$2x + 28 - x = 33$$
$$x + 28 = 33$$
$$x = 5$$

Then $28 - x = 28 - 5 = 23$. Thus, the man purchased 5 300-plate packages and 23 100-plate packages, giving him a total of $5(300) + 23(100) = 1500 + 2300 = 3800$ plates.

CHECK: 5 300-plate packages at $2 per package cost $10. 23 100-plate packages at $1 per package cost $23. $10 + $23 \overset{\checkmark}{=} $33.

43. Let t = # of seconds that a VF-44 printer needs.
   $120t = 80(1800)$ ← the number of seconds in 30 minutes

   $120t = 144000$

   $t = 1200$

   Thus, the VF-44 printer needs 1200 seconds or 20 minutes to print the document.

   CHECK: GL-70 printer prints $80(1800) = 144{,}000$ characters. VF-44 printer prints $120(1200) = 144{,}000$ characters during the same time.

45. Let W = width of the rectangle (in feet). Then 3W + 2 = length of the rectangle (in feet).

   Perimeter $= 2W + 2(3W + 2) = 2W + 6W + 4 = 8W + 4$
   $5 \le W \le 12$
   $40 \le 8w \le 96$
   $44 \le 8w + 4 \le 100$
   $44 \le \text{perimeter} \le 100$

   Thus, the perimeter varies from 44 to 100 feet.

   CHECK: If W = 5, then 3W + 2 = 17, and the perimeter $= 2(5) + 2(17) = 44$ ft. If W = 12, then 3W + 2 = 38, and the perimeter $= 2(12) + 2(38) = 100$ ft. For $5 < W < 12$, $44 < p < 100$.

47. degree = 5

49. $(3x^2 + 2x - 4) - (2x^2 - 3x + 5) = 3x^2 + 2x - 4 - 2x^2 + 3x - 5 = x^2 + 5x - 9$

51. $(x - y)(x - 3y) = (x)(x) + (x)(-3y) + (-y)(x) + (-y)(-3y)$
   $= x^2 - 3xy - xy + 3y^2 = x^2 - 4xy + 3y^2$

53. $(x + y - 2)(x + y) = x(x + y) + y(x + y) - 2(x + y)$
   $= x(x) + x(y) + y(x) + y(y) - 2(x) - 2(y)$
   $= x^2 + xy + xy + y^2 - 2x - 2y = x^2 + 2xy + y^2 - 2x - 2y$

55. $(3x - 5y)^2 = (3x)^2 - 2(3x)(5y) + (5y)^2 = 9x^2 - 30xy + 25y^2$

57. $(3x - 5y)(3x + 5y) = (3x)^2 - (5y)^2 = 9x^2 - 25y^2$

59. $(2x + y - 3)(2x + y + 3) = ((2x + y) - 3)((2x + y) + 3) = (2x + y)^2 - 3^2$
$$= (2x)^2 + 2(2x)(y) + y^2 - 9 = 4x^2 + 4xy + y^2 - 9$$

61. $x^2 - 5x - 24 = (x - 8)(x + 3)$

63. $y^2 - 12xy + 35x^2 = (y - 5x)(y - 7x)$

65. $4y^2 + 16yz + 15z^2 = (2y + 3z)(2y + 5z)$

67. $25a^2 + 20ab + 4b^2 = (5a)^2 + 2(5a)(2b) + (2b)^2 = (5a + 2b)^2$

69. $36x^2 - 9 = 9(4x^2 - 1) = 9((2x)^2 - 1^2) = 9(2x + 1)(2x - 1)$

71. $25a^2 + 20ab - 4b^2$ is not factorable

73. $3y^3 + 5y^2 - 2y = y(3y^2 + 5y - 2) = y(3y - 1)(y + 2)$

75. $49a^4 - 14a^2z - 3z^2 = (7a^2 - 3z)(7a^2 + z)$

77. $(x - y)^2 - 16 = (x - y)^2 - 4^2 = (x - y + 4)(x - y - 4)$

79. $16a^4 - b^4 = (4a^2)^2 - (b^2)^2 = (4a^2 + b^2)(4a^2 - b^2)$
$$= (4a^2 + b^2)((2a)^2 - b^2) = (4a^2 + b^2)(2a + b)(2a - b)$$

81. $8a^3 + 125b^3 = (2a)^3 + (5b)^3 = (2a + 5b)((2a)^2 - (2a)(5b) + (5b)^2)$
$$= (2a + 5b)(4a^2 - 10ab + 25b^2)$$

83.
$$\begin{array}{r}
x - 2 \\
x + 4 \overline{)x^2 + 2x + 3} \\
\underline{-(x^2 + 4x)} \\
-2x + 3 \\
\underline{-(-2x - 8)} \\
11
\end{array}$$

Answer: $x - 2$, R11 or $x - 2 + \dfrac{11}{x + 4}$

85.

$$2x + 3 \overline{)\begin{array}{r} 2x^2 - 3x + 6 \\ 4x^3 + 0x^2 + 3x + 1 \end{array}}$$

$$\underline{-(4x^3 + 6x^2)}$$
$$-6x^2 + 3x + 1$$
$$\underline{-(-6x^2 - 9x)}$$
$$12x + 1$$
$$\underline{-(12x + 18)}$$
$$-17$$

Answer: $2x^2 - 3x + 6$, R -17 or

$$2x^2 - 3x + 6 - \frac{17}{2x + 3}$$

Cumulative Practice Test: Chapters 1–3

1. A = {4,8,12,16,20}

    B = {2,4,6,8,10,12,14,16,18,20,22}

    (a) $A \cap B = \{4,8,12,16,20\} = A$

    (b) $A \cup B = \{2,4,6,8,10,12,14,16,18,20,22\} = B$

3. (a) $(-2)(-2) - (-2)^2(2) = (-2)(-2) - (4)(2) = 4 - 8 = -4$

    (b) $5 - \{6 + [2 - 3(4 - 9)]\} = 5 - \{6 + [2 - 3(-5)]\}$
    $$= 5 - \{6 + [2 + 15]\} = 5 - \{6 + 17\}$$
    $$= 5 - 23 = -18$$

5. (a) $3x - 2 = 5x + 4$ CHECK: $3x - 2 = 5x + 4$

    $-2 = 2x + 4$ $3(-3) - 2 \overset{?}{=} 5(-3) + 4$

    $-6 = 2x$ $-9 - 2 \overset{?}{=} -15 + 4$

    $-3 = x$ $-11 \overset{\checkmark}{=} -11$

    contradiction (no solution)

    (b) $3(x - 5) - 2(x - 5) = 3 - (5 - x)$

    $3x - 15 - 2x + 10 = 3 - 5 + x$

    $x - 5 = -2 + x$

    $-5 = -2$

    (c) $|x - 3| = 8$

    $x - 3 = 8$ or $x - 3 = -8$

    $x = 11$ or $x = -5$

CHECK x = 11:   CHECK x = −5:

$|x − 3| = 8$ $|x − 3| = 8$

$|11 − 3| \overset{?}{=} 8$ $|−5 − 3| \overset{?}{=} 8$

$|8| \overset{?}{=} 8$ $|−8| \overset{?}{=} 8$

$8 \overset{\checkmark}{=} 8$ $8 \overset{\checkmark}{=} 8$

7. Let x = # of 10-lb. packages on the truck. Then 170 − x = # of 30-lb. packages on the truck.

$$10x + 30(170 − x) = 3140$$
$$10x + 5100 − 30x = 3140$$
$$−20x + 5100 = 3140$$
$$−20x = −1960$$
$$x = 98$$

Then 170 − x = 170 − 98 = 72. Thus, there are 98 10-lb. packages and 72 30-lb. packages on the truck.

CHECK: 98 10-lb. packages weigh 980 lbs. 72 30-lb. packages weigh 2160 lbs.

$980 + 2160 \overset{\checkmark}{=} 3140$

9. (a) $(2x^2 − 3xy + 4y^2) − (5x^2 − 2xy + y^2) = 2x^2 − 3xy + 4y^2 − 5x^2 + 2xy − y^2$

$$= −3x^2 − xy + 3y^2$$

(b) $(3a − 2b)(5a + 3b) = (3a)(5a) + (3a)(3b) + (−2b)(5a) + (−2b)(3b)$

$$= 15a^2 + 9ab − 10ab − 6b^2 = 15a^2 − ab − 6b^2$$

(c) $(2x^2 − y)(2x^2 + y) = (2x^2)^2 − y^2 = 4x^4 − y^2$

(d) $(3y − 2z)^2 = (3y)^2 − 2(3y)(2z) + (2z)^2 = 9y^2 − 12yz + 4z^2$

(e) $(x + y − 3)^2 = ((x + y) − 3)^2 = (x + y)^2 − 2(x + y)(3) + 3^2$

$$= x^2 + 2xy + y^2 − 6(x + y) + 9$$
$$= x^2 + 2xy + y^2 − 6(x + y) + 9$$
$$= x^2 + 2xy + y^2 − 6x − 6y + 9$$

11.

$$\begin{array}{r} 2x^2 - 4x + 14 \\ x+2\overline{)2x^3 + 0x^2 + 6x + 5} \\ \underline{-(2x^3 + 4x^2)} \\ -4x^2 + 6x + 5 \\ \underline{-(-4x^2 - 8x)} \\ 14x + 5 \\ \underline{-(14x + 28)} \\ -23 \end{array}$$

Answer: $2x^2 - 4x + 14$, R -23 or

$$2x^2 - 4x + 14 - \frac{23}{x+2}$$

CHAPTER 4
RATIONAL EXPRESSIONS

Exercises 4.1

1. $\dfrac{x}{x-2}$ is undefined when $x - 2 = 0$. Hence, $x \neq 2$.

7. $\dfrac{3xy}{3x-y}$ is undefined when $3x - y = 0$. Hence, $y \neq 3x$.

9. $\dfrac{3x^2y}{15xy} = \dfrac{\cancel{3} \cdot \cancel{x} \cdot x \cdot \cancel{y}}{\cancel{3} \cdot 5 \cdot \cancel{x} \cdot \cancel{y}} = \dfrac{x}{5}$

13. $\dfrac{(x-5)\cancel{(x+4)}}{(x-3)\cancel{(x+4)}} = \dfrac{x-5}{x-3}$

17. $\dfrac{(2x-3)(x-5)}{(2x-3)(5-x)} = \dfrac{\cancel{(2x-3)}\cancel{(x-5)}}{\cancel{(2x-3)}(-1)\cancel{(x-5)}} = \dfrac{1}{-1} = -1$

21. $\dfrac{x^2-9}{x^2-6x+9} = \dfrac{\cancel{(x-3)}(x+3)}{\cancel{(x-3)}(x-3)} = \dfrac{x+3}{x-3}$

23. $\dfrac{w^2-8wz+7z^2}{w^2+8wz+7z^2} = \dfrac{(w-7z)(w-z)}{(w+7z)(w+z)}$

27. $\dfrac{x^2+x-12}{3-x} = \dfrac{(x+4)\cancel{(x-3)}}{-1\cancel{(x-3)}} = \dfrac{x+4}{-1} = -(x+4)$

29. $\dfrac{3a^2-13a-30}{15a^2+28a+5} = \dfrac{\cancel{(3a+5)}(a-6)}{\cancel{(3a+5)}(5a+1)} = \dfrac{a-6}{5a+1}$

31. $\dfrac{2x^3-2x^2-12x}{3x^2-6x} = \dfrac{2x(x^2-x-6)}{3x(x-2)} = \dfrac{2\cancel{x}(x-3)(x+2)}{3\cancel{x}(x-2)} = \dfrac{2(x-3)(x+2)}{3(x-2)}$

35. $\dfrac{3a^2-7a-6}{3+5a-2a^2} = \dfrac{3a^2-7a-6}{-1(2a^2-5a-3)} = \dfrac{(3a+2)\cancel{(a-3)}}{(-1)(2a+1)\cancel{(a-3)}}$

$\qquad\qquad = \dfrac{3a+2}{(-1)(2a+1)} = -\dfrac{3a+2}{2a+1}$

39. $\dfrac{a^3+b^3}{(a+b)^3} = \dfrac{\cancel{(a+b)}(a^2-ab+b^2)}{\cancel{(a+b)}(a+b)^2} = \dfrac{a^2-ab+b^2}{(a+b)^2}$

41. $\dfrac{ax + bx - 2ay - 2by}{4xy - 2x^2} = \dfrac{(a+b)x - 2y(a+b)}{2x(2y-x)} = \dfrac{(a+b)(x-2y)}{2x(2y-x)}$

$$= \dfrac{(a+b)\cancel{(x-2y)}}{(2x)(-1)\cancel{(x-2y)}} = \dfrac{a+b}{-2x} = -\dfrac{a+b}{2x}$$

45. $\dfrac{2x}{3y} = \dfrac{2x(3x^2)}{3y(3x^2)} = \dfrac{\boxed{6x^3}}{9x^2 y}$

49. $\dfrac{3x}{x-5} = \dfrac{\boxed{3x(x+5)}}{(x-5)(x+5)}$

53. $\dfrac{y-2}{2y-3} = \dfrac{(y-2)(y+4)(-1)}{(2y-3)(y+4)(-1)} = \dfrac{\boxed{-(y-2)(y+4)}}{12 - 5y - 2y^2}$

55. $\dfrac{x-y}{a^2 - b^2} = \dfrac{(x-y)(x+y)}{(a^2 - b^2)(x+y)} = \dfrac{\boxed{(x-y)(x+y)}}{a^2 x + a^2 y - b^2 x - b^2 y}$

57. (a) We cannot cancel terms, only factors.

(b) $a^3 + b^3$ is the sum of cubes, while $(a+b)^3$ is the cube of a sum. These two are in general different, and therefore cannot be cancelled.

(c) The cancellation of $a + b$ is improperly done. We must cross out the factor of $(a + b)$ in the second term of the numerator as well, leading to an answer of $x + 2y$.

61. $3 - [x - (2 - x)] = 4 - x$
$3 - [x - 2 + x] = 4 - x$
$3 - [2x - 2] = 4 - x$
$3 - 2x + 2 = 4 - x$
$5 - 2x = 4 - x$
$5 = 4 + x$
$1 = x$

63. Let n = # of nickels Valerie has. Then $28 - n$ = # of dimes Valerie has.
$5n + 10(28 - n) = 250$
$5n + 280 - 10n = 250$
$-5n + 280 = 250$
$-5n = -30$
$n = 6$

Then $28 - n = 28 - 6 = 22$. So Valerie has 6 nickels and 22 dimes.

Exercises 4.2

1. $\dfrac{10a^2b}{3xy^3} \cdot \dfrac{9xy^{4}}{5a^4b^7} = \dfrac{6y}{a^2b^6}$

   (overwritten: $\overset{2}{10a^2b}$, $\overset{3y}{9xy^4}$)

3. $\dfrac{17a^2b^3}{18yx} \div \dfrac{34a^2}{9xy} = \dfrac{17a^2b^3}{18yx} \cdot \dfrac{9xy}{34a^2} = \dfrac{b^3}{4}$

7. $\dfrac{32r^2s^3}{12a^2b} \div \left(\dfrac{15ab^2}{16r} \cdot \dfrac{24rs^2}{5ab} \right) = \dfrac{32r^2s^3}{12a^2b} \div \dfrac{9bs^2}{2} = \dfrac{12r^2s^5}{a^2}$

9. $\dfrac{32r^2s^3}{12a^2b} \div \left(\dfrac{15ab^2}{16r} \cdot \dfrac{24rs^2}{5ab} \right) = \dfrac{32r^2s^3}{12a^2b} \div \dfrac{9bs^2}{2} = \dfrac{32r^2s^3}{12a^2b} \cdot \dfrac{2}{9bs^2} = \dfrac{16r^2s}{27a^2b^2}$

13. $\dfrac{x^2 - x - 2}{x+3} \cdot \dfrac{3x+9}{2x+2} = \dfrac{(x-2)(x+1)}{x+3} \cdot \dfrac{3(x+3)}{2(x+1)} = \dfrac{3(x-2)}{2}$

15. $\dfrac{5a^3 - 5a^2b}{3a^2 + 3ab} \cdot (a+b) = \dfrac{5a^2(a-b)}{3a(a+b)} \cdot \dfrac{(a+b)}{1} = \dfrac{5a(a-b)}{3}$

19. $\dfrac{2x^2 + 3x - 5}{2x+5} \cdot \dfrac{1}{1-x} = \dfrac{(2x+5)(x-1)}{2x+5} \cdot \dfrac{1}{(-1)(x-1)} = \dfrac{1}{-1} = -1$

21. $\dfrac{2a^2 - 7a + 6}{4a^2 - 9} \cdot \dfrac{4a^2 + 12a + 9}{a^2 - a - 2} = \dfrac{(2a-3)(a-2)}{(2a+3)(2a-3)} \cdot \dfrac{(2a+3)(2a+3)}{(a-2)(a+1)}$

    $= \dfrac{2a+3}{a+1}$

25. $\dfrac{9x^2 + 3xy - 2y^2}{6x^2y - 2xy^2} \div \dfrac{3x+2y}{6x^2y} = \dfrac{9x^2 + 3xy - 2y^2}{6x^2y - 2xy^2} \cdot \dfrac{6x^2y}{3x+2y}$

    $= \dfrac{(3x+2y)(3x-y)}{2xy(3x-y)} \cdot \dfrac{6x^2y^{3x}}{3x+2y} = \dfrac{3x}{1} = 3x$

29. $\dfrac{6q^2 - pq - 2p^2}{8pq^2 + 4p^2q} \cdot \dfrac{8p^2q^2}{6pq^2 - 4p^2q} = \dfrac{(3q-2p)(2q+p)}{4pq(2q+p)} \cdot \dfrac{8p^2q^2}{2pq(3q-2p)} = \dfrac{1}{1} = 1$

33. $\left(\dfrac{2a^2 + 3ab + b^2}{3a^2 - ab - 2b^2} \right)\left(\dfrac{a-b}{2a-b} \right)\left(\dfrac{3a+b}{a+b} \right) = \dfrac{(2a+b)(a+b)}{(3a+2b)(a-b)} \cdot \dfrac{a-b}{2a-b} \cdot \dfrac{3a+b}{a+b}$

    $= \dfrac{(2a+b)(3a+b)}{(3a+2b)(2a-b)}$

35. $\dfrac{9a^2 + 9ab + 2b^2}{3a^2 - 2ab - b^2} \cdot \left(\dfrac{a-b}{3a^2 + 4ab + b^2} \div \dfrac{3a+2b}{a+b} \right) = \dfrac{9a^2 + 9ab + 2b^2}{3a^2 - 2ab - b^2} \cdot \left(\dfrac{a-b}{(3a+b)(a+b)} \cdot \dfrac{a+b}{3a+2b} \right)$

$\qquad = \dfrac{(3a+b)(3a+2b)}{(3a+b)(a-b)} \cdot \dfrac{a-b}{(3a+b)(3a+2b)} = \dfrac{1}{3a+b}$

39. $\dfrac{4ax + 6x - 2ay - 3y}{x^2 + xy} \div \left(\dfrac{2x^2 - xy - y^2}{3x^3 + 3x^2 y} \cdot \dfrac{2a+3}{x-y} \right)$

$\qquad = \dfrac{2x(2a+3) - y(2a+3)}{x(x+y)} \div \left(\dfrac{(2x+y)(x-y)}{3x^2(x+y)} \cdot \dfrac{2a+3}{x-y} \right) = \dfrac{(2a+3)(2x-y)}{x(x+y)} \div \dfrac{(2x+y)(2a+3)}{3x^2(x+y)}$

$\qquad = \dfrac{(2a+3)(2x-y)}{x(x+y)} \cdot \dfrac{3x^2(x+y)}{(2x+y)(2a+3)} = \dfrac{3x(2x-y)}{2x+y}$

41. $|-6-3| - |6| - |3| = |-9| - |6| - |3| = 9 - 6 - 3 = 0$

43. The distributive property

45. Let t = Bobby's running time, and $t + 1$ = Linda's running time.

$\qquad 6t = 5(t+1)$

$\qquad 6t = 5t + 5$

$\qquad t = 5$

Thus, it takes Bobby 5 hours to overtake Linda.

Exercises 4.3

1. $\dfrac{x+2}{x} - \dfrac{2-x}{x} = \dfrac{x+2 - (2-x)}{x} = \dfrac{x+2-2+x}{x} = \dfrac{2x}{x} = \dfrac{2}{1} = 2$

5. $\dfrac{3a}{3a^2 + a - 2} - \dfrac{2}{3a^2 + a - 2} = \dfrac{3a-2}{3a^2 + a - 2} = \dfrac{3a-2}{(3a-2)(a+1)} = \dfrac{1}{a+1}$

7. $\dfrac{x^2}{x^2 - y^2} + \dfrac{y^2}{x^2 - y^2} - \dfrac{2xy}{x^2 - y^2} = \dfrac{x^2 + y^2 - 2xy}{x^2 - y^2} = \dfrac{x^2 - 2xy + y^2}{x^2 - y^2} = \dfrac{(x-y)(x-y)}{(x-y)(x+y)} = \dfrac{x-y}{x+y}$

11. $\dfrac{3x}{2y^2} - \dfrac{4x^2}{9y} = \dfrac{3x(9)}{2y^2(9)} - \dfrac{4x^2(2y)}{9y(2y)} = \dfrac{27x}{18y^2} - \dfrac{8x^2 y}{18y^2} = \dfrac{27x - 8x^2 y}{18y^2}$

17. $\dfrac{36a^2}{b^2 c} + \dfrac{24}{bc^3} - \dfrac{3}{7bc} = \dfrac{36a^2(7c^2)}{b^2 c(7c^2)} + \dfrac{24(7b)}{bc^3(7b)} - \dfrac{3(bc^2)}{7bc(bc^2)}$

$\qquad = \dfrac{252a^2 c^2}{7b^2 c^3} + \dfrac{168b}{7b^2 c^3} - \dfrac{3bc^2}{7b^2 c^3} = \dfrac{252a^2 c^2 + 168b - 3bc^2}{7b^2 c^3}$

21.
$$\frac{a}{a-b} - \frac{b}{a} = \frac{a(a)}{(a-b)(a)} - \frac{b(a-b)}{a(a-b)} = \frac{a^2}{a(a-b)} - \frac{b(a-b)}{a(a-b)}$$

$$= \frac{a^2 - b(a-b)}{a(a-b)} = \frac{a^2 - ab + b^2}{a(a-b)}$$

25.
$$\frac{2}{x-7} + \frac{3x+1}{x+2} = \frac{2(x+2)}{(x-7)(x+2)} + \frac{(3x+1)(x-7)}{(x+2)(x-7)} = \frac{2(x+2) + (3x+1)(x-7)}{(x-7)(x+2)}$$

$$= \frac{2x+4+3x^2-20x-7}{(x-7)(x+2)} = \frac{3x^2-18x-3}{(x-7)(x+2)} = \frac{3(x^2-6x-1)}{(x-7)(x+2)}$$

27.
$$\frac{2r+s}{r-s} - \frac{r-2s}{r+s} = \frac{(2r+s)(r+s)}{(r-s)(r+s)} - \frac{(r-2s)(r-s)}{(r+s)(r-s)}$$

$$= \frac{(2r+s)(r+s) - (r-2s)(r-s)}{(r-s)(r+s)} = \frac{2r^2+3rs+s^2 - (r^2-3rs+2s^2)}{(r-s)(r+s)}$$

$$= \frac{2r^2+3rs+s^2-r^2+3rs-2s^2}{(r-s)(r+s)} = \frac{r^2+6rs-s^2}{(r-s)(r+s)}$$

$$2r^2 + 2rs + sr + s^2 -$$
$$r^2 - rs - 2rs + 2s^2$$
$$2r^2 \quad 3rs + s^2 - 3rs + s^2$$

31.
$$\frac{7a+3}{2a-1} + \frac{5a+4}{1-2a} = \frac{7a+3}{2a-1} + \frac{(5a+4)(-1)}{(1-2a)(-1)} = \frac{7a+3}{2a-1} + \frac{(5a+4)(-1)}{2a-1}$$

$$= \frac{7a+3-(5a+4)}{2a-1} = \frac{7a+3-5a-4}{2a-1} = \frac{2a-1}{2a-1} = 1$$

33.
$$\frac{a}{a-b} - \frac{b}{a^2-b^2} = \frac{a}{a-b} - \frac{b}{(a-b)(a+b)} = \frac{a(a+b)}{(a-b)(a+b)} - \frac{b}{(a-b)(a+b)}$$

$$= \frac{a(a+b)-b}{(a-b)(a+b)} = \frac{a^2+ab-b}{(a-b)(a+b)}$$

37.
$$\frac{2y+1}{y+2} + \frac{3y}{y+3} + y = \frac{(2y+1)(y+3)}{(y+2)(y+3)} + \frac{3y(y+2)}{(y+3)(y+2)} + \frac{y(y+2)(y+3)}{(y+2)(y+3)}$$

$$= \frac{2y^2+7y+3}{(y+2)(y+3)} + \frac{3y^2+6y}{(y+2)(y+3)} + \frac{y^3+5y^2+6y}{(y+2)(y+3)}$$

$$= \frac{2y^2+7y+3+3y^2+6y+y^3+5y^2+6y}{(y+2)(y+3)}$$

$$= \frac{y^3+10y^2+19y+3}{(y+2)(y+3)}$$

45.

$$\frac{a+3}{a-3}+\frac{a}{a+4}-\frac{3}{a^2+a-12}=\frac{a+3}{a-3}+\frac{a}{a+4}-\frac{3}{(a-3)(a+4)}$$

$$=\frac{(a+3)(a+4)}{(a-3)(a+4)}+\frac{a(a-3)}{(a+4)(a-3)}-\frac{3}{(a-3)(a+4)}$$

$$=\frac{a^2+7a+12}{(a-3)(a+4)}+\frac{a^2-3a}{(a-3)(a+4)}-\frac{3}{(a-3)(a+4)}$$

$$=\frac{a^2+7a+12+a^2-3a-3}{(a-3)(a+4)}=\frac{2a^2+4a+9}{(a-3)(a+4)}$$

49.

$$\frac{3x-4}{x-5}+\frac{4x}{10+3x-x^2}=\frac{3x-4}{x-5}+\frac{4x}{(-1)(x-5)(x+2)}$$

$$=\frac{(3x-4)(-1)(x+2)}{(-1)(x-5)(x+2)}+\frac{4x}{(-1)(x-5)(x+2)}$$

$$=\frac{-3x^2-2x+8}{(-1)(x-5)(x+2)}+\frac{4x}{(-1)(x-5)(x+2)}$$

$$=\frac{-3x^2-2x+8+4x}{(-1)(x-5)(x+2)}=\frac{-3x^2+2x+8}{(-1)(x-5)(x+2)}$$

$$=\frac{(-1)(3x+4)(x-2)}{(-1)(x-5)(x+2)}=\frac{(3x+4)(x-2)}{(x-5)(x+2)}$$

55.

$$\frac{2r+xs}{r+s}+\frac{2s+xr}{r+s}=\frac{2r+xs+2s+xr}{r+s}=\frac{2r+2s+xr+xs}{r+s}$$

$$=\frac{2(r+s)+x(r+s)}{r+s}=\frac{(r+s)(2+x)}{r+s}=2+x$$

57.

$$\frac{r^2}{r^3-s^3}+\frac{rs}{r^3-s^3}+\frac{s^2}{r^3-s^3}=\frac{r^2+rs+s^2}{r^3-s^3}=\frac{r^2+rs+s^2}{(r-s)(r^2+rs+s^2)}=\frac{1}{r-s}$$

61. (a)

$$\frac{x^2+4x}{4x}=\frac{x(x+4)}{4x}=\frac{x+4}{4}$$

(b)

$$\frac{x^2+4x}{4x}=\frac{x^2}{4x}+\frac{4x}{4x}=\frac{x}{4}+1\ \left(\text{Note that }\frac{x}{4}+1=\frac{x}{4}+\frac{4}{4}=\frac{x+4}{4}.\right)$$

65. (a)

$$\frac{6m^2n-4m^3n^2-9mn}{15mn^2}=\frac{mn(6m-4m^2n-9)}{15mn^2}=\frac{6m-4m^2n-9}{15n}$$

(b)

$$\frac{6m^2n-4m^3n^2-9mn}{15mn^2}=\frac{6m^2n}{15mn^2}-\frac{4m^3n^2}{15mn^2}-\frac{9mn}{15mn^2}=\frac{2m}{5n}-\frac{4m^2}{15}-\frac{3}{5n}$$

70

67. In all cases, the value of the expression is equal to 2, except when x = 1. This is true because for $x \neq 1$,

$$\frac{3x+5}{x-1} + \frac{x+7}{1-x} = \frac{3x+5}{x-1} - \frac{x+7}{x-1}$$

$$= \frac{3x+5-(x+7)}{x-1}$$

$$= \frac{3x+5-x-7}{x-1}$$

$$= \frac{2x-2}{x-1}$$

$$= \frac{2(x-1)}{x-1} = 2$$

When x = 1, the expression is not defined.

69. Let x = # of 25-lb. packages. Then 60 − x = # of 30-lb. packages.

$$25x + 30(60-x) = 1680$$
$$25x + 1800 - 30x = 1680$$
$$-5x + 1800 = 1680$$
$$-5x = -120$$
$$x = 24$$

Then 60 − x = 60 − 24 = 36. So there are 24 25-lb. packages and 36 30-lb. packages in the truck.

71.
$$\frac{2x^3y + x^2y - 6xy}{x^2 - 2x - 8} \cdot \frac{x-4}{2x^3y - x^2y^2 - 3xy^2} = \frac{xy(2x^2 + x - 6)}{(x-4)(x+2)} \cdot \frac{x-4}{xy(2x^2 - xy - 3y^2)}$$

$$= \frac{xy(2x-3)(x+2)}{(x-4)(x+2)} \cdot \frac{x-4}{xy(2x-3y)(x+y)} = \frac{xy(2x-3)(x+2)(x-4)}{(x-4)(x+2)xy(2x-3y)(x+y)}$$

$$= \frac{2x-3}{(2x-3y)(x+y)}$$

Exercises 4.4

1. $$\frac{\dfrac{3}{xy^2}}{\dfrac{15}{x^2y}} = \frac{3}{xy^2} \div \frac{15}{x^2y} = \frac{3}{xy^2} \cdot \frac{x^2y}{15} = \frac{x}{5y}$$

(handwritten:) $\dfrac{39}{1} \cdot \dfrac{9}{1} - \dfrac{39}{1} \cdot \dfrac{1}{3}$

$3a^2 - a' = a(3a-1)$

5. $$\frac{a - \dfrac{1}{3}}{\dfrac{9a^2 - 1}{3a}} = \frac{3a\left(a - \dfrac{1}{3}\right)}{3a(9a^2 - 1)} = \frac{3a^2 - a}{9a^2 - 1} = \frac{a(3a-1)}{(3a+1)(3a-1)} = \frac{a}{3a+1}$$

9. $$\left(1 - \frac{4}{x^2}\right) \div \left(\frac{1}{x} - \frac{2}{x^2}\right) = \left(\frac{x^2}{x^2} - \frac{4}{x^2}\right) \div \left(\frac{x}{x^2} - \frac{2}{x^2}\right)$$

$$= \frac{x^2 - 4}{x^2} \div \frac{x-2}{x^2} = \frac{(x+2)(x-2)}{x^2} \cdot \frac{x^2}{x-2} = x+2$$

13. $$\dfrac{1-\dfrac{4}{z}+\dfrac{4}{z^2}}{\dfrac{1}{z^2}-\dfrac{2}{z^3}}=\dfrac{z^3\left(1-\dfrac{4}{z}+\dfrac{4}{z^2}\right)}{z^3\left(\dfrac{1}{z^2}-\dfrac{2}{z^3}\right)}=\dfrac{z^3-4z^2+4z}{z-2}=\dfrac{z(z-2)(z-2)}{z-2}=z(z-2)$$

15. $$\dfrac{\dfrac{4}{y^2}-\dfrac{12}{xy}-\dfrac{9}{x^2}}{\dfrac{4}{y^2}-\dfrac{9}{x^2}}=\dfrac{x^2y^2\left(\dfrac{4}{y^2}-\dfrac{12}{xy}+\dfrac{9}{x^2}\right)}{x^2y^2\left(\dfrac{4}{y^2}-\dfrac{9}{x^2}\right)}=\dfrac{4x^2-12xy+9y^2}{4x^2-9y^2}$$

$$=\dfrac{(2x-3y)(2x-3y)}{(2x+3y)(2x-3y)}=\dfrac{2x-3y}{2x+3y}$$

19. $$\left(3+\dfrac{1}{2x-1}\right)\div\left(5+\dfrac{x}{2x-1}\right)=\left(\dfrac{3(2x-1)}{2x-1}+\dfrac{1}{2x-1}\right)\div\left(\dfrac{5(2x-1)}{2x-1}+\dfrac{x}{2x-1}\right)$$

$$=\left(\dfrac{6x-3}{2x-1}+\dfrac{1}{2x-1}\right)\div\left(\dfrac{10x-5}{2x-1}+\dfrac{x}{2x-1}\right)$$

$$=\left(\dfrac{6x-3+1}{2x-1}\right)\div\left(\dfrac{10x-5+x}{2x-1}\right)=\dfrac{6x-2}{2x-1}\div\dfrac{11x-5}{2x-1}$$

$$=\dfrac{6x-2}{2x-1}\cdot\dfrac{2x-1}{11x-5}=\dfrac{6x-2}{11x-5}$$

21. $$\dfrac{x+\dfrac{2}{x+3}}{x-5+\dfrac{12}{x+3}}=\dfrac{(x+3)\left(x+\dfrac{2}{x+3}\right)}{(x+3)\left(x-5+\dfrac{12}{x+3}\right)}=\dfrac{x(x+3)+2}{(x+3)(x-5)+12}$$

$$=\dfrac{x^2+3x+2}{x^2-2x-15+12}=\dfrac{x^2+3x+2}{x^2-2x-3}=\dfrac{(x+1)(x+2)}{(x+1)(x-3)}=\dfrac{x+2}{x-3}$$

25. $$\dfrac{\dfrac{2x}{x-3}}{\dfrac{x}{x+2}+\dfrac{3}{x-3}}=\dfrac{(x+2)(x-3)\left(\dfrac{2x}{x-3}\right)}{(x+2)(x+3)\left(\dfrac{x}{x+2}+\dfrac{3}{x-3}\right)}$$

$$=\dfrac{2x(x+2)}{x(x-3)+3(x+2)}=\dfrac{2x(x+2)}{x^2-3x+3x+6}=\dfrac{2x(x+2)}{x^2+6}$$

29. $$\dfrac{\dfrac{2a^2-6a}{a+2}-a}{\dfrac{a}{a-3}-\dfrac{4a}{a^2-a-6}}=\dfrac{\dfrac{2a^2-6a}{a+2}-a}{\dfrac{a}{a-3}-\dfrac{4a}{(a+2)(a-3)}}=\dfrac{(a+2)(a-3)\left(\dfrac{2a^2-6a}{a+2}-a\right)}{(a+2)(a-3)\left(\dfrac{a}{a-3}-\dfrac{4a}{(a+2)(a-3)}\right)}=$$

$$\dfrac{(a-3)(2a^2-6a)-a(a+2)(a-3)}{a(a+2)-4a}=\dfrac{2a(a-3)(a-3)-a(a+2)(a-3)}{a^2+2a-4a}$$

$$=\dfrac{a(a-3)(2(a-3)-(a+2))}{a^2-2a}=\dfrac{a(a-3)(a-8)}{a(a-2)}=\dfrac{(a-3)(a-8)}{a-2}$$

31. First note that $\dfrac{\dfrac{2}{3}}{a+2}=\dfrac{2}{1}\div\dfrac{3}{a+2}=\dfrac{2}{1}\cdot\dfrac{a+2}{3}=\dfrac{2(a+2)}{3}$.

Then $\dfrac{3a - \dfrac{2}{a}}{a - \dfrac{2}{\dfrac{3}{a+2}}} = \dfrac{3a - \dfrac{2}{a}}{a - \dfrac{2(a+2)}{3}} = \dfrac{3a\left(3a - \dfrac{2}{a}\right)}{3a\left(a - \dfrac{2(a+2)}{3}\right)} = \dfrac{9a^2 - 6}{3a^2 - 2a(a+2)}$

$$= \dfrac{9a^2 - 6}{3a^2 - 2a^2 - 4a} = \dfrac{9a^2 - 6}{a^2 - 4a} = \dfrac{3(3a^2 - 2)}{a(a - 4)}$$

33. First note that $3 - \dfrac{x}{3-x} = \dfrac{3(3-x)}{3-x} - \dfrac{x}{3-x} = \dfrac{9-3x}{3-x} - \dfrac{x}{3-x}$

$$= \dfrac{9 - 3x - x}{3 - x} = \dfrac{9 - 4x}{3 - x}$$

Then $\dfrac{x}{3 - \dfrac{x}{3-x}} = \dfrac{x}{\dfrac{9-4x}{3-x}} = \dfrac{x}{1} \div \dfrac{9-4x}{3-x} = \dfrac{x}{1} \cdot \dfrac{3-x}{9-4x} = \dfrac{x(3-x)}{9-4x}$

Finally, $3 - \dfrac{x}{3 - \dfrac{x}{3-x}} = 3 - \dfrac{x(3-x)}{9-4x} = \dfrac{3(9-4x)}{9-4x} - \dfrac{x(3-x)}{9-4x} = \dfrac{3(9-4x) - x(3-x)}{9-4x}$

$$= \dfrac{27 - 12x - 3x + x^2}{9 - 4x} = \dfrac{x^2 - 15x + 27}{9 - 4x}$$

37. The fundamental principle of fractions requires the multiplying factor of the numerator to be the same as the multiplying factor of the denominator.

Exercises 4.5

1. $\dfrac{x}{3} - \dfrac{x}{2} + \dfrac{x}{4} = 1$ CHECK x = 12:

$$12\left(\dfrac{x}{3} - \dfrac{x}{2} + \dfrac{x}{4}\right) = 12 \cdot 1$$

$$\dfrac{12}{1} \cdot \dfrac{x}{3} - \dfrac{12}{1} \cdot \dfrac{x}{2} + \dfrac{12}{1} \cdot \dfrac{x}{4} = 12$$

$$4x - 6x + 3x = 12$$

$$x = 12$$

$$\dfrac{x}{3} - \dfrac{x}{2} + \dfrac{x}{4} = 1$$

$$\dfrac{12}{3} - \dfrac{12}{2} + \dfrac{12}{4} \stackrel{?}{=} 1$$

$$4 - 6 + 3 \stackrel{?}{=} 1$$

$$1 \stackrel{\checkmark}{=} 1$$

5. $\dfrac{a-1}{6} + \dfrac{a+1}{10} = a - 3$

$$30\left(\dfrac{a-1}{6} + \dfrac{a+1}{10}\right) = 30(a-3)$$

$$\dfrac{\overset{5}{\cancel{30}}}{1} \cdot \dfrac{a-1}{\cancel{6}} + \dfrac{\overset{3}{\cancel{30}}}{1} \cdot \dfrac{a+1}{\cancel{10}} = 30(a-3)$$

$$5(a-1) + 3(a+1) = 30(a-3)$$

$$5a - 5 + 3a + 3 = 30a - 90$$

$$8a - 2 = 30a - 90$$

$$-2 = 22a - 90$$

$$88 = 22a$$

$$4 = a$$

CHECK a = 4:

$$\dfrac{a-1}{6} + \dfrac{a+1}{10} = a - 3$$

$$\dfrac{4-1}{6} + \dfrac{4+1}{10} \overset{?}{=} 4 - 3$$

$$\dfrac{3}{6} + \dfrac{5}{10} \overset{?}{=} 1$$

$$\dfrac{1}{2} + \dfrac{1}{2} \overset{?}{=} 1$$

$$1 \overset{\checkmark}{=} 1$$

9. $\dfrac{y-5}{2} \le \dfrac{y-2}{5} + 3$

$$10\left(\dfrac{y-5}{2}\right) \le 10\left(\dfrac{y-2}{5} + 3\right)$$

$$\dfrac{\overset{5}{\cancel{10}}}{1} \cdot \dfrac{y-5}{\cancel{2}} \le \dfrac{\overset{2}{\cancel{10}}}{1} \cdot \dfrac{y-2}{\cancel{5}} + 10 \cdot 3$$

$$5(y-5) \le 2(y-2) + 30$$

$$5y - 25 \le 2y - 4 + 30$$

$$5y - 25 \le 2y + 26$$

$$3y - 25 \le 26$$

$$3y \le 51$$

$$y \le 17$$

13. $\dfrac{x-3}{4} - \dfrac{x-4}{3} \ge 2$

$$12\left(\dfrac{x-3}{4} - \dfrac{x-4}{3}\right) \ge 12 \cdot 2$$

$$\dfrac{\overset{3}{\cancel{12}}}{1} \cdot \dfrac{x-3}{\cancel{4}} - \dfrac{\overset{4}{\cancel{12}}}{1} \cdot \dfrac{x-4}{\cancel{3}} \ge 24$$

$$3(x-3) - 4(x-4) \ge 24$$

$$3x - 9 - 4x + 16 \ge 24$$

$$-x + 7 \ge 24$$

$$-x \ge 17$$

$$x \le -17$$

15.

$$\frac{3x+11}{6} - \frac{2x+1}{3} = x+5$$

$$6\left(\frac{3x+11}{6} - \frac{2x+1}{3}\right) = 6(x+5)$$

$$\frac{\cancel{6}}{1} \cdot \frac{3x+11}{\cancel{6}} - \frac{\cancel{6}^{2}}{1} \cdot \frac{2x+1}{\cancel{3}} = 6(x+5)$$

$$3x+11 - 2(2x+1) = 6(x+5)$$

$$3x+11 - 4x - 2 = 6x + 30$$

$$-x + 9 = 6x + 30$$

$$9 = 7x + 30$$

$$-21 = 7x$$

$$-3 = x$$

CHECK x = −3:

$$\frac{3x+11}{6} - \frac{2x+1}{3} = x+5$$

$$\frac{3(-3)+11}{6} - \frac{2(-3)+1}{3} \stackrel{?}{=} -3 + 5$$

$$\frac{-9+11}{6} - \frac{-6+1}{3} \stackrel{?}{=} 2$$

$$\frac{2}{6} - \left(-\frac{5}{3}\right) \stackrel{?}{=} 2$$

$$\frac{1}{3} + \frac{5}{3} \stackrel{?}{=} 2$$

$$\frac{6}{3} \stackrel{?}{=} 2$$

$$2 \stackrel{\checkmark}{=} 2$$

19.

$$\frac{5}{x} - \frac{1}{2} = \frac{3}{x}$$

$$2x\left(\frac{5}{x} - \frac{1}{2}\right) = 2x\left(\frac{3}{x}\right)$$

$$\frac{2\cancel{x}}{1} \cdot \frac{5}{\cancel{x}} - \frac{\cancel{2}x}{1} \cdot \frac{1}{\cancel{2}} = \frac{2\cancel{x}}{1} \cdot \frac{3}{\cancel{x}}$$

$$10 - x = 6$$

$$-x = -4$$

$$x = 4$$

CHECK x = 4:

$$\frac{5}{x} - \frac{1}{2} = \frac{3}{x}$$

$$\frac{5}{4} - \frac{1}{2} \stackrel{?}{=} \frac{3}{4}$$

$$\frac{5}{4} - \frac{2}{4} \stackrel{?}{=} \frac{3}{4}$$

$$\frac{3}{4} \stackrel{\checkmark}{=} \frac{3}{4}$$

23.

$$\frac{1}{t-3} + \frac{2}{t} = \frac{5}{3t}$$

$$3t(t-3)\left(\frac{1}{t-3} + \frac{2}{t}\right) = 3t(t-3)\left(\frac{5}{3t}\right)$$

$$\frac{3t(\cancel{t-3})}{1} \cdot \frac{1}{\cancel{t-3}} + \frac{3\cancel{t}(t-3)}{1} \cdot \frac{2}{\cancel{t}} = \frac{\cancel{3t}(t-3)}{1} \cdot \frac{5}{\cancel{3t}}$$

$$3t + 6(t-3) = 5(t-3)$$

$$3t + 6t - 18 = 5t - 15$$

$$9t - 18 = 5t - 15$$

$$4t - 18 = -15$$

$$4t = 3$$

$$t = \frac{3}{4}$$

CHECK t = $\frac{3}{4}$:

$$\frac{1}{t-3} + \frac{2}{t} = \frac{5}{3t}$$

$$\frac{1}{\frac{3}{4}-3} + \frac{2}{\frac{3}{4}} \stackrel{?}{=} \frac{5}{3\left(\frac{3}{4}\right)}$$

$$\frac{1}{-\frac{9}{4}} + \frac{2}{\frac{3}{4}} \stackrel{?}{=} \frac{5}{\frac{9}{4}}$$

$$-\frac{4}{9} + \frac{8}{3} \stackrel{?}{=} \frac{20}{9}$$

$$-\frac{4}{9} + \frac{24}{9} \stackrel{?}{=} \frac{20}{9}$$

$$\frac{20}{9} \stackrel{\checkmark}{=} \frac{20}{9}$$

27.

$$\frac{7}{x-5} + 2 = \frac{x+2}{x-5}$$

$$(x-5)\left(\frac{7}{x-5} + 2\right) = (x-5)\left(\frac{x+2}{x-5}\right)$$

$$\frac{x-5}{1} \cdot \frac{7}{x-5} + 2(x-5) = \frac{x-5}{1} \cdot \frac{x+2}{x-5}$$

$$7 + 2(x-5) = x+2$$

$$7 + 2x - 10 = x+2$$

$$2x - 3 = x+2$$

$$x - 3 = 2$$

$$x = 5$$

CHECK x = 5:

$$\frac{7}{x-5} + 2 = \frac{x+2}{x-5}$$

$$\frac{7}{5-5} + 2 \overset{?}{=} \frac{5+2}{5-5}$$

$$\frac{7}{0} + 2 \neq \frac{7}{0},$$

since we are not allowed
to divide by 0. Therefore
the equation has no solution.

31.

$$\frac{5}{y^2 + 3y} - \frac{4}{3y} + \frac{1}{2} = \frac{5}{y(y+3)} - \frac{4}{3y} + \frac{1}{2}$$

$$= \frac{5(6)}{y(y+3)(6)} - \frac{4(2(y+3))}{3y(2(y+3))} + \frac{1(3y(y+3))}{2(3y(y+3))}$$

$$= \frac{30}{6y(y+3)} - \frac{8(y+3)}{6y(y+3)} + \frac{3y(y+3)}{6y(y+3)} = \frac{30 - 8(y+3) + 3y(y+3)}{6y(y+3)}$$

$$= \frac{30 - 8y - 24 + 3y^2 + 9y}{6y(y+3)} = \frac{3y^2 + y + 6}{6y(y+3)}$$

35.

$$\frac{1}{x^2 - x - 2} + \frac{2}{x^2 - 1} = \frac{1}{x^2 - 3x + 2}$$

$$\frac{1}{(x-2)(x+1)} + \frac{2}{(x-1)(x+1)} = \frac{1}{(x-1)(x-2)}$$

$$(x-1)(x+1)(x-2)\left(\frac{1}{(x-2)(x+1)} + \frac{2}{(x-1)(x+1)}\right) = (x-1)(x+1)(x-2)\left(\frac{1}{(x-1)(x-2)}\right)$$

$$\frac{(x-1)(x+1)(x-2)}{1} \cdot \frac{1}{(x-2)(x+1)} + \frac{(x-1)(x+1)(x-2)}{1} \cdot \frac{2}{(x-1)(x+1)}$$

$$= \frac{(x-1)(x+1)(x-2)}{1} \cdot \frac{1}{(x-1)(x-2)}$$

$$x - 1 + 2(x-2) = x+1$$

$$x - 1 + 2x - 4 = x+1$$

$$3x - 5 = x+1$$

$$2x - 5 = 1$$

$$2x = 6$$

$$x = 3$$

CHECK x = 3:

$$\frac{1}{x^2 - x - 2} + \frac{2}{x^2 - 1} = \frac{1}{x^2 - 3x + 2}$$

$$\frac{1}{3^2 - 3 - 2} + \frac{2}{3^2 - 1} \overset{?}{=} \frac{1}{3^2 - 3(3) + 2}$$

$$\frac{1}{9 - 3 - 2} + \frac{2}{9 - 1} \overset{?}{=} \frac{1}{9 - 9 + 2}$$

$$\frac{1}{4} + \frac{2}{8} \overset{?}{=} \frac{1}{2}$$

$$\frac{1}{4} + \frac{1}{4} \overset{?}{=} \frac{1}{2}$$

$$\frac{2}{4} \overset{?}{=} \frac{1}{2}$$

$$\frac{1}{2} \overset{\checkmark}{=} \frac{1}{2}$$

39.

$$\frac{n}{3n+2} + \frac{6}{9n^2-4} - \frac{2}{3n-2} = \frac{n}{3n+2} + \frac{6}{(3n+2)(3n-2)} - \frac{2}{3n-2}$$

$$= \frac{n(3n-2)}{(3n+2)(3n-2)} + \frac{6}{(3n+2)(3n-2)} - \frac{2(3n+2)}{(3n-2)(3n+2)}$$

$$= \frac{n(3n-2) + 6 - 2(3n+2)}{(3n+2)(3n-2)}$$

$$= \frac{3n^2 - 2n + 6 - 6n - 4}{(3n+2)(3n-2)} = \frac{3n^2 - 8n + 2}{(3n+2)(3n-2)}$$

41.

$$\frac{1}{3n+4} + \frac{8}{9n^2-16} = \frac{1}{3n-4}$$

$$\frac{1}{3n+4} + \frac{8}{(3n+4)(3n-4)} = \frac{1}{3n-4}$$

$$(3n+4)(3n-4)\left(\frac{1}{3n+4} + \frac{8}{(3n+4)(3n-4)}\right) = (3n+4)(3n-4)\left(\frac{1}{3n-4}\right)$$

$$\frac{(3n+4)(3n-4)}{1} \cdot \frac{1}{3n+4} + \frac{(3n+4)(3n-4)}{1} \cdot \frac{8}{(3n+4)(3n-4)} = \frac{(3n+4)(3n-4)}{1} \cdot \frac{1}{3n-4}$$

$$3n - 4 + 8 = 3n + 4$$

$$3n + 4 = 3n + 4$$

identity

This means that any value of n will satisfy the equation. However, we must exclude the values $n = \pm\frac{4}{3}$, since each of these values will cause a zero denominator to occur in the original equation.

45.

$$\frac{6}{3a+5} - \frac{2}{a-4} = \frac{10}{3a^2-7a-20}$$

$$\frac{6}{3a+5} - \frac{2}{a-4} = \frac{10}{(3a+5)(a-4)}$$

$$(3a+5)(a-4)\left(\frac{6}{3a+5} - \frac{2}{a-4}\right) = (3a+5)(a-4)\left(\frac{10}{(3a+5)(a-4)}\right)$$

$$\frac{(3a+5)(a-4)}{1} \cdot \frac{6}{3a+5} - \frac{(3a+5)(a-4)}{1} \cdot \frac{2}{a-4} = \frac{(3a+5)(a-4)}{1} \cdot \frac{10}{(3a+5)(a-4)}$$

$$6(a-4) - 2(3a+5) = 10$$

$$6a - 24 - 6a - 10 = 10$$

$$-34 = 10$$

contradiction (no solution)

49.

$$\frac{4}{4x^2-9} - \frac{5}{4x^2-8x+3} = \frac{8}{4x^2+4x-3}$$

77

$$\frac{4}{(2x+3)(2x-3)} - \frac{5}{(2x-3)(2x-1)} = \frac{8}{(2x+3)(2x-1)}$$

$$(2x+3)(2x-3)(2x-1)\left(\frac{4}{(2x+3)(2x-3)} - \frac{5}{(2x-3)(2x-1)}\right) = (2x+3)(2x-3)(2x-1)\left(\frac{8}{(2x+3)(2x-1)}\right)$$

$$\frac{\cancel{(2x+3)}\cancel{(2x-3)}(2x-1)}{1} \cdot \frac{4}{\cancel{(2x+3)}\cancel{(2x-3)}} - \frac{(2x+3)\cancel{(2x-3)}\cancel{(2x-1)}}{1} \cdot \frac{5}{\cancel{(2x-3)}\cancel{(2x-1)}}$$

$$= \frac{(2x+3)(2x-3)\cancel{(2x-1)}}{1} \cdot \frac{8}{(2x+3)\cancel{(2x-1)}}$$

$$4(2x-1) - 5(2x+3) = 8(2x-3)$$
$$8x - 4 - 10x - 15 = 16x - 24$$
$$-2x - 19 = 16x - 24$$
$$-19 = 18x - 24$$
$$5 = 18x$$
$$\frac{5}{18} = x$$

51. $(x-y)^2 - 9 = (x-y)^2 - 3^2 = (x-y-3)(x-y+3)$

53. $\dfrac{3x^2-6x}{x+4} \div \left(\dfrac{2y^2-xy^2}{x^2-9} \cdot \dfrac{x^2}{xy+4y}\right) = \dfrac{3x^2-6x}{x+4} \div \left(\dfrac{y^2(2-x)}{(x-3)(x+3)} \cdot \dfrac{x^2}{\cancel{y}(x+4)}\right)$

$$= \frac{3x^2-6x}{x+4} \div \frac{x^2y(2-x)}{(x-3)(x+3)(x+4)} = \frac{3\cancel{x}(x-2)}{\cancel{x+4}} \cdot \frac{(x-3)(x+3)\cancel{(x+4)}}{x^{\cancel{2}}y(2-x)}$$

$$= \frac{3(x-2)(x-3)(x+3)}{xy((-1)(x-2))} = \frac{-3(x-3)(x+3)}{xy}$$

Exercises 4.6

1. $$5x + 7y = 4$$
    $$5x + 7 - 7y = 4 - 7y$$
    $$5x = 4 - 7y$$
    $$\frac{5x}{5} = \frac{4-7y}{5}$$
    $$x = \frac{4-7y}{5}$$

5. $$w + 4z - 1 = 2w - z + 3$$
    $$w + 4z - 1 - w = 2w - z + 3 - w$$
    $$4z - 1 = w - z + 3$$
    $$4z - 1 + z = w - z + 3 + z$$
    $$5z - 1 = w + 3$$
    $$5z - 1 - 3 = w + 3 - 3$$
    $$5z - 4 = w$$

7.

$$2(6r - 5t) > 5(2r + t)$$

$$12r - 10t > 10r + 5t$$

$$12r - 10t - 10r > 10r + 5t - 10r$$

$$2r - 10t > 5t$$

$$2r - 10t + 10t > 5t + 10t$$

$$2r > 15t$$

$$\frac{2r}{2} > \frac{15t}{2}$$

$$r > \frac{15t}{2}$$

11.

$$\frac{a}{5} - \frac{b}{3} = \frac{a}{2} - \frac{b}{6}$$

$$30\left(\frac{a}{5} - \frac{b}{3}\right) = 30\left(\frac{a}{2} - \frac{b}{6}\right)$$

$$\frac{\overset{6}{\cancel{30}}}{1} \cdot \frac{a}{\cancel{5}} - \frac{\overset{10}{\cancel{30}}}{1} \cdot \frac{b}{\cancel{3}} = \frac{\overset{15}{\cancel{30}}}{1} \cdot \frac{a}{\cancel{2}} - \frac{\overset{5}{\cancel{30}}}{1} \cdot \frac{b}{\cancel{6}}$$

$$6a - 10b = 15a - 5b$$

$$6a - 10b - 6a = 15a - 5b - 6a$$

$$-10b = 9a - 5b$$

$$-10b + 5b = 9a - 5b + 5b$$

$$-5b = 9a$$

$$\frac{-5b}{9} = \frac{9a}{9}$$

$$\frac{-5b}{9} = a$$

15.

$$ax + b = cx + d$$

$$ax + b - cx = cx + d - cx$$

$$ax + b - cx = d$$

$$ax + b - cx - b = d - b$$

$$ax - cx = d - b$$

$$(a - c)x = d - b$$

$$\frac{(a - c)x}{a - c} = \frac{d - b}{a - c}$$

$$x = \frac{d - b}{a - c}$$

19. $(x + 3)(y + 7) = a$

$$\frac{(x + 3)(y + 7)}{y + 7} = \frac{a}{y + 7}$$

$$x + 3 = \frac{a}{y + 7}$$

$$x + 3 - 3 = \frac{a}{y + 7} - 3$$

$$x = \frac{a}{y + 7} - 3$$

23. $x = \dfrac{2t - 3}{3t - 2}$

$$(3t - 2)x = \left(\frac{3t - 2}{1}\right)\left(\frac{2t - 3}{3t - 2}\right)$$

$$(3t - 2)x = 2t - 3$$

$$3tx - 2x = 2t - 3$$

$$3tx - 2x - 2t = 2t - 3 - 2t$$

$$3tx - 2x - 2t = -3$$

$$2x + 2tx - 2x - 2t = 2x - 3$$

$$3tx - 2t = 2x - 3$$

$$(3x - 2)t = 2x - 3$$

$$\frac{(3x - 2)t}{3x - 2} = \frac{2x - 3}{3x - 2}$$

$$t = \frac{2x - 3}{3x - 2}$$

25. $A = \dfrac{1}{2} bh$

$$2A = 2\left(\frac{1}{2} bh\right)$$

$$2A = bh$$

$$\frac{2A}{h} = \frac{bh}{h}$$

$$\frac{2A}{h} = b$$

29.　$A = P(1 + rt)$

$$\frac{A}{P} = \frac{P(1 + rt)}{P}$$

$$\frac{A}{P} = 1 + rt$$

$$\frac{A}{P} - 1 = 1 + rt - 1$$

$$\frac{A}{P} - 1 = rt$$

$$\frac{\frac{A}{P} - 1}{t} = \frac{rt}{t}$$

$$\frac{\frac{A}{P} - 1}{t} = r$$

$$\frac{P\left(\frac{A}{P} - 1\right)}{Pt} = r$$

$$\frac{A - P}{Pt} = r$$

31.　$C = \dfrac{5}{9}(F - 32)$

$$9C = \frac{9}{1} \cdot \frac{5}{9}(F - 32)$$

$$9C = 5(F - 32)$$

$$\frac{9C}{5} = \frac{5(F - 32)}{5}$$

$$\frac{9C}{5} = F - 32$$

$$\frac{9C}{5} + 32 = F - 32 + 32$$

$$\frac{9C}{5} + 32 = F$$

35.　$S = s_0 + v_0 t + \dfrac{1}{2}gt^2$

$$S - s_0 = s_0 + v_0 t + \frac{1}{2} gt^2 - s_0$$

$$S - s_0 = v_0 t + \frac{1}{2} gt^2$$

$$S - s_0 - v_0 t = v_0 t + \frac{1}{2} gt^2 - v_0 t$$

$$S - s_0 - v_0 t = \frac{1}{2} gt^2$$

$$2(S - s_0 - v_0 t) = \frac{\cancel{2}}{1} \cdot \frac{1}{\cancel{2}} gt^2$$

$$2(S - s_0 - v_0 t) = gt^2$$

$$\frac{2(S - s_0 - v_0 t)}{t^2} = \frac{gt^2}{t^2}$$

$$\frac{2(S - s_0 - v_0 t)}{t^2} = g$$

39. $$\frac{1}{f} = \frac{1}{f_1} + \frac{1}{f_2}$$

$$ff_1 f_2 \left(\frac{1}{f} \right) = ff_1 f_2 \left(\frac{1}{f_1} + \frac{1}{f_2} \right)$$

$$\frac{\cancel{f}f_1 f_2}{1} \cdot \frac{1}{\cancel{f}} = \frac{f\cancel{f_1} f_2}{1} \cdot \frac{1}{\cancel{f_1}} + \frac{f f_1 \cancel{f_2}}{1} \cdot \frac{1}{\cancel{f_2}}$$

$$f_1 f_2 = ff_2 + ff_1$$

$$f_1 f_2 - ff_1 = ff_2 + ff_1 - ff_1$$

$$f_1 f_2 - ff_1 = ff_2$$

$$f_1 (f_2 - f) = ff_2$$

$$\frac{f_1 (f_2 - f)}{f_2 - f} = \frac{ff_2}{f_2 - f}$$

$$f_1 = \frac{ff_2}{f_2 - f}$$

43. $$|3x - 2| > 5$$

$$3x - 2 < -5 \text{ or } 3x - 2 > 5$$

$$3x < -3 \text{ or } 3x > 7$$

$$x < -1 \text{ or } x > \frac{7}{3}$$

45. $$\frac{1 + \dfrac{2}{x+3}}{\dfrac{3}{x+3} + 1} = \frac{(x+3)\left(1 + \dfrac{2}{x+3}\right)}{(x+3)\left(\dfrac{3}{x+3} + 1\right)} = \frac{(x+3) + 2}{3 + (x+3)} = \frac{x+5}{x+6}$$

47. Let x = width of the rectangle (in inches). $3x + 5$ = length of the rectangle (in inches).

$$2x + 2(3x + 5) = 42$$
$$2x + 6x + 10 = 42$$
$$8x + 10 = 42$$
$$8x = 32$$
$$x = 4$$

Then $3x + 5 = 3(4) + 5 = 17$. Thus, the dimensions of the rectangle are 4" by 17".

Exercises 4.7

1. Let n = the number.

$$\frac{3}{4}n = \frac{2}{5}n - 7$$
$$\overset{5}{\cancel{20}}\left(\frac{3}{\cancel{4}}n\right) = \overset{4}{\cancel{20}}\left(\frac{2}{\cancel{5}}n\right) - 20 \cdot 7$$
$$15n = 8n - 140$$
$$7n = -140$$
$$n = -20$$

Thus, the number is –20.

CHECK: $\frac{3}{4}(-20) = -15$, which is 7 less than $\frac{2}{5}(-20) = -8$, as required.

5. Let n = larger number. Then $n - 21$ = smaller number.

$$\frac{n - 21}{n} = \frac{5}{12}$$
$$12\cancel{n}\left(\frac{n - 21}{\cancel{n}}\right) = \cancel{12}n\left(\frac{5}{\cancel{12}}\right)$$
$$12n - 252 = 5n$$
$$-252 = -7n$$
$$36 = n$$

Then $n - 21 = 36 - 21 = 15$. Thus, the numbers are 15 and 36.

CHECK: 15 is 21 less than 36 and $\frac{15}{36} = \frac{\cancel{3}(5)}{\cancel{3}(12)} = \frac{5}{12}$.

7. Let x = # of inches in 52 cm.

$$\frac{x}{52} = \frac{1}{2.54}$$
$$\cancel{52}\left(\frac{x}{\cancel{52}}\right) = 52\left(\frac{1}{2.54}\right)$$
$$x = \frac{52}{2.54} = 20.47 \text{ inches}$$

15. Let x = distance between Mike's home and the ballfield (in miles).

$$\frac{3}{4}x = 6$$

$$\cancel{4}\left(\frac{3}{\cancel{4}}x\right) = 4 \cdot 6$$

$$3x = 24$$

$$x = 8$$

Thus, the distance in question is 8 miles.

CHECK: $\frac{1}{4}(8) = 2$ miles, leaving $8 - 2 \overset{\checkmark}{=} 6$ miles.

19. Let R_3 = third resistance (in ohms).

$$\frac{1}{R} = \frac{1}{R_1} + \frac{1}{R_2} + \frac{1}{R_3}$$

$$\frac{1}{\frac{5}{4}} = \frac{1}{2} + \frac{1}{5} + \frac{1}{R_3}$$

$$\frac{4}{5} = \frac{1}{2} + \frac{1}{5} + \frac{1}{R_3}$$

$$\frac{4}{5} - \frac{1}{2} - \frac{1}{5} = \frac{1}{R_3}$$

$$\frac{3}{5} - \frac{1}{2} = \frac{1}{R_3}$$

$$\overset{2}{\cancel{10}}R_3\left(\frac{3}{\cancel{5}}\right) - \overset{5}{\cancel{10}}R_3\left(\frac{1}{\cancel{2}}\right) = 10\cancel{R_3}\left(\frac{1}{\cancel{R_3}}\right)$$

$$6R_3 - 5R_3 = 10$$

$$R_3 = 10$$

Thus, the third resistance is 10 ohms.

CHECK: $\frac{1}{2} + \frac{1}{5} + \frac{1}{10} = \frac{5}{10} + \frac{2}{10} + \frac{1}{10} = \frac{8}{10} = \frac{4}{5}$. If $\frac{1}{R} = \frac{4}{5}$, then $5 = 4R$, so $R = \frac{5}{4}$, as stated.

21. Let x = amount invested in bonds (in dollars). Then x + 3000 = amount invested in certificate (in dollars).

$$0.10x + 0.06(x + 3000) = 580$$

$$10x + 6(x + 3000) = 58000$$

$$10x + 6x + 18000 = 58000$$

$$16x + 18000 = 58000$$

$$16x = 40000$$

$$x = 2500$$

Then x + 3000 = 2500 + 3000 = 5500. Thus, Cindy invested $2500 in the bond and $5500 in the certificate.

CHECK: 10% of $2500 = (0.10)(2500) = 250. 6% of $5500 = (0.06)(5500) = $330. Then 250 + 330 = 580.

25. Let x = amount invested at $8\frac{1}{2}\%$ (in dollars). $18000 - x$ = amount invested at 11% (in dollars).

$$0.085x + 0.11(18000 - x) = 0.10(18000)$$
$$85x + 110(18000 - x) = 100(18000)$$
$$85x + 1980000 - 110x = 1800000$$
$$-25x + 1980000 = 1800000$$
$$-25x = -180000$$
$$x = 7200$$

Then, $18000 - x = 18000 - 7200 = 10800$. Thus, Lew invests $7,200 at $8\frac{1}{2}\%$ and $10,800 at 11%.

CHECK: $8\frac{1}{2}\%$ of $7200 = (0.085)(7200)= $612. 11% of $10800 = (0.11)(10800) = $1188.

$612 + 1188 = 1800, which is 10% of $18,000.

29. Let x = # of hours that the two services work together.
$$\frac{x}{30} + \frac{x}{20} = 1$$
$$\overset{2}{60}\left(\frac{x}{30}\right) + \overset{3}{60}\left(\frac{x}{20}\right) = 60 \cdot 1$$
$$2x + 3x = 60$$
$$5x = 60$$
$$x = 12$$

Thus, the two services can clean the building in 12 hours if they work together.

CHECK: In 12 hours, Quickie cleans $\frac{12}{30} = \frac{2}{5}$ of the job. In 12 hours, Super Quickie cleans $\frac{12}{20} = \frac{3}{5}$ of the job.

$\frac{2}{5} + \frac{3}{5} = 1$ (the entire job).

33. Let x = # of hours that Super Quickie service needs to complete the job.
$$\frac{10}{30} + \frac{x}{20} = 1$$
$$\frac{1}{3} + \frac{x}{20} = 1$$
$$\overset{20}{60}\left(\frac{1}{3}\right) + \overset{3}{60}\left(\frac{x}{20}\right) = 60 \cdot 1$$
$$20 + 3x = 60$$
$$3x = 40$$
$$x = \frac{40}{3}$$

Thus, the cleaning job takes $10 + \frac{40}{3} = \frac{70}{3}$ hours altogether.

CHECK: In 10 hours, Quickie cleans $\frac{10}{30} = \frac{1}{3}$ of the job. In $\frac{40}{3}$ hours, Super Quickie cleans $\frac{\frac{40}{3}}{20} = \frac{40}{60} = \frac{2}{3}$ of the job. $\frac{1}{3} + \frac{2}{3} = 1$ (the entire job).

37. Let x = Bill's riding speed (in kph). Then x + 10 = Jill's riding speed (in kph).

$$\frac{10}{x} = \frac{15}{x + 10}$$

$$\cancel{x}(x + 10)\left(\frac{10}{\cancel{x}}\right) = x(\cancel{x+10})\left(\frac{15}{\cancel{x+10}}\right)$$

$$10(x + 10) = 15x$$
$$10x + 100 = 15x$$
$$100 = 5x$$
$$20 = x$$

Thus, Bill's riding speed is 20 kph.

CHECK: At 20 kph, Bill covers 10 km in $\frac{10}{20} = \frac{1}{2}$ hour. At 20 + 10 = 30 kph, Jill covers 15 km in $\frac{15}{30} = \frac{1}{2}$.

41. Let x = # of ounces of the 20% alcohol solution.
$$0.20x + 0.50(5) = 0.30(x + 5)$$
$$20x + 50(5) = 30(x + 5)$$
$$20x + 250 = 30x + 150$$
$$250 = 10x + 150$$
$$100 = 10x$$
$$10 = x$$

Thus, 10 oz. of the 20% solution are needed.

CHECK: 10 oz. of the 20% solution contains 0.20(10) = 2 oz. of pure alcohol. 5 oz. of the 50% solution contains 0.50(5) = 2.5 oz. of pure alcohol. 2 + 2.5 = 4.5, which is 30% of 10 + 5 = 15.

45. Let x = # of tons of 40% iron alloy. Then 80 − x = # of tons of 60% iron alloy.
$$0.40x + 0.60(80 - x) = 0.55(80)$$
$$40x + 60(80 - x) = 55(80)$$
$$40x + 4800 - 60x = 4400$$
$$-20x + 4800 = 4400$$
$$-20x = -400$$
$$x = 20$$

Then 80 − x = 80 − 20 = 60. Thus, 20 tons of the 40% alloy should be mixed with 60 tons of the 60% alloy.

CHECK: 20 tons of 40% alloy contains 0.40(20) = 8 tons of iron. 60 tons of 60% alloy contains 0.60(60) = 36 tons of iron. 8 + 36 = 44, which is 55% of 80.

49. Let x = # of gallons drained off.
$$0.30(3 - x) + 0x = 0.20(3)$$
$$30(3 - x) = 20(3)$$
$$90 - 30x = 60$$
$$-30x = -30$$
$$x = 1$$

Thus, Jack must drain off 1 gallon from his radiator.

CHECK: Originally, radiator contains 30% of 3 gallons. After draining, it contains 30% of 2 gallons, or $0.30(2) = 0.6$ gallons of antifreeze, which is 20% of 3 gallon capacity.

53. Let n = # of nickels that Sam had. Then n + 5 = # of dimes that Sam had and 2n = # of quarters that Sam had.
$$5n + 10(n + 5) + 25(2n) = 2000$$
$$5n + 10n + 50 + 50n = 2000$$
$$65n + 50 = 2000$$
$$65n = 1950$$
$$n = 30$$

Then $n + 5 = 30 + 5 = 35$ and $2n = 2(30) = 60$. Thus, Sam had 30 nickels, 35 dimes, and 60 quarters.

CHECK:
$$30 \text{ nickels} = 30(5\text{¢}) = 150\text{¢}$$
$$35 \text{ dimes} = 35(10\text{¢}) = 350\text{¢}$$
$$\underline{60 \text{ quarters} = 60(25\text{¢}) = 1500\text{¢}}$$
$$\text{total value} = 2000\text{¢}$$
$$\overset{\vee}{=} \$20.00$$

57. Let x = Ari's final exam score.
$$0.20(85) + 0.20(65) + 0.20(72) + 0.40x \geq 80$$
$$20(85) + 20(65) + 20(72) + 40x \geq 8000$$
$$1700 + 1300 + 1440 + 40x \geq 8000$$
$$4440 + 40x \geq 8000$$
$$40x \geq 3560$$
$$x \geq 89$$

Thus, Ari must get at least an 89 on the final exam.

63. Since Joe travels 100 miles (half the distance) at the rate of 25 mph, this part of the trip takes him $\frac{100}{25} = 4$ hours to complete. He now must cover the remaining 100 miles, but has no time left to do so. Therefore, it is impossible for Joe to achieve an average speed of 50 mph, no matter how fast he decides to go.

65. $3x^3y - 6x^2y^2 + 3xy^3 = 3xy(x^2 - 2xy + y^2) = 3xy(x - y)^2$

67. $|2x + 8| \le 10$

$\qquad -10 \le 2x + 8 \le 10$

$\qquad -18 \le 2x \le 2$

$\qquad -9 \le x \le 1$

Chapter 4 Review Exercises

1. $\dfrac{4x^2y^3}{16xy^5} = \dfrac{\cancel{4xy^3}(x)}{\cancel{4xy^3}(4y^2)} = \dfrac{x}{4y^2}$

3. $\dfrac{x^2 + 2x - 8}{x^2 + 3x - 10} = \dfrac{(x + 4)\cancel{(x - 2)}}{(x + 5)\cancel{(x - 2)}} = \dfrac{x + 4}{x + 5}$

5. $\dfrac{x^4 - 2x^3 + 3x^2}{x^2} = \dfrac{\cancel{x^4}}{\cancel{x^2}} - \dfrac{2\cancel{x^3}}{\cancel{x^2}} + \dfrac{3\cancel{x^2}}{\cancel{x^2}} = x^2 - 2x + 3$

7. $\dfrac{5xa - 7a + 5xb - 7b}{3xa - 2a + 3xb - 2b} = \dfrac{a(5x - 7) + b(5x - 7)}{a(3x - 2) + b(3x - 2)} = \dfrac{(5x - 7)\cancel{(a + b)}}{(3x - 2)\cancel{(a + b)}} = \dfrac{5x - 7}{3x - 2}$

9. $\dfrac{\cancel{4x^2}y^3z^2}{12xy^4} \cdot \dfrac{\cancel{24xy^3}}{16xy^4} = \dfrac{xy^3z^2}{2}$

11. $\dfrac{5}{3x^2y} + \dfrac{1}{3x^2y} = \dfrac{5 + 1}{3x^2y} = \dfrac{\cancel{6}}{\cancel{3}x^2y} = \dfrac{2}{x^2y}$

13. $\dfrac{3x}{x - 1} + \dfrac{3}{x - 1} = \dfrac{3x + 3}{x - 1} = \dfrac{3(x + 1)}{x - 1}$

15. $\dfrac{2x^2}{x^2 + x - 6} + \dfrac{2x}{x^2 + x - 6} - \dfrac{12}{x^2 + x - 6} = \dfrac{2x^2 + 2x - 12}{x^2 + x - 6} = \dfrac{2\cancel{(x^2 + x - 6)}}{\cancel{x^2 + x - 6}} = \dfrac{2}{1} = 2$

17. $\dfrac{5}{3a^2b} - \dfrac{8}{4ab^4} = \dfrac{5(4b^3)}{3a^2b(4b^3)} - \dfrac{8(3a)}{4ab^4(3a)} = \dfrac{20b^3}{12a^2b^4} - \dfrac{24a}{12a^2b^4} = \dfrac{20b^3 - 24a}{12a^2b^4}$

$\qquad\qquad\qquad = \dfrac{4(5b^3 - 6a)}{4(3a^2b^4)} = \dfrac{5b^3 - 6a}{3a^2b^4}$

19. $\dfrac{3x + 1}{2x^2} - \dfrac{3x - 2}{5x} = \dfrac{(3x + 1)(5)}{2x^2(5)} - \dfrac{(3x - 2)(2x)}{5x(2x)} = \dfrac{15x + 5}{10x^2} - \dfrac{6x^2 - 4x}{10x^2}$

$\qquad\qquad = \dfrac{15x + 5 - (6x^2 - 4x)}{10x^2} = \dfrac{15x + 5 - 6x^2 + 4x}{10x^2} = \dfrac{-6x^2 + 19x + 5}{10x^2}$

21. $\dfrac{x - 7}{5 - x} + \dfrac{3x + 3}{x - 5} = \dfrac{(x - 7)(-1)}{(5 - x)(-1)} + \dfrac{3x + 3}{x - 5} = \dfrac{-x + 7}{x - 5} + \dfrac{3x + 3}{x - 5}$

$\qquad\qquad = \dfrac{-x + 7 + 3x + 3}{x - 5} = \dfrac{2x + 10}{x - 5} = \dfrac{2(x + 5)}{x - 5}$

23. $\dfrac{x^2 + x - 6}{x + 4} \cdot \dfrac{2x^2 + 8x}{x^2 + x - 6} = \dfrac{(x + 3)\cancel{(x - 2)}}{\cancel{x + 4}} \cdot \dfrac{2x\cancel{(x + 4)}}{(x + 3)\cancel{(x - 2)}} = \dfrac{2x}{1} = 2x$

25. $\dfrac{a^2 - 2ab + b^2}{a + b} \div \dfrac{(a - b)^3}{a + b} = \dfrac{(a - b)^2}{a + b} \div \dfrac{(a - b)^3}{a + b} = \dfrac{\cancel{(a - b)^2}}{\cancel{a + b}} \cdot \dfrac{\cancel{a + b}}{\cancel{(a - b)^3}} = \dfrac{1}{a - b}$

88

27. $\dfrac{3x}{2x+3} - \dfrac{5}{x-4} = \dfrac{3x(x-4)}{(2x+3)(x-4)} - \dfrac{5(2x+3)}{(x-4)(2x+3)} = \dfrac{3x^2-12x}{(2x+3)(x-4)} - \dfrac{10x+15}{(2x+3)(x-4)}$

$\qquad = \dfrac{3x^2-12x-(10x+15)}{(2x+3)(x-4)} = \dfrac{3x^2-12x-10x-15}{(2x+3)(x-4)} = \dfrac{3x^2-22x-15}{(2x+3)(x-4)}$

29. $\dfrac{3x-2}{2x-7} + \dfrac{5x+2}{2x-3} = \dfrac{(3x-2)(2x-3)}{(2x-7)(2x-3)} + \dfrac{(5x+2)(2x-7)}{(2x-3)(2x-7)}$

$\qquad = \dfrac{6x^2-13x+6}{(2x-7)(2x-3)} + \dfrac{10x^2-31x-14}{(2x-7)(2x-3)}$

$\qquad = \dfrac{6x^2-13x+6+10x^2-31x-14}{(2x-7)(2x-3)}$

$\qquad = \dfrac{16x^2-44x-8}{(2x-7)(2x-3)} = \dfrac{4(4x^2-11x-2)}{(2x-7)(2x-3)}$

31. $\dfrac{5a}{a^2-3a} + \dfrac{2}{4a^3+4a^2} = \dfrac{\cancel{5a}}{\cancel{a}(a-3)} + \dfrac{\cancel{2}}{2\cancel{4}a^2(a+1)} = \dfrac{5}{a-3} + \dfrac{1}{2a^2(a+1)}$

$\qquad = \dfrac{5(2a^2(a+1))}{(a-3)(2a^2(a+1))} + \dfrac{a-3}{2a^2(a+1)(a-3)}$

$\qquad = \dfrac{10a^3+10a^2}{2a^2(a+1)(a-3)} + \dfrac{a-3}{2a^2(a+1)(a-3)} = \dfrac{10a^3+10a^2+a-3}{2a^2(a+1)(a-3)}$

33. $\dfrac{5}{x^2-4x+4} + \dfrac{3}{x^2-4} = \dfrac{5}{(x-2)^2} + \dfrac{3}{(x+2)(x-2)}$

$\qquad = \dfrac{5(x+2)}{(x-2)^2(x+2)} + \dfrac{3(x-2)}{(x+2)(x-2)(x-2)}$

$\qquad = \dfrac{5x+10+3x-6}{(x-2)^2(x+2)} = \dfrac{8x+4}{(x-2)^2(x+2)} = \dfrac{4(2x+1)}{(x-2)^2(x+2)}$

35. $\dfrac{2x}{7x^2-14x-21} + \dfrac{2x}{14x-42} = \dfrac{2x}{7(x^2-2x-3)} + \dfrac{\cancel{2}x}{7\cancel{14}(x-3)} = \dfrac{2x}{7(x-3)(x+1)} + \dfrac{x}{7(x-3)}$

$\qquad = \dfrac{2x}{7(x-3)(x+1)} + \dfrac{x(x+1)}{7(x-3)(x+1)}$

$\qquad = \dfrac{2x}{7(x-3)(x+1)} + \dfrac{x^2+x}{7(x-3)(x+1)}$

$\qquad = \dfrac{2x+x^2+x}{7(x-3)(x+1)} = \dfrac{x^2+3x}{7(x-3)(x+1)} = \dfrac{x(x+3)}{7(x-3)(x+1)}$

37. $\dfrac{5x}{x-2} + \dfrac{3x}{x+2} - \dfrac{2x+3}{x^2-4} = \dfrac{5x}{x-2} + \dfrac{3x}{x+2} - \dfrac{2x+3}{(x+2)(x-2)}$

$\qquad = \dfrac{5x(x+2)}{(x-2)(x+2)} + \dfrac{3x(x-2)}{(x+2)(x-2)} - \dfrac{2x+3}{(x+2)(x-2)}$

$\qquad = \dfrac{5x(x+2)+3x(x-2)-(2x+3)}{(x+2)(x-2)}$

$\qquad = \dfrac{5x^2+10x+3x^2-6x-2x-3}{(x+2)(x-2)} = \dfrac{8x^2+2x-3}{(x+2)(x-2)} = \dfrac{(2x-1)(4x+3)}{(x+2)(x-2)}$

39. $\dfrac{2x+y}{5x^2y-xy^2} \cdot \dfrac{25x^2-y^2}{10x^2+3xy-y^2} \cdot \dfrac{5x^2-xy}{5x+y} = \dfrac{\cancel{2x+y}}{xy(5x-y)} \cdot \dfrac{(5x-y)(5x+y)}{(5x-y)(2x+y)} \cdot \dfrac{x(5x-y)}{5x+y} = \dfrac{1}{y}$

41. $\dfrac{4x+11}{x^2+x-6} - \dfrac{x+2}{x^2+4x+3} = \dfrac{4x+11}{(x+3)(x-2)} - \dfrac{x+2}{(x+3)(x+1)}$

$$= \dfrac{(4x+11)(x+1)}{(x+3)(x-2)(x+1)} - \dfrac{(x+2)(x-2)}{(x+3)(x+1)(x-2)}$$

$$= \dfrac{4x^2+15x+11}{(x+3)(x-2)(x+1)} - \dfrac{x^2-4}{(x+3)(x-2)(x+1)}$$

$$= \dfrac{4x^2+15x+11-(x^2-4)}{(x+3)(x-2)(x+1)} = \dfrac{4x^2+15x+11-x^2+4}{(x+3)(x-2)(x+1)}$$

$$= \dfrac{3x^2+15x+15}{(x+3)(x-2)(x+1)} = \dfrac{3(x^2+5x+5)}{(x+3)(x-2)(x+1)}$$

43. $\dfrac{5x}{x^2-x-2} + \dfrac{4x+3}{x^3+x^2} - \dfrac{x-6}{x^3-2x^2} = \dfrac{5x}{(x-2)(x+1)} + \dfrac{4x+3}{x^2(x+1)} - \dfrac{x-6}{x^2(x-2)}$

$$= \dfrac{5x(x^2)}{(x-2)(x+1)(x^2)} + \dfrac{(4x+3)(x-2)}{x^2(x+1)(x-2)} - \dfrac{(x-6)(x+1)}{x^2(x-2)(x+1)}$$

$$= \dfrac{5x^3}{x^2(x-2)(x+1)} + \dfrac{4x^2-5x-6}{x^2(x-2)(x+1)} - \dfrac{x^2-5x-6}{x^2(x-2)(x+1)}$$

$$= \dfrac{5x^3+4x^2-5x-6-(x^2-5x-6)}{x^2(x-2)(x+1)}$$

$$= \dfrac{5x^3+4x^2-5x-6-x^2+5x+6}{x^2(x-2)(x+1)} = \dfrac{5x^3+3x^2}{x^2(x-2)(x+1)}$$

$$= \dfrac{\cancel{x^2}(5x+3)}{\cancel{x^2}(x-2)(x+1)} = \dfrac{5x+3}{(x-2)(x+1)}$$

45. $\dfrac{4x^2+12x+9}{8x^3+27} \cdot \dfrac{12x^3-18x^2+27x}{4x^2-9} = \dfrac{\cancel{(2x+3)}(2x+3)}{\cancel{(2x+3)}\cancel{(4x^2-6x+9)}} \cdot \dfrac{3x\cancel{(4x^2-6x+9)}}{(2x+3)(2x-3)} = \dfrac{3x}{2x-3}$

47. $4x \div \left(\dfrac{8x^2-8xy}{2ax+bx-2ay-by} \div \dfrac{2ax+2bx+3ay+3by}{2a^2+3ab+b^2} \right)$

$$= 4x \div \left(\dfrac{8x^2-8xy}{2ax+bx-2ay-by} \cdot \dfrac{2a^2+3ab+b^2}{2ax+2bx+3ay+3by} \right)$$

$$= 4x \div \left(\dfrac{8x(x-y)}{x(2a+b)-y(2a+b)} \cdot \dfrac{(2a+b)(a+b)}{2x(a+b)+3y(a+b)} \right)$$

$$= 4x \div \left(\dfrac{8x\cancel{(x-y)}}{\cancel{(2a+b)}\cancel{(x-y)}} \cdot \dfrac{\cancel{(2a+b)}\cancel{(a+b)}}{\cancel{(a+b)}(2x+3y)} \right)$$

$$= 4x \div \dfrac{8x}{2x+3y} = \dfrac{\cancel{4x}}{1} \cdot \dfrac{2x+3y}{\cancel{8x}_2} = \dfrac{2x+3y}{2}$$

49. $\left(\dfrac{x}{2} + \dfrac{3}{x} \right) \cdot \dfrac{x+1}{x} = \left(\dfrac{x(x)}{2(x)} + \dfrac{3(2)}{x(2)} \right) \cdot \dfrac{x+1}{x}$

$$= \left(\dfrac{x^2}{2x} + \dfrac{6}{2x} \right) \cdot \dfrac{x+1}{x} = \dfrac{x^2+6}{2x} \cdot \dfrac{x+1}{x} = \dfrac{(x^2+6)(x+1)}{2x^2}$$

51.

$$\frac{\dfrac{3x^2y}{2ab}}{\dfrac{9x}{16a^2}} = \frac{3x^2y}{2ab} \div \frac{9x}{16^2} = \frac{3x^2y}{2ab} \cdot \frac{16a^2}{9x} = \frac{8axy}{3b}$$

53.

$$\frac{\dfrac{3}{a} - \dfrac{2}{a}}{\dfrac{5}{a}} = \frac{\dfrac{3-2}{a}}{\dfrac{5}{a}} = \frac{\dfrac{1}{a}}{\dfrac{5}{a}} = \frac{1}{a} \div \frac{5}{a} = \frac{1}{a} \cdot \frac{a}{5} = \frac{1}{5}$$

55.

$$\frac{\dfrac{3}{b+1} + 2}{\dfrac{2}{b-1} + b} = \frac{(b+1)(b-1)\left(\dfrac{3}{b+1} + 2\right)}{(b+1)(b-1)\left(\dfrac{2}{b-1} + b\right)}$$

$$= \frac{\dfrac{(b+1)(b-1)}{1} \cdot \dfrac{3}{b+1} + 2(b+1)(b-1)}{\dfrac{(b+1)(b-1)}{1} \cdot \dfrac{2}{b-1} + b(b+1)(b-1)}$$

$$= \frac{3(b-1) + 2(b+1)(b-1)}{2(b+1) + b(b+1)(b-1)} = \frac{(b-1)(3 + 2(b+1))}{(b+1)(2 + b(b-1))}$$

$$= \frac{(b-1)(3 + 2b + 2)}{(b+1)(2 + b^2 - b)} = \frac{(b-1)(2b+5)}{(b+1)(b^2 - b + 2)}$$

57.

$$\frac{x}{3} + \frac{x-1}{2} = \frac{7}{6}$$

$$6\left(\frac{x}{3} + \frac{x-1}{2}\right) = 6\left(\frac{7}{6}\right)$$

$$\frac{6}{1} \cdot \frac{x}{3} + \frac{6}{1} \cdot \frac{x-1}{2} = \frac{6}{1} \cdot \frac{7}{6}$$

$$2x + 3(x-1) = 7$$

$$2x + 3x - 3 = 7$$

$$5x - 3 = 7$$

$$5x = 10$$

$$x = 2$$

CHECK x = 2:

$$\frac{x}{3} + \frac{x-1}{2} = \frac{7}{6}$$

$$\frac{2}{3} + \frac{2-1}{2} \overset{?}{=} \frac{7}{6}$$

$$\frac{2}{3} + \frac{1}{2} \overset{?}{=} \frac{7}{6}$$

$$\frac{4}{6} + \frac{3}{6} \overset{?}{=} \frac{7}{6}$$

$$\frac{7}{6} \overset{\checkmark}{=} \frac{7}{6}$$

59.

$$\frac{x}{5} - \frac{x+1}{3} < \frac{1}{3}$$

$$15\left(\frac{x}{5} - \frac{x+1}{3}\right) < 15\left(\frac{1}{3}\right)$$

$$\frac{15}{1} \cdot \frac{x}{5} - \frac{15}{1} \cdot \frac{x+1}{3} < \frac{15}{1} \cdot \frac{1}{3}$$

$$3x - 5(x+1) < 5$$

$$3x - 5x - 5 < 5$$

$$-2x - 5 < 5$$

$$-2x < 10$$

$$x > -5$$

61. $\dfrac{5}{x} - \dfrac{1}{3} = \dfrac{11}{3x}$

$$3x\left(\dfrac{5}{x} - \dfrac{1}{3}\right) = 3x\left(\dfrac{11}{3x}\right)$$

$$\dfrac{3x}{1} \cdot \dfrac{5}{x} - \dfrac{3x}{1} \cdot \dfrac{1}{3} = \dfrac{3x}{1} \cdot \dfrac{11}{3x}$$

$$15 - x = 11$$

$$-x = -4$$

$$x = 4$$

CHECK x = 4:

$$\dfrac{5}{x} - \dfrac{1}{3} = \dfrac{11}{3x}$$

$$\dfrac{5}{4} - \dfrac{1}{3} \overset{?}{=} \dfrac{11}{3(4)}$$

$$\dfrac{5}{4} - \dfrac{1}{3} \overset{?}{=} \dfrac{11}{12}$$

$$\dfrac{15}{12} - \dfrac{4}{12} \overset{?}{=} \dfrac{11}{12}$$

$$\dfrac{11}{12} \overset{\checkmark}{=} \dfrac{11}{12}$$

63. $\dfrac{x+1}{3} - \dfrac{x}{2} > 4$

$$6\left(\dfrac{x+1}{3} - \dfrac{x}{2}\right) > 6 \cdot 4$$

$$\dfrac{\overset{2}{6}}{1} \cdot \dfrac{x+1}{3} - \dfrac{\overset{3}{6}}{1} \cdot \dfrac{x}{2} > 24$$

$$2(x+1) - 3x > 24$$

$$2x + 2 - 3x > 24$$

$$-x + 2 > 24$$

$$-x > 22$$

$$x < -22$$

65. $-\dfrac{7}{x} + 1 = -13$

$$x\left(-\dfrac{7}{x} + 1\right) = (x)(-13)$$

$$\dfrac{x}{1}\left(-\dfrac{7}{x}\right) + x \cdot 1 = -13x$$

$$-7 + x = -13x$$

$$-7 = -14x$$

$$\dfrac{1}{2} = x$$

CHECK $x = \dfrac{1}{2}$:

$$-\dfrac{7}{x} + 1 = -13$$

$$\dfrac{-7}{\dfrac{1}{2}} + 1 \overset{?}{=} -13$$

$$-14 + 1 \overset{?}{=} -13$$

$$-13 \overset{\checkmark}{=} -13$$

67. $\dfrac{5}{x-2} - 1 = 0$

$$\dfrac{5}{x-2} = 1$$

$$(x-2)\left(\dfrac{5}{x-2}\right) = (x-2)(1)$$

$$5 = x - 2$$

$$7 = x$$

CHECK x = 7:

$$\dfrac{5}{x-2} - 1 = 0$$

$$\dfrac{5}{7-2} - 1 \overset{?}{=} 0$$

$$\dfrac{5}{5} - 1 \overset{?}{=} 0$$

$$1 - 1 \overset{?}{=} 0$$

$$0 \overset{\checkmark}{=} 0$$

69.
$$\frac{x-2}{5} - \frac{3-x}{15} > \frac{1}{9}$$
$$45\left(\frac{x-2}{5} - \frac{3-x}{15}\right) > 45\left(\frac{1}{9}\right)$$
$$\frac{\overset{9}{\cancel{45}}}{1} \cdot \frac{x-2}{\cancel{5}} - \frac{\overset{3}{\cancel{45}}}{1} \cdot \frac{3-x}{\cancel{15}} > \frac{\overset{5}{\cancel{45}}}{1} \cdot \frac{1}{\cancel{9}}$$
$$9(x-2) - 3(3-x) > 5$$
$$9x - 18 - 9 + 3x > 5$$
$$12x - 27 > 5$$
$$12x > 32$$
$$x > \frac{8}{3}$$

71.
$$\frac{7}{x-1} + 4 = \frac{x+6}{x-1}$$
$$(x-1)\left(\frac{7}{x-1} + 4\right) = (x-1)\left(\frac{x+6}{x-1}\right)$$
$$\frac{\cancel{x-1}}{1} \cdot \frac{7}{\cancel{x-1}} + 4(x-1) = \frac{\cancel{x-1}}{1} \cdot \frac{x+6}{\cancel{x-1}}$$
$$7 + 4(x-1) = x+6$$
$$7 + 4x - 4 = x+6$$
$$4x + 3 = x+6$$
$$3x + 3 = 6$$
$$3x = 3$$
$$x = 1$$

CHECK x = 1:

$$\frac{7}{x-1} + 4 = \frac{x+6}{x-1}$$
$$\frac{7}{1-1} + 4 \overset{?}{=} \frac{1+6}{1-1}$$
$$\frac{7}{0} + 4 \neq \frac{7}{0},$$

since we cannot divide by 0. Therefore, the equation has no solution.

73.
$$\frac{4x+1}{x^2 - x - 6} = \frac{2}{x-3} + \frac{5}{x+2}$$
$$\frac{4x+1}{(x-3)(x+2)} = \frac{2}{x-3} + \frac{5}{x+2}$$
$$(x-3)(x+2)\left(\frac{4x+1}{(x-3)(x+2)}\right) = (x-3)(x+2)\left(\frac{2}{x-3} + \frac{5}{x+2}\right)$$
$$\frac{(x-3)(x+2)}{1} \cdot \frac{4x+1}{(x-3)(x+2)} = \frac{(x-3)(x+2)}{1} \cdot \frac{2}{x-3} + \frac{(x-3)(x+2)}{1} \cdot \frac{5}{x+2}$$

$$4x + 1 = 2(x + 2) + 5(x - 3)$$

$$4x + 1 = 2x + 4 + 5x - 15$$
$$4x + 1 = 7x - 11$$
$$1 = 3x - 11$$
$$12 = 3x$$
$$4 = x$$

CHECK $x = 4$:

$$\frac{4x + 1}{x^2 - x - 6} = \frac{2}{x - 3} + \frac{5}{x + 2}$$

$$\frac{4(4) + 1}{4^2 - 4 - 6} \overset{?}{=} \frac{2}{4 - 3} + \frac{5}{4 + 2}$$

$$\frac{16 + 1}{16 - 4 - 6} \overset{?}{=} \frac{2}{1} + \frac{5}{6}$$

$$\frac{17}{6} \overset{?}{=} \frac{12}{6} + \frac{5}{6}$$

$$\frac{17}{6} \overset{\checkmark}{=} \frac{17}{6}$$

75.
$$5x - 3y = 2x + 7y$$
$$5x - 3y - 2x = 2x + 7y - 2x$$
$$3x - 3y = 7y$$
$$3x - 3y + 3y = 7y + 3y$$
$$3x = 10y$$
$$\frac{3x}{3} = \frac{10y}{3}$$
$$x = \frac{10y}{3}$$

77.
$$3xy = 2xy + 4$$
$$3xy - 2xy = 2xy + 4 - 2xy$$
$$xy = 4$$
$$\frac{xy}{x} = \frac{4}{x}$$
$$y = \frac{4}{x}$$

79.
$$\frac{2x + 1}{y} = x$$
$$y\left(\frac{2x + 1}{y}\right) = y(x)$$
$$2x + 1 = yx$$
$$\frac{2x + 1}{x} = \frac{yx}{x}$$
$$\frac{2x + 1}{x} = y$$

81.
$$\frac{ax + b}{cx + d} = y$$

$$(cx + d)\left(\frac{ax + b}{cx + d}\right) = (cx + d)y$$

$$ax + b = cxy + dy$$

$$ax + b - cxy = cxy + dy - cxy$$

$$ax + b - cxy = dy$$

$$ax + b - cxy - b = dy - b$$

$$ax - cxy = dy - b$$

$$x(a - cy) = dy - b$$

$$\frac{x(a - cy)}{a - cy} = \frac{dy - b}{a - cy}$$

$$x = \frac{dy - b}{a - cy}$$

83.
$$\frac{1}{a} + \frac{1}{b} + \frac{1}{c} = \frac{1}{d}$$

$$abcd\left(\frac{1}{a} + \frac{1}{b} + \frac{1}{c}\right) = abcd\left(\frac{1}{d}\right)$$

$$\frac{abcd}{1} \cdot \frac{1}{a} + \frac{abcd}{1} \cdot \frac{1}{b} + \frac{abcd}{1} \cdot \frac{1}{c} = \frac{abcd}{1} \cdot \frac{1}{d}$$

$$bcd + acd + abd = abc$$

$$bcd + acd + abd - bcd = abc - bcd$$

$$acd + abd = abc - bcd$$

$$acd + abd - abd = abc - bcd - abd$$

$$acd = abc - bcd - abd$$

$$acd = b(ac - cd - ad)$$

$$\frac{acd}{ac - cd - ad} = \frac{b(ac - cd - ad)}{ac - cd - ad}$$

$$\frac{acd}{ac - cd - ad} = b$$

85. Let x = # of inches in one centimeter.

$$\frac{x}{1} = \frac{1}{2.54}$$

$$x = \frac{1}{2.54} \approx .39 \text{ inches}$$

87. Let x = # of miles that Carol walked.

$$\frac{1}{2}x + \frac{1}{3}\left(\frac{1}{2}x\right) + \frac{3}{2} = x$$

$$\frac{1}{2}x + \frac{1}{6}x + \frac{3}{2} = x$$

$$6\left(\frac{1}{2}x\right) + 6\left(\frac{1}{6}x\right) + 6\left(\frac{3}{2}\right) = 6x$$

$$3x + x + 9 = 6x$$

CHECK: $\frac{1}{2}\left(\frac{9}{2}\right) = \frac{9}{4}$

$$\frac{1}{3}\left(\frac{9}{2} - \frac{9}{4}\right) = \frac{1}{3}\left(\frac{9}{4}\right) = \frac{3}{4}$$

$$\frac{9}{4} + \frac{3}{4} + \frac{3}{2} = \frac{9}{4} + \frac{3}{4} + \frac{6}{4}$$

$$= \frac{18}{4} \stackrel{\checkmark}{=} \frac{9}{2}$$

$$4x + 9 = 6x$$
$$9 = 2x$$
$$\frac{9}{2} = x$$

Thus, Carol walked $\frac{9}{2}$ miles.

89. Let x = # of days required by Charles and Ellen if they work together.

$$\frac{x}{2\frac{1}{2}} + \frac{x}{2\frac{1}{3}} = 1$$

$$\frac{x}{\frac{5}{2}} + \frac{x}{\frac{7}{3}} = 1$$

$$\frac{2x}{5} + \frac{3x}{7} = 1$$

$$\overset{7}{35}\left(\frac{2x}{\cancel{5}}\right) + \overset{5}{\cancel{35}}\left(\frac{3x}{\cancel{7}}\right) = 35 \cdot 1$$

$$14x + 15x = 35$$
$$29x = 35$$
$$x = \frac{35}{29}$$

Thus, it takes Charles and Ellen $\frac{35}{29}$ days to refinish the room if they work together.

91. Let x = # of liters of 35% alcohol solution needed.

$$0.35x + 0.70(5) = 0.60(x + 5)$$
$$35x + 70(5) = 60(x + 5)$$
$$35x + 350 = 60x + 300$$
$$350 = 25x + 300$$
$$50 = 25x$$
$$2 = x$$

Thus, 2 liters of the 35% alcohol solution are needed.

CHECK: 2 liters of 35% solution contains 0.7 liters of pure alcohol. 5 liters of 70% solution contains 3.5 liters.

$0.7 + 3.5 = 4.2$ liters, and 4.2 is 60% of $2 + 5 = 7$.

93. Let x = # of children's tickets sold. Then 980 − x = # of adult's tickets sold.

$$1.50x + 4.25(980 - x) = 3010$$
$$150x + 425(980 - x) = 301000$$
$$150x + 416500 - 425x = 301000$$
$$-275x + 416500 = 301000$$
$$-275x = -115500$$
$$x = 420$$

Then 980 − x = 980 − 420 = 560. Thus, 420 children's tickets and 560 adult's tickets were sold.

CHECK: 420 tickets at $1.50 each = $630

<u>560 tickets at $4.25 each = $2380</u>

total $\overset{\vee}{=}$ $3010

Chapter 4 Practice Test

1. (a) $\dfrac{24x^2y^4}{64x^3y} = \dfrac{8x^2y(3y^3)}{8x^2y(8x)} = \dfrac{3y^3}{8x}$

(b) $\dfrac{x^2 - 9}{x^2 - 6x + 9} = \dfrac{(x-3)(x+3)}{(x-3)(x-3)} = \dfrac{x+3}{x-3}$

(c) $\dfrac{6x^3 - 9x^2 - 6x}{5x^3 - 10x^2} = \dfrac{3x(2x^2 - 3x - 2)}{5x^2(x-2)} = \dfrac{3x(2x+1)(x-2)}{5x^2(x-2)} = \dfrac{3(2x+1)}{5x}$

3. $\dfrac{\dfrac{3}{x+1} - 2}{\dfrac{5}{x} + 1} = \dfrac{x(x+1)\left(\dfrac{3}{x+1} - 2\right)}{x(x+1)\left(\dfrac{5}{x} + 1\right)} = \dfrac{\dfrac{x(x+1)}{1} \cdot \dfrac{3}{x+1} - 2x(x+1)}{\dfrac{x(x+1)}{1} \cdot \dfrac{5}{x} + 1x(x+1)}$

$= \dfrac{3x - 2x(x+1)}{5(x+1) + x(x+1)} = \dfrac{x(3 - 2(x+1))}{(x+1)(5+x)}$

$= \dfrac{x(3 - 2x - 2)}{(x+1)(5+x)} = \dfrac{x(1 - 2x)}{(x+1)(5+x)}$

5.
$$y = \frac{x-2}{2x+1}$$

$$(2x+1)y = (2x+1)\left(\frac{x-2}{2x+1}\right)$$

$$2xy + y = x - 2$$

$$2xy + y - 2xy = x - 2 - 2xy$$

$$y = x - 2 - 2xy$$

$$y + 2 = x - 2 - 2xy + 2$$

$$y + 2 = x - 2xy$$

$$y + 2 = x(1 - 2y)$$

$$\frac{y+2}{1-2y} = \frac{x(1-2y)}{1-2y}$$

$$\frac{y+2}{1-2y} = x$$

7.   Let x = # of hours needed if they work together.

$$\frac{x}{3\frac{1}{2}} + \frac{x}{2} = 1$$

$$\frac{x}{\frac{7}{2}} + \frac{x}{2} = 1$$

$$\frac{2x}{7} + \frac{x}{2} = 1$$

$$\overset{2}{14}\left(\frac{2x}{7}\right) + \overset{7}{14}\left(\frac{x}{2}\right) = 14 \cdot 1$$

$$4x + 7x = 14$$

$$11x = 14$$

$$x = \frac{14}{11}$$

Thus, it takes Jackie and Eleanor $\frac{14}{11}$ hours to complete the job if they work together.

CHAPTER 5
GRAPHING LINEAR EQUATIONS AND INEQUALITIES

Exercises 5.1

1. $$3x - 5y = 17$$
    $$3(4) - 5(1) \overset{?}{=} 17$$
    $$12 - 5 \overset{?}{=} 17$$
    $$7 \neq 17$$

    So the ordered pair (4, 1) does not satisfy the equation.

5. $$2x + 3y = 2$$
    $$\cancel{2}\left(\frac{3}{\cancel{7}}\right) + \cancel{3}\left(-\frac{1}{\cancel{3}}\right) \overset{?}{=} 2$$
    $$3 - 1 \overset{?}{=} 2$$
    $$2 \overset{\checkmark}{=} 2$$

    So the ordered pair $\left(\frac{3}{2}, -\frac{1}{3}\right)$ satisfies the equation.

9. $x + y = 8$

    $(-1, \underline{9})$, $(0, \underline{8})$, $(1, \underline{7})$, $(\underline{10}, -2)$, $(\underline{8}, 0)$, $(\underline{4}, 4)$

13. $\dfrac{x}{3} + \dfrac{y}{4} = 1$

    $(-3, \underline{8})$, $(0, \underline{4})$, $(3, \underline{0})$, $(\underline{6}, -4)$, $(\underline{3}, 0)$, $(\underline{0}, 4)$

15. To find the x-intercept, set y = 0:

    $$x + y = 6$$
    $$x + 0 = 6$$
    $$x = 6$$

    Hence, the graph crosses the x-axis at (6, 0). To find the y-intercept, set x = 0:

    $$x + y = 6$$
    $$0 + y = 6$$
    $$y = 6$$

    Hence, the graph crosses the y-axis at (0, 6).

17. To find the x-intercept, set y = 0:

    $$x - y = 6$$
    $$x - 0 = 6$$

$$x = 6$$

Hence, the graph crosses the x-axis at (6, 0). To find the y-intercept, set x = 0:

$$x - y = 6$$
$$0 - y = 6$$
$$-y = 6$$
$$y = -6$$

Hence the graph crosses the y-axis at (0, –6)

19. To find the x-intercept, set y = 0:

$$y - x = 6$$
$$0 - x = 6$$
$$-x = 6$$
$$x = -6$$

Hence, the graph crosses the x-axis at (–6, 0). To find the y-intercept, set x = 0:

$$y - x = 6$$
$$y - 0 = 6$$
$$y = 6$$

Hence, the graph crosses the y-axis at (0, 6).

23. To find the x-intercept, set y = 0:

$$3y + 4x = 12$$
$$3(0) + 4x = 12$$
$$4x = 12$$
$$x = 3$$

Hence, the graph crosses the x-axis at (3, 0). To find the y-intercept, set x = 0:

$$3y + 4x = 12$$
$$3y + 4(0) = 12$$
$$3y = 12$$
$$y = 4$$

Hence, the graph crosses the y-axis at (0, 4).

29. To find the x-intercept, set y = 0:

$$4x + 3y = 0$$
$$4x + 3(0) = 0$$
$$4x + 0 = 0$$
$$4x = 0$$
$$x = 0$$

Hence, the graph crosses the x-axis at (0, 0). This is also the point

at which the graph crosses the y-axis. To find a second point,

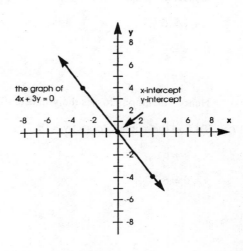

choose y = 4:

$$4x + 3y = 0$$

$$4x + 3(4) = 0$$

$$4x + 12 = 0$$

$$4x = -12$$

$$x = -3$$

Hence, the graph passes through (–3, 4). To find a check point,

choose y = –4:

$$4x + 3y = 0$$

$$4x + 3(-4) = 0$$

$$4x - 12 = 0$$

$$4x = 12$$

$$x = 3$$

Hence, the graph passes through (3, –4).

33. To find the x-intercept, set y = 0:

$$\frac{x}{2} - \frac{y}{3} = 1$$

$$\frac{x}{2} - \frac{0}{3} = 1$$

$$\frac{x}{2} - 0 = 1$$

$$\frac{x}{2} = 1$$

$$x = 2$$

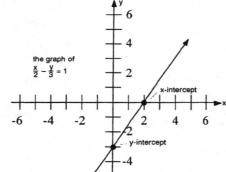

the graph of
$\frac{x}{2} - \frac{y}{3} = 1$

x-intercept

y-intercept

Hence, the graph crosses the x-axis at (2, 0). To find the y-intercept, set x = 0:

$$\frac{x}{2} - \frac{y}{3} = 1$$

$$\frac{0}{2} - \frac{y}{3} = 1$$

$$0 - \frac{y}{3} = 1$$

$$-\frac{y}{3} = 1$$

$$-y = 3$$

$$y = -3$$

Hence, the graph crosses the y-axis at (0, –3). To find a check point, choose y = –6:

$$\frac{x}{2} - \frac{y}{3} = 1$$

$$\frac{x}{2} - \frac{-6}{3} = 1$$

$$\frac{x}{2} - (-2) = 1$$

$$\frac{x}{2} + 2 = 1$$

$$\frac{x}{2} = -1$$

$$x = -2$$

Hence, the graph passes through $(-2, -6)$.

37. To find the x-intercept, set $y = 0$:

$$y = -\frac{2}{3}x + 4$$

$$0 = -\frac{2}{3}x + 4$$

$$\frac{2}{3}x = 4$$

$$2x = 12$$

$$x = 6$$

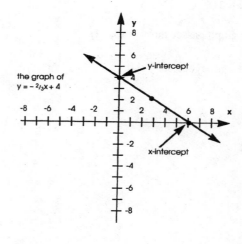

the graph of
$y = -{}^{2}/_{3}x + 4$

Hence, the graph crosses the x-axis at $(6, 0)$.

To find the y-intercept, set $x = 0$:

$$y = -\frac{2}{3}x + 4$$

$$y = -\frac{2}{3}(0) + 4$$

$$y = 0 + 4$$

$$y = 4$$

Hence, the graph crosses the y-axis at $(0, 4)$.

To find a check point, choose $x = 3$:

$$y = -\frac{2}{3}x + 4$$

$$y = -\frac{2}{\cancel{3}}(\cancel{3}) + 4$$

$$y = -2 + 4$$

$$y = 2$$

Hence, the graph passes through $(3, 2)$.

102

41. To find the x-intercept, set y = 0:

$$5x - 7y = 30$$

$$5x - 7(0) = 30$$

$$5x - 0 = 30$$

$$5x = 30$$

$$x = 6$$

Hence, the graph crosses the x-axis at (6, 0).

To find the y-intercept, set x = 0:

$$5x - 7y = 30$$

$$5(0) - 7y = 30$$

$$0 - 7y = 30$$

$$-7y = 30$$

$$y = -\frac{30}{7}$$

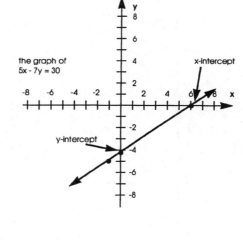

the graph of
5x - 7y = 30

x-intercept

y-intercept

Hence, the graph crosses the y-axis at $\left(0, -\dfrac{30}{7}\right)$.

To find a check point, choose y = –5:

$$5x - 7y = 30$$

$$5x - 7(-5) = 30$$

$$5x + 35 = 30$$

$$5x = -5$$

$$x = -1$$

Hence, the graph passes through (–1, –5).

45. The graph of x = 5 is a vertical line five units to the right of the y-axis.

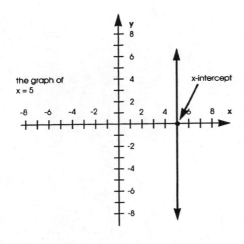

the graph of
x = 5

x-intercept

103

51.   The graph of y + 5 = 0, or equivalently, the graph of y = –5, is a horizontal line five units below the x-axis.

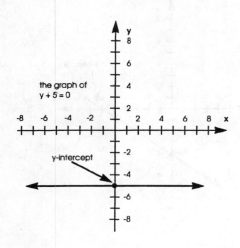

the graph of
y + 5 = 0

y-intercept

53.   To find the x-intercept, set y = 0:

$$5x - 4y = 0$$

$$5x - 4(0) = 0$$

$$5x - 0 = 0$$

$$5x = 0$$

$$x = 0$$

Hence, the graph crosses the x-axis at (0, 0). This is also the point

at which the graph crosses the y-axis. To find a second point,

choose y = 5:

$$5x - 4y = 0$$

$$5x - 4(5) = 0$$

$$5x - 20 = 0$$

$$5x = 20$$

$$x = 4$$

Hence, the graph passes through (4, 5). To find a check point,

choose y = –5:

$$5x - 4y = 0$$

$$5x - 4(-5) = 0$$

$$5x + 20 = 0$$

$$5x = -20$$

$$x = -4$$

Hence, the graph passes through (–4, –5).

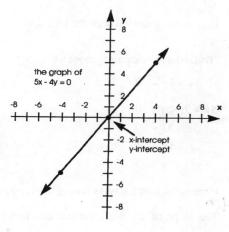

the graph of
5x - 4y = 0

x-intercept
y-intercept

104

57. The x-intercept of a graph is the x-coordinate of the point at which the graph crosses the x-axis. To find an x-intercept, we set y = 0 and solve the resulting equation for x. The y-intercept of a graph is the y-coordinate of the point at which the graph crosses the y-axis. To find a y-intercept, we set x = 0 and solve the resulting equation for y.

58. To find the t-intercept, set s = 0:

$$2s + 3t = 6$$

$$2(0) + 3t = 6$$

$$0 + 3t = 6$$

$$3t = 6$$

$$t = 2$$

To find the s-intercept, set t = 0:

$$2s + 3t = 6$$

$$2s + 3(0) = 6$$

$$2s + 0 = 6$$

$$2s = 6$$

$$s = 3$$

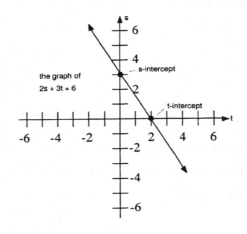

the graph of 2s + 3t = 6

59. The s- and t-intercepts are the same as in Exercise 58. Reversing the labeling of the axes does affect the graph. The two graphs are mirror images of one another in the line s = t.

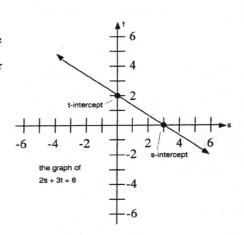

the graph of 2s + 3t = 6

60. (a) To find the x-intercept, set y = 0:

$$y = 200x + 400$$

$$0 = 200x + 400$$

$$-400 = 200x$$

$$-2 = x$$

Hence the graph crosses the x-axis at (−2, 0).

To find the y-intercept, set x = 0:

$$y = 200x + 400$$

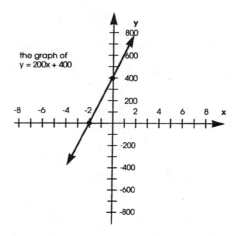

the graph of y = 200x + 400

y = 200(0) + 400

y = 0 + 400

y = 400

Hence, the graph crosses the y-axis at (0, 400).

(b) To find the x-intercept, set y = 0:

y = 0.04x

0 = 0.04x

0 = x

Hence, the graph crosses the x-axis at (0, 0). This is also
the point at which the graph crosses the y-axis. To find
a second point, choose x = 100:

y = 0.04x

y = 0.04(100)

y = 4

Hence, the graph passes through (100, 4).

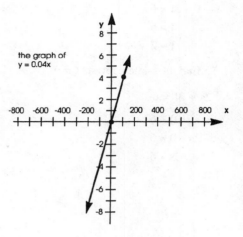

the graph of
y = 0.04x

61. $-3 - (-4) + (-5) - (-8) = -3 + (+4) + (-5) + (+8) = 1 + (-5) + (+8) = (-4) + (+8) = +4$

63. $\dfrac{4}{x-2} - \dfrac{2x}{x-2} = \dfrac{4-2x}{x-2} = \dfrac{-2(x-2)}{x-2} = -2$

65.   Let x = length of the second side of the triangle (in inches). 2x = length of the first side of the triangle (in inches).

2x + x + 24 = 75

3x + 24 = 75

3x = 51

x = 17

Then 2x = 34

Thus, the first side is 34 inches and the second side is 17 inches.

Exercises 5.2

1. $m = \dfrac{y_2 - y_1}{x_2 - x_1} = \dfrac{1-(-2)}{-3-1}$

    $= \dfrac{3}{-4} = -\dfrac{3}{4}$

5. $m = \dfrac{y_2 - y_1}{x_2 - x_1} = \dfrac{-5-(-4)}{-2-(-3)}$

    $= \dfrac{-5+4}{-2+3} = \dfrac{-1}{1}$

    $= -1$

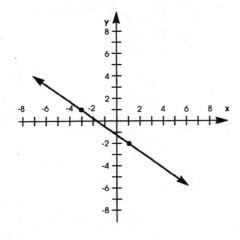

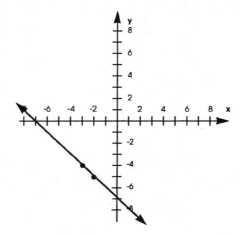

7. $m = \dfrac{y_2 - y_1}{x_2 - x_1} = \dfrac{4-4}{-3-2} = \dfrac{0}{-5} = 0$

9. $m = \dfrac{y_2 - y_1}{x_2 - x_1} = \dfrac{-3-2}{4-4} = -\dfrac{5}{0}$, which is undefined.

13. $m = \dfrac{y_2 - y_1}{x_2 - x_1} = \dfrac{b^2 - a^2}{b-a} = \dfrac{\cancel{(b-a)}(b+a)}{\cancel{b-a}} = b+a$

17. $m = \dfrac{y_2 - y_1}{x_2 - x_1} = \dfrac{0.06 - 7}{0.14 - (-0.2)}$

    $= \dfrac{-6.94}{1.86} \approx -3.73$

21. $m \approx \dfrac{7}{2}$

23. The line passes through the points (5,0) and (0,-3).

    $m = \dfrac{y_2 - y_1}{x_2 - x_1} = \dfrac{0-(-3)}{5-0} = \dfrac{3}{5}$

27.

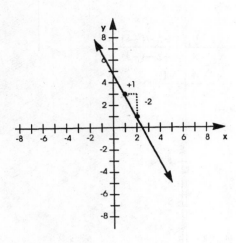

31.

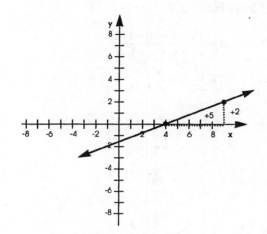

33.

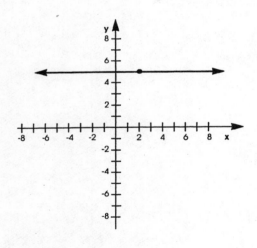

35.

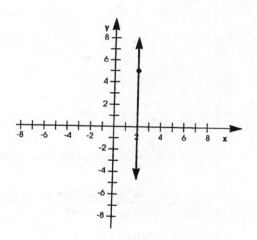

37.

$$m_{P_1P_2} = \frac{y_2 - y_1}{x_2 - x_1} = \frac{4 - 2}{3 - 1} = \frac{2}{2} = 1$$

$$m_{P_3P_4} = \frac{y_2 - y_1}{x_2 - x_1} = \frac{-4 - (-2)}{-3 - (-1)} = \frac{-4 + 2}{-3 + 1} = \frac{-2}{-2} = 1$$

Thus, the lines are parallel, since their slopes are equal.

39. $m_{P_1 P_2} = \dfrac{y_2 - y_1}{x_2 - x_1} = \dfrac{2 - 4}{-1 - 0} = \dfrac{-2}{-1} = 2$

$m_{P_3 P_4} = \dfrac{y_2 - y_1}{x_2 - x_1} = \dfrac{7 - 5}{1 - (-3)} = \dfrac{2}{4} = \dfrac{1}{2}$

Thus, the lines are neither parallel nor perpendicular.

43. $m = \dfrac{y_2 - y_1}{x_2 - x_1}$

$-5 = \dfrac{h - (-2)}{1 - 4}$

$-5 = \dfrac{h + 2}{3}$

$(-3)(-5) = \dfrac{(\cancel{-3})}{1}\left(\dfrac{h + 2}{\cancel{-3}}\right)$

$15 = h + 2$

$13 = h$

45. The line through $(0, 1)$ and (c, c) has slope $= \dfrac{c - 1}{c}$. The line through $(0, 2)$ and $(-c, c)$ has slope $= \dfrac{c - 2}{-c}$ Since these lines are perpendicular, it must be true that $\left(\dfrac{c - 1}{c}\right)\left(\dfrac{c - 2}{-c}\right) = -1$

Then $\dfrac{(c - 1)(c - 1)}{-c^2} = -1$

$(c - 1)(c - 2) = c^2$

$c^2 - 3c + 2 = c^2$

$-3c + 2 = 0$

$2 = 3c$

$\dfrac{2}{3} = c$

47. Label the four given points as follows: A $(0, 0)$, B $(2, 1)$, C $(-2, 5)$, and D $(0, 6)$. Then

$m_{AB} = \dfrac{y_2 - y_1}{x_2 - x_1} = \dfrac{1 - 0}{2 - 0} = \dfrac{1}{2}$ and

$m_{CD} = \dfrac{y_2 - y_1}{x_2 - x_1} = \dfrac{6 - 5}{0 - (-2)} = \dfrac{1}{2}$

Since these slopes are equal, it follows that AB is parallel to CD. Also,

$m_{AC} = \dfrac{y_2 - y_1}{x_2 - x_1} = \dfrac{5 - 0}{-2 - 0} = \dfrac{5}{-2} = -\dfrac{5}{2}$ and

$m_{BD} = \dfrac{y_2 - y_1}{x_2 - x_1} = \dfrac{6 - 1}{0 - 2} = \dfrac{5}{-2} = -\dfrac{5}{2}$

Since these slopes are equal, it follows that AC is parallel to BD. Finally, since the opposite sides of quadrilateral ABDC are parallel, ABDC is a parallelogram.

51. Label the three given points as follows: A $(-2, -1)$, B $(0, 4)$, and, C $(2, 9)$. Then

$$m_{AB} = \frac{y_2 - y_1}{x_2 - x_1} = \frac{4 - (-1)}{0 - (-2)} = \frac{5}{2} \text{ and}$$

$$m_{BC} = \frac{y_2 - y_1}{x_2 - x_1} = \frac{9 - 4}{2 - 0} = \frac{5}{2}$$

Since these slopes are equal, the lines AB and BC are parallel. However, B is a point that is on both lines. This implies that A, B, and C are collinear.

52. A line with a positive slope rises; a line with a negative slope falls; a line with a zero slope is horizontal; a line with undefined slope is vertical.

53. 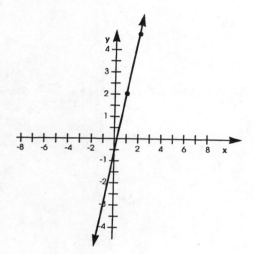 54.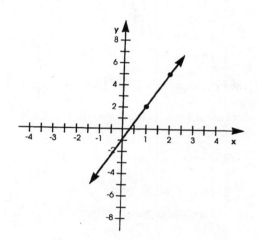

55. $162a^2y^2 - 32 = 2(81a^2y^2 - 16) = 2(9ay - 4)(9ay + 4)$

57. $\dfrac{\dfrac{3}{x+1} + 2}{\dfrac{2}{x-1} + x} = \dfrac{(x+1)(x-1)\left(\dfrac{3}{x+1} + 2\right)}{(x+1)(x-1)\left(\dfrac{2}{x-1} + x\right)} = \dfrac{3(x-1) + 2(x+1)(x-1)}{2(x+1) + x(x+1)(x-1)}$

$$= \dfrac{(x-1)(3 + 2(x+1))}{(x+1)(2 + x(x-1))} = \dfrac{(x-1)(2x+5)}{(x+1)(x^2 - x + 2)}$$

59. Let x = number of ounces of the 30% solution.

$$0.30x + 0.70(6) = 0.60(x + 6)$$

110

$$3x + 42 = 6(x + 6)$$
$$3x + 42 = 6x + 36$$
$$6 = 3x$$
$$2 = x$$

Thus, 2 ounces of the 30% solution of alcohol must be used.

Exercises 5.3

1.
$$y - y_1 = m(x - x_1)$$
$$y - (-3) = 5(x - 1)$$
$$y + 3 = 5(x - 1)$$
$$y + 3 = 5x - 5$$
$$y = 5x - 8$$

5.
$$y - y_1 = m(x - x_1)$$
$$y - 1 = \frac{2}{3}(x - 6)$$
$$y - 1 = \frac{2}{3}x - 4$$
$$y = \frac{2}{3}x - 3$$

9.
$$y = mx + b$$
$$y = \frac{3}{4}x + 5$$

11. Since L has slope 0, L is horizontal. Equation: $y = -4$.

13. Since L has undefined slope, L is vertical. Equation: $x = -4$.

17.
$$m = \frac{y_2 - y_1}{x_2 - x_1} = \frac{-5 - (-1)}{-3 - (-2)} = \frac{-5 + 1}{-3 + 2} = \frac{-4}{-1} = 4$$
$$y - y_1 = m(x - x_1)$$
$$y - (-1) = 4(x - (-2))$$
$$y + 1 = 4(x + 2)$$
$$y + 1 = 4x + 8$$
$$y = 4x + 7$$

21.
$$y = mx + b$$
$$y = 4x + 6$$

23.
$$y - y_1 = m(x - x_1)$$
$$y - 0 = -2(x - (-3))$$
$$y = -2(x + 3)$$
$$y = -2x - 6$$

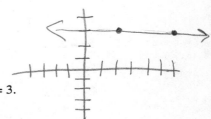

25. Since the two points have the same y-coordinate, L must be horizontal. Equation: $y = 3$.

29.   L passes through $(-3, 0)$ and $(0, 2)$.

$$m = \frac{y_2 - y_1}{x_2 - x_1} = \frac{2 - 0}{0 - (-3)} = \frac{2}{3}$$

$$y = mx + b$$

$$y = \frac{2}{3}x + 2$$

31.   The line $y = 3x + 7$ has slope 3 (by comparison with $y = mx + b$). So L has slope 3, because the lines are parallel.

$$y - y_1 = m(x - x_1)$$

$$y - 2 = 3(x - 2)$$
$$y - 2 = 3x - 6$$
$$y = 3x - 4$$

33.   The line $y = -\frac{2}{3}x - 1$ has slope $-\frac{2}{3}$ (by comparison with $y = mx + b$). So L has slope $\frac{3}{2}$, because the lines are

perpendicular.

$$y - y_1 = m(x - x_1)$$

$$y - (-3) = \frac{3}{2}(x - (-3))$$

$$y + 3 = \frac{3}{2}(x + 3)$$

$$y + 3 = \frac{3}{2}x + \frac{9}{2}$$

$$y = \frac{3}{2}x + \frac{3}{2}$$

37. $2y - 3x = 12$
$$2y = 3x + 12$$

$$y = \frac{3}{2}x + 6$$

So the given line has slope $\frac{3}{2}$. Therefore, L has slope $\frac{3}{2}$, because the lines are parallel.

$$y - y_1 = m(x - x_1)$$

$$y - (-2) = \frac{3}{2}(x - (-1))$$

$$y + 2 = \frac{3}{2}(x + 1)$$

$$y + 2 = \frac{3}{2}x + \frac{3}{2}$$

$$y = \frac{3}{2}x - \frac{1}{2}$$

41. $8x - 5y = 20$

$$-5y = -8x + 20$$
$$y = \frac{8}{5}x - 4$$

So the given line has slope $\dfrac{8}{5}$ and y-intercept –4. Therefore, L has slope $-\dfrac{5}{8}$ and y-intercept –4.

$y = mx + b$

$y = -\dfrac{5}{8}x - 4$

43. Slope of given line $= \dfrac{y_2 - y_1}{x_2 - x_1} = \dfrac{2 - (-6)}{-1 - 3} = \dfrac{8}{-4} = -2$. Therefore, L has slope –2.

$y = mx + b$

$y = -2x + 4$

47. A line perpendicular to the x-axis is vertical. Equation: $x = -1$.

49. $3x - 2y = 5$ $3x - 2y = 6$

    $-2y = -3x + 5$ $-2y = -3x + 6$

        $y = \dfrac{3}{2}x - \dfrac{5}{2}$ $y = \dfrac{3}{2}x - 3$

        $m = \dfrac{3}{2}$ $m = \dfrac{3}{2}$

Since these lines have equal slopes, they are parallel.

53. $5x + y = 2$ $5y = x + 3$

                                      $y = \dfrac{1}{5}x + \dfrac{3}{5}$

      $y = -5x + 2$

      $m = -5$ $m = \dfrac{1}{5}$

Since the product of these slopes is –1, the lines are perpendicular.

57. (18, 200), (100, 2660)

$m = \dfrac{2660 - 200}{100 - 18} = \dfrac{2460}{82} = 30$

$P - P_1 = m(x - x_1)$

$P - 200 = 30(x - 18)$

$P - 200 = 30x - 540$

$\quad\quad P = 30x - 340$

When x = 200,

$P = 30(200) - 340 = 6000 - 340 = \$5660$

61. (35,000, 70), (20,000, 85)

$m = \dfrac{85 - 70}{20,000 - 35,000} = \dfrac{15}{-15,000} = -\dfrac{1}{1000}$

$$E - E_1 = m(V - V_1)$$

$$E - 70 = -\frac{1}{1000}(V - 35,000)$$

$$E - 70 = -\frac{1}{1000}V + 35$$

$$E = -\frac{1}{1000}V + 105$$

When V = 15,000,

$$E = -\frac{1}{1000}(15,000) + 105 = -\frac{15,000}{1,000} + 105 = -15 + 105 = 90$$

63. If (x_1, y_1) satisfies the equation $Ax + By = C$, then $Ax_1 + By_1 = C$. Then

$$By_1 = C - Ax_1$$

$$y_1 = \frac{C - Ax_1}{B}$$

So $\left(x_1, \dfrac{C - Ax_1}{B}\right)$ satisfies the equation. Similarly, $\left(x_2, \dfrac{C - Ax_2}{B}\right)$ satisfies the equation. Using these two points and the slope formula, we get,

$$m = \frac{\left(\dfrac{C - Ax_2}{B} - \dfrac{C - Ax_1}{B}\right)}{x_2 - x_1} = \frac{\dfrac{(C - Ax_2) - (C - Ax_1)}{B}}{x_2 - x_1}$$

$$= \frac{\dfrac{C - Ax_2 - C + Ax_1}{B}}{x_2 - x_1} = \frac{\dfrac{-Ax_2 + Ax_1}{B}}{x_2 - x_1} = \frac{-\dfrac{A}{B}(x_2 - x_1)}{x_2 - x_1} = -\frac{A}{B}$$

which is independent of x_1 and x_2.

64. $Ax + By = C$

$$By = -Ax + C$$

$$y = -\frac{A}{B}x + \frac{C}{B}$$

Then this line has slope $-\dfrac{A}{B}$, by comparison with $y = mx + b$.

65. $5 - [3 - (a - 2)] = a - 2$

$$5 - [3 - a + 2] = a - 2$$

$$5 - [5 - a] = a - 2$$

$$5 - 5 + a = a - 2$$

$$a = a - 2$$

$$0 = -2$$

So there is no solution.

67. $|9 - 3a| \geq 6$

$9 - 3a \leq -6$ or $9 - 3a \geq 6$

$-3a \leq -15$ or $-3a \geq -3$

$a \geq 5$ or $a \leq 1$

69. $\dfrac{1}{a} - \dfrac{1}{b} = \dfrac{1}{c}$

$abc\left(\dfrac{1}{a} - \dfrac{1}{b}\right) = abc\left(\dfrac{1}{c}\right)$

$bc - ac = ab$

$bc = ab + ac$

$bc = a(b + c)$

$\dfrac{bc}{b + c} = a$

Exercises 5.4

1. (a) 7,500 calls
   (b) 4 a.m.
   (c) 28,000 calls
   (d) 4 p.m. to 4 a.m.
   (e) 10 a.m. to 4 p.m.
   (f) 10 a.m. to 4 p.m.

3. (a) This graph denies the belief that temperature drops steadily as altitude increases. Notice that the temperature increases as the altitude increases from 10 km to 50 km and again when the altitude is greater than 80 km.

   (b) The temperature decreases when the altitude is between 0 km and 10 km and also when it is between 50 km and 80 km.

   (c) For the first 5 km that the balloon rises, the temperature will increase; for the last 5 km, it will decrease.

5. (a) 30%
   (b) about 4 days
   (c) As time goes on, forgetting increases. The most forgetting occurs during the first day. After the first day, forgetting continues to increase but does so at a much more gradual pace.

7. (a) 14 trials
   (b) 3 trials
   (c) The skill of keeping a pointer on a moving target is learned more rapidly with spaced practice as opposed to massed practice.

11. During the treatment period, Treatment C appears to be the most effective therapy because it shows the greatest decrease in the number of aggressive behaviors. During this period, Treatment B seems to be the least effective. However, these conclusions are reversed in the post-treatment period. That is, in the long run, Treatment B is the most effective therapy and Treatment C is the least effective.

1. $\begin{cases} 2x + y = 12 \\ 3x - y = 12 \end{cases}$

   Add: $5x = 25$

   $x = 5$

   Substitute:

   $2x + y = 12$

   $2(5) + y = 12$

   $10 + y = 12$

   $y = 2$

   Solution: $x = 5$, $y = 2$; independent

   CHECK $x = 5$, $y = 2$:

   $2x + y = 12$

   $2(5) + 2 \overset{?}{=} 12$

   $10 + 2 \overset{?}{=} 12$

   $12 \overset{\checkmark}{=} 12$

   $3x - y = 13$

   $3(5) - 2 \overset{?}{=} 13$

   $15 - 2 \overset{?}{=} 13$

   $13 \overset{\checkmark}{=} 13$

5. $\begin{cases} 3x - y = 0 \\ 2x + 3y = 11 \end{cases}$

   Since $3x - y = 0$, it follows that $y = 3x$. Substitute this result into the second equation to find

   $2x + 3(3x) = 11$

   $2x + 9x = 11$

   $11x = 11$

   $x = 1$

   CHECK $x = 1$, $y = 3$:

   $3x - y = 0$

   $3(1) - 3 \overset{?}{=} 0$

   $3 - 3 \overset{?}{=} 0$

   $0 \overset{\checkmark}{=} 0$

   Since $y = 3x$, $y = 3(1) = 3$.

   $2x + 3y = 11$

   $2(1) + 3(3) \overset{?}{=} 11$

   $2 + 9 \overset{?}{=} 11$

   Solution: $x = 1$, $y = 3$; independent

   $11 \overset{\checkmark}{=} 11$

9. $\begin{cases} 4x + 5y = 0 \\ 2x + 3y = -2 \end{cases}$

   Multiply the second equation by -2, then add:

   $4x + 5y = 0$

   $\underline{-4x - 6y = 4}$

   $-y = 4$

   $y = -4$

   Substitute: $4x + 5y = 0$

   $4x + 5(-4) = 0$

   $4x - 20 = 0$

   $4x = 20$

   $x = 5$

   Solution: $x = 5$, $y = -4$; independent

   CHECK: $x = 5$, $y = -4$:

   $4x + 5y = 0$

   $4(5) + 5(-4) \overset{?}{=} 0$

   $20 - 20 \overset{?}{=} 0$

   $0 \overset{\checkmark}{=} 0$

   $2x + 3y = -2$

   $2(5) + 3(-4) \overset{?}{=} -2$

   $10 - 12 \overset{?}{=} -2$

   $-2 \overset{\checkmark}{=} -2$

11. $\begin{cases} 2x + 3y = 7 \\ 4x + 6y = 14 \end{cases}$

Multiply the first equation by -2, then add:

$-4x - 6y = -14$

$\underline{4x + 6y = 14}$

$0 = 0,$ an identity.

Thus, every ordered pair that satisfies one of the equations also satisfies the other. There are infinitely many solutions (x, y), where $2x + 3y = 7$. The system is dependent.

13. $\begin{cases} 5x - 6y = 3 \\ 10x - 12y = 5 \end{cases}$

Multiply the first equation by -2, then add:

$-10x + 12y = -6$

$\underline{10x - 12y = 5}$

$0 = -1,$ a contradiction

Thus, there are no solutions common to both equations. The system is inconsistent.

17. $\begin{cases} y = 2x + 3 \\ 2x + y = -1 \end{cases}$
$\qquad\qquad$ CHECK $\ x = -1, y = 1$:

Substitute the result of the first
$\qquad\qquad\qquad\qquad\qquad\quad$ $y = 2x + 3$

equation into the second:
$\qquad\qquad\qquad\qquad\quad$ $1 \overset{?}{=} 2(-1) + 3$

$\qquad\qquad\qquad\qquad\qquad\qquad\qquad\qquad$ $1 \overset{?}{=} -2 + 3$

$\qquad\qquad\qquad\qquad\qquad\qquad\qquad\qquad$ $1 \overset{\checkmark}{=} 1$

$\qquad\qquad 2x + y = -1$ $\qquad\qquad\qquad$ $2x + y = -1$

$\qquad 2x + (2x + 3) = -1$ $\qquad\qquad$ $2(-1) + 1 \overset{?}{=} -1$

$\qquad\qquad\quad 4x + 3 = -1$ $\qquad\qquad\qquad$ $-2 + 1 \overset{?}{=} -1$

$\qquad\qquad\qquad\quad 4x = -4$ $\qquad\qquad\qquad\qquad$ $-1 \overset{\checkmark}{=} -1$

$\qquad\qquad\qquad\qquad x = -1$

Since $y = 2x + 3$, $y = 2(-1) + 3 = -2 + 3 = 1$. Solution: $x = -1$, $y = 1$; independent.

21. $\begin{cases} s = 3t - 5 \\ t = 3s - 5 \end{cases}$
$\qquad\qquad$ CHECK $\ s = \dfrac{5}{2}, t = \dfrac{5}{2}$:

Substitute the result of the

first equation into the second:
$\qquad\qquad\qquad\qquad$ $s = 3t - 5$

$\qquad\qquad\qquad\qquad\qquad\qquad\qquad\qquad$ $\dfrac{5}{2} \overset{?}{=} 3\left(\dfrac{5}{2}\right) - 5$

$t = 3s - 5$
$\qquad\qquad\qquad\qquad\qquad\qquad$ $\dfrac{5}{2} \overset{?}{=} \dfrac{15}{2} - \dfrac{10}{2}$

$t = 3(3t - 5) - 5$
$\qquad\qquad\qquad\qquad\quad$ $\dfrac{5}{2} \overset{\checkmark}{=} \dfrac{5}{2}$

$t = 9t - 15 - 5$
$\qquad\qquad\qquad\qquad\qquad$ $t = 3s - 5$

$$t = 9t - 20$$

$$-8t = -20$$

$$t = \frac{5}{2}$$

$$\frac{5}{2} \overset{?}{=} 3\left(\frac{5}{2}\right) - 5$$

$$\frac{5}{2} \overset{?}{=} \frac{15}{2} - \frac{10}{2}$$

$$\frac{5}{2} \overset{\checkmark}{=} \frac{5}{2}$$

Since $s = 3t - 5$, $s = 3\left(\frac{5}{2}\right) - 5 = \frac{15}{2} - 5 = \frac{5}{2}$. Solution: $s = \frac{5}{2}$, $t = \frac{5}{2}$

25. $\begin{cases} 3p - 4q = 5 \\ 3q - 4p = -9 \end{cases}$

CHECK $p = 3$, $q = 1$:

Rewrite this system as:

$$3p - 4q = 5$$

$$-4p + 3q = -9$$

$$3p - 4q = 5$$

$$3(3) - 4(1) \overset{?}{=} 5$$

$$9 - 4 \overset{?}{=} 5$$

$$5 \overset{\checkmark}{=} 5$$

Multiply the first equation by 4,

the second by 3, and add:

$$12p - 16q = 20$$

$$\underline{-12p + 9q = -27}$$

$$-7q = -7$$

$$q = 1$$

$$3q - 4p = -9$$

$$3(1) - 4(3) \overset{?}{=} -9$$

$$3 - 12 \overset{?}{=} -9$$

$$-9 \overset{\checkmark}{=} -9$$

Substitute: $3p - 4q = 5$

$$3p - 4(1) = 5$$

$$3p - 4 = 5$$

$$3p = 9$$

$$p = 3$$

Solution: $p = 3$, $q = 1$: independent.

29. $\begin{cases} \dfrac{w}{4} + \dfrac{z}{6} = 4 \\ \dfrac{w}{2} - \dfrac{z}{3} = 4 \end{cases}$

CHECK $w = 12$, $z = 6$:

Multiply the first equation by 12 and the

second by 6, in order to clear fractions.

$$3w + 2z = 48$$

$$3w - 2z = 24$$

Add: $6w = 72$

$$w = 12$$

$$\frac{w}{4} + \frac{z}{6} = 4$$

$$\frac{12}{4} + \frac{6}{6} \overset{?}{=} 4$$

$$3 + 1 \overset{?}{=} 4$$

$$4 \overset{\checkmark}{=} 4$$

$$\frac{w}{2} - \frac{z}{3} = 4$$

$$\frac{12}{2} - \frac{6}{3} \overset{?}{=} 4$$

118

Substitute: $\dfrac{w}{4} + \dfrac{z}{6} = 4$

$\dfrac{12}{4} + \dfrac{z}{6} = 4$

$3 + \dfrac{z}{6} = 4$

$\dfrac{z}{6} = 1$

$z = 6$

$6 - 2 \overset{?}{=} 4$

$4 \overset{\checkmark}{=} 4$

Solution: $w = 12$, $z = 6$; independent

33. $\begin{cases} \dfrac{x+3}{2} + \dfrac{y-4}{3} = \dfrac{19}{6} \\[2mm] \dfrac{x-2}{3} + \dfrac{y-2}{2} = 2 \end{cases}$

CHECK $x = 2$, $y = 6$:

Multiply both equations by 6,

in order to clear fractions.

$3(x+3) + 2(y-4) = 19$

$2(x-2) + 3(y-2) = 12$

Simplify: $3x + 2y = 18$

$2x + 3y = 22$

Multiply the first equation by 2,

the second by -3, and add:

$6x + 4y = 36$

$\underline{-6x - 9y = -66}$

$-5y = -30$

$y = 6$

Substitute: $3x + 2y = 18$

$3x + 2(6) = 18$

$3x + 12 = 18$

$3x = 6$

$x = 2$

Solution: $x = 2$, $y = 6$; independent

$\dfrac{x+3}{2} + \dfrac{y-4}{3} = \dfrac{19}{6}$

$\dfrac{2+3}{2} + \dfrac{6-4}{3} \overset{?}{=} \dfrac{19}{6}$

$\dfrac{5}{2} + \dfrac{2}{3} \overset{?}{=} \dfrac{19}{6}$

$\dfrac{15}{6} + \dfrac{4}{6} \overset{?}{=} \dfrac{19}{6}$

$\dfrac{19}{6} \overset{\checkmark}{=} \dfrac{19}{6}$

$\dfrac{x-2}{3} + \dfrac{y-2}{2} = 2$

$\dfrac{2-2}{3} + \dfrac{6-2}{2} \overset{?}{=} 2$

$\dfrac{0}{3} + \dfrac{4}{2} \overset{?}{=} 2$

$0 + 2 \overset{?}{=} 2$

$2 \overset{\checkmark}{=} 2$

37. $\dfrac{x}{2} + 0.05y = 0.35$

$\qquad 0.3x + \dfrac{y}{4} = 0.65$

Multiply both equations by 100,

in order to clear decimals.

$50x + 5y = 35$
$30x + 25y = 65$

Multiply the first equation by -5 and add:

$-250x - 25y = -175$

$\underline{30x + 25y = 65}$
$-220x = -110$

$\qquad x = \dfrac{1}{2}$

Substitute: $50x + 5y = 35$

$\qquad 50\left(\dfrac{1}{2}\right) + 5y = 35$

$\qquad\quad 25 + 5y = 35$

$\qquad\qquad\quad 5y = 10$

$\qquad\qquad\quad\ y = 2$

Solution: $x = \dfrac{1}{2}$, $y = 2$; independent

CHECK $\quad x = \dfrac{1}{2}, y = 2$:

$\qquad \dfrac{x}{2} + 0.05y = 0.35$

$\qquad \dfrac{\frac{1}{2}}{2} + 0.05(2) \overset{?}{=} 0.35$

$\qquad\quad \dfrac{1}{4} + 0.10 \overset{?}{=} 0.35$

$\qquad 0.25 + 0.10 \overset{?}{=} 0.35$

$\qquad\qquad\quad 0.35 \overset{\checkmark}{=} 0.35$

$\qquad\quad 0.3x + \dfrac{y}{4} = 0.65$

$\qquad 0.3\left(\dfrac{1}{2}\right) + \dfrac{2}{4} \overset{?}{=} 0.65$

$\qquad\quad 0.15 + 0.50 \overset{?}{=} 0.65$

$\qquad\qquad\quad 0.65 \overset{\checkmark}{=} 0.65$

39. Let x = amount of money Susan should invest at 5%. y = amount of money Susan should invest at 8%.

$x + y = 14,000$

$0.05x + 0.08y = 835$

Multiply the first equation by -5,
the second by 100, and add:

$-5x - 5y = -70,000$

$\underline{5x + 8y = 83,500}$
$\qquad\ 3y = 13,500$
$\qquad\quad y = 4,500$

Substitute: $\qquad x + y = 14,000$

$\qquad\qquad x + 4,500 = 14,000$

$\qquad\qquad\qquad\quad x = 9,500$

CHECK:

$\$9,500 + \$4,500 \overset{\checkmark}{=} \$14,000$

5% of $9,500 = $475

$\underline{\text{8% of \$4,500} = \$360}$

Total interest $\overset{\checkmark}{=}$ $835

Conclude: Susan should invest $9,500 at 5% and $4,500 at 8%.

120

43. Let w = width of the rectangle (in cm). ℓ = length of the rectangle (in cm).

$2\ell + 2w = 36$

$\ell = w + 2$

Substitute the result of the second

equation into the first:

$2(w+2) + 2w = 36$

$2w + 4 + 2w = 36$

$4w + 4 = 36$

$4w = 32$

$w = 8$

CHECK: $2(10) + 2(8) = 36$

$20 + 16 \overset{?}{=} 36$

$36 \overset{\checkmark}{=} 36$

10 is 2 more than 8.

Then $\ell = w + 2 = 8 + 2 = 10$. Conclude: The width of the rectangle is 8 cm and the length of the rectangle is 10 cm.

47. Let x = cost of a single roll of 35 mm film (in dollars). y = cost of a single roll of movie film (in dollars).

$5x + 3y = 35.60$

$3x + 5y = 43.60$

Multiply the first equation by –3,

the second by 5, and add:

$-15x - 9y = -106.80$

$\underline{15x + 25y = 218.00}$

$16y = 111.20$

$y = 6.95$

Substitute: $5x + 3y = 35.60$

$5x + 3(6.95) = 35.60$

$5x + 20.85 = 35.60$

$5x = 14.75$

$x = 2.95$

CHECK:

5 rolls of 35 mm film costs 5($2.95) = $14.75

3 rolls of movie film costs 3($6.95) = $20.85

Albert's total cost = $35.60

3 rolls of 35 mm film costs 3($2.95) = $8.85

5 rolls of movie film costs 5($6.95) = $34.75

Audrey's total cost = $43.60

Conclude: A single roll of 35 mm film costs $2.95, and a single roll of movie film costs $6.95.

51. Let x = # of more expensive models that can be produced. y = # of less expensive models that can be produced.

$6x + 5y = 730$

$3x + 2y = 340$

Multiply the second equation

by – 2 and add:

$6x + 5y = 730$

$\underline{-6x - 4y = -680}$

$y = 50$

Substitute: $3x + 2y = 340$

$3x + 2(50) = 340$

$3x + 100 = 340$

$3x = 240$

$x = 80$

CHECK: Manufacture:

80 more expensive models at 6 hours

per model = 480 hours

50 less expensive models at 5 hours

per model = 250 hours

Total $\overset{\checkmark}{=}$ 730 hours

Assembly:

80 more expensive models at 3 hours

per model = 240 hours

500 less expensive models at 2 hours

per model = 100 hours

Total $\overset{\checkmark}{=}$ 340 hours

Conclude: The manufacturer can produce 80 of the more expensive models and 50 of the less expensive models.

55. Let x = flat rate (in dollars). y = charge per mile (in dollars).

$x + 85y = 44.30$

$x + 125y = 51.50$

Multiply the first equation

by −1 and add:

$-x - 85y = -44.30$

$\underline{x + 125y = 51.50}$

$40y = 7.20$

$y = 0.18$

Substitute: $x + 85y = 44.30$

$x + 85(0.18) = 44.30$

$x + 15.30 = 44.30$

$x = 29$

CHECK:

One-day rental with 85 miles is

$29 + 85(0.18) =$

$29 + 15.30 \overset{\checkmark}{=} \$44.30$

One-day rental with 125 miles is

$29 + 125(0.18) =$

$29 + 22.50 \overset{\checkmark}{=} 51.50$

Conclude: The flat rate is $29 and the charge per mile is 18¢.

59. $\dfrac{y-3}{x-2} = -1 --\rightarrow y - 3 = -1(x-2)$ or $y = -x + 5$

$\dfrac{y+2}{x-1} = 2 --\rightarrow y + 2 = 2(x-1)$ or $y = 2x - 4$

Substitute the result of the second equation into the first:

$y = -x + 5$

$2x - 4 = -x + 5$

$3x = 9$

$x = 3$

Substitute: $y = 2x - 4$

$y = 2(3) - 4$

$y = 6 - 4$

$y = 2$

CHECK:

Slope of first line =

$\dfrac{3-2}{2-3} \overset{\checkmark}{=} 1$

Slope of second line =

$\dfrac{-2-2}{1-3} \overset{\checkmark}{=} 2$

Conclude: The point in question is (3, 2).

61. If it is easy to solve one of the equations of the system explicitly for one of the variables, then the substitution method should be considered. This occurs when some variable in the system has a coefficient of either 1 or −1. In all other cases, it is probably easier to use the elimination method.

62. $\begin{cases} 4x - 9y = 3 \\ 10x - 6y = 7 \end{cases}$

Multiply the first equation by −2, the second by 3, and add:

$$-8x + 18y = -6$$

$$\underline{30x - 18y = 21}$$

$$22x = 15$$

$$x = \frac{15}{22}$$

Substitute: $\quad 4x - 9y = 3$

$$4\left(\frac{15}{22}\right) - 9y = 3$$

$$\frac{30}{11} - 9y = 3$$

$$-9y = 3 - \frac{30}{11}$$

$$-9y = \frac{3}{11}$$

$$y = -\frac{1}{33}$$

The variation of the elimination method that is described would be easier to use for the system just solved, since we could avoid most of the calculations that involve fractions in this way.

63. (a) $\begin{cases} \dfrac{1}{u} + \dfrac{1}{v} = 5 \\ \dfrac{1}{u} - \dfrac{1}{v} = 1 \end{cases}$

Letting $x = \dfrac{1}{u}$ and $y = \dfrac{1}{v}$, this system becomes

$x + y = 5$

$x - y = 1$

Add: $2x = 6$

$\qquad x = 3$

CHECK $u = \dfrac{1}{3}$, $v = \dfrac{1}{2}$:

$$\frac{1}{u} + \frac{1}{v} = 5 \qquad \frac{1}{u} - \frac{1}{v} = 1$$

$$\frac{1}{\frac{1}{3}} + \frac{1}{\frac{1}{2}} \overset{?}{=} 5 \qquad \frac{1}{\frac{1}{3}} - \frac{1}{\frac{1}{2}} \overset{?}{=} 1$$

$$3 + 2 \overset{?}{=} 5 \qquad 3 - 2 \overset{?}{=} 1$$

$$5 \overset{\checkmark}{=} 5 \qquad 1 \overset{\checkmark}{=} 1$$

Substitute: $x + y = 5$

$$3 + y = 5$$

$$y = 2$$

Then $3 = \dfrac{1}{u}$ and $2 = \dfrac{1}{v}$, so that $u = \dfrac{1}{3}$ and $v = \dfrac{1}{2}$.

(b) $\begin{cases} \dfrac{2}{u} + \dfrac{1}{v} = 3 \\ \dfrac{6}{u} + \dfrac{1}{v} = 5 \end{cases}$

Letting $x = \dfrac{1}{u}$ and $y = \dfrac{1}{v}$, this system becomes

$2x + y = 3$

$6x + y = 5$

Multiply the first equation by −1 and add:

$-2x - y = -3$

$$\underline{6x + y = 5}$$

$4x = 2$

$x = \dfrac{1}{2}$

Substitute: $2x + y = 3$

$2\left(\dfrac{1}{2}\right) + y = 2$

$1 + y = 3$

$y = 2$

Then $\dfrac{1}{2} = \dfrac{1}{u}$ and $2 = \dfrac{1}{v}$, so that $u = 2$ and $v = \dfrac{1}{2}$.

CHECK $u = 2$, $v = \dfrac{1}{2}$:

$\dfrac{2}{u} + \dfrac{1}{v} = 3$

$\dfrac{2}{2} + \dfrac{1}{\frac{1}{2}} \overset{?}{=} 3$

$1 + 2 \overset{?}{=} 3$

$3 \overset{\checkmark}{=} 3$

$\dfrac{6}{u} + \dfrac{1}{v} = 5$

$\dfrac{6}{2} + \dfrac{1}{\frac{1}{2}} \overset{?}{=} 5$

$3 + 2 \overset{?}{=} 5$

$5 \overset{\checkmark}{=} 5$

Exercises 5.6

1. boundary: solid line

    x-intercept: −3

    y-intercept: 3

    test point: (0, 0)

    $y \le x + 3$

    $0 \overset{?}{\le} 0 + 3$

    $0 \overset{\checkmark}{\le} 3$

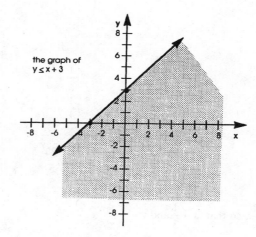

the graph of
$y \le x + 3$

5.   boundary: dotted line

     x-intercept: 3

     y-intercept: 3

     test point: (0, 0)

     $x + y < 3$

     $0 + 0 \overset{?}{<} 3$

       $0 \overset{\checkmark}{<} 3$

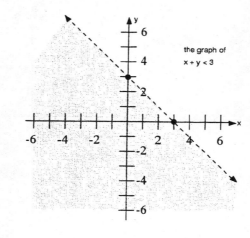

the graph of $x + y < 3$

7.   boundary: solid line

     x-intercept: 3

     y-intercept: 3

     test point: (0, 0)

     $x + y \geq 3$

     $0 + 0 \overset{?}{\geq} 3$

       $0 \not\geq 3$

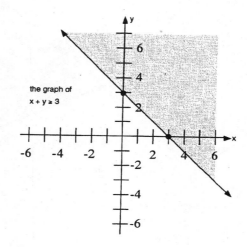

the graph of $x + y \geq 3$

11.   boundary: dotted line

     x-intercept: 6

     y-intercept: 3

     test point: (0, 0)

     $x + 2y > 6$

     $0 + 2(0) \overset{?}{>} 6$

      $0 + 0 \overset{?}{>} 6$

        $0 \not> 6$

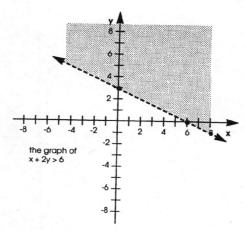

the graph of $x + 2y > 6$

125

15. boundary: dotted line

x-intercept: 5

y-intercept: 2

test point: (0, 0)

$2x + 5y < 10$

$2(0) + 5(0) \overset{?}{<} 10$

$0 + 0 \overset{?}{<} 10$

$0 \overset{\checkmark}{<} 10$

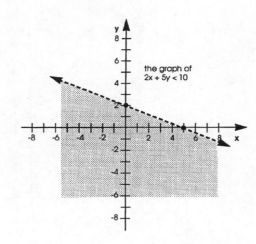

the graph of
$2x + 5y < 10$

19. boundary: solid line

x-intercept: 5

y-intercept: 2

test point: (0, 0)

$2x + 5y \geq 10$

$2(0) + 5(0) \overset{?}{\geq} 10$

$0 + 0 \overset{?}{\geq} 10$

$0 \not\geq 10$

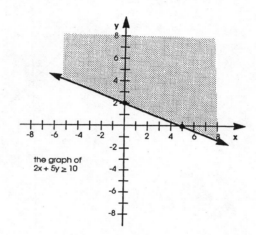

the graph of
$2x + 5y \geq 10$

23. boundary: solid line

x-intercept: 0

y-intercept: 0

second point: (1,1)

test point: (4, 0)

$y \leq x$

$0 \overset{\checkmark}{\leq} 4$

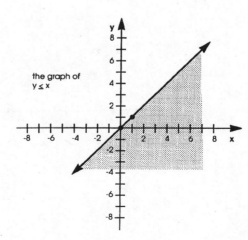

the graph of
$y \leq x$

27. boundary: solid line

x-intercept: 2

y-intercept: −8

test point: (0, 0)

$$4x - y \geq 8$$

$$4(0) - 0 \overset{?}{\geq} 8$$

$$0 - 0 \overset{?}{\geq} 8$$

$$0 \ngeq 8$$

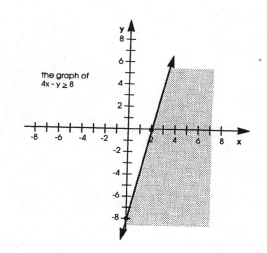

the graph of
$4x - y \geq 8$

31. boundary: dotted line

x-intercept: $\dfrac{15}{7}$

y-intercept: −5

test point: (0, 0)

$$7x - 3y < 15$$

$$7(0) - 3(0) \overset{?}{<} 15$$

$$0 - 0 \overset{?}{<} 15$$

$$0 \overset{\checkmark}{<} 15$$

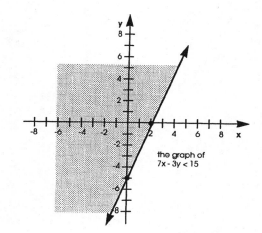

the graph of
$7x - 3y < 15$

35. boundary: dotted line

x-intercept: 8

y-intercept: 12

test point: (0, 0)

$$\frac{x}{2} + \frac{y}{3} < 4$$

$$\frac{0}{2} + \frac{0}{3} \overset{?}{<} 4$$

$$0 + 0 \overset{?}{<} 4$$

$$0 \overset{\checkmark}{<} 4$$

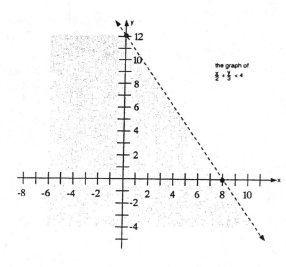

the graph of
$\frac{x}{2} + \frac{y}{3} < 4$

39. boundary: dotted line

horizontal, 3 units above

x-axis

test point: (0, 0)

y < 3

0 $\overset{y}{<}$ 3

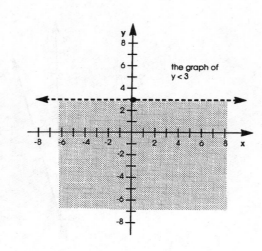

43. boundary: dotted line

vertical, 2 units to left of

y-axis

test point: (0, 0)

x < −2

0 $\not<$ − 2

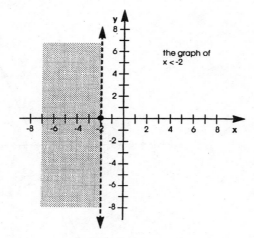

47. boundary: solid line

the y-axis

test point: (1, 0)

x ≤ 0

1 ≤ 0

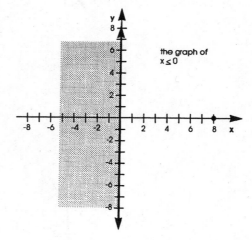

51.   boundary: dotted line

the x-axis

test point: (0, 1)

$$\frac{y}{2} > 0$$

$$\frac{1}{2} \not> 0$$

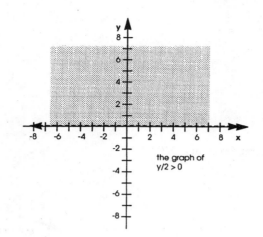

the graph of
y/2 > 0

Chapter 5 Review Exercises

1.   To find the x-intercept, set y = 0:

2x + y = 6

2x + (0) = 6

2x = 6

x = 3

Hence, the graph crosses the x-axis at (3, 0).

To find the y-intercept, set x = 0:

2x + y = 6

2(0) + y = 6

0 + y = 6

y = 6

Hence, the graph crosses the y-axis at (0, 6).

To find a check point, choose y = 2:

2x + y = 6

2x + (2) = 6

2x + 2 = 6

2x = 4

x = 2

Hence, the graph passes through (2, 2).

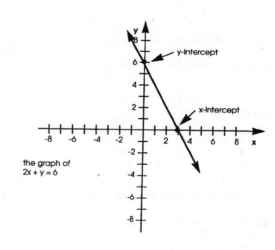

y-intercept

x-intercept

the graph of
2x + y = 6

3. To find the x-intercept, set y = 0:

$$2x - 6y = -6$$
$$2x - 6(0) = -6$$
$$2x - 0 = -6$$
$$2x = -6$$
$$x = -3$$

Hence, the graph crosses the x-axis at (-3, 0).

To find the y-intercept, set x = 0:

$$2x - 6y = -6$$
$$2(0) - 6y = -6$$
$$0 - 6y = -6$$
$$-6y = -6$$
$$y = 1$$

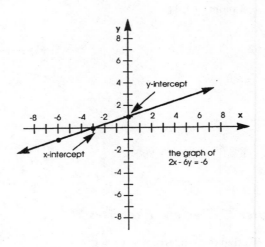

the graph of
2x - 6y = -6

Hence, the graph crosses the y-axis at (0, 1).

To find a check point, choose y = -1:

$$2x - 6y = -6$$
$$2x - 6(-1) = -6$$
$$2x + 6 = -6$$
$$2x = -12$$
$$x = -6$$

Hence, the graph passes through (-6, -1).

5. To find the x-intercept, set y = 0:

$$5x - 3y = 10$$
$$5x - 3(0) = 10$$
$$5x - 0 = 10$$
$$5x = 10$$
$$x = 2$$

Hence, the graph crosses the x-axis at (2, 0).

To find the y-intercept, set x = 0:

$$5x - 3y = 10$$
$$5(0) - 3y = 10$$
$$0 - 3y = 10$$
$$-3y = 10$$
$$y = -\frac{10}{3}$$

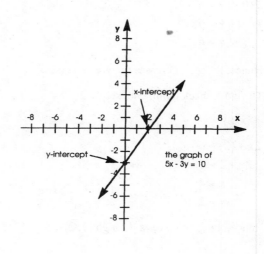

the graph of
5x - 3y = 10

Hence, the graph crosses the y-axis at $\left(0, -\frac{10}{3}\right)$.

130

To find a check point, choose y = 5:

$$5x - 3y = 10$$

$$5x - 3(5) = 10$$

$$5x - 15 = 10$$

$$5x = 25$$

$$x = 5$$

Hence, the graph passes through (5, 5).

7. To find the x-intercept, set y = 0:

$$2x + 5y = 7$$

$$2x + 5(0) = 7$$

$$2x + 0 = 7$$

$$2x = 7$$

$$x = \frac{7}{2}$$

Hence, the graph crosses the x-axis at $\left(\frac{7}{2}, 0\right)$.

To find the y-intercept, set x = 0:

$$2x + 5y = 7$$

$$2(0) + 5y = 7$$

$$0 + 5y = 7$$

$$5y = 7$$

$$y = \frac{7}{5}$$

Hence, the graph crosses the y-axis at $\left(0, \frac{7}{5}\right)$.

To find a check point, choose y = 1:

$$2x + 5y = 7$$

$$2x + 5(1) = 7$$

$$2x + 5 = 7$$

$$2x = 2$$

$$x = 1$$

Hence, the graph passes through (1, 1).

9. To find the x-intercept, set y = 0:

$$3x - 8y = 11$$

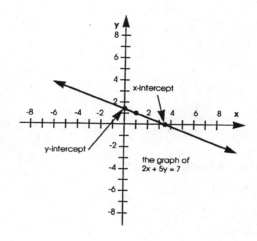

131

$$3x - 8(0) = 11$$

$$3x - 0 = 11$$

$$3x = 11$$

$$x = \frac{11}{3}$$

Hence, the graph crosses the x-axis at $\left(\frac{11}{3}, 0\right)$.

To find the y-intercept, set x = 0:

$$3x - 8y = 11$$

$$3(0) - 8y = 11$$

$$0 - 8y = 11$$

$$-8y = 11$$

$$y = -\frac{11}{8}$$

Hence, the graph crosses the y-axis at $\left(0, -\frac{11}{8}\right)$.

To find a check point, choose y = −1:

$$3x - 8y = 11$$

$$3x - 8(-1) = 11$$

$$3x + 8 = 11$$

$$3x = 3$$

$$x = 1$$

Hence, the graph passes through (1, −1).

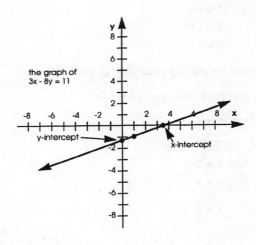

the graph of
3x − 8y = 11

y-intercept

x-intercept

11. To find the x-intercept, set y = 0:

$$5x + 7y = 21$$

$$5x + 7(0) = 21$$

$$5x + 0 = 21$$

$$5x = 21$$

$$x = \frac{21}{5}$$

Hence, the graph crosses the x-axis at $\left(\frac{21}{5}, 0\right)$.

To find the y-intercept, set x = 0:

$$5x + 7y = 21$$

$$5(0) + 7y = 21$$

$$0 + 7y = 21$$

$$7y = 21$$

$$y = 3$$

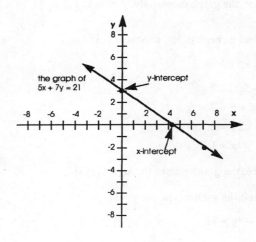

the graph of
5x + 7y = 21

y-intercept

x-intercept

132

Hence, the graph crosses the y-axis at (0, 3).

To find a check point, choose y = –2:

$$5x – 7y = 21$$

$$5x + 7(–2) = 21$$

$$5x – 14 = 21$$

$$5x = 35$$

$$x = 7$$

Hence, the graph passes through (7, –2).

13. To find the x-intercept, set y = 0:

$$y = x$$

$$0 = x$$

Hence, the graph crosses the x-axis at (0, 0).

This is also the point at which the graph crosses the

y-axis. To find a second point, choose x = 2:

$$y = x$$

$$y = 2$$

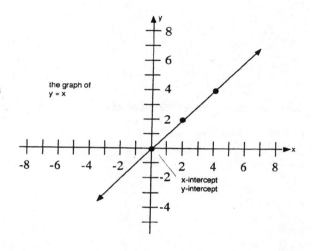

Hence, the graph passes through (2, 2).

To find a check point, choose x = 4:

$$y = x$$

$$y = 4$$

Hence, the graph passes through (4, 4).

15. To find the x-intercept, set y = 0:

$$y = –2x$$

$$0 = –2x$$

$$0 = x$$

Hence, the graph crosses the x-axis at (0, 0).

This is also the point at which the graph crosses the y-axis.

To find a second point, choose x = 1:

$$y = –2x$$

$$y = –2(1)$$

$$y = –2$$

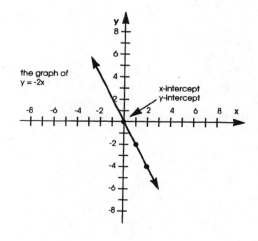

Hence, the graph passes through (1, –2).

To find a check point, choose x = 2:

$$y = -2x$$

$$y = -2(2)$$

$$y = -4$$

Hence, the graph passes through (2, –4).

17.   To find the x-intercept, set y = 0:

$$y = \frac{2}{3}x + 2$$

$$0 = \frac{2}{3}x + 2$$

$$-2 = \frac{2}{3}x$$

$$-6 = 2x$$

$$-3 = x$$

Hence, the graph crosses the x-axis at (–3, 0).

To find the y-intercept, set x = 0:

$$y = \frac{2}{3}x + 2$$

$$y = \frac{2}{3}(0) + 2$$

$$y = 0 + 2$$

$$y = 2$$

Hence, the graph crosses the y-axis at (0, 2).

To find a check point, choose x = 3:

$$y = \frac{2}{3}x + 2$$

$$y = \frac{2}{3}(3) + 2$$

$$y = 2 + 2$$

$$y = 4$$

Hence, the graph passes through (3, 4).

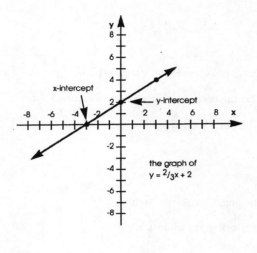

the graph of
$y = {}^{2}/_{3}x + 2$

19. To find the x-intercept, set y = 0:

$$\frac{x}{3} + \frac{y}{2} = 12$$

$$\frac{x}{3} + \frac{0}{2} = 12$$

$$\frac{x}{3} + 0 = 12$$

$$\frac{x}{3} = 12$$

$$x = 36$$

Hence, the graph crosses the x-axis at (36, 0).

To find the y-intercept, set x = 0:

$$\frac{x}{3} + \frac{y}{2} = 12$$

$$\frac{0}{3} + \frac{y}{2} = 12$$

$$0 + \frac{y}{2} = 12$$

$$\frac{y}{2} = 12$$

$$y = 24$$

Hence, the graph crosses the y-axis at (0, 24).

To find a check point, choose y = 12:

$$\frac{x}{3} + \frac{y}{2} = 12$$

$$\frac{x}{3} + \frac{12}{2} = 12$$

$$\frac{x}{3} + 6 = 12$$

$$\frac{x}{3} = 6$$

$$x = 18$$

Hence, the graph passes through (18, 12).

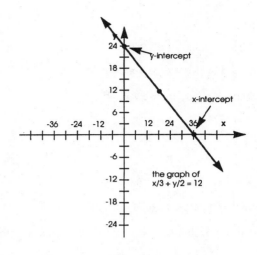

the graph of
x/3 + y/2 = 12

21. To find the x-intercept, set y = 0:

$$x - 2y = 8$$

$$x - 2(0) = 8$$

$$x - 0 = 8$$

$$x = 8$$

Hence, the graph crosses the x-axis at (8, 0).

To find the y-intercept, set x = 0:

x – 2y = 8

0 – 2y = 8

–2y = 8

y = –4

Hence, the graph crosses the y-axis at (0, –4).

To find a check point, choose y = –2:

x – 2y = 8

x – 2(–2) = 8

x + 4 = 8

x = 4

Hence, the graph passes through (4, –2).

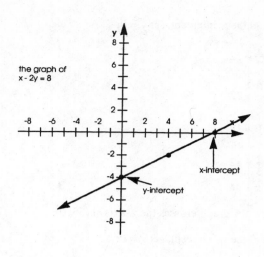

23.　x – 2 = 0, or equivalently, x = 2, is a vertical line two units to the right of the y-axis.

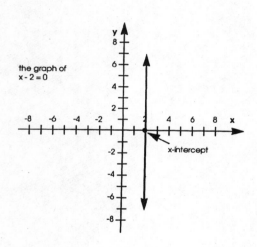

136

25. $2y = 5$, or equivalently, $y = \dfrac{5}{2}$, is a horizontal line $\dfrac{5}{2}$ units above the x-axis.

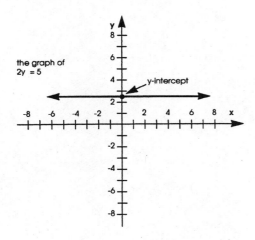

the graph of
$2y = 5$

y-intercept

27. $\quad m = \dfrac{y_2 - y_1}{x_2 - x_1} = \dfrac{-2 - 0}{3 - (-1)} = \dfrac{-2}{4} = -\dfrac{1}{2}$

29. Since $y = 3x - 5$ is an equation of the form $y = mx + b$, it follows that $m = 3$.

31. $\quad 4y - 3x = 1$

$\qquad 4y = 3x + 1$

$\qquad y = \dfrac{3}{4}x + \dfrac{1}{4}$

By comparison with $y = mx + b$, find that $m = \dfrac{3}{4}$.

33. The line passing through $(3, 5)$ and $(1, 4)$ has slope $\dfrac{4 - 5}{1 - 3} = \dfrac{-1}{-2} = \dfrac{1}{2}$. Therefore, the line in question has slope $\dfrac{1}{2}$.

35. The line passing through $(4, 7)$ and $(4, 9)$ is vertical. Therefore, the line in question is horizontal, so has slope 0.

37. The line passing through $(-7, 6)$ and $(2, 6)$ is horizontal, hence has slope 0.

39. Since the line $y = 3x - 7$ has slope 3, the line in question has slope 3 as well.

41. $\quad 3y - 5x + 6 = 0$

$\qquad 3y = 5x - 6$

$\qquad y = \dfrac{5}{3}x - 2$

So the given line has slope $\dfrac{5}{3}$. This means that the line in question has slope $-\dfrac{3}{5}$.

43. The line $x = 3$ is vertical. Hence, the line in question is vertical and has no slope.

45.
$$\frac{2-a}{1-4} = 4$$

$$\frac{2-a}{-3} = 4$$

$$(-3)\left(\frac{2-a}{-3}\right) = (-3)(4)$$

$$2 - a = -12$$

$$-a = -14$$

$$a = 14$$

47.
$$\frac{a-3}{-3-0} = \frac{a-0}{7-0}$$

$$\frac{a-3}{-3} = \frac{a}{7}$$

$$\overset{7}{(-21)}\left(\frac{a-3}{-3}\right) = \overset{-3}{(-21)}\left(\frac{a}{7}\right)$$

$$7(a-3) = -3a$$

$$7a - 21 = -3a$$

$$-21 = -10a$$

$$\frac{21}{10} = a$$

49.
$$m = \frac{y_2 - y_1}{x_2 - x_1} = \frac{-4-3}{1-(-2)} = \frac{-7}{3} = -\frac{7}{3}$$

$$y - y_1 = m(x - x_1)$$

$$y - 3 = -\frac{7}{3}(x - (-2))$$

$$y - 3 = -\frac{7}{3}(x + 2)$$

$$3(y - 3) = 3\left(-\frac{7}{3}(x+2)\right)$$

$$3y - 9 = -7x - 14$$

$$3y = -7x - 5$$

$$y = -\frac{7}{3}x - \frac{5}{3}$$

51. $m = \dfrac{y_2 - y_1}{x_2 - x_1} = \dfrac{-5 - 5}{-3 - 3} - \dfrac{-10}{-6} = \dfrac{5}{3}$

$$y - y_1 = m(x - x_1)$$

$$y - 5 = \dfrac{5}{3}(x - 3)$$

$$3(y - 5) = \cancel{3}\left(\dfrac{5}{\cancel{3}}(x - 3)\right)$$

$$3y - 15 = 5x - 15$$

$$3y = 5x$$

$$y = \dfrac{5}{3}x$$

53. $y - y_1 = m(x - x_1)$

$$y - 5 = \dfrac{2}{5}(x - 2)$$

$$5(y - 5) = \cancel{5}\left(\dfrac{2}{\cancel{5}}(x - 2)\right)$$

$$5y - 25 = 2x - 4$$

$$5y = 2x + 21$$

$$y = \dfrac{2}{5}x + \dfrac{21}{5}$$

55. $y - y_1 = m(x - x_1)$

$$y - 7 = 5(x - 4)$$

$$y - 7 = 5x - 20$$

$$y = 5x - 13$$

57. $y = mx + b$

$$y = 5x + 3$$

59. $y = 3$

61. The given line has slope $\dfrac{3}{2}$, so the line in question has slope $\dfrac{3}{2}$ as well.

$$y = mx + b$$

$$y = \dfrac{3}{2}x + 0$$

$$y = \dfrac{3}{2}x$$

63. $2y - 5x = 1$

$$2y = 5x + 1$$

$$y = \dfrac{5}{2}x + \dfrac{1}{2}$$

Therefore, the given line has slope $\frac{5}{2}$. So the line in question has slope $-\frac{2}{5}$.

$$y = mx + b$$

$$y = -\frac{2}{5}x + 6$$

65. $$3x = -5y$$

$$-\frac{3}{5}x = y$$

Therefore, the given line has slope $-\frac{3}{5}$. So the line in question has slope $\frac{5}{3}$.

$$y = mx + b$$

$$y = \frac{5}{3}x + 0$$

$$y = \frac{5}{3}x$$

67.   Since the line passes through $(3,0)$ and $(0, -5)$, its slope is $\frac{-5-0}{0-3} = \frac{-5}{-3} = \frac{5}{3}$.

$$y = mx + b$$

$$y = \frac{5}{3}x - 5$$

69. $3x - 2y = 5$ $5y = x + 3$

      $-2y = -3x + 5$

$$y = \frac{3}{2}x - \frac{5}{2}$$ $$y = \frac{1}{5}x + \frac{3}{5}$$

By comparison with $y = mx + b$, the first line has slope $\frac{3}{2}$ and the second has y-intercept $\frac{3}{5}$. Then the line in question has equation $y = \frac{3}{2}x + \frac{3}{5}$.

71. $(250, 12000)$, $(300, 20000)$

$$m = \frac{20000 - 12000}{300 - 250} = \frac{8000}{500} = 160$$

$$P - P_1 = m(x - x_1)$$

$$P - 12000 = 160(x - 250)$$

$$P - 12000 = 160x - 40000$$

$$P = 160x - 28000$$

When $x = 400$,

$$P = 160(400) - 28000 = 64000 - 28000 = \$36000$$

73.

hour	change in distance	rate of speed	direction
first	from 0 to 40	40 mph	away from home
second	from 40 to 90	50 mph	away from home
third	from 90 to 90	0 mph	—
fourth	from 90 to 30	60 mph	toward home
fifth	from 30 to 50	20 mph	away from home
sixth	from 50 to 0	50 mph	toward home

75. $\begin{cases} x - y = 4 \\ 2x - 3y = 7 \end{cases}$

CHECK $x = 5$, $y = 1$:

From the first equation, $x = y + 4$.

Substitute this result into the second equation:

$2(y + 4) - 3y = 7$

$2y + 8 - 3y = 7$

$-y + 8 = 7$

$-y = -1$

$y = 1$

$$x - y = 4$$
$$5 - 1 \overset{?}{=} 4$$
$$4 \overset{\checkmark}{=} 4$$

$$2x - 3y = 7$$
$$2(5) - 3(1) \overset{?}{=} 7$$
$$10 - 3 \overset{?}{=} 7$$
$$7 \overset{\checkmark}{=} 7$$

When $y = 1$, $x = 1 + 4 = 5$. Solution: $x = 5$, $y = 1$.

77. $\begin{cases} \dfrac{x}{6} - \dfrac{y}{4} = \dfrac{4}{3} \xrightarrow{\text{multiply by } 12} 2x - 3y = 16 \\ \dfrac{x}{5} - \dfrac{y}{2} = \dfrac{8}{5} \xrightarrow{\text{multiply by } 10} 2x - 5y = 16 \end{cases}$

CHECK $x = 8$, $y = 0$:

Subtract the second equation from the first:

$2y = 0$

$y = 0$

Substitute: $2x - 3y = 16$

$2x - 3(0) = 16$

$2x - 0 = 16$

$2x = 16$

$x = 8$

Solution: $x = 8$, $y = 0$

$$\frac{x}{6} - \frac{y}{4} = \frac{4}{3}$$
$$\frac{8}{6} - \frac{0}{4} \overset{?}{=} \frac{4}{3}$$
$$\frac{8}{6} - 0 \overset{?}{=} \frac{4}{3}$$
$$\frac{8}{6} \overset{?}{=} \frac{4}{3}$$
$$\frac{4}{3} \overset{\checkmark}{=} \frac{4}{3}$$

$$\frac{x}{5} - \frac{y}{2} = \frac{8}{5}$$
$$\frac{8}{5} - \frac{0}{2} \overset{?}{=} \frac{8}{5}$$
$$\frac{8}{5} - 0 \overset{?}{=} \frac{8}{5}$$
$$\frac{8}{5} \overset{\checkmark}{=} \frac{8}{5}$$

79.
$$\begin{cases} 3x - \dfrac{y}{4} = 2 \xrightarrow{\text{multiply by 4}} 12x - y = 8 \\ 6x - \dfrac{y}{2} = 4 \xrightarrow{\text{multiply by 2}} 12x - y = 8 \end{cases}$$

Since these equations are identical, the system is dependent. There are infinitely many solutions (x, y), where

$12x - y = 8$.

81.
$$\begin{cases} x = 2y - 3 \\ y = 3x + 2 \end{cases}$$

CHECK $x = -\dfrac{1}{5},\ y = \dfrac{7}{5}$:

Substitute the result of the second

equation into the first:

$x = 2(3x + 2) - 3$

$x = 6x + 4 - 3$

$x = 6x + 1$

$-5x = 1$

$x = -\dfrac{1}{5}$

$x = 2y - 3$

$-\dfrac{1}{5} \overset{?}{=} 2\left(\dfrac{7}{5}\right) - 3$

$-\dfrac{1}{5} \overset{?}{=} \dfrac{14}{5} - 3$

$-\dfrac{1}{5} \overset{?}{=} \dfrac{14}{5} - \dfrac{15}{5}$

$-\dfrac{1}{5} \overset{\checkmark}{=} -\dfrac{1}{5}$

$y = 3x + 2$

$\dfrac{7}{5} \overset{?}{=} 3\left(-\dfrac{1}{5}\right) + 2$

$\dfrac{7}{5} \overset{?}{=} -\dfrac{3}{5} + 2$

$\dfrac{7}{5} \overset{?}{=} -\dfrac{3}{5} + \dfrac{10}{5}$

$\dfrac{7}{5} \overset{\checkmark}{=} \dfrac{7}{5}$

When $x = -\dfrac{1}{5},\ y = 3\left(-\dfrac{1}{5}\right) + 2 = -\dfrac{3}{5} + 2 = \dfrac{7}{5}$

Solution: $x = -\dfrac{1}{5},\ y = \dfrac{7}{5}$

83. Let x = amount invested at rate of 4.75%

y = amount invested at rate of 6.65%

$x + y = 8,500$

$0.0475x + 0.0665y = 512.05$

Multiply the first equation by –475, the second by 10,000, and add the resulting equations:

$-475x - 475y = -4,037,500$

$\underline{475x + 665y = 5,120,500}$

$190y = 1,083,000$

$y = 5,700$

Substitute: $x + y = 8,500$

$x + 5,700 = 8,500$

$x = 2,800$

Conclude: $2,800 is invested at 4.75% and $5,700 is invested at 6.65%.

CHECK: $\$5,700 + \$2,800 \overset{\checkmark}{=} \$8,500$

$4.75\% \text{ of } \$2,800 = \$133$

$\underline{6.65\% \text{ of } \$5,700 = \$379.05}$

Total interest $\overset{\checkmark}{=} \$512.05$

85.   boundary: dotted line

x-intercept: -2

y-intercept: 4

test point: $(0, 0)$

$y - 2x < 4$

$0 - 2(0) \overset{?}{<} 4$

$0 - 0 \overset{?}{<} 4$

$0 \overset{\checkmark}{<} 4$

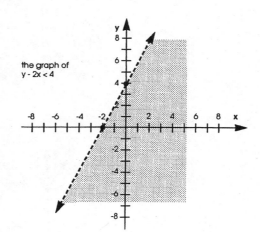

the graph of
y - 2x < 4

87.   boundary: dotted line

x-intercept: -2

y-intercept: 3

test point: $(0, 0)$

$2y - 3x > 6$

$2(0) - 3(0) \overset{?}{>} 6$

$0 - 0 \overset{?}{>} 6$

$0 \not> 6$

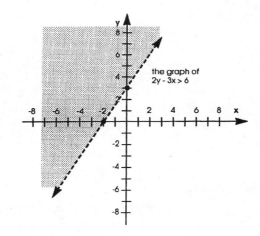

the graph of
2y - 3x > 6

89.   boundary: solid line

x-intercept: $-\dfrac{5}{2}$

y-intercept: 4

test point: (0, 0)

$$5y - 8x \leq 20$$

$$5(0) - 8(0) \overset{?}{\leq} 20$$

$$0 - 0 \overset{?}{\leq} 20$$

$$0 \overset{\checkmark}{\leq} 20$$

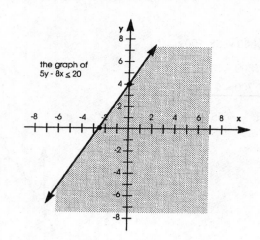

the graph of
5y - 8x ≤ 20

91. boundary: solid line

x-intercept: 12

y-intercept: 18

test point: (0, 0)

$$\frac{x}{2} + \frac{y}{3} \geq 6$$

$$\frac{0}{2} + \frac{0}{3} \overset{?}{\geq} 6$$

$$0 + 0 \overset{?}{\geq} 6$$

$$0 \ngeq 6$$

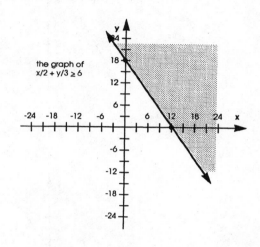

the graph of
x/2 + y/3 ≥ 6

93. boundary: dotted line

horizontal, 5 units above

x-axis

test point: (0, 0)

$$y < 5$$

$$0 \overset{\checkmark}{<} 5$$

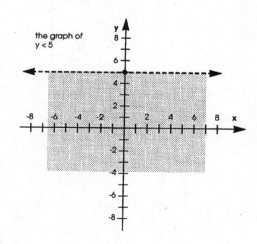

the graph of
y < 5

(c)   As in part (b), a negative answer would be impossible. Here, $8^{-\frac{1}{3}} = \dfrac{1}{8^{\frac{1}{3}}} = \dfrac{1}{2}$, from part (a).

78.   Negative exponents involve reciprocals of the base, whereas fractional exponents involve roots of the base.

79.   The same rules of exponents apply for rational, integer, and natural number exponents.

80. $9^{-2} = \dfrac{1}{9^2} = \dfrac{1}{81}$ (negative exponent involves a reciprocal)

$9^{\frac{1}{2}} = 3$ (fractional exponent involves a root)

$9^{-\frac{1}{2}} = \dfrac{1}{9^{\frac{1}{2}}} = \dfrac{1}{3}$ (negative fractional exponent involves a reciprocal of a root)

81. $m = \dfrac{5 - (-6)}{3 - 2} = \dfrac{5 + 6}{1} = \dfrac{11}{1} = 11$

83.   boundary: solid line
x-intercept: 4
y-intercept: −2
test point: (0,0)
$3x - 6y \leq 12$
$3(0) - 6(0) \overset{?}{\leq} 12$
$0 - 0 \overset{?}{\leq} 12$
$0 \overset{\checkmark}{=} 12$

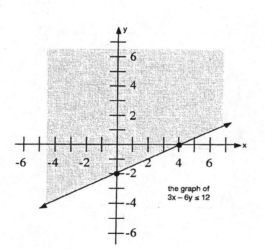

the graph of
$3x - 6y \leq 12$

Chapter 6 Review Exercises

1. $(x^2 x^5)(x^4 x) = x^7 x^5 = x^{12}$

3. $(-3x^2 y)(-2xy^4)(-x) = (-3)(-2)(-1)x^2 xxyy^4 = -6x^4 y^5$

5. $(a^3)^4 = a^{12}$

7. $(a^2 b^3)^7 = (a^2)^7 (b^3)^7 = a^{14} b^{21}$

9. $(a^2 b^3)^2 (a^2 b)^3 = (a^2)^2 (b^3)^2 (a^2)^3 b^3 = a^4 b^6 a^6 b^3 = a^{10} b^9$

11. $(2^2 \cdot 3^2)^5 = (2^2)^5(3^2)^5 = 2^{10}3^{10}$

13. $(a^2bc^2)^2(ab^2c)^3 = (a^2)^2b^2(c^2)^2a^3(b^2)^3c^3 = a^4b^2c^4a^3b^6c^3 = a^7b^8c^7$

15. $\dfrac{a^5}{a^6} = a^{-1} = \dfrac{1}{a}$

17. $\dfrac{x^2x^5}{x^4x^3} = \dfrac{x^7}{x^7} = 1$

19. $\dfrac{(x^3y^2)^3}{(x^5y^4)^5} = \dfrac{(x^3)^3(y^2)^3}{(x^5)^5(y^4)^5} = \dfrac{x^9y^6}{x^{25}y^{20}} = x^{-16}y^{-14} = \dfrac{1}{x^{16}} \cdot \dfrac{1}{y^{14}} = \dfrac{1}{x^{16}y^{14}}$

21. $\left(\dfrac{a^2\not{b}}{a\not{b}}\right)^4 = a^4$

23. $\dfrac{(2ax^2)^2(3ax)^2}{(-2x)^2} = \dfrac{4a^2(x^2)^2 9a^2x^2}{4x^2} = \dfrac{4a^2x^4 \cdot 9a^2x^2}{4x^2} = \dfrac{36a^4x^6}{4x^2} = 9a^4x^4$

25. $\left(\dfrac{-3\not{x}y}{x^2\not{x}}\right)^2\left(\dfrac{-2\not{x}y^2}{\not{x}}\right)^3 = \left(\dfrac{-3y}{x}\right)^2\left(\dfrac{-2y^2}{1}\right)^3 = \left(\dfrac{9y^2}{x^2}\right)\left(\dfrac{-8y^6}{1}\right) = \dfrac{-72y^8}{x^2}$

27. $2x^2y\left(\dfrac{-3\not{x}y^2}{x^2\not{y}}\right)^2 = 2x^2y\left(\dfrac{-3y}{x}\right)^2 = \dfrac{2\not{x}^2y}{1} \cdot \dfrac{9y^2}{\not{x}^2} = 18y^3$

29. $\left(\dfrac{(-3ab^2c)^2}{-6a^2b}\right)^2 = \left(\dfrac{\overset{3}{\not{9}}a^2b^{\overset{3}{4}}c^2}{\underset{2}{-\not{6}a^2\not{b}}}\right)^2 = \left(\dfrac{3b^3c^2}{-2}\right)^2 = \dfrac{3^2(b^3)^2(c^2)^2}{(-2)^2} = \dfrac{9b^6c^4}{4}$

31. $a^{-3}a^{-4}a^5 = a^{-3-4+5} = a^{-2} = \dfrac{1}{a^2}$

33. $(x^{-2}y^5)^{-4} = (x^{-2})^{-4}(y^5)^{-4} = x^8y^{-20} = \dfrac{x^8}{1} \cdot \dfrac{1}{y^{20}} = \dfrac{x^8}{y^{20}}$

35. $(a^{-2}b^2)^{-3}(a^3b^{-4})^2 = (a^{-2})^{-3}(b^2)^{-3}(a^3)^2(b^{-4})^2 = a^6b^{-6}a^6b^{-8} = a^{12}b^{-14} = \dfrac{a^{12}}{1} \cdot \dfrac{1}{b^{14}} = \dfrac{a^{12}}{b^{14}}$

37. $(-3)^{-4}(-2)^{-1} = \dfrac{1}{(-3)^4} \cdot \dfrac{1}{-2} = \left(\dfrac{1}{81}\right)\left(-\dfrac{1}{2}\right) = -\dfrac{1}{162}$

39. $(-3x^{-2}yx^{-3})^{-2}(9x^{-2}y)^{-3} = (-3x^{-5}y)^{-2}(9x^{-2}y)^{-3} = (-3)^{-2}(x^{-5})^{-2}y^{-2} \cdot 9^{-3}(x^{-2})^{-3}y^{-3}$

$$= \frac{1}{(-3)^2}x^{10}y^{-2} \cdot \frac{1}{9^3}x^6y^{-3} = \frac{1}{9}x^{10}y^{-2} \cdot \frac{1}{9^3}x^6y^{-3} = \frac{1}{9^4}x^{16}y^{-5}$$

$$= \frac{1}{9^4} \cdot \frac{x^{16}}{1} \cdot \frac{1}{y^5} = \frac{x^{16}}{9^4y^5}$$

41. $\left(\dfrac{3x^{-5}y^2z^{-4}}{2x^{-7}y^{-4}}\right)^0 = 1$, by definition of zero exponents.

43. $\dfrac{x^{-3}x^{-6}}{x^{-5}x^0} = \dfrac{x^{-9}}{x^{-5}} = x^{-9-(-5)} = x^{-4} = \dfrac{1}{x^4}$

45. $\dfrac{x^{-3}y^{-5}x^{-2}}{x^4y^{-3}} = \dfrac{x^{-5}y^{-5}}{x^4y^{-3}} = x^{-9}y^{-2} = \dfrac{1}{x^9} \cdot \dfrac{1}{y^2} = \dfrac{1}{x^9y^2}$

47. $\left(\dfrac{x^{-2}y^{-3}}{y^{-3}x^2}\right)^{-2} = \left(\dfrac{x^{-2}}{x^2}\right)^{-2} = (x^{-4})^{-2} = x^8$

49. $\left(\dfrac{r^{-2}s^{-3}r^{-2}}{s^{-4}}\right)^{-2}\left(\dfrac{r^{-1}}{s^{-1}}\right)^{-3} = (r^{-4}s)^{-2}\left(\dfrac{r^{-1}}{s^{-1}}\right)^{-3} = (r^{-4})^{-2}s^{-2}\left(\dfrac{(r^{-1})^{-3}}{(s^{-1})^{-3}}\right) = r^8s^{-2}\left(\dfrac{r^3}{s^3}\right)$

$$= r^{11}s^{-5} = \dfrac{r^{11}}{1} \cdot \dfrac{1}{s^5} = \dfrac{r^{11}}{s^5}$$

51. $\left(\dfrac{2}{5}\right)^{-2} = \left(\dfrac{5}{2}\right)^2 = \dfrac{5^2}{2^2} = \dfrac{25}{4}$

53. $\dfrac{(2x^2y^{-1}z)^{-2}}{(3xy^2)^{-3}} = \dfrac{2^{-2}(x^2)^{-2}(y^{-1})^{-2}z^{-2}}{3^{-3}x^{-3}(y^2)^{-3}} = \dfrac{\frac{1}{4}x^{-4}y^2z^{-2}}{\frac{1}{27}x^{-3}y^{-6}} = \dfrac{27}{4}x^{-1}y^8z^{-2} = \dfrac{27}{4} \cdot \dfrac{1}{x} \cdot \dfrac{y^8}{1} \cdot \dfrac{1}{z^2} = \dfrac{27y^8}{4xz^2}$

55. $\left(\dfrac{(2x^{-2}y)^{-2}}{(3x^{-1}y^{-1})^2}\right)^{-2} = \left(\dfrac{2^{-2}(x^{-2})^{-2}y^{-2}}{3^2(x^{-1})^2(y^{-1})^2}\right)^{-2} = \left(\dfrac{\frac{1}{4}x^4y^{-2}}{9x^{-2}y^{-2}}\right)^{-2} = \left(\dfrac{1}{36}x^6\right)^{-2} = \left(\dfrac{x^6}{36}\right)^{-2}$

$$= \left(\dfrac{36}{x^2}\right)^2 = \dfrac{36^2}{(x^6)^2} = \dfrac{1296}{x^{12}}$$

57. $(x^{-1} + y^{-1})(x - y) = x^{-1}x - x^{-1}y + y^{-1}x - y^{-1}y = \dfrac{1}{x} \cdot \dfrac{x}{1} - \dfrac{1}{x} \cdot \dfrac{y}{1} + \dfrac{1}{y} \cdot \dfrac{x}{1} - \dfrac{1}{y} \cdot \dfrac{y}{1}$

$$= 1 - \dfrac{y}{x} + \dfrac{x}{y} - 1 = -\dfrac{y}{x} + \dfrac{x}{y} = \dfrac{x^2 - y^2}{xy}$$

59. $\dfrac{x^{-1} + y^{-3}}{x^{-1}y^2} = \dfrac{x^{-1}}{x^{-1}y^2} + \dfrac{y^{-3}}{x^{-1}y^2} = \dfrac{1}{y^2} + \dfrac{y^{-5}}{x^{-1}} = \dfrac{1}{y^2} + \dfrac{\frac{1}{y^5}}{\frac{1}{x}} = \dfrac{1}{y^2} + \dfrac{x}{y^5} = \dfrac{y^3}{y^5} + \dfrac{x}{y^5} = \dfrac{y^3 + x}{y^5}$

61. $\dfrac{x^{-2} - y^{-1}}{x^{-1} + y^{-2}} = \dfrac{\frac{1}{x^2} - \frac{1}{y}}{\frac{1}{x} + \frac{1}{y^2}} = \dfrac{x^2y^2\left(\frac{1}{x^2} - \frac{1}{y}\right)}{x^2y^2\left(\frac{1}{x} + \frac{1}{y^2}\right)} = \dfrac{\frac{x^2y^2}{1} \cdot \frac{1}{x^2} - \frac{x^2y^2}{1} \cdot \frac{1}{y}}{\frac{x^2y^2}{1} \cdot \frac{1}{x} + \frac{x^2y^2}{1} \cdot \frac{1}{y^2}} = \dfrac{y^2 - x^2y}{xy^2 + x^2} = \dfrac{y(y - x^2)}{x(y^2 + x)}$

63. $2.83 \times 10^4 = 28,300$

65. $7.96 \times 10^{-5} = 0.0000796$

67. $7.936 = 7.936 \times 10^0$

69. $0.00578 = 5.78 \times 10^{-3}$

71. $625,897 = 6.25897 \times 10^5$

73. $\dfrac{(0.0014)(9,000)}{(20,000)(63,000)} = \dfrac{(1.4 \times 10^{-3})(9 \times 10^3)}{(2 \times 10^4)(6.3 \times 10^4)} = \dfrac{(1.4)(9)}{(2)(6.3)} \times \dfrac{10^{-3}10^3}{10^4 10^4} = 1 \times \dfrac{10^0}{10^8}$

$= 1 \times 10^{-8} = 0.00000001$

75. $25^{\frac{1}{2}} = 5$

77. $(-243)^{-\frac{3}{5}} = \dfrac{1}{(-243)^{\frac{3}{5}}} = \dfrac{1}{\left((-243)^{\frac{1}{5}}\right)^3} = \dfrac{1}{(-3)^3} = \dfrac{1}{-27} = -\dfrac{1}{27}$

79. $\left(\dfrac{64}{27}\right)^{\frac{1}{3}} = \dfrac{64^{\frac{1}{3}}}{27^{\frac{1}{3}}} = \dfrac{4}{3}$

81. $\left(-\dfrac{27}{125}\right)^{-\frac{1}{3}} = \left(-\dfrac{125}{27}\right)^{\frac{1}{3}} = \dfrac{(-125)^{\frac{1}{3}}}{(27)^{\frac{1}{3}}} = \dfrac{-5}{3} = -\dfrac{5}{3}$

83. $x^{\frac{1}{2}}x^{\frac{1}{3}} = x^{\frac{1}{2} + \frac{1}{3}} = x^{\frac{5}{6}}$

85. $\left(x^{-\frac{1}{2}}x^{\frac{1}{3}}\right)^{-6} = \left(x^{-\frac{1}{2} + \frac{1}{3}}\right)^{-6} = \left(x^{-\frac{1}{6}}\right)^{-6} = x$

87. $\dfrac{x^{\frac{1}{2}}x^{\frac{1}{3}}}{x^{\frac{2}{5}}} = \dfrac{x^{\frac{1}{2}+\frac{1}{3}}}{x^{\frac{2}{5}}} = \dfrac{x^{\frac{5}{6}}}{x^{\frac{2}{5}}} = x^{\frac{5}{6}-\frac{2}{5}} = x^{\frac{13}{30}}$

89. $\dfrac{r^{-\frac{1}{2}}s^{-\frac{1}{3}}}{r^{\frac{1}{3}}s^{-\frac{2}{3}}} = r^{-\frac{1}{2}-\frac{1}{3}}s^{-\frac{1}{3}-\left(-\frac{2}{3}\right)} = r^{-\frac{5}{6}}s^{\frac{1}{3}} = \dfrac{1}{r^{\frac{5}{6}}}\cdot\dfrac{s^{\frac{1}{3}}}{1} = \dfrac{s^{\frac{1}{3}}}{r^{\frac{5}{6}}}$

91. $\left(\dfrac{a^{\frac{1}{2}}a^{-\frac{1}{3}}}{a^{\frac{1}{2}}b^{\frac{1}{5}}}\right)^{-15} = \dfrac{\left(a^{-\frac{1}{3}}\right)^{-15}}{\left(b^{\frac{1}{5}}\right)^{-15}} = \dfrac{a^5}{b^{-3}} = \dfrac{a^5}{\frac{1}{b^3}} = \dfrac{a^5}{1}\cdot\dfrac{b^3}{1} = a^5b^3$

93. $\dfrac{\left(x^{-\frac{1}{2}}y^{\frac{1}{2}}\right)^{-2}}{\left(x^{-\frac{1}{3}}y^{-\frac{1}{3}}\right)^{-\frac{1}{2}}} = \dfrac{\left(x^{-\frac{1}{2}}\right)^{-2}\left(y^{\frac{1}{2}}\right)^{-2}}{\left(x^{-\frac{1}{3}}\right)^{-\frac{1}{2}}\left(y^{-\frac{1}{3}}\right)^{-\frac{1}{2}}} = \dfrac{xy^{-1}}{x^{\frac{1}{6}}y^{\frac{1}{6}}} = x^{\frac{5}{6}}y^{-\frac{7}{6}} = \dfrac{x^{\frac{5}{6}}}{1}\cdot\dfrac{1}{y^{\frac{7}{6}}} = \dfrac{x^{\frac{5}{6}}}{y^{\frac{7}{6}}}$

95. $\left(\dfrac{4^{-\frac{1}{2}}16^{-\frac{3}{4}}}{8^{\frac{1}{3}}}\right)^2 = \left(\dfrac{\frac{1}{4^{\frac{1}{2}}}\cdot\frac{1}{\left(16^{\frac{1}{4}}\right)^3}}{8^{\frac{1}{3}}}\right)^{-2} = \left(\dfrac{\frac{1}{2}\cdot\frac{1}{2^3}}{2}\right)^{-2} = \left(\dfrac{\frac{1}{2^4}}{2}\right)^{-2} = \left(\dfrac{2^{-4}}{2}\right)^{-2} = (2^{-5})^{-2} = 2^{10}$

97. $\left(a^{\frac{1}{2}}-b^{\frac{1}{2}}\right)a^{-\frac{2}{3}} = \dfrac{a^{\frac{1}{2}}-b^{\frac{1}{2}}}{a^{\frac{2}{3}}} = \dfrac{a^{\frac{1}{2}}}{a^{\frac{2}{3}}}-\dfrac{b^{\frac{1}{2}}}{a^{\frac{2}{3}}} = a^{\frac{1}{2}-\frac{2}{3}}-\dfrac{b^{\frac{1}{2}}}{a^{\frac{2}{3}}}$

$= a^{-\frac{1}{6}}-\dfrac{b^{\frac{1}{2}}}{a^{\frac{2}{3}}} = \dfrac{1}{a^{\frac{1}{6}}}-\dfrac{b^{\frac{1}{2}}}{a^{\frac{2}{3}}}$

99. $\left(a^{-\frac{1}{2}}+2b^{\frac{1}{2}}\right)^2 = \left(a^{-\frac{1}{2}}+2b^{\frac{1}{2}}\right)\left(a^{-\frac{1}{2}}+2b^{\frac{1}{2}}\right) = a^{-\frac{1}{2}}a^{-\frac{1}{2}}+2a^{-\frac{1}{2}}b^{\frac{1}{2}}+2a^{-\frac{1}{2}}b^{\frac{1}{2}}+4b^{\frac{1}{2}}b^{\frac{1}{2}}$

$= a^{-1}+4a^{-\frac{1}{2}}b^{\frac{1}{2}}+4b = \dfrac{1}{a}+\dfrac{4b^{\frac{1}{2}}}{a^{\frac{1}{2}}}+4b$

101. $x^{\frac{2}{5}} = \sqrt[5]{x^2}$ or $\left(\sqrt[5]{x}\right)^2$

103. $3y^{\frac{2}{5}} = 3\left(\sqrt[5]{y^2}\right)$ or $3\left(\sqrt[5]{y}\right)^2$

Chapter 6 Practice Test

1. $(a^2b^3)(ab^4)^2 = a^2b^3a^2(b^4)^2 = a^2b^3a^2b^8 = a^4b^{11}$

3. $(-3ab^{-2})^{-1}(-2x^{-1}y)^2 = (-3)^{-1}a^{-1}(b^{-2})^{-1}(-2)^2(x^{-1})^2y^2 = \frac{1}{-3}a^{-1}b^2 \cdot 4x^{-2}y^2$

$$= -\frac{4}{3}a^{-1}b^2x^{-2}y^2 = -\frac{4}{3} \cdot \frac{1}{a} \cdot \frac{b^2}{1} \cdot \frac{1}{x^2} \cdot \frac{y^2}{1} = -\frac{4b^2y^2}{3ax^2}$$

5. $\left(\frac{5r^{-1}s^{-3}}{3rs^2}\right)^{-2} = \left(\frac{5r^{-2}s^{-5}}{3}\right)^{-2} = \frac{5^{-2}(r^{-2})^{-2}(s^{-5})^{-2}}{3^{-2}} = \frac{\frac{1}{5^2}r^4s^{10}}{\frac{1}{3^2}} = \frac{\frac{1}{25}r^4s^{10}}{\frac{1}{9}} = \frac{9}{25}r^4s^{10}$

7. $\frac{27^{\frac{2}{3}}3^{-4}}{9^{-\frac{1}{2}}} = \frac{(3^3)^{\frac{2}{3}}3^{-4}}{(3^2)^{-\frac{1}{2}}} = \frac{3^2 3^{-4}}{3^{-1}} = \frac{3^{-2}}{3^{-1}} = 3^{-1} = \frac{1}{3}$

9. $\frac{x^{-3}+x^{-1}}{yx^{-2}} = \frac{\frac{1}{x^3}+\frac{1}{x}}{\frac{y}{x^2}} = \frac{x^3\left(\frac{1}{x^3}+\frac{1}{x}\right)}{x^3\left(\frac{y}{x^2}\right)} = \frac{\frac{x^3}{1}\cdot\frac{1}{x^3}+\frac{x^3}{1}\cdot\frac{1}{x}}{\frac{x^3}{1}\cdot\frac{y}{x^2}} = \frac{1+x^2}{xy}$

11. $\left(x^{-\frac{1}{2}}+3x^{\frac{1}{2}}\right)^2 = \left(x^{-\frac{1}{2}}+3x^{\frac{1}{2}}\right)\left(x^{-\frac{1}{2}}+3x^{\frac{1}{2}}\right)$

$$= x^{-\frac{1}{2}}x^{-\frac{1}{2}}+3x^{\frac{1}{2}}x^{-\frac{1}{2}}+3x^{\frac{1}{2}}x^{-\frac{1}{2}}+9x^{\frac{1}{2}}x^{\frac{1}{2}} = x^{-1}+3+3+9x$$

$$= x^{-1}+6+9x = \frac{1}{x}+6+9x = \frac{1}{x}+\frac{6x}{x}+\frac{9x^2}{x} = \frac{1+6x+9x^2}{x} = \frac{(1+3x)^2}{x}$$

13. $3a^{\frac{2}{3}} = 3\left(\sqrt[3]{a^2}\right)$ or $3\left(\sqrt[3]{a}\right)^2$

15. There are 3,600 seconds in one hour. Therefore, in one hour light travels $(186,000)(3,600)$ miles.

$$(186,000)(3,600) = (1.86 \times 10^5)(3.6 \times 10^3)$$
$$= (1.86)(3.6) \times (10^5)(10^3)$$
$$= 6.696 \times 10^8$$
$$= 669,600,000 \text{ miles}$$

Cumulative Review: Chapters 4–6

1. $\frac{18a^3b^2}{16a^5b} = \frac{9b}{8a^2}$

162

3. $\dfrac{2a^3 - 5a^2b - 3ab^2}{a^4 - 2a^3b - 3a^2b^2} = \dfrac{a(2a^2 - 5ab - 3b^2)}{a^2(a^2 - 2ab - 3b^2)} = \dfrac{(2a+b)(a-3b)}{a(a+b)(a-3b)} = \dfrac{2a+b}{a(a+b)}$

5. $\dfrac{3xy^2}{5a^3b} \div \dfrac{21x^3y}{25ab^3} = \dfrac{3xy^2}{5a^3b} \cdot \dfrac{25ab^3}{21x^3y} = \dfrac{5yb^2}{7a^2x^2}$

7. $\dfrac{3x}{2y} + \dfrac{2y}{3x} = \dfrac{3x(3x)}{2y(3x)} + \dfrac{2y(2y)}{3x(2y)} = \dfrac{9x^2}{6xy} + \dfrac{4y^2}{6xy} = \dfrac{9x^2 + 4y^2}{6xy}$

9. $\left(\dfrac{2x^3 - 2x^2 - 24x}{x+2}\right)\left(\dfrac{x+2}{4x^2 + 12x}\right) = \dfrac{2x(x^2 - x - 12)}{x+2} \cdot \dfrac{x+2}{4x(x+3)}$

$= \dfrac{2x(x-4)(x+3)}{x+2} \cdot \dfrac{x+2}{4x(x+3)} = \dfrac{x-4}{2}$

11. $\dfrac{2}{x-5} - \dfrac{3}{x-2} = \dfrac{2(x-2)}{(x-5)(x-2)} - \dfrac{3(x-5)}{(x-2)(x-5)} = \dfrac{2x-4}{(x-5)(x-2)} - \dfrac{3x-15}{(x-5)(x-2)}$

$= \dfrac{2x - 4 - (3x - 15)}{(x-5)(x-2)} = \dfrac{2x - 4 - 3x + 15}{(x-5)(x-2)} = \dfrac{-x + 11}{(x-5)(x-2)}$

13. $\dfrac{x^3 + x^2y}{2x^2 + xy} \div \left[\left(\dfrac{x^2 + 2xy - 3y^2}{2x^2 - xy - y^2}\right)(x+y)\right] = \dfrac{x^3 + x^2y}{2x^2 + xy} \div \left[\dfrac{(x-y)(x+3y)}{(x-y)(2x+y)} \cdot \dfrac{x+y}{1}\right]$

$= \dfrac{x^2(x+y)}{x(2x+y)} \div \dfrac{(x+3y)(x+y)}{2x+y}$

$= \dfrac{x^2(x+y)}{x(2x+y)} \cdot \dfrac{2x+y}{(x+3y)(x+y)} = \dfrac{x}{x+3y}$

15. $\dfrac{x - \dfrac{2}{x}}{\dfrac{1}{2} - x} = \dfrac{2x\left(x - \dfrac{2}{x}\right)}{2x\left(\dfrac{1}{2} - x\right)} = \dfrac{2x^2 - \dfrac{2x}{1} \cdot \dfrac{2}{x}}{\dfrac{2x}{1} \cdot \dfrac{1}{2} - 2x^2} = \dfrac{2x^2 - 4}{x - 2x^2} = \dfrac{2(x^2 - 2)}{x(1 - 2x)}$

17. $\dfrac{2}{x} - \dfrac{1}{2} = 2 - \dfrac{1}{x}$

$2x\left(\dfrac{2}{x} - \dfrac{1}{2}\right) = 2x\left(2 - \dfrac{1}{x}\right)$

$\dfrac{2x}{1} \cdot \dfrac{2}{x} - \dfrac{2x}{1} \cdot \dfrac{1}{2} = 2x(2) - \dfrac{2x}{1} \cdot \dfrac{1}{x}$

$4 - x = 4x - 2$

$4 = 5x - 2$

CHECK $x = \dfrac{6}{5}$:

$\dfrac{2}{x} - \dfrac{1}{2} = 2 - \dfrac{1}{x}$

$\dfrac{2}{\frac{6}{5}} - \dfrac{1}{2} \overset{?}{=} 2 - \dfrac{1}{\frac{6}{5}}$

$\dfrac{2}{1} \cdot \dfrac{5}{6} - \dfrac{1}{2} \overset{?}{=} 2 - 1 \cdot \dfrac{5}{6}$

$\dfrac{10}{6} - \dfrac{1}{2} \overset{?}{=} 2 - \dfrac{5}{6}$

$$6 = 5x$$

$$\frac{6}{5} = x$$

$$\frac{10}{6} - \frac{3}{6} \overset{?}{=} \frac{12}{6} - \frac{5}{6}$$

$$\frac{7}{6} \overset{\checkmark}{=} \frac{7}{6}$$

19. $\quad \dfrac{x}{2} - \dfrac{x+1}{3} > \dfrac{2}{3}$

$$6\left(\frac{x}{2} - \frac{x+1}{3}\right) > 6 \cdot \frac{2}{3}$$

$$\overset{3}{\frac{\cancel{6}}{1}} \cdot \frac{x}{\cancel{2}} - \overset{2}{\frac{\cancel{6}}{1}} \cdot \frac{x+1}{\cancel{3}} > \overset{2}{\frac{\cancel{6}}{1}} \cdot \frac{2}{\cancel{3}}$$

$$3x - 2(x+1) > 4$$

$$3x - 2x - 2 > 4$$

$$x - 2 > 4$$

$$x > 6$$

21. $\quad \dfrac{3}{x-2} + \dfrac{5}{x+1} = \dfrac{1}{x^2 - x - 2}$

$$\frac{3}{x-2} + \frac{5}{x+1} = \frac{1}{(x-2)(x+1)}$$

$$(x-2)(x+1)\left(\frac{3}{x-2} + \frac{5}{x+1}\right) = (x-2)(x+1)\left(\frac{1}{(x-2)(x+1)}\right)$$

$$\frac{\cancel{(x-2)}(x+1)}{1} \cdot \frac{3}{\cancel{x-2}} + \frac{(x-2)\cancel{(x+1)}}{1} \cdot \frac{5}{\cancel{x+1}} = \frac{\cancel{(x-2)(x+1)}}{1} \cdot \frac{1}{\cancel{(x-2)(x+1)}}$$

$$3(x+1) + 5(x-2) = 1$$

$$3x + 3 + 5x - 10 = 1$$

$$8x - 7 = 1$$

$$8x = 8$$

$$x = 1$$

CHECK x = 1:

$$\frac{3}{x-2} + \frac{5}{x+1} = \frac{1}{x^2 - x - 2}$$

$$\frac{3}{1-2} + \frac{5}{1+1} \overset{?}{=} \frac{1}{1^2 - 1 - 2}$$

$$\frac{3}{-1} + \frac{5}{2} \overset{?}{=} \frac{1}{1 - 1 - 2}$$

$$-\frac{6}{2} + \frac{5}{2} \overset{?}{=} \frac{1}{-2}$$

$$-\frac{1}{2} \overset{\checkmark}{=} -\frac{1}{2}$$

23. $\quad 2a + 3b = 5b - 4a$

$$2a + 3b + 4a = 5b - 4a + 4a$$

$$6a + 3b = 5b$$

$$6a + 3b - 3b = 5b - 3b$$

$$6a = 2b$$

$$\frac{6a}{6} = \frac{\cancel{2}b}{\cancel{6}3}$$

$$a = \frac{b}{3}$$

164

25.
$$\frac{x-y}{y} = x$$

$$\cancel{y}\left(\frac{x-y}{\cancel{y}}\right) = yx$$

$$x - y = yx$$

$$x - y + y = yx + y$$

$$x = yx + y$$

$$x = y(x+1)$$

$$\frac{x}{x+1} = \frac{y\cancel{(x+1)}}{\cancel{x+1}}$$

$$\frac{x}{x+1} = y$$

27. Let x = # of foreign-made cars in the town.

$$\frac{x}{1,200} = \frac{5}{6}$$

$$\cancel{1,200}\left(\frac{x}{\cancel{1,200}}\right) = \overset{200}{\cancel{1,200}}\left(\frac{5}{\cancel{6}}\right) \qquad \text{CHECK: } \frac{1,000}{1,200} = \frac{5\cancel{(200)}}{6\cancel{(200)}} \overset{\checkmark}{=} \frac{5}{6}$$

$$x = 1,000$$

Thus, there are 1,000 foreign-made cars in the town. Since there are 1,200 American-made cars in the town, there are 1,000 + 1,200 = 2,200 cars in the town altogether.

29. Let x = # of hours that they work together.

$$\frac{x}{4} + \frac{x}{4\frac{1}{2}} = 1$$

$$\frac{x}{4} + \frac{x}{\frac{9}{2}} = 1$$

$$\frac{x}{4} + \frac{2x}{9} = 1$$

$$36\left(\frac{x}{4} + \frac{2x}{9}\right) = 36 \cdot 1$$

$$\overset{9}{\cancel{\frac{36}{1}}} \cdot \frac{x}{\cancel{4}} + \overset{4}{\cancel{\frac{36}{1}}} \cdot \frac{2x}{\cancel{9}} = 36$$

$$9x + 8x = 36$$

$$17x = 36$$

$$x = \frac{36}{17}$$

Thus, it would take Carmen and Judy $\frac{36}{17}$ hours to paint the room if they worked together.

31. $5x - 4y = 10$

To find the x-intercept, set y = 0:

$5x - 4(0) = 10$, so

$5x = 10$ or $x = 2$.

To find the y-intercept, set x = 0:

5(0) − 4y = 10, so

−4y = 10 or $y = -\dfrac{5}{2}$

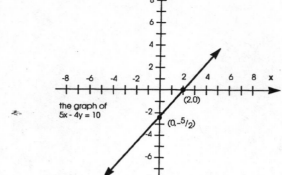

the graph of
5x - 4y = 10

(2,0)

(0,−5/2)

33.   5y = 3x − 10

To find the x-intercept, set y = 0:

5(0) = 3x − 10, so

0 = 3x − 10 or $x = \dfrac{10}{3}$.

To find the y-intercept, set x = 0:

5y = 3(0) − 10, so

5y = −10 or y = −2.

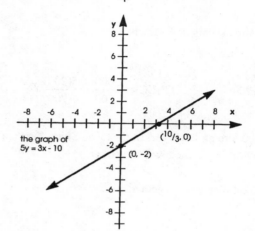

the graph of
5y = 3x - 10

($^{10}/_3$, 0)

(0, -2)

35.   3x − 2 = 8

3x = 10

$x = \dfrac{10}{3}$

The graph of this equation is a vertical line, $\dfrac{10}{3}$ units to the right of the y-axis.

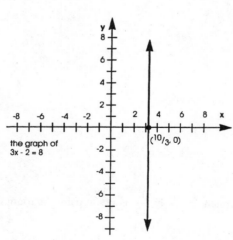

the graph of
3x - 2 = 8

($^{10}/_3$, 0)

37. $m = \dfrac{y_2 - y_1}{x_2 - x_1} = \dfrac{5 - (-3)}{3 - 2} = \dfrac{8}{1} = 8$

39. $y = 5x - 8$. By comparison with $y = mx + b$, it follows that $m = 5$.

41. The slope of the line passing through the points $(2, -1)$ and $(6, 4)$ is $m = \dfrac{4 - (-1)}{6 - 2} = \dfrac{5}{4}$. Therefore, any line parallel to

this one also has $m = \dfrac{5}{4}$.

43. $\dfrac{-2 - a}{a - 2} = 3$

$(a - 2)\left(\dfrac{-2 - a}{a - 2}\right) = (a - 2)(3)$

$\qquad -2 - a = 3a - 6$

$\qquad\qquad -2 = 4a - 6$

$\qquad\qquad\quad 4 = 4a$

$\qquad\qquad\quad 1 = a$

45. $y - y_1 = m(x - x_1)$

$\quad y - 7 = 3(x - (-2))$

$\quad y - 7 = 3(x + 2)$

$\quad y - 7 = 3x + 6$

$\qquad\; y = 3x + 13$

47. $m = \dfrac{1 - 7}{3 - 2} = -\dfrac{6}{1} = -6$

$\qquad y - y_1 = m(x - x_1)$

$\qquad\; y - 7 = -6(x - 2)$

$\qquad\; y - 7 = -6x + 12$

$\qquad\qquad y = -6x + 19$

49. $y = mx + b$

$\quad y = 4x + 2$

51. $3x + 5y = 4$

$\quad 5y = -3x + 4$

$\quad\; y = -\dfrac{3}{5}x + \dfrac{4}{5}$

So the given line has slope $-\dfrac{3}{5}$, which means that the line in question also has slope $-\dfrac{3}{5}$.

$$y - y_1 = m(x - x_1)$$

$$y - (-3) = -\frac{3}{5}(x - 2)$$

$$y + 3 = -\frac{3}{5}(x - 2)$$

$$y + 3 = -\frac{3}{5}x + \frac{6}{5}$$

$$y = -\frac{3}{5}x - \frac{9}{5}$$

$+5$

$\dfrac{-2}{-10} + \dfrac{5}{5}$ $\quad \dfrac{6}{5} - \dfrac{3}{1} = \dfrac{6}{5} - \dfrac{15}{5} = -\dfrac{9}{5}$

-3

$\dfrac{6}{5} + 3 = \dfrac{-9}{5}$

53. (86,450), (80,325)

$$m = \frac{325 - 450}{80 - 86} = \frac{-125}{-6} = \frac{125}{6}$$

$$P - P_1 = m(T - T_1)$$

$$P - 450 = \frac{125}{6}(T - 86)$$

$$P - 450 = \frac{125}{6}T - \frac{5,375}{3}$$

$$P = \frac{125}{6}T - \frac{4,025}{3}$$

When T = 90,

$$P = \frac{125}{6}(90) - \frac{4,025}{3}$$

$$= 1,875 - \frac{4,025}{3}$$

$$= \frac{1,600}{3}$$

$$= 533.\overline{33}$$

Conclude: The daily profit on a 90° F day is $533.33.

55. $\begin{cases} 3x - 2y = 8 \ (\text{as is}) \\ 5x + \ y = 9 \xrightarrow{\text{multiply by 2}} \end{cases}$

$$\begin{array}{r} 3x - 2y = \ 8 \\ 10x + 2y = 18 \\ \hline \text{Add: } 13x \qquad = 26 \\ x = 2 \end{array}$$

Substitute: $3x - 2y = 8$

$$3(2) - 2y = 8$$

$$6 - 2y = 8$$

$$-2y = 2$$

$$y = -1$$

Solution: x = 2, y = -1

CHECK: $5x + y = 9$

$$5(2) + (-1) \overset{?}{=} 9$$

$$10 - 1 \overset{?}{=} 9$$

$$9 \overset{\checkmark}{=} 9$$

57. $\begin{cases} 7u + 5v = 23 \xrightarrow{\text{multiply by 9}} & 63u + 45v = 207 \\ 8u + 9v = 23 \xrightarrow{\text{multiply by } -5} & \underline{-40u - 45v = -115} \end{cases}$

$$\text{Add:} \quad 23u = 92$$
$$u = 4$$

Substitute: $7u + 5v = 23$

$7(4) + 5v = 23$

$28 + 5v = 23$

$5v = -5$

$v = -1$

Solution: $u = 4, v = -1$

CHECK: $8u + 9v = 23$

$8(4) + 9(-1) \overset{?}{=} 23$

$32 - 9 \overset{?}{=} 23$

$23 \overset{\checkmark}{=} 23$

59. $\begin{cases} 2m = 3n - 5 \\ 3n = 2m - 5 \end{cases}$

Substitute the value of $2m$ from the first equation into the second:

$3n = 3n - 5 - 5$

$3n = 3n - 10$

$0 = -10$

Thus, the system is inconsistent.

61. $\begin{cases} \dfrac{2}{3}y - \dfrac{1}{2}x = 6 \xrightarrow{\text{multiply by 30}} \\ \dfrac{4}{5}y - \dfrac{3}{4}x = 6 \xrightarrow{\text{multiply by } -20} \end{cases}$

$$20y - 15x = 180$$
$$\underline{-16y + 15x = -120}$$
$$\text{Add:} \quad 4y = 60$$
$$y = 15$$

Substitute: $\dfrac{2}{3}y - \dfrac{1}{2}x = 6$

$\dfrac{2}{3}(15) - \dfrac{1}{2}x = 6$

$10 - \dfrac{1}{2}x = 6$

$-\dfrac{1}{2}x = -4$

$x = 8$

CHECK: $\dfrac{4}{5}y - \dfrac{3}{4}x = 6$

$\dfrac{4}{5}(15) - \dfrac{3}{4}(8) \overset{?}{=} 6$

$12 - 6 \overset{?}{=} 6$

$6 \overset{\checkmark}{=} 6$

Solution: $x = 8, y = 15$

63. boundary: solid line

x-intercept: 9

y-intercept: 6

test point: (0,0)

$$2x + 3y \geq 18$$

$$2(0) + 3(0) \overset{?}{\geq} 18$$

$$0 + 0 \overset{?}{\geq} 18$$

$$0 \not\geq 18$$

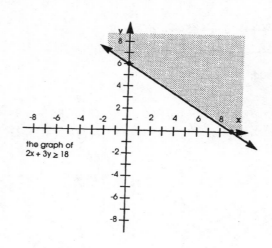

the graph of
$2x + 3y \geq 18$

65. boundary: dotted line

x-intercept: 4

y-intercept: −10

test point: (0,0)

$$2y < 5x - 20$$

$$2(0) \overset{?}{<} 5(0) - 20$$

$$0 \overset{?}{<} 0 - 20$$

$$0 \not< -20$$

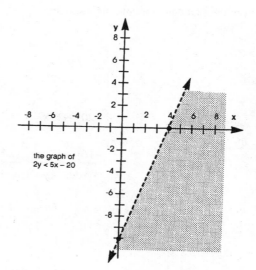

the graph of
$2y < 5x - 20$

67. $(-2x^2y)(-3xy^2)^3 = (-2x^2y)(-3)^3x^3(y^2)^3 = -2x^2y(-27x^3y^6) = (-2)(-27)x^2x^3yy^6 = 54x^5y^7$

69. $\left(\dfrac{\cancel{2xy^2}}{\cancel{4x^2y^3}}\right)^2 = \left(\dfrac{1}{2xy}\right)^2 = \dfrac{1^2}{(2xy)^2} = \dfrac{1}{2^2x^2y^2} = \dfrac{1}{4x^2y^2}$

(handwritten: 2xy)

71. $(x^{-5}y^{-2})(x^{-7}y) = x^{-5}x^{-7}y^{-2}y = x^{-5+(-7)}y^{-2+1} = x^{-12}y^{-1} = \dfrac{1}{x^{12}} \cdot \dfrac{1}{y} = \dfrac{1}{x^{12}y}$

73. $\dfrac{r^{-3}s^{-2}}{r^{-2}s^0} = \dfrac{r^{-3}}{r^{-2}} \cdot \dfrac{s^{-2}}{s^0} = r^{-3-(-2)}s^{-2-0} = r^{-1}s^{-2} = \dfrac{1}{r} \cdot \dfrac{1}{s^2} = \dfrac{1}{rs^2}$

75. $(-2r^{-3}s)^{-1}(-4r^{-2}s^{-3})^2 = (-2)^{-1}(r^{-3})^{-1}s^{-1}(-4)^2(r^{-2})^2(s^{-3})^2 = \dfrac{1}{-2}r^3s^{-1} \cdot 16r^{-4}s^{-6}$

$$= \dfrac{1}{-2} \cdot 16r^3r^{-4}s^{-1}s^{-6} = -8r^{3+(-4)}s^{-1-6} = -8r^{-1}s^{-7} = -\dfrac{8}{1} \cdot \dfrac{1}{r} \cdot \dfrac{1}{s^7} = \dfrac{-8}{rs^7}$$

77. $(x^{-1} - y^{-1})(x + y) = x^{-1}x + x^{-1}y - y^{-1}x - y^{-1}y = 1 + \dfrac{1}{x} \cdot y - \dfrac{1}{y} \cdot x - 1$

$$= \dfrac{y}{x} - \dfrac{x}{y} = \dfrac{y^2}{xy} - \dfrac{x^2}{xy} = \dfrac{y^2 - x^2}{xy}$$

79. $56,429.32 = 5.642932 \times 10^4$

81. 1 day = 24 hours

24 hours = 24(60) minutes

24(60) minutes = 24(60)(60) seconds

So 1 day = 24(60)(60) = 8.64×10^4 seconds. Since light travels 186,000 miles per second, light travels

$(186,000)(8.64 \times 10^4) = (1.86 \times 10^5)(8.64 \times 10^4)$

$$= (1.86)(8.64) \times 10^5 10^4 = 16.0704 \times 10^9$$

$$= 1.60704 \times 10^{10} \text{ miles in one day}$$

83. $(-1,000)^{\frac{1}{3}} = \sqrt[3]{-1,000} = -10$

85. $a^{\frac{2}{3}}a^{-\frac{1}{2}} = a^{\frac{2}{3} - \frac{1}{2}} = a^{\frac{1}{6}}$

87. $\left[(81)^{-\frac{1}{3}} 3^2 \right]^{-3} = \left(81^{-\frac{1}{3}} \right)^{-3} (3^2)^{-3} = 81 \cdot 3^{-6} = 3^4 3^{-6} = 3^{-2} = \dfrac{1}{3^2} = \dfrac{1}{9}$

89. $x^{\frac{3}{4}} = \sqrt[4]{x^3} \text{ or } \left(\sqrt[4]{x} \right)^3$

Cumulative Practice Test: Chapters 4–6

1. (a) $\left(\dfrac{25x^2 - 9y^2}{x - y} \right) \left(\dfrac{2x^2 + xy - 3y^2}{10x^2 + 9xy - 9y^2} \right)$

$$= \dfrac{(5x - 3y)(5x + 3y)}{x - y} \cdot \dfrac{(2x + 3y)(x - y)}{(2x + 3y)(5x - 3y)} = 5x + 3y$$

(b) $\dfrac{2y}{2x - y} + \dfrac{4x}{y - 2x} = \dfrac{2y}{2x - y} + \dfrac{4x(-1)}{(y - 2x)(-1)} = \dfrac{2y}{2x - y} - \dfrac{4x}{2x - y}$

$$= \dfrac{2y - 4x}{2x - y} = \dfrac{2(y - 2x)}{2x - y} = \dfrac{2(-1)(2x - y)}{2x - y} = -2$$

171

(c) $\dfrac{3x}{x^2-10x+21} - \dfrac{2}{x^2-8x+15} = \dfrac{3x}{(x-3)(x-7)} - \dfrac{2}{(x-3)(x-5)}$

$$= \dfrac{3x(x-5)}{(x-3)(x-7)(x-5)} - \dfrac{2(x-7)}{(x-3)(x-5)(x-7)}$$

$$= \dfrac{3x^2-15x}{(x-3)(x-7)(x-5)} - \dfrac{2x-14}{(x-3)(x-5)(x-7)}$$

$$= \dfrac{3x^2-15x-(2x-14)}{(x-3)(x-7)(x-5)} = \dfrac{3x^2-15x-2x+14}{(x-3)(x-7)(x-5)}$$

$$= \dfrac{3x^2-17x+14}{(x-3)(x-7)(x-5)} = \dfrac{(3x-14)(x-1)}{(x-3)(x-7)(x-5)}$$

3. (a) $\dfrac{2}{x+1} + \dfrac{3}{2x} = \dfrac{6}{x^2+x}$

$\dfrac{2}{x+1} + \dfrac{3}{2x} = \dfrac{6}{x(x+1)}$

$2x(x+1)\left(\dfrac{2}{x+1} + \dfrac{3}{2x}\right) = 2x(x+1)\left(\dfrac{6}{x(x+1)}\right)$

$\dfrac{2x(x+1)}{1} \cdot \dfrac{2}{x+1} + \dfrac{2x(x+1)}{1} \cdot \dfrac{3}{2x} = \dfrac{2x(x+1)}{1} \cdot \dfrac{6}{x(x+1)}$

$4x + 3(x+1) = 12$ CHECK $x = \dfrac{9}{7}$:

$4x + 3x + 3 = 12$ $\dfrac{2}{x+1} + \dfrac{3}{2x} = \dfrac{6}{x^2+x}$

$7x + 3 = 12$ $\dfrac{2}{\frac{9}{7}+1} + \dfrac{3}{2\left(\frac{9}{7}\right)} \overset{?}{=} \dfrac{6}{\left(\frac{9}{7}\right)^2 + \frac{9}{7}}$

$7x = 9$ $\dfrac{2}{\frac{16}{7}} + \dfrac{3}{\frac{18}{7}} \overset{?}{=} \dfrac{6}{\frac{81}{49} + \frac{9}{7}}$

     $\dfrac{14}{16} + \dfrac{21}{18} \overset{?}{=} \dfrac{6}{\frac{144}{49}}$

$x = \dfrac{9}{7}$ $\dfrac{7}{8} + \dfrac{7}{6} \overset{?}{=} \dfrac{49}{24}$

     $\dfrac{21}{24} + \dfrac{28}{24} \overset{?}{=} \dfrac{49}{24}$

     $\dfrac{49}{24} \overset{\checkmark}{=} \dfrac{49}{24}$

(b)
$$\frac{x}{3} - \frac{x+2}{7} < 4$$

$$21\left(\frac{x}{3} - \frac{x+2}{7}\right) < 21 \cdot 4$$

$$\frac{\overset{7}{\cancel{21}}}{1} \cdot \frac{x}{\cancel{3}} - \frac{\overset{3}{\cancel{21}}}{1} \cdot \frac{x+2}{\cancel{7}} < 84$$

$$7x - 3(x+2) < 84$$

$$7x - 3x - 6 < 84$$

$$4x - 6 < 84$$

$$4x < 90$$

$$x < \frac{45}{2}$$

(c)
$$\frac{5}{x+3} + 2 = \frac{x+8}{x+3}$$

$$(x+3)\left(\frac{5}{x+3} + 2\right) = (x+3)\left(\frac{x+8}{x+3}\right)$$

$$\frac{\cancel{x+3}}{1} \cdot \frac{5}{\cancel{x+3}} + (x+3)(2) = \frac{\cancel{x+3}}{1} \cdot \frac{x+8}{\cancel{x+3}}$$

$$5 + 2(x+3) = x + 8$$

$$5 + 2x + 6 = x + 8$$

$$2x + 11 = x + 8$$

$$x + 11 = 8$$

$$x = -3$$

CHECK $x = -3$:

$$\frac{5}{x+3} + 2 = \frac{x+8}{x+3}$$

$$\frac{5}{-3+3} + 2 \overset{?}{=} \frac{-3+8}{-3+3}$$

$$\frac{5}{0} + 2 \neq \frac{5}{0}$$

since we cannot divide by 0. Thus, the equation has no solution.

5. Let x = # of hours Carol and Joe need if they work together.

$$\frac{x}{6} + \frac{x}{5\frac{1}{2}} = 1$$

$$\frac{x}{6} + \frac{x}{\frac{11}{2}} = 1$$

$$\frac{x}{6} + \frac{2x}{11} = 1$$

$$66\left(\frac{x}{6} + \frac{2x}{11}\right) = 66 \cdot 1$$

$$\frac{\overset{11}{\cancel{66}}}{1} \cdot \frac{x}{\cancel{6}} + \frac{\overset{6}{\cancel{66}}}{1} \cdot \frac{2x}{\cancel{11}} = 66$$

$$11x + 12x = 66$$

$$23x = 66$$

$$x = \frac{66}{23}$$

Thus, they can process the forms in $\frac{66}{23}$ hours if they work together.

7. $$m = \frac{y_2 - y_1}{x_2 - x_1} = \frac{-4 - (-3)}{3 - 2} = \frac{-4 + 3}{3 - 2} = \frac{-1}{1} = -1$$

9. (40, 30), (60, 25)
$$m = \frac{25 - 30}{60 - 40} = \frac{-5}{20} = -\frac{1}{4}$$

$$B - B_1 = m(A - A_1)$$

$$B - 30 = -\frac{1}{4}(A - 40)$$

$$B - 30 = -\frac{1}{4}A + 10$$

$$B = -\frac{1}{4}A + 40$$

When A = 48,

$$B = -\frac{1}{4}(48) + 40$$

$$= -12 + 40$$

$$= 28$$

11. boundary: dotted line
x-intercept: 6
y-intercept: 3
test point: (0,0)
$$3x + 6y > 18$$
$$3(0) + 6(0) \overset{?}{>} 18$$
$$0 + 0 \overset{?}{>} 18$$
$$0 \not> 18$$

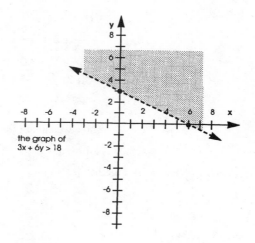

the graph of
$3x + 6y > 18$

13. $0.000034 = 3.4 \times 10^{-5}$

15. $$(-128)^{-\frac{3}{7}} = \frac{1}{(-128)^{\frac{3}{7}}} = \frac{1}{\left(\sqrt[7]{-128}\right)^3} = \frac{1}{(-2)^3} = \frac{1}{-8} = -\frac{1}{8}$$

17. Let's assume that Kyle leaves his home at 9:00 a.m. He travels at 50 mph and reaches his first destination, 50 miles away from his home, at 10:00 a.m. He stays at this first meeting until 12:00 noon. Then he travels for an hour at 40

mph to pick up his partner at a point 10 miles from Kyle's home. Together, they travel at 60 mph for an hour to attend another meeting at 2:00 p.m., 70 miles from Kyle's home. When this meeting ends at 3:00 p.m., they travel at $46\frac{2}{3}$ mph and return to Kyle's home, where they arrive at 4:30 p.m.

CHAPTER 7
RADICAL EXPRESSIONS

Exercises 7.1

1. $\sqrt[3]{64} = 4$

5. $\sqrt[8]{-1}$ is not a real number

7. $-\sqrt[9]{-1} = -(-1) = 1$

11. $-\sqrt[4]{1296} = -6$

13. $\sqrt[8]{256} = 2$

17. $9^{\frac{1}{2}} = \sqrt{9} = 3$

19. $(-125)^{\frac{1}{3}} = \sqrt[3]{-125} = -5$

23. $\sqrt[3]{8^2} = \left(\sqrt[3]{8}\right)^2 = 2^2 = 4$

25. $\left(\sqrt[4]{81}\right)^3 = 3^3 = 3$

29. $\sqrt{(-16)^2} = \sqrt{256} = 16$

31. $\sqrt{-16}$ is not a real number

33. $-\left(\sqrt{16}\right)^2 = -16$

35. $\sqrt[n]{3^{2n}} = \sqrt[n]{\left(3^2\right)^n} = \sqrt[n]{9^n} = 9$

39. $\sqrt{x^2 + y^2} = \left(x^2 + y^2\right)^{\frac{1}{2}}$

41. $\sqrt[5]{5a^2b^3} = \left(5a^2b^3\right)^{\frac{1}{5}}$

45. $5\sqrt[3]{(x-y)^2} = 5(x-y)^{\frac{2}{3}}$

49. $\sqrt[n]{x^{5n+1}y^{2n-1}} = \left(x^{5n+1}y^{2n-1}\right)^{\frac{1}{n}}$

51. $x^{\frac{1}{3}} = \sqrt[3]{x}$

55. $(-a)^{\frac{2}{3}} = \sqrt[3]{(-a)^2}$

57. $-a^{\frac{2}{3}} = -\sqrt[3]{a^2}$

61. $\left(x^2 + y^2\right)^{\frac{1}{2}} = \sqrt{x^2 + y^2}$

65. $\sqrt[m]{\sqrt[n]{a}} = \sqrt[m]{a^{\frac{1}{n}}} = \left(a^{\frac{1}{n}}\right)^{\frac{1}{m}} = a^{\frac{1}{mn}}$

$\sqrt[mn]{a} = a^{\frac{1}{mn}}$

So $\sqrt[m]{\sqrt[n]{a}} = \sqrt[mn]{a}$, since each is equal to $a^{\frac{1}{mn}}$.

66. (a) $\sqrt[3]{\sqrt[4]{6}} = \sqrt[3 \cdot 4]{6} = \sqrt[12]{6}$

(b) $\sqrt{\sqrt[3]{x^2y}} = \sqrt[2 \cdot 3]{x^2y} = \sqrt[6]{x^2y}$

(c) $\sqrt[4]{\sqrt[3]{\sqrt{x}}} = \sqrt[4 \cdot 3]{\sqrt{x}} = \sqrt[12]{\sqrt{x}} = \sqrt[12 \cdot 2]{x} = \sqrt[24]{x}$

67. $\sqrt{(-4)^2}$ is a real number (it equals 4), but $\left(\sqrt{-4}\right)^2$ is the square of a number that is not real. Therefore, the two cannot be equal.

69. $\dfrac{x^2y^{-3}}{x^{-2}y^3} = x^{2-(-2)}y^{-3-3} = x^4y^{-6} = x^4 \cdot \dfrac{1}{y^6} = \dfrac{x^4}{y^6}$

71. $\quad x = \dfrac{2-y}{3y}$

$3y(x) = \cancel{3y}\left(\dfrac{2-y}{\cancel{3y}}\right)$

$\quad 3xy = 2 - y$

$\quad 3xy + y = 2$

$\quad (3x+1)y = 2$

$\quad y = \dfrac{2}{3x+1}$

73. $(1.6)(186{,}000) = (1.6)(1.86 \times 10^5) = 2.976 \times 10^5$ km/sec

Exercises 7.2

1. $\sqrt{56} = \sqrt{4 \cdot 14} = \sqrt{4}\sqrt{14} = 2\sqrt{14}$

3. $\sqrt{48} = \sqrt{16 \cdot 3} = \sqrt{16}\sqrt{3} = 4\sqrt{3}$

7. $\sqrt{8}\sqrt{18} = \sqrt{8 \cdot 18} = \sqrt{144} = 12$

9. $\sqrt{64x^8} = \sqrt{64}\sqrt{x^8} = 8x^4$

13. $\sqrt{128x^{60}} = \sqrt{64x^{60}}\sqrt{2} = 8x^{30}\sqrt{2}$

15. $\sqrt[4]{128x^{60}} = \sqrt[4]{16x^{60}}\sqrt[4]{8} = 2x^{15}\sqrt[4]{8}$

17. $\sqrt[5]{128x^{60}} = \sqrt[5]{32x^{60}}\sqrt[5]{4} = 2x^{12}\sqrt[5]{4}$

21. $\sqrt{32a^2b^4} = \sqrt{16a^2b^4}\sqrt{2} = 4ab^2\sqrt{2}$

25. $\sqrt{x^3y}\sqrt{xy^3} = \sqrt{x^3y \cdot xy^3} = \sqrt{x^4y^4} = x^2y^2$

27. $\sqrt{\dfrac{1}{2}} = \dfrac{\sqrt{1}}{\sqrt{2}} = \dfrac{1}{\sqrt{2}} = \dfrac{1}{\sqrt{2}} \cdot \dfrac{\sqrt{2}}{\sqrt{2}} = \dfrac{\sqrt{2}}{2}$

29. $\dfrac{\sqrt{x}}{\sqrt{5}} = \dfrac{\sqrt{x}\sqrt{5}}{\sqrt{5}\sqrt{5}} = \dfrac{\sqrt{5x}}{5}$

33. $\dfrac{1}{\sqrt{75}} = \dfrac{1}{\sqrt{75}} \cdot \dfrac{\sqrt{75}}{\sqrt{75}} = \dfrac{\sqrt{75}}{75} = \dfrac{\sqrt{25}\sqrt{3}}{75} = \dfrac{\cancel{5}\sqrt{3}}{\cancel{75}_{15}} = \dfrac{\sqrt{3}}{15}$

37. $\sqrt[3]{81x^8y^7} = \sqrt[3]{27x^6y^6}\sqrt[3]{3x^2y} = 3x^2y^2\sqrt[3]{3x^2y}$

41. $\sqrt[6]{(x+y^2)^6} = x + y^2$

43. $\sqrt[4]{x^4 - y^4}$ cannot be simplified

47. $\left(3a\sqrt[3]{2b^4}\right)\left(2a^2\sqrt[3]{4b^2}\right) = (3a)(2a^2)\sqrt[3]{2b^4 \cdot 4b^2} = 6a^3\sqrt[3]{8b^6} = 6a^3 \cdot 2b^2 = 12a^3b^2$

49. $\sqrt[3]{\dfrac{x^3y^6}{8}} = \dfrac{\sqrt[3]{x^3y^6}}{\sqrt[3]{8}} = \dfrac{xy^2}{2}$

53. $\dfrac{\sqrt{54xy}}{\sqrt{2xy}} = \sqrt{\dfrac{\cancel{54xy}^{27}}{\cancel{2xy}}} = \sqrt{27} = \sqrt{9 \cdot 3} = \sqrt{9}\sqrt{3} = 3\sqrt{3}$

57. $\sqrt{\dfrac{3\cancel{xy}}{5x\cancel{y}}} = \sqrt{\dfrac{3}{5x}} = \dfrac{\sqrt{3}}{\sqrt{5x}} = \dfrac{\sqrt{3}\sqrt{5x}}{\sqrt{5x}\sqrt{5x}} = \dfrac{\sqrt{15x}}{5x}$

61. $\sqrt[3]{\dfrac{3}{2}} = \dfrac{\sqrt[3]{3}}{\sqrt[3]{2}} = \dfrac{\sqrt[3]{3}\sqrt[3]{4}}{\sqrt[3]{2}\sqrt[3]{4}} = \dfrac{\sqrt[3]{12}}{\sqrt[3]{8}} = \dfrac{\sqrt[3]{12}}{2}$

65. $\sqrt[4]{\dfrac{9}{4}} = \dfrac{\sqrt[4]{9}}{\sqrt[4]{4}} = \dfrac{\sqrt[4]{9}\,\sqrt[4]{4}}{\sqrt[4]{4}\,\sqrt[4]{4}} = \dfrac{\sqrt[4]{36}}{\sqrt[4]{16}} = \dfrac{\sqrt[4]{36}}{2}$

69. $\dfrac{3a^2\sqrt{a^2x^5}}{3a^3\;9a^5\sqrt[6]{a^6x}} = \dfrac{1}{3a^3}\sqrt[6]{\dfrac{a^2x^5\,x^4}{a^6\,x}\;^{4}_{a}} = \dfrac{1}{3a^3}\sqrt{\dfrac{x^4}{a^4}} = \dfrac{1}{3a^3}\dfrac{\sqrt{x^4}}{\sqrt{a^4}} = \dfrac{1}{3a^3}\cdot\dfrac{x^2}{a^2} = \dfrac{x^2}{3a^5}$

73. $\sqrt[12]{a^6} = (a^6)^{\frac{1}{12}} = a^{\frac{6}{12}} = a^{\frac{1}{2}} = \sqrt{a}$

75. $\left(\sqrt[3]{x}\right)\left(\sqrt[4]{x^3}\right) = x^{\frac{1}{3}}x^{\frac{3}{4}} = x^{\frac{1}{3}+\frac{3}{4}} = x^{\frac{13}{12}} = \left(\sqrt[12]{x^{13}}\right) = \sqrt[12]{x^{12}}\;\sqrt[12]{x} = x\sqrt[12]{x}$

79. $\sqrt[n]{x^{5n}y^{3n}} = \left(x^{5n}y^{3n}\right)^{\frac{1}{n}} = \left(x^{5n}\right)^{\frac{1}{n}}\left(y^{3n}\right)^{\frac{1}{n}} = x^{\frac{5n}{n}}y^{\frac{3n}{n}} = x^5y^3$

81. (a)   The step $\left(\sqrt{-2}\right)^2 = \sqrt{(-2)^2}$ is wrong since $\sqrt{-2}$ is not a real number and $\sqrt{a^n} = \left(\sqrt{a}\right)^n$ only applies if $\sqrt{a}$ is real. (Note $\sqrt{(-2)^2}$ is a real number.)

(b)   The step $\sqrt[6]{(-8)^2} = \sqrt[3]{-8}$ is wrong since -8 is negative. Property 3 of radicals requires the radicand to be nonnegative in order to factor 2 from both the index and the exponent of the radicand.

83. the distributive property

85. Let x = the number

$$\frac{2}{3}x = \frac{3}{5}x + 5$$

$$15\left(\frac{2}{3}x\right) = 15\left(\frac{3}{5}x + 5\right)$$

$$10x = 9x + 75$$

$$x = 75$$

Exercises 7.3

1. $5\sqrt{3} - \sqrt{3} = (5-1)\sqrt{3} = 4\sqrt{3}$

5. $8\sqrt{3} - \left(4\sqrt{3} - 2\sqrt{6}\right) = 8\sqrt{3} - 4\sqrt{3} + 2\sqrt{6} = (8-4)\sqrt{3} + 2\sqrt{6} = 4\sqrt{3} + 2\sqrt{6}$

7. $2\sqrt{3} - 2\sqrt{5} - \left(\sqrt{3} - \sqrt{5}\right) = 2\sqrt{3} - 2\sqrt{5} - \sqrt{3} + \sqrt{5} = (2-1)\sqrt{3} + (-2+1)\sqrt{5} = \sqrt{3} - \sqrt{5}$

11. $5a\sqrt{b} - 3a^3\sqrt{b} + 2a\sqrt{b} = (5+2)a\sqrt{b} - 3a^3\sqrt{b} = 7a\sqrt{b} - 3a^3\sqrt{b} = (7a - 3a^3)\sqrt{b}$

15. $\left(7 - 3\sqrt[3]{a}\right) - \left(6 - \sqrt[3]{a}\right) = 7 - 3\sqrt[3]{a} - 6 + \sqrt[3]{a} = 7 - 6 + (-3+1)\sqrt[3]{a} = 1 - 2\sqrt[3]{a}$

17. $\sqrt{12} - \sqrt{27} = \sqrt{4\cdot3} - \sqrt{9\cdot3} = \sqrt{4}\sqrt{3} - \sqrt{9}\sqrt{3} = 2\sqrt{3} - 3\sqrt{3} = (2-3)\sqrt{3} = -\sqrt{3}$

21. $6\sqrt{3} - 4\sqrt{81} = 6\sqrt{3} - 4 \cdot 9 = 6\sqrt{3} - 36$

23. $3\sqrt{24} - 5\sqrt{48} - \sqrt{6} = 3\sqrt{4 \cdot 6} - 5\sqrt{16 \cdot 3} - \sqrt{6} = 3\sqrt{4}\sqrt{6} - 5\sqrt{16}\sqrt{3} - \sqrt{6}$

$= 3\left(2\sqrt{6}\right) - 5\left(4\sqrt{3}\right) - \sqrt{6} = 6\sqrt{6} - 20\sqrt{3} - \sqrt{6} = (6-1)\sqrt{6} - 20\sqrt{3}$

$= 5\sqrt{6} - 20\sqrt{3}$

25. $3\sqrt[3]{24} - 5\sqrt[3]{48} - \sqrt[3]{6} = 3\sqrt[3]{8 \cdot 3} - 5\sqrt[3]{8 \cdot 6} - \sqrt[3]{6} = 3\sqrt[3]{8}\sqrt[3]{3} - 5\sqrt[3]{8}\sqrt[3]{6} - \sqrt[3]{6}$

$= 3\left(2\sqrt[3]{3}\right) - 5\left(2\sqrt[3]{6}\right) - \sqrt[3]{6}$

$= 6\sqrt[3]{3} - 10\sqrt[3]{6} - \sqrt[3]{6} = 6\sqrt[3]{3} + (-10-1)\sqrt[3]{6} = 6\sqrt[3]{3} - 11\sqrt[3]{6}$

29. $\sqrt[3]{x^4} - x\sqrt[3]{x} = \sqrt[3]{x^3 x} - x\sqrt[3]{x} = \sqrt[3]{x^3}\sqrt[3]{x} - x\sqrt[3]{x} = x\sqrt[3]{x} - x\sqrt[3]{x} = 0$

31. $\sqrt{20x^9 y^8} + 2xy\sqrt{5x^7 y^6} = \sqrt{4x^8 y^8 \cdot 5x} + 2xy\sqrt{x^6 y^6 \cdot 5x} = \sqrt{4x^8 y^8}\sqrt{5x} + 2xy\sqrt{x^6 y^6}\sqrt{5x}$

$= 2x^4 y^4 \sqrt{5x} + 2xy\left(x^3 y^3\right)\sqrt{5x} = 2x^4 y^4 \sqrt{5x} + 2x^4 y^4 \sqrt{5x} = 4x^4 y^4 \sqrt{5x}$

35. $4\sqrt[4]{16x} - 7\sqrt[4]{x^5} + x\sqrt[4]{81x} = 4\sqrt[4]{16}\sqrt[4]{x} - 7\sqrt[4]{x^4}\sqrt[4]{x} + x\sqrt[4]{81}\sqrt[4]{x}$

$= 4\left(2\sqrt[4]{x}\right) - 7x\sqrt[4]{x} + x\left(3\sqrt[4]{x}\right) = 8\sqrt[4]{x} - 7x\sqrt[4]{x} + 3x\sqrt[4]{x}$

$= 8\sqrt[4]{x} - 4x\sqrt[4]{x} = (8 - 4x)\sqrt[4]{x}$

39. $\dfrac{12}{\sqrt{6}} - 2\sqrt{6} = \dfrac{12}{\sqrt{6}} \cdot \dfrac{\sqrt{6}}{\sqrt{6}} - 2\sqrt{6} = \dfrac{12\sqrt{6}}{6} - 2\sqrt{6} = 2\sqrt{6} - 2\sqrt{6} = 0$

43. $\sqrt{\dfrac{5}{2}} + \sqrt{\dfrac{2}{5}} = \dfrac{\sqrt{5}}{\sqrt{2}} + \dfrac{\sqrt{2}}{\sqrt{5}} = \dfrac{\sqrt{5}\sqrt{2}}{\sqrt{2}\sqrt{2}} + \dfrac{\sqrt{2}\sqrt{5}}{\sqrt{5}\sqrt{5}} = \dfrac{\sqrt{10}}{2} + \dfrac{\sqrt{10}}{5} = \dfrac{5\sqrt{10}}{10} + \dfrac{2\sqrt{10}}{10} = \dfrac{7\sqrt{10}}{10}$

47. $\dfrac{1}{\sqrt[3]{2}} - 6\sqrt[3]{4} = \dfrac{1}{\sqrt[3]{2}} \cdot \dfrac{\sqrt[3]{4}}{\sqrt[3]{4}} - 6\sqrt[3]{4} = \dfrac{\sqrt[3]{4}}{\sqrt[3]{8}} - 6\sqrt[3]{4} = \dfrac{\sqrt[3]{4}}{2} - 6\sqrt[3]{4}$

$= \dfrac{\sqrt[3]{4}}{2} - \dfrac{12\sqrt[3]{4}}{2} = \dfrac{-11\sqrt[3]{4}}{2}$

51. $\boxed{\dfrac{1}{\sqrt[3]{9}} - \dfrac{1}{\sqrt[3]{3}}} = \boxed{\dfrac{1}{\sqrt[3]{9}} \cdot \dfrac{\sqrt[3]{3}}{\sqrt[3]{3}}} \boxed{\dfrac{3}{\sqrt[3]{3}} \cdot \dfrac{\sqrt[3]{9}}{\sqrt[3]{3}}} = \dfrac{\sqrt[3]{3}}{\sqrt[3]{27}} - \dfrac{3\sqrt[3]{9}}{\sqrt[3]{27}} = \dfrac{\sqrt[3]{3}}{3} - \dfrac{3\sqrt[3]{9}}{3} = \boxed{\dfrac{\sqrt[3]{3} - 3\sqrt[3]{9}}{3}}$

55. $3\sqrt{10} - \dfrac{4}{\sqrt{10}} + \dfrac{2}{\sqrt{10}} = 3\sqrt{10} - \dfrac{2}{\sqrt{10}} = 3\sqrt{10} - \dfrac{2\sqrt{10}}{\sqrt{10}\sqrt{10}} = 3\sqrt{10} - \dfrac{2\sqrt{10}}{\cancel{10}5}$

$= 3\sqrt{10} - \dfrac{\sqrt{10}}{5} = \dfrac{15\sqrt{10}}{5} - \dfrac{\sqrt{10}}{5} = \boxed{\dfrac{14\sqrt{10}}{5}}$

59. $\left(xy^{-1}\right)^{-2}\left(x^{-2}y\right)^3 = \left(x^{-2}\left(y^{-1}\right)^{-2}\right)\left(\left(x^{-2}\right)^3 y^3\right) = x^{-2}y^2 x^{-6} y^3 = x^{-2-6} y^{2+3} = x^{-8} y^5 = \dfrac{1}{x^8} \cdot y^5 = \dfrac{y^5}{x^8}$

61. $\dfrac{5x+1}{x^2+x-6}-\dfrac{2}{2-x}=\dfrac{5x+1}{(x+3)(x-2)}+\dfrac{2}{x-2}=\dfrac{5x+1}{(x+3)(x-2)}+\dfrac{2(x+3)}{(x+3)(x-2)}$

$$=\dfrac{5x+1+2(x+3)}{(x+3)(x-2)}=\dfrac{7x+7}{(x+3)(x-2)}\text{ or }\dfrac{7(x+1)}{(x+3)(x-2)}$$

63. Let t = number of hours it takes the faster car to catch up

$50t = 40(t+1)$

$50t = 40t + 40$

$10t = 40$

$t = 4$

Thus, the faster car needs 4 hours to catch the slower car.

Exercises 7.4

3. $2\left(\sqrt{3}-\sqrt{5}\right)-4\left(\sqrt{3}+\sqrt{5}\right)=2\sqrt{3}-2\sqrt{5}-4\sqrt{3}-4\sqrt{5}=-2\sqrt{3}-6\sqrt{5}$

5. $\sqrt{a}\left(\sqrt{a}+\sqrt{b}\right)=\sqrt{a}\sqrt{a}+\sqrt{a}\sqrt{b}=a+\sqrt{ab}$

9. $3\sqrt{5}\left(2\sqrt{3}-4\sqrt{5}\right)=\left(3\sqrt{5}\right)\left(2\sqrt{3}\right)-\left(3\sqrt{5}\right)\left(4\sqrt{5}\right)=6\sqrt{15}-12\cdot5=6\sqrt{15}-60$

13. $\left(\sqrt{5}-2\right)\left(\sqrt{3}+1\right)=\left(\sqrt{5}\right)\left(\sqrt{3}\right)+\left(\sqrt{5}\right)(1)-(2)\left(\sqrt{3}\right)-(2)(1)=\sqrt{15}+\sqrt{5}-2\sqrt{3}-2$

17. $\left(\sqrt{5}-\sqrt{3}\right)^2=\left(\sqrt{5}-\sqrt{3}\right)\left(\sqrt{5}-\sqrt{3}\right)=\sqrt{5}\sqrt{5}-\sqrt{5}\sqrt{3}-\sqrt{3}\sqrt{5}+\sqrt{3}\sqrt{3}$

$$=5-\sqrt{15}-\sqrt{15}+3=8-2\sqrt{15}$$

21. $\left(5\sqrt{2}-3\sqrt{5}\right)\left(5\sqrt{2}+3\sqrt{5}\right)=\left(5\sqrt{2}\right)\left(5\sqrt{2}\right)+\left(5\sqrt{2}\right)\left(3\sqrt{5}\right)-\left(3\sqrt{5}\right)\left(5\sqrt{2}\right)-\left(3\sqrt{5}\right)\left(3\sqrt{5}\right)$

$$=25\cdot2+15\sqrt{10}-15\sqrt{10}-9\cdot5=50-45=5$$

27. $\left(\sqrt{x}-3\right)^2=\left(\sqrt{x}-3\right)\left(\sqrt{x}-3\right)=\sqrt{x}\sqrt{x}-3\sqrt{x}-3\sqrt{x}+3\cdot3=x-6\sqrt{x}+9$

29. $\left(\sqrt{x-3}\right)^2=x-3$

33. $\left(\sqrt[3]{2}-\sqrt[3]{3}\right)\left(\sqrt[3]{4}+\sqrt[3]{6}+\sqrt[3]{9}\right)=\sqrt[3]{2}\left(\sqrt[3]{4}+\sqrt[3]{6}+\sqrt[3]{9}\right)-\sqrt[3]{3}\left(\sqrt[3]{4}+\sqrt[3]{6}+\sqrt[3]{9}\right)$

$$=\sqrt[3]{2}\sqrt[3]{4}+\sqrt[3]{2}\sqrt[3]{6}+\sqrt[3]{2}\sqrt[3]{9}-\sqrt[3]{3}\sqrt[3]{4}-\sqrt[3]{3}\sqrt[3]{6}-\sqrt[3]{3}\sqrt[3]{9}$$

$$=\sqrt[3]{8}+\sqrt[3]{12}+\sqrt[3]{18}-\sqrt[3]{12}-\sqrt[3]{18}-\sqrt[3]{27}=\sqrt[3]{8}-\sqrt[3]{27}=2-3=-1$$

37. $\dfrac{5\sqrt{8}-2\sqrt{7}}{8}=\dfrac{5\left(\sqrt{4\cdot2}\right)-2\sqrt{7}}{8}=\dfrac{5\sqrt{4}\sqrt{2}-2\sqrt{7}}{8}=\dfrac{5\left(2\sqrt{2}\right)-2\sqrt{7}}{8}$

$$=\dfrac{2\left(5\sqrt{2}-\sqrt{7}\right)}{8}=\dfrac{5\sqrt{2}-\sqrt{7}}{4}$$

43. $\dfrac{1}{\sqrt{2}-3} = \dfrac{1}{\sqrt{2}-3} \cdot \dfrac{\sqrt{2}+3}{\sqrt{2}+3} = \dfrac{\sqrt{2}+3}{2-9} = \dfrac{\sqrt{2}+3}{-7} = -\dfrac{\sqrt{2}+3}{7}$

45. $\dfrac{10}{\sqrt{5}+1} = \dfrac{10}{\sqrt{5}+1} \cdot \dfrac{\sqrt{5}-1}{\sqrt{5}-1} = \dfrac{10(\sqrt{5}-1)}{5-1} = \dfrac{\overset{5}{\cancel{10}}(\sqrt{5}-1)}{\cancel{4}2} = \dfrac{5(\sqrt{5}-1)}{2}$

49. $\dfrac{\sqrt{x}}{\sqrt{x}-\sqrt{y}} = \dfrac{\sqrt{x}}{\sqrt{x}-\sqrt{y}} \cdot \dfrac{\sqrt{x}+\sqrt{y}}{\sqrt{x}+\sqrt{y}} = \dfrac{\sqrt{x}\left(\sqrt{x}+\sqrt{y}\right)}{x-y} = \dfrac{\sqrt{x}\sqrt{x}+\sqrt{x}\sqrt{y}}{x-y} = \dfrac{x+\sqrt{xy}}{x-y}$

53. $\dfrac{2\sqrt{2}}{2\sqrt{5}-\sqrt{2}} = \dfrac{2\sqrt{2}}{2\sqrt{5}-\sqrt{2}} \cdot \dfrac{2\sqrt{5}+\sqrt{2}}{2\sqrt{5}+\sqrt{2}} = \dfrac{2\sqrt{2}\left(2\sqrt{5}+\sqrt{2}\right)}{20-2} = \dfrac{\cancel{2}\sqrt{2}\left(2\sqrt{5}+\sqrt{2}\right)}{\cancel{18}\,9}$

$= \dfrac{2\sqrt{2}\sqrt{5}+\sqrt{2}\sqrt{2}}{9} = \dfrac{2\sqrt{10}+2}{9}$

57. $\dfrac{\sqrt{3}+\sqrt{2}}{\sqrt{3}-\sqrt{2}} = \dfrac{\sqrt{3}+\sqrt{2}}{\sqrt{3}-\sqrt{2}} \cdot \dfrac{\sqrt{3}+\sqrt{2}}{\sqrt{3}+\sqrt{2}} = \dfrac{\sqrt{3}\sqrt{3}+\sqrt{3}\sqrt{2}+\sqrt{2}\sqrt{3}+\sqrt{2}\sqrt{2}}{3-2}$

$= \dfrac{3+\sqrt{6}+\sqrt{6}+2}{1} = 5+2\sqrt{6}$

59. $\dfrac{3\sqrt{5}-2\sqrt{2}}{2\sqrt{5}-3\sqrt{2}} = \dfrac{3\sqrt{5}-2\sqrt{2}}{2\sqrt{5}-3\sqrt{2}} \cdot \dfrac{2\sqrt{5}+3\sqrt{2}}{2\sqrt{5}+3\sqrt{2}}$

$= \dfrac{\left(3\sqrt{5}\right)\left(2\sqrt{5}\right)+\left(3\sqrt{5}\right)\left(3\sqrt{2}\right)-\left(2\sqrt{2}\right)\left(2\sqrt{5}\right)-\left(2\sqrt{2}\right)\left(3\sqrt{2}\right)}{20-18}$

$= \dfrac{6\cdot 5+9\sqrt{10}-4\sqrt{10}-6\cdot 2}{2} = \dfrac{30+5\sqrt{10}-12}{2} = \dfrac{18+5\sqrt{10}}{2}$

63. $\dfrac{x^2-x-2}{\sqrt{x}-\sqrt{2}} = \dfrac{(x-2)(x+1)}{\sqrt{x}-\sqrt{2}} = \dfrac{(x-2)(x+1)}{\sqrt{x}-\sqrt{2}} \cdot \dfrac{\sqrt{x}+\sqrt{2}}{\sqrt{x}+\sqrt{2}}$

$= \dfrac{\cancel{(x-2)}(x+1)\left(\sqrt{x}+\sqrt{2}\right)}{\cancel{x-2}} = (x+1)\left(\sqrt{x}+\sqrt{2}\right)$

65. $\dfrac{12}{\sqrt{6}-2} - \dfrac{36}{\sqrt{6}} = \dfrac{12}{\sqrt{6}-2} \cdot \dfrac{\sqrt{6}+2}{\sqrt{6}+2} - \dfrac{36}{\sqrt{6}} \cdot \dfrac{\sqrt{6}}{\sqrt{6}} = \dfrac{12\left(\sqrt{6}+2\right)}{6-4} - \dfrac{36\sqrt{6}}{6}$

$= \dfrac{\overset{6}{\cancel{12}}\left(\sqrt{6}+2\right)}{\cancel{2}} - \dfrac{\overset{6}{\cancel{36}}\sqrt{6}}{\cancel{6}} = 6\sqrt{6}+12-6\sqrt{6} = 12$

69. For exercise 65, the calculator gives $26.69694 - 14.696938 = 12.000002$. This differs very slightly from the actual answer of 12, as a result of rounding off the value of $\sqrt{6}$.

For exercise 66, the calculator gives $2.0501256 + 9.9498746 = 12.0000002$. Once again, the actual answer is 12, and the difference occurs because the calculator has rounded off the value of $\sqrt{11}$.

71. $-4 \le 3x-2 < 5$

$-2 \le 3x < 7$

$-\dfrac{2}{3} \le x < \dfrac{7}{3}$

181

73. $\left(\dfrac{64x^{\frac{1}{2}}y^{\frac{1}{5}}}{x^{-\frac{1}{2}}}\right)^{-\frac{1}{2}} = \left(64xy^{\frac{1}{5}}\right)^{-\frac{1}{2}} = (64)^{-\frac{1}{2}}x^{-\frac{1}{2}}\left(y^{\frac{1}{5}}\right)^{-\frac{1}{2}} = \dfrac{1}{\sqrt{64}}x^{-\frac{1}{2}}y^{-\frac{1}{10}}$

$= \dfrac{1}{8} \cdot \dfrac{1}{x^{\frac{1}{2}}} \cdot \dfrac{1}{y^{\frac{1}{10}}} = \dfrac{1}{8x^{\frac{1}{2}}y^{\frac{1}{10}}}$

Exercises 7.5

3.　$\sqrt{x-3} = 5$

　　$x - 3 = 25$

　　　$x = 28$

CHECK x = 28:

$\sqrt{x-3} = 5$

$\sqrt{28-3} \overset{?}{=} 5$

$\sqrt{25} \overset{?}{=} 5$

$5 \overset{\checkmark}{=} 5$

7.　$\sqrt{3a} + 4 = 5$

　　$\sqrt{3a} = 1$

　　$3a = 1$

　　$a = \dfrac{1}{3}$

CHECK a = $\dfrac{1}{3}$:

$\sqrt{3a} + 4 = 5$

$\sqrt{3\left(\dfrac{1}{3}\right)} + 4 \overset{?}{=} 5$

$\sqrt{1} + 4 \overset{?}{=} 5$

$1 + 4 \overset{?}{=} 5$

$5 \overset{\checkmark}{=} 5$

9.　$\sqrt{3a+4} = 5$

　　$3a + 4 = 25$

　　　$3a = 21$

　　　$a = 7$

CHECK a = 7:

$\sqrt{3a+4} = 5$

$\sqrt{3(7)+4} \overset{?}{=} 5$

$\sqrt{21+4} \overset{?}{=} 5$

$\sqrt{25} \overset{?}{=} 5$

$5 \overset{\checkmark}{=} 5$

13.　$2 = 5 - \sqrt{3x-1}$

　　$\sqrt{3x-1} + 2 = 5$

　　$\sqrt{3x-1} = 3$

　　$3x - 1 = 9$

　　$3x = 10$

　　$x = \dfrac{10}{3}$

CHECK x = $\dfrac{10}{3}$:

$2 = 5 - \sqrt{3x-1}$

$2 \overset{?}{=} 5 - \sqrt{3\left(\dfrac{10}{3}\right)-1}$

$2 \overset{?}{=} 5 - \sqrt{10-1}$

$2 \overset{?}{=} 5 - \sqrt{9}$

$2 \overset{?}{=} 5 - 3$

$2 \overset{\checkmark}{=} 2$

17. $3\sqrt{x} = 6$

  $\sqrt{x} = 2$

  $x = 4$

CHECK x = 4:

  $3\sqrt{x} = 6$

  $3\sqrt{4} \overset{?}{=} 6$

  $3 \cdot 2 \overset{?}{=} 6$

  $6 \overset{\checkmark}{=} 6$

23. $5\sqrt{y} - 2 = 7$

  $5\sqrt{y} = 9$

  $\sqrt{y} = \dfrac{9}{5}$

  $y = \dfrac{81}{25}$

CHECK $y = \dfrac{81}{25}$:

  $5\sqrt{y} - 2 = 7$

  $5\sqrt{\dfrac{81}{25}} - 2 \overset{?}{=} 7$

  $5\left(\dfrac{9}{5}\right) - 2 \overset{?}{=} 7$

  $9 - 2 \overset{?}{=} 7$

  $7 \overset{\checkmark}{=} 7$

27. $3 + \sqrt{x-1} = 7$

  $\sqrt{x-1} = 4$

  $x - 1 = 16$

  $x = 17$

CHECK x = 17:

  $3 + \sqrt{x-1} = 7$

  $3 + \sqrt{17-1} \overset{?}{=} 7$

  $3 + \sqrt{16} \overset{?}{=} 7$

  $3 + 4 \overset{?}{=} 7$

  $7 \overset{\checkmark}{=} 7$

31. $\sqrt{2x-3} - \sqrt{x+5} = 0$

  $\sqrt{2x-3} = \sqrt{x+5}$

  $2x - 3 = x + 5$

  $x - 3 = 5$

  $x = 8$

CHECK x = 8:

  $\sqrt{2x-3} - \sqrt{x+5} = 0$

  $\sqrt{2(8)-3} - \sqrt{8+5} \overset{?}{=} 0$

  $\sqrt{16-3} - \sqrt{8+5} \overset{?}{=} 0$

  $\sqrt{13} - \sqrt{13} \overset{?}{=} 0$

  $0 \overset{\checkmark}{=} 0$

33. $\sqrt{2x+1} + \sqrt{x+1} = 0$

  $\sqrt{2x+1} = -\sqrt{x+1}$

  $2x + 1 = x + 1$

  $x + 1 = 1$

  $x = 0$

CHECK x = 0:

  $\sqrt{2x+1} + \sqrt{x+1} = 0$

  $\sqrt{2(0)+1} + \sqrt{0+1} \overset{?}{=} 0$

  $\sqrt{0+1} + \sqrt{0+1} \overset{?}{=} 0$

  $\sqrt{1} + \sqrt{1} \overset{?}{=} 0$

  $1 + 1 \overset{?}{=} 0$

  $2 \neq 0$

Thus, the equation has no solution.

37. $4 = \sqrt[3]{x}$

  $4^3 = \left(\sqrt[3]{x}\right)^3$

  $64 = x$

CHECK x = 64:

  $4 = \sqrt[3]{x}$

  $4 \overset{?}{=} \sqrt[3]{64}$

  $4 \overset{\checkmark}{=} 4$

41. $\sqrt[5]{s} = -3$

        $\left(\sqrt[5]{s}\right)^5 = (-3)^5$

          $s = -243$

CHECK s = −243:

       $\sqrt[5]{s} = -3$

      $\sqrt[5]{-243} \overset{?}{=} -3$

       $-3 \overset{\checkmark}{=} -3$

45. $-4 = \sqrt[3]{2x + 3}$

       $(-4)^3 = \left(\sqrt[3]{2x + 3}\right)^3$

        $-64 = 2x + 3$

        $-67 = 2x$

      $-\dfrac{67}{2} = x$

CHECK $x = -\dfrac{67}{2}$:

      $-4 = \sqrt[3]{2x + 3}$

      $-4 \overset{?}{=} \sqrt[3]{2\left(-\dfrac{67}{2}\right) + 3}$

      $-4 \overset{?}{=} \sqrt[3]{-67 + 3}$

      $-4 \overset{?}{=} \sqrt[3]{-64}$

      $-4 \overset{\checkmark}{=} -4$

47. $x^{\frac{1}{3}} = 5$

       $\left(x^{\frac{1}{3}}\right)^3 = 5^3$

        $x = 125$

CHECK x = 125:

       $x^{\frac{1}{3}} = 5$

      $125^{\frac{1}{3}} \overset{?}{=} 5$

      $\sqrt[3]{125} \overset{?}{=} 5$

       $5 \overset{\checkmark}{=} 5$

49. $x^{\frac{1}{3}} + 7 = 5$

         $x^{\frac{1}{3}} = -2$

      $\left(x^{\frac{1}{3}}\right)^3 = (-2)^3$

           $x = -8$

CHECK x = −8:

       $x^{\frac{1}{3}} + 7 = 5$

    $(-8)^{\frac{1}{3}} + 7 \overset{?}{=} 5$

    $\sqrt[3]{-8} + 7 \overset{?}{=} 5$

      $-2 + 7 \overset{?}{=} 5$

        $5 \overset{\checkmark}{=} 5$

53. $x^{-\frac{1}{4}} = 4$

      $\left(x^{-\frac{1}{4}}\right)^{-4} = 4^{-4}$

          $x = \dfrac{1}{4^4}$

          $x = \dfrac{1}{256}$

CHECK $x = \dfrac{1}{256}$:

       $x^{-\frac{1}{4}} = 4$

    $\left(\dfrac{1}{256}\right)^{-\frac{1}{4}} \overset{?}{=} 4$

      $256^{\frac{1}{4}} \overset{?}{=} 4$

      $\sqrt[4]{256} \overset{?}{=} 4$

       $4 \overset{\checkmark}{=} 4$

55.

$$(x+3)^{\frac{1}{3}} + 4 = 2$$

$$(x+3)^{\frac{1}{3}} = -2$$

$$\left((x+3)^{\frac{1}{3}}\right)^3 = (-2)^3$$

$$x + 3 = -8$$

$$x = -11$$

CHECK x = –11:

$$(x+3)^{\frac{1}{3}} + 4 = 2$$

$$(-11+3)^{\frac{1}{3}} + 4 \overset{?}{=} 2$$

$$(-8)^{\frac{1}{3}} + 4 \overset{?}{=} 2$$

$$\sqrt[3]{-8} + 4 \overset{?}{=} 2$$

$$-2 + 4 \overset{?}{=} 2$$

$$2 \overset{\checkmark}{=} 2$$

57. When we square $\sqrt{x+4}$, we just get the quantity under the square root sign, namely x + 4. To square $\sqrt{x} + 4$, we use the FOIL method to obtain $x + 8\sqrt{x} + 16$.

59. $\dfrac{\sqrt{3}}{\sqrt{3}-\sqrt{2}} = \dfrac{\sqrt{3}}{\sqrt{3}-\sqrt{2}} \cdot \dfrac{\sqrt{3}+\sqrt{2}}{\sqrt{3}+\sqrt{2}} = \dfrac{\sqrt{3}\left(\sqrt{3}+\sqrt{2}\right)}{3-2} = \dfrac{3+\sqrt{6}}{1} = 3 + \sqrt{6}$

Exercises 7.6

1. $i^3 = -i$

3. $-i^{21} = -(i^4)^5 i = -(1)^5 i = -i$

5. $(-i)^{29} = (-1)^{29} i^{29} = -1(i^4)^7 i = -1(1)^7 i = -i$

9. $i^{67} = (i^4)^{16} i^3 = (1)^{16} i^3 = 1i^3 = i^3 = -i$

13. $\sqrt{-5} = \sqrt{5}i = 0 + \sqrt{5}\,i$

17. $2\sqrt{-4} - \sqrt{28} = 2\sqrt{4}\,i - \sqrt{4}\sqrt{7} = 2\cdot 2i - 2\sqrt{7} = -2\sqrt{7} + 4i$

21. $\dfrac{6-\sqrt{-3}}{3} = \dfrac{6-\sqrt{3}\,i}{3} = \dfrac{6}{3} - \dfrac{\sqrt{3}\,i}{3} = 2 - \dfrac{\sqrt{3}}{3}i$

23. $(3+2i) - (2+3i) = 3 + 2i - 2 - 3i = (3-2) + (2-3)i = 1 - i$

27. $5(3-7i) = 5(3) - 5(7i) = 15 - 35i$

31. $(5i)(3i) = 5 \cdot 3 \cdot i \cdot i = 15i^2 = 15(-1) = -15$

35. $2i(3+2i) = 2i(3) + 2i(2i) = 6i + 4i^2 = 6i + 4(-1) = -4 + 6i$

41. $(7-4i)(3+i) = 7(3) + 7(i) - 4i(3) - 4i(i) = 21 + 7i - 12i - 4i^2$
$= 21 - 5i - 4i^2 = 21 - 5i - 4(-1) = 21 - 5i + 4 = 25 - 5i$

45. $(2-7i)(2+7i) = 2(2) + 2(7i) - 7i(2) - 7i(7i) = 4 + 14i - 14i - 49i^2$

$$= 4 - 49i^2 = 4 - 49(-1) = 4 + 49 = 53$$

49. $(2-i)^2 - 4(2-i) = (2-i)(2-i-4) = (2-i)(-2-i) = 2(-2) - 2i - i(-2) + i(i)$

$$= -4 - 2i + 2i + i^2 = -4 + i^2 = -4 - 1 = -5$$

53. $\dfrac{6-5i}{3} = \dfrac{6}{3} - \dfrac{5}{3}i = 2 - \dfrac{5}{3}i$

55. $\dfrac{3-i}{i} = \dfrac{3-i}{i} \cdot \dfrac{i}{i} = \dfrac{(3-i)i}{i^2} = \dfrac{3i - i^2}{-1} = \dfrac{3i - (-1)}{-1} = \dfrac{3i + 1}{-1} = -3i - 1 = -1 - 3i$

59. $\dfrac{2i}{5-2i} = \dfrac{2i}{5-2i} \cdot \dfrac{5+2i}{5+2i} = \dfrac{2i(5+2i)}{5(5) + 5(2i) - 2i(5) - 2i(2i)} = \dfrac{10i + 4i^2}{25 + 10i - 10i - 4i^2}$

$$= \dfrac{10i + 4(-1)}{25 - 4(-1)} = \dfrac{10i - 4}{25 + 4} = \dfrac{10i - 4}{29} = -\dfrac{4}{29} + \dfrac{10}{29}i$$

63. $\dfrac{2+i}{2-i} = \dfrac{2+i}{2-i} \cdot \dfrac{2+i}{2+i} = \dfrac{2(2) + 2(i) + i(2) + i(i)}{2(2) + 2(i) - i(2) - i(i)} = \dfrac{4 + 4i + i^2}{4 - i^2} = \dfrac{4 + 4i - 1}{4 - (-1)}$

$$= \dfrac{3 + 4i}{4 + 1} = \dfrac{3 + 4i}{5} = \dfrac{3}{5} + \dfrac{4}{5}i$$

67. $\dfrac{3-7i}{5+2i} = \dfrac{3-7i}{5+2i} \cdot \dfrac{5-2i}{5-2i} = \dfrac{3(5) - 3(2i) - 7i(5) - 7i(-2i)}{5(5) + 5(-2i) + 2i(5) + 2i(-2i)} = \dfrac{15 - 6i - 35i + 14i^2}{25 - 10i + 10i - 4i^2}$

$$= \dfrac{15 - 41i + 14(-1)}{25 - 4(-1)} = \dfrac{15 - 41i - 14}{25 + 4} = \dfrac{1 - 41i}{29} = \dfrac{1}{29} - \dfrac{41}{29}i$$

69. $\dfrac{5-\sqrt{-4}}{3+\sqrt{-25}} = \dfrac{5-2i}{3+5i} = \dfrac{5-2i}{3+5i} \cdot \dfrac{3-5i}{3-5i} = \dfrac{5(3) + 5(-5i) - 2i(3) - 2i(-5i)}{3(3) + 3(-5i) + 5i(3) + 5i(-5i)} = \dfrac{15 - 25i - 6i + 10i^2}{9 - 15i + 15i - 25i^2}$

$$= \dfrac{15 - 31i + 10(-1)}{9 - 25(-1)} = \dfrac{15 - 31i - 10}{9 + 25} = \dfrac{5 - 31i}{34} = \dfrac{5}{34} - \dfrac{31}{34}i$$

73. $x^2 - 4x + 5 = 0$

$(2-i)^2 - 4(2-i) + 5 \overset{?}{=} 0$

$4 - 4i + i^2 - 8 + 4i + 5 \overset{?}{=} 0$

$1 + i^2 \overset{?}{=} 0$

$1 + (-1) \overset{?}{=} 0$

$0 \overset{\checkmark}{=} 0$

Thus, $2 - i$ is a solution of the given equation.

75. The irrational numbers are not closed under multiplication. Both $\sqrt{2}$ and $\sqrt{8}$ are irrational numbers, but their product, $\sqrt{2}\sqrt{8} = \sqrt{16} = 4$, is not an irrational number.

76. $\sqrt[6]{-1} = \sqrt{\sqrt[3]{-1}} = \sqrt{-1} = i$

$\sqrt[10]{-1} = \sqrt{\sqrt[5]{-1}} = \sqrt{-1} = i$

$\sqrt[14]{-1} = \sqrt{\sqrt[7]{-1}} = \sqrt{-1} = i$

77.

$$i^{4n} = (i^4)^n = 1^n = 1$$

$$i^{4n+1} = i^{4n} \cdot i = 1 \cdot i = i$$

$$i^{4n+2} = i^{4n} \cdot i^2 = 1 \cdot i^2 = -1$$

$$i^{4n+3} = i^{4n} \cdot i^3 = 1 \cdot i^3 = -i$$

Chapter 7 Review Exercises

1. $x^{\frac{1}{2}} = \sqrt{x}$

3. $xy^{\frac{1}{2}} = x\sqrt{y}$

5. $m^{\frac{2}{3}} = \sqrt[3]{m^2}$ or $\left(\sqrt[3]{m}\right)^2$

7. $(5x)^{\frac{3}{4}} = \sqrt[4]{(5x)^3}$ or $\left(\sqrt[4]{5x}\right)^3$

9. $s^{-\frac{4}{5}} = \dfrac{1}{s^{\frac{4}{5}}} = \dfrac{1}{\sqrt[5]{s^4}} = \dfrac{1}{\sqrt[5]{s^4}} \cdot \dfrac{\sqrt[5]{s}}{\sqrt[5]{s}} = \dfrac{\sqrt[5]{s}}{\sqrt[5]{s^5}} = \dfrac{\sqrt[5]{s}}{s}$

11. $\sqrt[3]{a} = a^{\frac{1}{3}}$

13. $-\sqrt[5]{n^4} = -n^{\frac{4}{5}}$

15. $\left(\sqrt[3]{t}\right)^5 = t^{\frac{5}{3}}$

17. $\dfrac{1}{\sqrt[5]{t^7}} = \dfrac{1}{t^{\frac{7}{5}}} = t^{-\frac{7}{5}}$

19. $\sqrt{54} = \sqrt{9 \cdot 6} = \sqrt{9}\sqrt{6} = 3\sqrt{6}$

21. $\sqrt{x^{60}} = (x^{60})^{\frac{1}{2}} = x^{30}$

23. $\sqrt[5]{x^{60}} = \left(x^{60}\right)^{\frac{1}{5}} = x^{12}$

25. $\sqrt[3]{48x^4y^8} = \sqrt[3]{8x^3y^6 \cdot 6xy^2} = \sqrt[3]{8x^3y^6}\,\sqrt[3]{6xy^2} = 2xy^2\sqrt[3]{6xy^2}$

27. $\sqrt[4]{81x^9y^{10}} = \sqrt[4]{81x^8y^8 \cdot xy^2} = \sqrt[4]{81x^8y^8}\,\sqrt[4]{xy^2} = 3x^2y^2\sqrt[4]{xy^2}$

29. $\sqrt{75xy}\,\sqrt{3x} = \sqrt{75xy \cdot 3x} = \sqrt{225x^2y} = \sqrt{225x^2 \cdot y} = \sqrt{225x^2}\,\sqrt{y} = 15x\sqrt{y}$

31. $\left(x\sqrt{xy}\right)\left(2x^2y\sqrt{xy^2}\right) = x \cdot 2x^2y\sqrt{xy}\,\sqrt{xy^2} = 2x^3y\sqrt{xy \cdot xy^2} = 2x^3y\sqrt{x^2y^3} = 2x^3y\sqrt{x^2y^2 \cdot y}$

$$= 2x^3y\sqrt{x^2y^2}\,\sqrt{y} = 2x^3y \cdot xy\sqrt{y} = 2x^4y^2\sqrt{y}$$

33. $\dfrac{\sqrt{28}}{\sqrt{63}} = \sqrt{\dfrac{28}{63}} = \sqrt{\dfrac{7(4)}{7(9)}} = \sqrt{\dfrac{4}{9}} = \dfrac{\sqrt{4}}{\sqrt{9}} = \dfrac{2}{3}$

35. $\dfrac{y}{x\sqrt{y}} = \dfrac{y}{x\sqrt{y}} \cdot \dfrac{\sqrt{y}}{\sqrt{y}} = \dfrac{y\sqrt{y}}{xy} = \dfrac{\sqrt{y}}{x}$

37. $\sqrt{\dfrac{48a^2b}{3a^5b^2}}^{\,16}_{\,a^3b} = \sqrt{\dfrac{16}{a^3b}} = \dfrac{\sqrt{16}}{\sqrt{a^3b}} = \dfrac{4}{\sqrt{a^2 \cdot ab}} = \dfrac{4}{\sqrt{a^2}\,\sqrt{ab}} = \dfrac{4}{a\sqrt{ab}} = \dfrac{4}{a\sqrt{ab}} \cdot \dfrac{\sqrt{ab}}{\sqrt{ab}}$

$$= \dfrac{4\sqrt{ab}}{a(ab)} = \dfrac{4\sqrt{ab}}{a^2b}$$

39. $\dfrac{\sqrt{x^3y^2}\,\sqrt{2xy}}{\sqrt{xy^5}} = \dfrac{\sqrt{x^3y^2 \cdot 2xy}}{\sqrt{xy^5}} = \sqrt{\dfrac{x^3y^2 \cdot 2xy}{xy^5}} = \sqrt{\dfrac{2\cancel{x}\cancel{y^2}x^3}{\cancel{xy^5}y^2}} = \sqrt{\dfrac{2x^3}{y^2}} = \dfrac{\sqrt{2x^3}}{\sqrt{y^2}}$

$= \dfrac{\sqrt{x^2 \cdot 2x}}{y} = \dfrac{\sqrt{x^2}\,\sqrt{2x}}{y} = \dfrac{x\sqrt{2x}}{y}$

41. $\sqrt{\dfrac{5}{a}} = \dfrac{\sqrt{5}}{\sqrt{a}} = \dfrac{\sqrt{5}}{\sqrt{a}} \cdot \dfrac{\sqrt{a}}{\sqrt{a}} = \dfrac{\sqrt{5a}}{a}$

43. $\dfrac{4}{\sqrt[3]{2a}} = \dfrac{4}{\sqrt[3]{2a}} \cdot \dfrac{\sqrt[3]{4a^2}}{\sqrt[3]{4a^2}} = \dfrac{4\sqrt[3]{4a^2}}{\sqrt[3]{8a^3}} = \dfrac{\cancel{4}^{\,2}\sqrt[3]{4a^2}}{\cancel{2}a} = \dfrac{2\sqrt[3]{4a^2}}{a}$

45. $\sqrt[6]{x^4} = (x^4)^{\frac{1}{6}} = x^{\frac{4}{6}} = x^{\frac{2}{3}} = \sqrt[3]{x^2} \ \text{or} \ \left(\sqrt[3]{x}\right)^2$

47. $\sqrt{2}\,\sqrt[3]{2} = 2^{\frac{1}{2}}2^{\frac{1}{3}} = 2^{\frac{1}{2}+\frac{1}{3}} = 2^{\frac{5}{6}} = \sqrt[6]{2^5} \ \text{or} \ \left(\sqrt[6]{2}\right)^5$

49. $8\sqrt{3} - 3\sqrt{3} - 6\sqrt{3} = (8 - 3 - 6)\sqrt{3} = -\sqrt{3}$

51. $7\sqrt{54} + 6\sqrt{24} = 7\left(\sqrt{9 \cdot 6}\right) + 6\left(\sqrt{4 \cdot 6}\right) = 7\left(\sqrt{9}\sqrt{6}\right) + 6\left(\sqrt{4}\sqrt{6}\right) = 7\left(3\sqrt{6}\right) + 6\left(2\sqrt{6}\right)$

$= 21\sqrt{6} + 12\sqrt{6} = 33\sqrt{6}$

53. $b\sqrt{a^3b} + a\sqrt{ab^3} = b\sqrt{a^2 \cdot ab} + a\sqrt{b^2 \cdot ab} = b\sqrt{a^2}\sqrt{ab} + a\sqrt{b^2}\sqrt{ab} = ba\sqrt{ab} + ab\sqrt{ab} = 2ab\sqrt{ab}$

55. $\sqrt{\dfrac{3}{2}} + \sqrt{\dfrac{5}{3}} = \dfrac{\sqrt{3}}{\sqrt{2}} + \dfrac{\sqrt{5}}{\sqrt{3}} = \dfrac{\sqrt{3}}{\sqrt{2}} \cdot \dfrac{\sqrt{2}}{\sqrt{2}} + \dfrac{\sqrt{5}}{\sqrt{3}} \cdot \dfrac{\sqrt{3}}{\sqrt{3}} = \dfrac{\sqrt{6}}{2} + \dfrac{\sqrt{15}}{3} = \dfrac{3\sqrt{6}}{6} + \dfrac{2\sqrt{15}}{6}$

$= \dfrac{3\sqrt{6} + 2\sqrt{15}}{6}$

57. $2\sqrt[3]{\dfrac{1}{9}} - 3\sqrt[3]{3} = 2 \cdot \dfrac{\sqrt[3]{1}}{\sqrt[3]{9}} - 3\sqrt[3]{3} = 2 \cdot \dfrac{1}{\sqrt[3]{9}} \cdot \dfrac{\sqrt[3]{3}}{\sqrt[3]{3}} - 3\sqrt[3]{3} = \dfrac{2\sqrt[3]{3}}{\sqrt[3]{27}} - 3\sqrt[3]{3}$

$= \dfrac{2\sqrt[3]{3}}{3} - 3\sqrt[3]{3} = \dfrac{2\sqrt[3]{3}}{3} - \dfrac{9\sqrt[3]{3}}{3} = -\dfrac{7\sqrt[3]{3}}{3}$

59. $\sqrt{3}\left(\sqrt{6} - \sqrt{2}\right) + \sqrt{2}\left(\sqrt{3} - 3\right) = \sqrt{3}\sqrt{6} - \sqrt{3}\sqrt{2} + \sqrt{2}\sqrt{3} + \sqrt{2}(-3)$

$= \sqrt{18} - \sqrt{6} + \sqrt{6} - 3\sqrt{2} = \sqrt{18} - 3\sqrt{2} = \sqrt{9 \cdot 2} - 3\sqrt{2}$

$= \sqrt{9}\sqrt{2} - 3\sqrt{2} = 3\sqrt{2} - 3\sqrt{2} = 0$

61. $\left(3\sqrt{7} - \sqrt{3}\right)\left(\sqrt{7} - 2\sqrt{3}\right) = \left(3\sqrt{7}\right)\left(\sqrt{7}\right) + \left(3\sqrt{7}\right)\left(-2\sqrt{3}\right) - \left(\sqrt{3}\right)\left(\sqrt{7}\right) - \left(\sqrt{3}\right)\left(-2\sqrt{3}\right)$

$= 3 \cdot 7 - 6\sqrt{21} - \sqrt{21} + 2 \cdot 3 = 21 - 7\sqrt{21} + 6 = 27 - 7\sqrt{21}$

63. $\left(\sqrt{x} - 5\right)^2 = \left(\sqrt{x} - 5\right)\left(\sqrt{x} - 5\right) = \sqrt{x}\sqrt{x} - 5\sqrt{x} - 5\sqrt{x} + (-5)(-5) = x - 10\sqrt{x} + 25$

65. $\left(\sqrt{a+7}\right)^2 - \left(\sqrt{a}+7\right)^2 = a+7 - \left(\sqrt{a}+7\right)\left(\sqrt{a}+7\right) = a+7 - \left(\sqrt{a}\sqrt{a} + 7\sqrt{a} + 7\sqrt{a} + 7\cdot 7\right)$

$$= a+7 - \left(a + 14\sqrt{a} + 49\right) = a+7 - a - 14\sqrt{a} - 49 = -42 - 14\sqrt{a}$$

67. $\dfrac{20}{\sqrt{3}-\sqrt{5}} = \dfrac{20}{\sqrt{3}-\sqrt{5}} \cdot \dfrac{\sqrt{3}+\sqrt{5}}{\sqrt{3}+\sqrt{5}} = \dfrac{20\left(\sqrt{3}+\sqrt{5}\right)}{\left(\sqrt{3}-\sqrt{5}\right)\left(\sqrt{3}+\sqrt{5}\right)}$

$$= \dfrac{20\left(\sqrt{3}+\sqrt{5}\right)}{\sqrt{3}\sqrt{3} + \sqrt{3}\sqrt{5} - \sqrt{5}\sqrt{3} - \sqrt{5}\sqrt{5}} = \dfrac{20\left(\sqrt{3}+\sqrt{5}\right)}{3 + \sqrt{15} - \sqrt{15} - 5}$$

$$= \dfrac{\overset{10}{20}\left(\sqrt{3}+\sqrt{5}\right)}{\underset{-1}{-2}} = -10\left(\sqrt{3}+\sqrt{5}\right)$$

69. $\dfrac{m-n^2}{\sqrt{m}-n} = \dfrac{m-n^2}{\sqrt{m}-n} \cdot \dfrac{\sqrt{m}+n}{\sqrt{m}+n} = \dfrac{(m-n^2)\left(\sqrt{m}+n\right)}{\left(\sqrt{m}-n\right)\left(\sqrt{m}+n\right)} = \dfrac{(m-n^2)\left(\sqrt{m}+n\right)}{\sqrt{m}\sqrt{m} + n\sqrt{m} - n\sqrt{m} - nn}$

$$= \dfrac{(m-n^2)\left(\sqrt{m}+n\right)}{m-n^2} = \sqrt{m}+n$$

71. $\dfrac{20}{3-\sqrt{5}} - \dfrac{50}{\sqrt{5}} = \dfrac{20}{3-\sqrt{5}} \cdot \dfrac{3+\sqrt{5}}{3+\sqrt{5}} - \dfrac{50}{\sqrt{5}} \cdot \dfrac{\sqrt{5}}{\sqrt{5}} = \dfrac{20\left(3+\sqrt{5}\right)}{\left(3-\sqrt{5}\right)\left(3+\sqrt{5}\right)} - \dfrac{50\sqrt{5}}{5}$

$$= \dfrac{20\left(3+\sqrt{5}\right)}{9 + 3\sqrt{5} - 3\sqrt{5} - 5} - \dfrac{50\sqrt{5}}{5} = \dfrac{\overset{5}{20}\left(3+\sqrt{5}\right)}{\cancel{4}} - \dfrac{\overset{10}{50}\sqrt{5}}{\cancel{5}}$$

$$= 5\left(3+\sqrt{5}\right) - 10\sqrt{5} = 15 + 5\sqrt{5} - 10\sqrt{5} = 15 - 5\sqrt{5}$$

73. $\left(\sqrt[3]{5} - \sqrt[3]{2}\right)\left(\sqrt[3]{25} + \sqrt[3]{10} + \sqrt[3]{4}\right) = \sqrt[3]{5}\left(\sqrt[3]{25} + \sqrt[3]{10} + \sqrt[3]{4}\right) - \sqrt[3]{2}\left(\sqrt[3]{25} + \sqrt[3]{10} + \sqrt[3]{4}\right)$

$$= \sqrt[3]{5}\sqrt[3]{25} + \sqrt[3]{5}\sqrt[3]{10} + \sqrt[3]{5}\sqrt[3]{4} - \sqrt[3]{2}\sqrt[3]{25} - \sqrt[3]{2}\sqrt[3]{10} - \sqrt[3]{2}\sqrt[3]{4}$$

$$= \sqrt[3]{125} + \sqrt[3]{50} + \sqrt[3]{20} - \sqrt[3]{50} - \sqrt[3]{20} - \sqrt[3]{8} = \sqrt[3]{125} - \sqrt[3]{8} = 5 - 2 = 3$$

75. $\sqrt{2x} - 5 = 7$ CHECK x = 72:

   $\sqrt{2x} = 12$ $\sqrt{2x} - 5 = 7$

   $2x = 144$ $\sqrt{2(72)} - 5 \overset{?}{=} 7$

   $x = 72$ $\sqrt{144} - 5 \overset{?}{=} 7$

   $12 - 5 \overset{?}{=} 7$

   $7 \overset{\checkmark}{=} 7$

77. $\sqrt{2x-5} = 7$ CHECK x = 27:

   $2x - 5 = 49$ $\sqrt{2x-5} = 7$

   $2x = 54$ $\sqrt{2(27)-5} \overset{?}{=} 7$

   $x = 27$ $\sqrt{54-5} \overset{?}{=} 7$

   $\sqrt{49} \overset{?}{=} 7$

   $7 \overset{\checkmark}{=} 7$

79. $\sqrt[4]{x-4} = 3$

$\left(\sqrt[4]{x-4}\right)^4 = 3^4$

$x - 4 = 81$

$x = 85$

CHECK $x = 85$:

$\sqrt[4]{x-4} = 3$

$\sqrt[4]{85-4} \overset{?}{=} 3$

$\sqrt[4]{81} \overset{?}{=} 3$

$3 \overset{\checkmark}{=} 3$

81. $x^{\frac{1}{4}} + 1 = 3$

$x^{\frac{1}{4}} = 2$

$\left(x^{\frac{1}{4}}\right)^4 = 2^4$

$x = 16$

CHECK $x = 16$:

$x^{\frac{1}{4}} + 1 = 3$

$16^{\frac{1}{4}} + 1 \overset{?}{=} 3$

$\sqrt[4]{16} + 1 \overset{?}{=} 3$

$2 + 1 \overset{?}{=} 3$

$3 \overset{\checkmark}{=} 3$

83. $i^{11} = (i^4)^2 i^3 = 1^2 i^3 = 1(-i) = -i$

85. $-i^{14} = -(i^4)^3 i^2 = -1^3 i^2 = -1(-1) = 1$

87. $(5 + i) + (4 - 2i) = 5 + 4 + i - 2i = 9 - i$

89. $(7 - 2i)(2 - 3i) = 7(2) + 7(-3i) - 2i(2) - 2i(-3i) = 14 - 21i - 4i + 6i^2$

$\qquad = 14 - 25i + 6(-1) = 14 - 25i - 6 = 8 - 25i$

91. $2i(3i - 4) = 2i(3i) - 2i(4) = 6i^2 - 8i = 6(-1) - 8i = -6 - 8i$

93. $(3 - 2i)^2 = (3 - 2i)(3 - 2i) = 9 - 3(2i) - 3(2i) - 2i(2i) = 9 - 6i - 6i + 4i^2$

$\qquad = 9 - 12i + 4i^2 = 9 - 12i + 4(-1) = 9 - 12i - 4 = 5 - 12i$

95. $(6 - i)^2 - 12(6 - i) = (6 - i)(6 - i - 12) = (6 - i)(-6 - i) = 6(-6) + 6(-i) - i(-6) - i(-i)$

$\qquad = -36 - 6i + 6i + i^2 = -36 + i^2 = -36 + (-1) = -37$

97. $\dfrac{4 - 3i}{3 + i} = \dfrac{4 - 3i}{3 + i} \cdot \dfrac{3 - i}{3 - i} = \dfrac{4(3) - 4i - 3i(3) - 3i(-i)}{(3)(3) - 3i + 3i - i^2} = \dfrac{12 - 4i - 9i + 3i^2}{9 - i^2}$

$\qquad = \dfrac{12 - 13i + 3(-1)}{9 - (-1)} = \dfrac{12 - 13i - 3}{10} = \dfrac{9 - 13i}{10} = \dfrac{9}{10} - \dfrac{13}{10} i$

99. $x^2 - 2x + 5 = 0$

$(1 - 2i)^2 - 2(1 - 2i) + 5 \overset{?}{=} 0$

$(1 - 2i)(1 - 2i) - 2(1 - 2i) + 5 \overset{?}{=} 0$

$1 - 2i - 2i + 4i^2 - 2 + 4i + 5 \overset{?}{=} 0$

$1 + 4i^2 - 2 + 5 \overset{?}{=} 0$

$1 + 4(-1) - 2 + 5 \overset{?}{=} 0$

$1 - 4 - 2 + 5 \overset{?}{=} 0$

$0 \overset{\checkmark}{=} 0$

Thus, $1 - 2i$ is a solution for $x^2 - 2x + 5 = 0$.

Chapter 7 Practice Test

1. $\sqrt[3]{27x^6y^9} = \sqrt[3]{27}\sqrt[3]{x^6}\sqrt[3]{y^9} = 3x^2y^3$

3. $\left(2x^2\sqrt{x}\right)\left(3x\sqrt{xy^2}\right) = 2x^2 \cdot 3x\sqrt{x}\sqrt{xy^2} = 6x^3\sqrt{x \cdot xy^2} = 6x^3\sqrt{x^2y^2} = 6x^3 \cdot xy = 6x^4y$

5. $\sqrt{\dfrac{5}{7}} = \dfrac{\sqrt{5}}{\sqrt{7}} = \dfrac{\sqrt{5}}{\sqrt{7}} \cdot \dfrac{\sqrt{7}}{\sqrt{7}} = \dfrac{\sqrt{35}}{7}$

7. $\dfrac{\left(xy\sqrt{2xy}\right)\left(3x\sqrt{y}\right)}{\sqrt{4x^3}} = \dfrac{(xy \cdot 3x)\sqrt{2xy}\sqrt{y}}{\sqrt{4x^3}} = \dfrac{3x^2y\sqrt{2xy \cdot y}}{\sqrt{4x^3}} = \dfrac{3x^2y\sqrt{2xy^2}}{\sqrt{4x^3}} = \dfrac{3x^2y\sqrt{y^2 \cdot 2x}}{\sqrt{4x^2 \cdot x}}$

$= \dfrac{3x^2y\sqrt{y^2}\sqrt{2x}}{\sqrt{4x^2}\sqrt{x}} = \dfrac{3x^2y \cdot y\sqrt{2x}}{2x\sqrt{x}} = \dfrac{3\cancel{x}y^2}{2\cancel{x}} \cdot \sqrt{\dfrac{2\cancel{x}}{\cancel{x}}} = \dfrac{3xy^2\sqrt{2}}{2}$

9. $\sqrt{50} - 3\sqrt{8} + 2\sqrt{18} = \sqrt{25 \cdot 2} - 3\sqrt{4 \cdot 2} + 2\sqrt{9 \cdot 2} = \sqrt{25}\sqrt{2} - 3\sqrt{4}\sqrt{2} + 2\sqrt{9}\sqrt{2}$

$= 5\sqrt{2} - 3\left(2\sqrt{2}\right) + 2\left(3\sqrt{2}\right) = 5\sqrt{2} - 6\sqrt{2} + 6\sqrt{2} = (5 - 6 + 6)\sqrt{2} = 5\sqrt{2}$

11. $\sqrt{24} - 4\sqrt{\dfrac{2}{3}} = \sqrt{4 \cdot 6} - \dfrac{4\sqrt{2}}{\sqrt{3}} = \sqrt{4}\sqrt{6} - \dfrac{4\sqrt{2}}{\sqrt{3}} = 2\sqrt{6} - \dfrac{4\sqrt{2}}{\sqrt{3}} \cdot \dfrac{\sqrt{3}}{\sqrt{3}} = 2\sqrt{6} - \dfrac{4\sqrt{6}}{3}$

$= \left(2 - \dfrac{4}{3}\right)\sqrt{6} = \dfrac{2}{3}\sqrt{6}$

13. $\sqrt{x - 3} + 4 = 8$ CHECK x = 19:

     $\sqrt{x - 3} = 4$ $\sqrt{x - 3} + 4 = 8$

      $x - 3 = 16$ $\sqrt{19 - 3} + 4 \overset{?}{=} 8$

        $x = 19$ $\sqrt{16} + 4 \overset{?}{=} 8$

                               $4 + 4 \overset{?}{=} 8$

                                $8 \overset{\checkmark}{=} 8$

15. $\sqrt[3]{x - 3} + 4 = 2$ CHECK x = −5:

      $\sqrt[3]{x - 3} = -2$ $\sqrt[3]{x - 3} + 4 = 2$

   $\left(\sqrt[3]{x - 3}\right)^3 = (-2)^3$ $\sqrt[3]{-5 - 3} + 4 \overset{?}{=} 2$

         $x - 3 = -8$ $\sqrt[3]{-8} + 4 \overset{?}{=} 2$

           $x = -5$ $-2 + 4 \overset{?}{=} 2$

                                $2 \overset{\checkmark}{=} 2$

17. $i^{51} = (i^4)^{12}i^3 = 1^{12}i^3 = 1(-i) = -i$

19. $\dfrac{2i+3}{i-2} = \dfrac{2i+3}{i-2} \cdot \dfrac{i+2}{i+2} = \dfrac{2i(i)+2i(2)+3i+6}{i^2+2i-2i-4} = \dfrac{2i^2+4i+3i+6}{i^2-4}$

$= \dfrac{2(-1)+7i+6}{(-1)-4} = \dfrac{4+7i}{-5} = -\dfrac{4}{5}-\dfrac{7}{5}i$

CHAPTER 8
SECOND DEGREE EQUATIONS
AND INEQUALITIES

Exercises 8.1

1. $(x+2)(x-3) = 0$

   $x+2 = 0$ or $x-3 = 0$

   $x = -2$ or $x = 3$

   CHECK $x = -2$:

   $(x+2)(x-3) = 0$

   $(-2+2)(-2-3) \overset{?}{=} 0$

   $(0)(-5) \overset{?}{=} 0$

   $0 \overset{\checkmark}{=} 0$

   CHECK $x = 3$:

   $(x+2)(x-3) = 0$

   $(3+2)(3-3) \overset{?}{=} 0$

   $(5)(0) \overset{?}{=} 0$

   $0 \overset{\checkmark}{=} 0$

5. $x^2 = 25$

   $x = \pm\sqrt{25}$

   $x = \pm 5$

   CHECK $x = 5$:

   $x^2 = 25$

   $5^2 \overset{?}{=} 25$

   $25 \overset{\checkmark}{=} 25$

   CHECK $x = -5$:

   $x^2 = 25$

   $(-5)^2 \overset{?}{=} 25$

   $25 \overset{\checkmark}{=} 25$

9. $12 = x(x-4)$

   $12 = x^2 - 4x$

   $0 = x^2 - 4x - 12$

   $0 = (x-6)(x+2)$

   $0 = x-6$ or $0 = x+2$

   $6 = x$ or $-2 = x$

   CHECK $x = 6$:

   $12 = x(x-4)$

   $12 \overset{?}{=} 6(6-4)$

   $12 \overset{?}{=} 6(2)$

   $12 \overset{\checkmark}{=} 12$

   CHECK $x = -2$:

   $12 = x(x-4)$

   $12 \overset{?}{=} -2(-2-4)$

   $12 \overset{?}{=} -2(-6)$

   $12 \overset{\checkmark}{=} 12$

13. $x^2 - 16 = 0$

   $x^2 = 16$

   $x = \pm\sqrt{16}$

   $x = \pm 4$

   CHECK $x = 4$:

   $x^2 - 16 = 0$

   $4^2 - 16 \overset{?}{=} 0$

   $16 - 16 \overset{?}{=} 0$

   $0 \overset{\checkmark}{=} 0$

   CHECK $x = -4$:

   $x^2 - 16 = 0$

   $(-4)^2 - 16 \overset{?}{=} 0$

   $16 - 16 \overset{?}{=} 0$

   $0 \overset{\checkmark}{=} 0$

17.　$8a^2 - 18 = 0$

$$8a^2 = 18$$

$$a^2 = \frac{18}{8}$$

$$a^2 = \frac{9}{4}$$

$$a = \pm\sqrt{\frac{9}{4}}$$

$$a = \pm\frac{3}{2}$$

CHECK $a = \frac{3}{2}$:

$$8a^2 - 18 = 0$$

$$8\left(\frac{3}{2}\right)^2 - 18 \stackrel{?}{=} 0$$

$$\overset{2}{8}\left(\frac{9}{\cancel{4}}\right) - 18 \stackrel{?}{=} 0$$

$$18 - 18 \stackrel{?}{=} 0$$

$$0 \stackrel{\checkmark}{=} 0$$

CHECK $a = -\frac{3}{2}$:

$$8a^2 - 18 = 0$$

$$8\left(-\frac{3}{2}\right)^2 - 18 \stackrel{?}{=} 0$$

$$\overset{2}{8}\left(\frac{9}{\cancel{4}}\right) - 18 \stackrel{?}{=} 0$$

$$18 - 18 \stackrel{?}{=} 0$$

$$0 = 0$$

19.　$0 = x^2 - x - 6$

$$0 = (x - 3)(x + 2)$$

$$0 = x - 3 \text{ or } 0 = x + 2$$

$$3 = x \text{ or } -2 = x$$

CHECK $x = 3$:

$$0 = x^2 - x - 6$$

$$0 \stackrel{?}{=} 3^2 - 3 - 6$$

$$0 \stackrel{?}{=} 9 - 3 - 6$$

$$0 \stackrel{\checkmark}{=} 0$$

CHECK $x = -2$:

$$0 = x^2 - x - 6$$

$$0 \stackrel{?}{=} (-2)^2 - (-2) - 6$$

$$0 \stackrel{?}{=} 4 + 2 - 6$$

$$0 \stackrel{\checkmark}{=} 0$$

23.　$0 = 8x^2 + 4x - 112$

$$0 = 4\left(2x^2 + x - 28\right)$$

$$0 = 2x^2 + x - 28$$

$$0 = (2x - 7)(x + 4)$$

$$0 = 2x - 7 \text{ or } 0 = x + 4$$

CHECK $x = \frac{7}{2}$:

$$0 = 8x^2 + 4x - 112$$

$$0 \stackrel{?}{=} 8\left(\frac{7}{2}\right)^2 + 4\left(\frac{7}{2}\right) - 112$$

$$0 \stackrel{?}{=} \overset{2}{8}\left(\frac{49}{\cancel{4}}\right) + \overset{2}{\cancel{4}}\left(\frac{7}{\cancel{2}}\right) - 112$$

$$0 \stackrel{?}{=} 98 + 14 - 112$$

$7 = 2x$ or $-4 = x$

$\dfrac{7}{2} = x$ or $-4 = x$

$0 \overset{\checkmark}{=} 0$

CHECK $x = -4$:

$0 = 8x^2 + 4x - 112$

$0 \overset{?}{=} 8(-4)^2 + 4(-4) - 112$

$0 \overset{?}{=} 8(16) + 4(-4) - 112$

$0 \overset{?}{=} 128 - 16 - 112$

$0 \overset{\checkmark}{=} 0$

27. $7a^2 = 19$

$a^2 = \dfrac{19}{7}$

$a = \pm\sqrt{\dfrac{19}{7}}$

$a = \dfrac{\pm\sqrt{19}}{\sqrt{7}}$

$a = \dfrac{\pm\sqrt{19}\sqrt{7}}{\sqrt{7}\sqrt{7}}$

$a = \pm\dfrac{\sqrt{133}}{7}$

CHECK $a = \dfrac{\sqrt{133}}{7}$:

$7a^2 = 19$

$7\left(\dfrac{\sqrt{133}}{7}\right)^2 \overset{?}{=} 19$

$\cancel{7}\left(\dfrac{133}{\cancel{49}\,7}\right) \overset{?}{=} 19$

$19 \overset{\checkmark}{=} 19$

CHECK $a = -\dfrac{\sqrt{133}}{7}$:

$7a^2 = 19$

$7\left(-\dfrac{\sqrt{133}}{7}\right)^2 \overset{?}{=} 19$

$\cancel{7}\left(\dfrac{133}{\cancel{49}\,7}\right) \overset{?}{=} 19$

$19 \overset{\checkmark}{=} 19$

31. $6a^2 + 3a - 1 = 4a$

$6a^2 - a - 1 = 0$

$(3a + 1)(2a - 1) = 0$

$3a + 1 = 0$ or $2a - 1 = 0$

$3a = -1$ or $2a = 1$

$a = -\dfrac{1}{3}$ or $a = \dfrac{1}{2}$

CHECK $a = -\dfrac{1}{3}$:

$6a^2 + 3a - 1 = 4a$

$6\left(-\dfrac{1}{3}\right)^2 + 3\left(-\dfrac{1}{3}\right) - 1 \overset{?}{=} 4\left(-\dfrac{1}{3}\right)$

$6\left(\dfrac{1}{9}\right) + 3\left(-\dfrac{1}{3}\right) - 1 \overset{?}{=} 4\left(-\dfrac{1}{3}\right)$

$\dfrac{2}{3} - 1 - 1 \overset{?}{=} -\dfrac{4}{3}$

$-\dfrac{4}{3} \overset{\checkmark}{=} -\dfrac{4}{3}$

$$\text{CHECK } a = \frac{1}{2}:$$

$$6a^2 + 3a - 1 = 4a$$

$$6\left(\frac{1}{2}\right)^2 + 3\left(\frac{1}{2}\right) - 1 \overset{?}{=} 4\left(\frac{1}{2}\right)$$

$$\overset{3}{\cancel{6}}\left(\frac{1}{\cancel{4}_2}\right) + 3\left(\frac{1}{2}\right) - 1 \overset{?}{=} \overset{2}{\cancel{4}}\left(\frac{1}{\cancel{2}}\right)$$

$$\frac{3}{2} + \frac{3}{2} - 1 \overset{?}{=} 2$$

$$2 \overset{\checkmark}{=} 2$$

33. $3y^2 - 5y + 8 = 9y^2 - 10y + 2$

$$0 = 6y^2 - 5y - 6$$

$$0 = (3y + 2)(2y - 3)$$

$$3y + 2 = 0 \text{ or } 2y - 3 = 0$$

$$3y = -2 \text{ or } 2y = 3$$

$$y = -\frac{2}{3} \text{ or } y = \frac{3}{2}$$

$$\text{CHECK } y = -\frac{2}{3}:$$

$$3y^2 - 5y + 8 = 9y^2 - 10y + 2$$

$$3\left(-\frac{2}{3}\right)^2 - 5\left(-\frac{2}{3}\right) + 8 \overset{?}{=} 9\left(-\frac{2}{3}\right)^2 - 10\left(-\frac{2}{3}\right) + 2$$

$$\overset{1}{\cancel{3}}\left(\frac{4}{\cancel{9}_3}\right) - 5\left(-\frac{2}{3}\right) + 8 \overset{?}{=} \cancel{9}\left(-\frac{4}{\cancel{9}}\right) - 10\left(-\frac{2}{3}\right) + 2$$

$$\frac{4}{3} + \frac{10}{3} + 8 \overset{?}{=} 4 + \frac{20}{3} + 2$$

$$\frac{38}{3} \overset{\checkmark}{=} \frac{38}{3}$$

$$\text{CHECK } y = \frac{3}{2}:$$

$$3y^2 - 5y + 8 = 9y^2 - 10y + 2$$

$$3\left(\frac{3}{2}\right)^2 - 5\left(\frac{3}{2}\right) + 8 \overset{?}{=} 9\left(\frac{3}{2}\right)^2 - \overset{5}{\cancel{10}}\left(\frac{3}{\cancel{2}}\right)_1 + 2$$

$$3\left(\frac{9}{4}\right) - 5\left(\frac{3}{2}\right) + 8 \overset{?}{=} 9\left(\frac{9}{4}\right) - 15 + 2$$

$$\frac{27}{4} - \frac{15}{2} + 8 \overset{?}{=} \frac{81}{4} - 15 + 2$$

$$\frac{29}{4} \overset{\checkmark}{=} \frac{29}{4}$$

37. $(2s - 3)(3s + 1) = 7$

$$6s^2 - 7s - 3 = 7$$

$$6s^2 - 7s - 10 = 0$$

$$(6s + 5)(s - 2) = 0$$

$$6s + 5 = 0 \text{ or } s - 2 = 0$$

$$6s = -5 \text{ or } s = 2$$

$$s = -\frac{5}{6} \text{ or } s = 2$$

$$\text{CHECK } s = -\frac{5}{6}:$$

$$(2s - 3)(3s + 1) = 7$$

$$\left(\overset{}{\cancel{2}}\left(-\frac{5}{\cancel{6}_3}\right) - 3\right)\left(\overset{1}{\cancel{3}}\left(-\frac{5}{\cancel{6}_2}\right) + 1\right) \overset{?}{=} 7$$

$$\left(-\frac{5}{3} - 3\right)\left(-\frac{5}{2} + 1\right) \overset{?}{=} 7$$

$$\left(-\frac{14}{\cancel{3}}\right)\left(-\frac{\cancel{3}}{\cancel{2}}\right) \overset{?}{=} 7$$

$$(-7)(-1) \overset{?}{=} 7$$

$$7 \overset{\checkmark}{=} 7$$

CHECK s = 2:

$$(2s - 3)(3s + 1) = 7$$

$$(2(2) - 3)(3(2) + 1) \stackrel{?}{=} 7$$

$$(4 - 3)(6 + 1) \stackrel{?}{=} 7$$

$$(1)(7) \stackrel{?}{=} 7$$

$$7 \stackrel{\checkmark}{=} 7$$

39. $(x + 2)^2 = 25$

$$x + 2 = \pm\sqrt{25}$$

$$x + 2 = \pm 5$$

$$x + 2 = 5 \text{ or } x + 2 = -5$$

$$x = 3 \text{ or } x = -7$$

CHECK x = 3:

$$(x + 2)^2 = 25$$

$$(3 + 2)^2 \stackrel{?}{=} 25$$

$$5^2 \stackrel{?}{=} 25$$

$$25 \stackrel{\checkmark}{=} 25$$

CHECK x = -7:

$$(x + 2)^2 = 25$$

$$(-7 + 2)^2 \stackrel{?}{=} 25$$

$$(-5)^2 \stackrel{?}{=} 25$$

$$25 \stackrel{\checkmark}{=} 25$$

41. $(x - 2)(3x - 1) = (2x - 3)(x + 1)$

$$3x^2 - 7x + 2 = 2x^2 - x - 3$$

$$x^2 - 6x + 5 = 0$$

$$(x - 1)(x - 5) = 0$$

$$x - 1 = 0 \text{ or } x - 5 = 0$$

$$x = 1 \text{ or } x = 5$$

CHECK x = 1:

$$(x - 2)(3x - 1) = (2x - 3)(x + 1)$$

$$(1 - 2)(3(1) - 1) \stackrel{?}{=} (2(1) - 3)(1 + 1)$$

$$(-1)(2) \stackrel{?}{=} (-1)(2)$$

$$-2 \stackrel{\checkmark}{=} -2$$

CHECK x = 5:

$$(x - 2)(3x - 1) = (2x - 3)(x + 1)$$

$$(5 - 2)(3(5) - 1) \stackrel{?}{=} (2(5) - 3)(5 + 1)$$

$$(3)(14) \stackrel{?}{=} (7)(6)$$

$$42 \stackrel{\checkmark}{=} 42$$

47. $x^2 + 3 = 1$

$$x^2 = -2$$

$$x = \pm\sqrt{-2}$$

$$x = \pm\sqrt{2}\, i$$

CHECK x = $\sqrt{2}$i:

$$x^2 + 3 = 1$$

$$\left(\sqrt{2}i\right)^2 + 3 \stackrel{?}{=} 1$$

$$2i^2 + 3 \stackrel{?}{=} 1$$

$$2(-1) + 3 \stackrel{?}{=} 1$$

$$-2 + 3 \stackrel{?}{=} 1$$

$$1 \stackrel{\checkmark}{=} 1$$

CHECK $x = -\sqrt{2}i$:

$$x^2 + 3 = 1$$

$$\left(-\sqrt{2}i\right)^2 + 3 \stackrel{?}{=} 1$$

$$2i^2 + 3 \stackrel{?}{=} 1$$

$$2(-1) + 3 \stackrel{?}{=} 1$$

$$-2 + 3 \stackrel{?}{=} 1$$

$$1 \stackrel{\checkmark}{=} 1$$

49.　　　$8 = (x - 8)^2$

$$\pm\sqrt{8} = x - 8$$

$$\pm 2\sqrt{2} = x - 8$$

$$8 \pm 2\sqrt{2} = x$$

CHECK $x = 8 + 2\sqrt{2}$:

$$8 = (x - 8)^2$$

$$8 \stackrel{?}{=} \left(8 + 2\sqrt{2} - 8\right)^2$$

$$8 \stackrel{?}{=} \left(2\sqrt{2}\right)^2$$

$$8 \stackrel{\checkmark}{=} 8$$

CHECK $x = 8 - 2\sqrt{2}$:

$$8 = (x - 8)^2$$

$$8 \stackrel{?}{=} \left(8 - 2\sqrt{2} - 8\right)^2$$

$$8 \stackrel{?}{=} \left(-2\sqrt{2}\right)^2$$

$$8 \stackrel{\checkmark}{=} 8$$

53.　$\dfrac{2}{x - 1} + x = 4$

$$(x - 1)\left(\frac{2}{x - 1} + x\right) = (x - 1)4$$

$$(x - 1)\left(\frac{2}{x - 1}\right) + (x - 1)x = (x - 1)4$$

$$2 + x^2 - x = 4x - 4$$

$$x^2 - 5x + 6 = 0$$

$$(x - 2)(x - 3) = 0$$

$$x - 2 = 0 \text{ or } x - 3 = 0$$

$$x = 2 \text{ or } x = 3$$

CHECK $x = 2$:

$$\frac{2}{x - 1} + x = 4$$

$$\frac{2}{2 - 1} + 2 \stackrel{?}{=} 4$$

$$\frac{2}{1} + 2 \stackrel{?}{=} 4$$

$$2 + 2 \stackrel{?}{=} 4$$

$$4 \stackrel{\checkmark}{=} 4$$

$$\frac{2}{x-1} + x = 4$$

$$\frac{2}{3-1} + 3 \overset{?}{=} 4$$

$$\frac{2}{2} + 3 \overset{?}{=} 4$$

$$1 + 3 \overset{?}{=} 4$$

$$4 \overset{\checkmark}{=} 4$$

57.

$$\frac{5x+2}{x-2} = \frac{2x+1}{x-2}$$

$$(x-2)\left(\frac{5x+2}{x-2}\right) = (x-2)\left(\frac{2x+1}{x-2}\right)$$

$$5x + 2 = 2x + 1$$

$$3x + 2 = 1$$

$$3x = -1$$

$$x = -\frac{1}{3}$$

CHECK $x = -\frac{1}{3}$:

$$\frac{5x+2}{x-2} = \frac{2x+1}{x-2}$$

$$\frac{5\left(-\frac{1}{3}\right)+2}{-\frac{1}{3}-2} \overset{?}{=} \frac{2\left(-\frac{1}{3}\right)+1}{-\frac{1}{3}-2}$$

$$\frac{-\frac{5}{3}+2}{-\frac{1}{3}-2} \overset{?}{=} \frac{-\frac{2}{3}+1}{-\frac{1}{3}-2}$$

$$\frac{\frac{1}{3}}{-\frac{7}{3}} \overset{?}{=} \frac{\frac{1}{3}}{-\frac{7}{3}}$$

$$-\frac{1}{7} \overset{\checkmark}{=} -\frac{1}{7}$$

63.

$$\frac{3}{x-2} + \frac{7}{x+2} = \frac{x+1}{x-2}$$

$$(x-2)(x+2)\left(\frac{3}{x-2} + \frac{7}{x+2}\right) = (x-2)(x+2)\left(\frac{x+1}{x-2}\right)$$

$$(x-2)(x+2)\left(\frac{3}{x-2}\right) + (x-2)(x+2)\left(\frac{7}{x+2}\right) = (x-2)(x+2)\left(\frac{x+1}{x-2}\right)$$

$$(x+2)(3) + (x-2)(7) = (x+2)(x+1)$$

$$3x + 6 + 7x - 14 = x^2 + 3x + 2$$

$$10x - 8 = x^2 + 3x + 2$$

$$0 = x^2 - 7x + 10$$

$$0 = (x-2)(x-5)$$

$$0 = x - 2 \text{ or } 0 = x - 5$$

$$2 = x \text{ or } 5 = x$$

CHECK x = 5:

$$\frac{3}{x-2} + \frac{7}{x+2} = \frac{x+1}{x-2}$$

$$\frac{3}{5-2} + \frac{7}{5+2} \overset{?}{=} \frac{5+1}{5-2}$$

CHECK x = 2:

$$\frac{3}{x-2} + \frac{7}{x+2} = \frac{x+1}{x-2}$$

$$\frac{3}{2-2} + \frac{7}{2+2} \overset{?}{=} \frac{2+1}{2-2}$$

$$\frac{3}{3} + \frac{7}{7} \overset{?}{=} \frac{6}{3} \qquad\qquad \frac{3}{0} + \frac{7}{4} \neq \frac{3}{0}$$

$$1 + 1 \overset{?}{=} 2 \qquad\qquad \text{since we cannot divide by 0.}$$

$$2 \overset{\checkmark}{=} 2$$

Hence, the only solution is x = 5.

67. $8a^2 + 3b = 5b$

$8a^2 = 2b$

$a^2 = \dfrac{2b}{8}$

$a^2 = \dfrac{b}{4}$

$a = \pm\sqrt{\dfrac{b}{4}}$

$a = \pm\dfrac{\sqrt{b}}{2}$

71. $V = \dfrac{2}{3}\pi r^2, \ r > 0$

$3V = 2\pi r^2$

$\dfrac{3V}{2\pi} = r^2$

$\sqrt{\dfrac{3V}{2\pi}} = r$

$\dfrac{\sqrt{3V}}{\sqrt{2\pi}} = r$

$\dfrac{\sqrt{3V}\,\sqrt{2\pi}}{\sqrt{2\pi}\,\sqrt{2\pi}} = r$

$\dfrac{\sqrt{6\pi V}}{2\pi} = r$

75. $x^2 - xy - 6y^2 = 0$

$(x - 3y)(x + 2y) = 0$

$x - 3y = 0 \ \text{or} \ x + 2y = 0$

$x = 3y \ \text{or} \ x = -2y$

77. The factoring method is based on the zero factor property, which tells us that if the product of two quantities is zero, then either one or both of the quantities must be zero.

78. (a) If the product of two quantities is equal to 7, then we cannot draw any conclusion about the value of either quantity.

(b) The "solution" x = 3 is impossible. In fact, we can divide both sides of the equation by 3, obtaining $x(x - 2) = 0$, from which we get x = 0 and x = 2 as solutions.

200

79. (a) <u>Factoring method</u>

$$x^2 = 9$$
$$x^2 - 9 = 0$$
$$(x - 3)(x + 3) = 0$$
$$x - 3 = 0 \text{ or } x + 3 = 0$$
$$x = 3 \text{ or } x = -3$$

<u>Square root method</u>

$$x^2 = 9$$
$$x = \pm\sqrt{9}$$
$$x = \pm 3$$

(b) $9x^2 - 16 = 0$
$$(3x - 4)(3x + 4) = 0$$
$$3x - 4 = 0 \text{ or } 3x + 4 = 0$$
$$3x = 4 \text{ or } 3x = -4$$
$$x = \frac{4}{3} \text{ or } x = -\frac{4}{3}$$

$9x^2 - 16 = 0$
$$9x^2 = 16$$
$$x^2 = \frac{16}{9}$$
$$x = \pm\sqrt{\frac{16}{9}}$$
$$x = \pm\frac{4}{3}$$

(c) $7x^2 - 5 = 3x^2 + 4$
$$4x^2 - 5 = 4$$
$$4x^2 - 9 = 0$$
$$(2x - 3)(2x + 3) = 0$$
$$2x - 3 = 0 \text{ or } 2x + 3 = 0$$
$$2x = 3 \text{ or } 2x = -3$$
$$x = \frac{3}{2} \text{ or } x = -\frac{3}{2}$$

$7x^2 - 5 = 3x^2 + 4$
$$4x^2 - 5 = 4$$
$$4x^2 = 9$$
$$x^2 = \frac{9}{4}$$
$$x = \pm\sqrt{\frac{9}{4}}$$
$$x = \pm\frac{3}{2}$$

80. (1) given

(2) addition property of equality

(3) addition property of equality

(4) arithmetic fact $(7 + 4 = 11)$

(5) factorization of $x^2 + 4x + 4$

(6) square root method

(7) subtraction property of equality

81.
$$x^2 - 4x + 1 = 0$$
$$(2 + \sqrt{3})^2 - 4(2 + \sqrt{3}) + 1 \overset{?}{=} 0$$
$$4 + 4\sqrt{3} + (\sqrt{3})^2 - 8 - 4\sqrt{3} + 1 \overset{?}{=} 0$$
$$4 + 4\sqrt{3} + 3 - 8 - 4\sqrt{3} + 1 \overset{?}{=} 0$$
$$0 \overset{\checkmark}{=} 0$$

Thus, $2 + \sqrt{3}$ is a solution to the equation $x^2 - 4x + 1 = 0$.

$$x^2 - 4x + 1 = 0$$
$$(2 - \sqrt{3})^2 - 4(2 - \sqrt{3}) + 1 \overset{?}{=} 0$$
$$4 - 4\sqrt{3} + (\sqrt{3})^2 - 8 + 4\sqrt{3} + 1 \overset{?}{=} 0$$
$$4 - 4\sqrt{3} + 3 - 8 + 4\sqrt{3} + 1 \overset{?}{=} 0$$
$$0 \overset{\checkmark}{=} 0$$

Thus, $2 - \sqrt{3}$ is a solution to the equation $x^2 - 4x + 1 = 0$.

[handwritten in left margin: $(2+\sqrt{3})(2+\sqrt{3})$ $+2\sqrt{3} + 2\sqrt{3} + \sqrt{3} + \sqrt{3}$ $+4\sqrt{3} + (\sqrt{3})^2$]

Exercises 8.2

3.

$$0 = c^2 - 2c - 5$$

$$5 = c^2 - 2c$$

$$5 + 1 = c^2 - 2c + 1$$

$$6 = (c - 1)^2$$

$$\pm\sqrt{6} = c - 1$$

$$1 \pm \sqrt{6} = c$$

$$\left(\frac{1}{2}(-2)\right)^2 = (-1)^2 = 1$$

7.

$$2x^2 + 3x - 1 = x^2 - 2$$

$$x^2 + 3x - 1 = -2$$

$$x^2 + 3x = -1$$

$$x^2 + 3x + \frac{9}{4} = -1 + \frac{9}{4}$$

$$\left(x + \frac{3}{2}\right)^2 = \frac{5}{4}$$

$$x + \frac{3}{2} = \pm\sqrt{\frac{5}{4}}$$

$$x + \frac{3}{2} = \pm\frac{\sqrt{5}}{\sqrt{4}}$$

$$x + \frac{3}{2} = \pm\frac{\sqrt{5}}{2}$$

$$x = -\frac{3}{2} \pm \frac{\sqrt{5}}{2} = \frac{-3 \pm \sqrt{5}}{2}$$

$$\left(\frac{1}{3}(3)\right)^2 = \left(\frac{3}{2}\right)^2 = \frac{9}{4}$$

9.

$$(a - 2)(a + 1) = 2$$

$$a^2 - a - 2 = 2$$

$$\left(\frac{1}{2}(-1)\right)^2 = \left(-\frac{1}{2}\right)^2 = \frac{1}{4}$$

$$a^2 - a = 4$$

$$a^2 - a + \frac{1}{4} = 4 + \frac{1}{4}$$

$$\left(a - \frac{1}{2}\right)^2 = \frac{17}{4}$$

$$a - \frac{1}{2} = \pm\sqrt{\frac{17}{4}}$$

$$a - \frac{1}{2} = \pm\frac{\sqrt{17}}{\sqrt{4}}$$

$$a - \frac{1}{2} = \pm\frac{\sqrt{17}}{2}$$

$$a = \frac{1}{2} \pm \frac{\sqrt{17}}{2} = \frac{1 \pm \sqrt{17}}{2}$$

13.
$$3x^2 + 3x = x^2 - 5x + 4$$
$$2x^2 + 3x = -5x + 4$$
$$2x^2 + 8x = 4$$
$$x^2 + 4x = 2$$
$$x^2 + 4x + 4 = 2 + 4$$
$$(x + 2)^2 = 6$$
$$x + 2 = \pm\sqrt{6}$$
$$x = -2 \pm \sqrt{6}$$

$$\left(\frac{1}{2}(4)\right)^2 = 2^2 = 4$$

17.
$$2x - 7 = x^2 - 3x + 4$$
$$-7 = x^2 - 5x + 4$$
$$-11 = x^2 - 5x$$
$$-11 + \frac{25}{4} = x^2 - 5x + \frac{25}{4}$$
$$-\frac{19}{4} = \left(x - \frac{5}{2}\right)^2$$
$$\pm\sqrt{-\frac{19}{4}} = x - \frac{5}{2}$$
$$\pm\frac{\sqrt{19}}{\sqrt{4}}i = x - \frac{5}{2}$$
$$\pm\frac{\sqrt{19}}{2}i = x - \frac{5}{2}$$
$$\frac{5}{2} \pm \frac{\sqrt{19}}{2}i = x$$
$$\frac{5 \pm \sqrt{19}i}{2} = x$$

$$\left(\frac{1}{2}(-5)\right)^2 = \left(-\frac{5}{2}\right)^2 = \frac{25}{4}$$

19. $5x^2 + 10x - 14 = 20$

$$5x^2 + 10x = 34$$

$$x^2 + 2x = \frac{34}{5}$$

$$x^2 + 2x + 1 = \frac{34}{5} + 1$$

$$\left(\frac{1}{2}(2)\right)^2 = 1^2 = 1$$

$$(x+1)^2 = \frac{39}{5}$$

$$x + 1 = \pm\sqrt{\frac{39}{5}}$$

$$x + 1 = \pm\frac{\sqrt{39}}{\sqrt{5}}$$

$$x + 1 = \pm\frac{\sqrt{39}\sqrt{5}}{\sqrt{5}\sqrt{5}}$$

$$x + 1 = \pm\frac{\sqrt{195}}{5}$$

$$x = -1 \pm \frac{\sqrt{195}}{5} = \frac{-5 \pm \sqrt{195}}{5}$$

23. $(2x+5)(x-3) = (x+4)(x-1)$

$2x^2 - 6x + 5x - 15$
$2x^2 - x - 15$

$$2x^2 - x - 15 = x^2 + 3x - 4$$

$$x^2 - x - 15 = 3x - 4$$

$$x^2 - 4x - 15 = -4$$

$$x^2 - 4x = 11$$

$$\left(\frac{1}{2}(-4)\right)^2 = (-2)^2 = 4$$

$$x^2 - 4x + 4 = 11 + 4$$

$$(x-2)^2 = 15$$

$$x - 2 = \pm\sqrt{15}$$

$$x = 2 \pm \sqrt{15}$$

25. $\dfrac{2x}{2x-3} = \dfrac{3x-1}{x+1}$

$$(2x-3)(x+1)\left(\frac{2x}{2x-3}\right) = (2x-3)(x+1)\left(\frac{3x-1}{x+1}\right) =$$

$2x(x+1) = (2x-3)(3x-1)$

$2x^2 + 2x = 6x^2 - 2x - 9x + 3$

$$2x^2 + 2x = 6x^2 - 11x + 3$$

$$2x = 4x^2 - 11x + 3$$

$$0 = 4x^2 - 13x + 3$$

$$\frac{-3}{4} = \frac{4x^2 - 13x}{4}$$

$$-\frac{3}{4} = x^2 - \frac{13}{4}x$$

$$\left(\frac{1}{2}\left(-\frac{13}{4}\right)\right)^2 = \left(-\frac{13}{8}\right)^2 = \frac{169}{64}$$

$-\frac{3}{4} = \frac{48}{64} + \frac{169}{64}$

$-\frac{3}{4} + \frac{169}{64} = x^2 - \frac{13}{4}x + \frac{169}{64}$

$\frac{121}{64} = \left(x - \frac{13}{8}\right)^2 = \left(x - \frac{13}{8}\right)\left(x - \frac{13}{8}\right) = x^2 - \frac{13}{8}x - \frac{13}{8} + \frac{169}{64}$

$\frac{\sqrt{121}}{\sqrt{64}} = \frac{11}{8}$

$\pm\sqrt{\frac{121}{64}} = x - \frac{13}{8}$

$\pm\frac{11}{8} = x - \frac{13}{8}$

$\frac{13}{8} \pm \frac{11}{8} = x$

add $\rightarrow \frac{13}{8} + \frac{11}{8} = x$ or $\frac{13}{8} - \frac{11}{8} = x \checkmark$ subtract

divide $\rightarrow \frac{24}{8} = x$ or $\frac{2}{8} = x \leftarrow$ divide

$\boxed{3 = x \text{ or } \frac{1}{4} = x}$

29.
$0 = 2y^2 + 2y + 5$

$-5 = 2y^2 + 2y$

$-\frac{5}{2} = y^2 + y$

$-\frac{5}{2} + \frac{1}{4} = y^2 + y + \frac{1}{4}$

$-\frac{9}{4} = \left(y + \frac{1}{2}\right)^2$

$\pm\sqrt{-\frac{9}{4}} = y + \frac{1}{2}$

$\pm\frac{3}{2}i = y + \frac{1}{2}$

$-\frac{1}{2} \pm \frac{3}{2}i = y$

$\frac{-1 \pm 3i}{2} = y$

$\left(\frac{1}{2}(1)\right)^2 = \left(\frac{1}{2}\right)^2 = \frac{1}{4}$

31.
$5n^2 - 3n = 2n^2 - 6$

$3n^2 - 3n = -6$

$n^2 - n = -2$

$n^2 - n + \frac{1}{4} = -2 + \frac{1}{4}$

$\left(\frac{1}{2}(-1)\right)^2 = \left(-\frac{1}{2}\right)^2 = \frac{1}{4}$

$$\left(n - \frac{1}{2}\right)^2 = -\frac{7}{4}$$

$$n - \frac{1}{2} = \pm\sqrt{-\frac{7}{4}}$$

$$n - \frac{1}{2} = \pm\frac{\sqrt{7}}{\sqrt{4}}\,i$$

$$n - \frac{1}{2} = \pm\frac{\sqrt{7}}{2}\,i$$

$$n = \frac{1}{2} \pm \frac{\sqrt{7}}{2}\,i = \frac{1 \pm \sqrt{7}i}{2}$$

35.
$$\frac{3}{x+2} - \frac{2}{x-1} = 5$$

$$(x+2)(x-1)\left(\frac{3}{x+2} - \frac{2}{x-1}\right) = (x+2)(x-1)(5)$$

$$(x+2)(x-1)\left(\frac{3}{x+2}\right) - (x+2)(x-1)\left(\frac{2}{x-1}\right) = (x+2)(x-1)(5)$$

$$3x - 3 - 2x - 4 = 5x^2 + 5x + 10$$

$$x - 7 = 5x^2 + 5x + 10$$

$$-7 = 5x^2 + 4x + 10$$

$$3 = 5x^2 + 4x$$

$$\frac{3}{5} = x^2 + \frac{4}{5}x \qquad\qquad \left(\frac{1}{2}\left(\frac{4}{5}\right)\right)^2 = \left(\frac{2}{5}\right)^2 = \frac{4}{25}$$

$$\frac{3}{5} + \frac{4}{25} = x^2 + \frac{4}{5}x + \frac{4}{25}$$

$$\frac{19}{25} = \left(x + \frac{2}{5}\right)^2$$

$$\pm\sqrt{\frac{19}{25}} = x + \frac{2}{5}$$

$$\pm\frac{\sqrt{19}}{\sqrt{25}} = x + \frac{2}{5}$$

$$\pm\frac{\sqrt{19}}{5} = x + \frac{2}{5}$$

$$-\frac{2}{5} \pm \frac{\sqrt{19}}{5} = x$$

$$\frac{-2 \pm \sqrt{19}}{5} = x$$

37. Assuming that the perfect square has a leading coefficient of 1, the numerical term is the square of one-half the middle term coefficient.

38.
$$3x^2 + 11x = 4$$

$$x^2 + \frac{11}{3}x = \frac{4}{3} \qquad\qquad \left(\frac{1}{2}\left(\frac{11}{3}\right)^2\right) = \left(\frac{11}{6}\right)^2 = \frac{121}{36}$$

206

$$x^2 + \frac{11}{3}x + \frac{121}{36} = \frac{4}{3} + \frac{121}{36}$$

$$\left(x + \frac{11}{6}\right)^2 = \frac{169}{36}$$

$$x + \frac{11}{6} = \pm\sqrt{\frac{169}{36}}$$

$$x + \frac{11}{6} = \pm\frac{13}{6}$$

$$x = -\frac{11}{6} \pm \frac{13}{6}$$

$$x = -\frac{11}{6} + \frac{13}{6} \text{ or } x = -\frac{11}{6} - \frac{13}{6}$$

$$x = \frac{2}{6} \text{ or } x = -\frac{24}{6}$$

$$x = \frac{1}{3} \text{ or } x = -4$$

The fact that the answers are rational tells us that the original equation could have been solved by factoring.

$$3x^2 + 11x = 4$$
$$3x^2 + 11x - 4 = 0$$
$$(3x - 1)(x + 4) = 0$$
$$3x - 1 = 0 \text{ or } x + 4 = 0$$
$$3x = 1 \text{ or } x = -4$$
$$x = \frac{1}{3} \text{ or } x = -4$$

The second method is easier than the first in this example.

39.　　$x^2 + rx + q = 0$

　　　　$x^2 + rx = -q$

$\left(\frac{1}{2}r\right)^2 = \left(\frac{r}{2}\right)^2 = \frac{r^2}{4}$

$x^2 + rx + \frac{r^2}{4} = -q + \frac{r^2}{4}$

$\frac{1}{2}\left(\frac{r}{1}\right) = \left(\frac{r}{2}\right)^2 \frac{r^2}{4}$

$\left(x + \frac{r}{2}\right)^2 = \frac{-4q + r^2}{4} = \frac{r^2 - 4q}{4}$

$x + \frac{r}{2} = \pm\sqrt{\frac{r^2 - 4q}{4}} = \pm\frac{\sqrt{r^2 - 4q}}{2}$

$x = -\frac{r}{2} \pm \frac{\sqrt{r^2 - 4q}}{2} = \frac{-r \pm \sqrt{r^2 - 4q}}{2}$

40.　　$Ax^2 + Bx + C = 0$

　　　　$Ax^2 + Bx = -C$

　　　　$x^2 + \frac{B}{A}x = -\frac{C}{A}$

$\left(\frac{1}{2}\left(\frac{B}{A}\right)\right)^2 = \left(\frac{B}{2A}\right)^2 = \frac{B^2}{4A^2}$

$$x^2 + \frac{B}{A}x + \frac{B^2}{4A^2} = -\frac{C}{A} + \frac{B^2}{4A^2}$$

$$\left(x + \frac{B}{2A}\right)^2 = \frac{-4AC + B^2}{4A^2} = \frac{B^2 - 4AC}{4A^2}$$

$$x + \frac{B}{2A} = \pm\sqrt{\frac{B^2 - 4AC}{4A^2}} = \pm\frac{\sqrt{B^2 - 4AC}}{\sqrt{4A^2}}$$

$$x + \frac{B}{2A} = \pm\frac{\sqrt{B^2 - 4AC}}{2A}$$

$$x = -\frac{B}{2A} \pm \frac{\sqrt{B^2 - 4AC}}{2A}$$

$$x = \frac{-B \pm \sqrt{B^2 - 4AC}}{2A}$$

41. $3\left\{6 - \left[(-4-5)^2 + 5\right] - 9\right\} = 3\left\{6 - \left[(-9)^2 + 5\right] - 9\right\} = 3\left\{6 - [81 + 5] - 9\right\}$
$$= 3\{6 - 86 - 9\} = 3\{-89\} = -267$$

43. $\dfrac{2x+1}{x^2 - 9} \div \dfrac{3x}{3-x} = \dfrac{2x+1}{x^2-9} \cdot \dfrac{3-x}{3x} = \dfrac{2x+1}{(x-3)(x+3)} \cdot \dfrac{(-1)(x-3)}{3x} = -\dfrac{2x+1}{3x(x+3)}$

45. (a) $\dfrac{8 + \sqrt{28}}{10} = \dfrac{8 + 2\sqrt{7}}{10} = \dfrac{2(4 + \sqrt{7})}{10\,5} = \dfrac{4 + \sqrt{7}}{5}$

(b) $\dfrac{2\sqrt{5}}{\sqrt{6} - \sqrt{2}} = \dfrac{2\sqrt{5}}{\sqrt{6} - \sqrt{2}} \cdot \dfrac{\sqrt{6} + \sqrt{2}}{\sqrt{6} + \sqrt{2}} = \dfrac{2\sqrt{5}(\sqrt{6} + \sqrt{2})}{6 - 2} = \dfrac{2\sqrt{5}(\sqrt{6} + \sqrt{2})}{4\,2} = \dfrac{\sqrt{30} + \sqrt{10}}{2}$

Exercises 8.3

1. $x^2 - 4x - 5 = 0 \qquad A = 1,\ B = -4,\ C = -5$

$$x = \frac{-B \pm \sqrt{B^2 - 4AC}}{2A} = \frac{-(-4) \pm \sqrt{(-4)^2 - 4(1)(-5)}}{2(1)} = \frac{4 \pm \sqrt{16 + 20}}{2} = \frac{4 \pm \sqrt{36}}{2} = \frac{4 \pm 6}{2}$$

$$x = \frac{4 + 6}{2} = \frac{10}{2} = 5 \text{ or } x = \frac{4 - 6}{2} = \frac{-2}{2} = -1$$

5. $(3y - 1)(2y - 3) = y$
$$6y^2 - 11y + 3 = y$$
$$6y^2 - 12y + 3 = 0$$
$$3(2y^2 - 4y + 1) = 0$$
$$2y^2 - 4y + 1 = 0 \qquad A = 2,\ B = -4,\ C = 1$$

$$y = \frac{-B \pm \sqrt{B^2 - 4AC}}{2A} = \frac{-(-4) \pm \sqrt{(-4)^2 - 4(2)(1)}}{2(2)} = \frac{4 \pm \sqrt{16 - 8}}{4}$$

$$= \frac{4 \pm \sqrt{8}}{4} = \frac{4 \pm 2\sqrt{2}}{4} = \frac{\cancel{2}(2 \pm \sqrt{2})}{\cancel{4}2} = \frac{2 \pm \sqrt{2}}{2}$$

7. $y^2 - 3y + 4 = 2y^2 + 4y - 3$

$ -3y + 4 = y^2 + 4y - 3$

$ 4 = y^2 + 7y - 3$

$ 0 = y^2 + 7y - 7 \qquad A = 2, B = 7, C = -7$

$$y = \frac{-B \pm \sqrt{B^2 - 4AC}}{2A} = \frac{-7 \pm \sqrt{7^2 - 4(1)(-7)}}{2(1)} = \frac{-7 \pm \sqrt{49 + 28}}{2} = \frac{-7 \pm \sqrt{77}}{2}$$

9. $2x^2 - 3x - 7 = 0 \qquad A = 2, B = -3, C = -7$

$$x = \frac{-B \pm \sqrt{B^2 - 4AC}}{2A} = \frac{-(-3) \pm \sqrt{(-3)^2 - 4(2)(-7)}}{2(2)} = \frac{3 \pm \sqrt{9 + 56}}{4} = \frac{3 \pm \sqrt{65}}{4}$$

13. $(s - 3)(s + 4) = (2s - 1)(s + 2)$

$ s^2 + s - 12 = 2s^2 + 3s - 2$

$ s - 12 = s^2 + 3s - 2$

$ -12 = s^2 + 2s - 2$

$ 0 = s^2 + 2s + 10 \qquad A = 1, B = 2, C = 10$

$$s = \frac{-B \pm \sqrt{B^2 - 4AC}}{2A} = \frac{-2 \pm \sqrt{2^2 - 4(1)(10)}}{2(1)}$$

$$= \frac{-2 \pm \sqrt{4 - 40}}{2} = \frac{-2 \pm \sqrt{-36}}{2} = \frac{-2 \pm 6i}{2} = -1 \pm 3i$$

17. $ x^2 - 10x = -25$

$x^2 - 10x + 25 = 0 \qquad A = 1, B = -10, C = 25$

$B^2 - 4AC = (-10)^2 - 4(1)(25) = 100 - 100 = 0$

Thus, the roots are real and equal.

21. $2a^2 + 4a = 0 \qquad A = 2, B = 4, C = 0$

$B^2 - 4AC = 4^2 - 4(2)(0) = 16 - 0 = 16 > 0$

Thus, the roots are real and distinct.

25. $(2y + 3)(y - 1) = y + 5$

$ 2y^2 + y - 3 = y + 5$

$ 2y^2 - 3 = 5$

$ 2y^2 - 8 = 0 \qquad A = 2, B = 0, C = -8$

$B^2 - 4AC = 0^2 - 4(2)(-8) = 64 > 0$

Thus, the roots are real and distinct.

27. $a^2 - 3a - 4 = 0$
$(a - 4)(a + 1) = 0$
$a - 4 = 0$ or $a + 1 = 0$
$a = 4$ or $a = -1$

31. $(5x - 4)(2x - 3) = 0$
$5x - 4 = 0$ or $2x - 3 = 0$
$5x = 4$ or $2x = 3$
$x = \dfrac{4}{5}$ or $x = \dfrac{3}{2}$

35. $(a - 1)(a + 2) = -2$
$a^2 + a - 2 = -2$
$a^2 + a = 0$
$a(a + 1) = 0$
$a = 0$ or $a + 1 = 0$
$a = 0$ or $a = -1$

37. $(y - 3)(y + 3) = 12$
$y^2 - 9 = 12$
$y^2 = 21$
$y = \pm\sqrt{21}$

41. $x^2 - 3x + 5 = 0$ $\qquad A = 1, B = -3, C = 5$

$x = \dfrac{-B \pm \sqrt{B^2 - 4AC}}{2A} = \dfrac{-(-3) \pm \sqrt{(-3)^2 - 4(1)(5)}}{2(1)} = \dfrac{3 \pm \sqrt{9 - 20}}{2} = \dfrac{3 \pm \sqrt{-11}}{2} = \dfrac{3 \pm \sqrt{11}i}{2}$

43. $2s^2 - 5s - 12 = -5s$
$2s^2 - 12 = 0$
$2s^2 = 12$
$s^2 = 6$
$s = \pm\sqrt{6}$

47. $x^2 + 3x - 8 = x^2 - x + 11$
$3x - 8 = -x + 11$
$4x - 8 = 11$
$4x = 19$
$x = \dfrac{19}{4}$

51. $3a^2 - 4a + 2 = 0$ $\qquad$ A = 3, B = -4, C = 2

$$a = \frac{-B \pm \sqrt{B^2 - 4AC}}{2A} = \frac{-(-4) \pm \sqrt{(-4)^2 - 4(3)(2)}}{2(3)} = \frac{4 \pm \sqrt{16 - 24}}{6} = \frac{4 \pm \sqrt{-8}}{6}$$

$$= \frac{4 \pm \sqrt{8}i}{6} = \frac{4 \pm 2\sqrt{2}i}{6} = \frac{\cancel{2}(2 \pm \sqrt{2}i)}{\cancel{6}\,3} = \frac{2 \pm \sqrt{2}i}{3}$$

53. $(x - 4)(2x + 3) = x^2 - 4$

$\qquad 2x^2 - 5x - 12 = x^2 - 4$

$\qquad\quad x^2 - 5x - 12 = -4$

$\qquad\qquad x^2 - 5x - 8 = 0 \qquad$ A = 1, B = -5, C = -8

$$x = \frac{-B \pm \sqrt{B^2 - 4AC}}{2A} = \frac{-(-5) \pm \sqrt{(-5)^2 - 4(1)(-8)}}{2(1)} = \frac{5 \pm \sqrt{25 + 32}}{2} = \frac{5 \pm \sqrt{57}}{2}$$

55. $(a + 1)(a - 3) = (3a + 1)(a - 2)$

$\qquad a^2 - 2a - 3 = 3a^2 - 5a - 2$

$\qquad\quad -2a - 3 = 2a^2 - 5a - 2$

$\qquad\qquad -3 = 2a^2 - 3a - 2$

$\qquad\qquad\quad 0 = 2a^2 - 3a + 1$

$\qquad\qquad\quad 0 = (2a - 1)(a - 1)$

$\qquad\qquad 0 = 2a - 1 \text{ or } 0 = a - 1$

$\qquad\qquad 1 = 2a \text{ or } 1 = a$

$\qquad\qquad \frac{1}{2} = a \text{ or } 1 = a$

59. $(x - 3)(x + 4) + 2 = x - 3$

$\qquad x^2 + x - 12 + 2 = x - 3$

$\qquad\quad x^2 + x - 10 = x - 3$

$\qquad\qquad\quad x^2 - 10 = -3$

$\qquad\qquad\qquad x^2 = 7$

$\qquad\qquad\qquad x = \pm\sqrt{7}$

61.
$$\frac{y}{y - 2} = \frac{y - 3}{y}$$

$$(y - 2)(y)\left(\frac{y}{y - 2}\right) = (y - 2)(y)\left(\frac{y - 3}{y}\right)$$

$\qquad\qquad y^2 = y^2 - 5y + 6$

$\qquad\qquad 0 = -5y + 6$

$\qquad\qquad 5y = 6$

$\qquad\qquad y = \frac{6}{5}$

65.
$$\frac{3a}{a+1} + \frac{2}{a-2} = 5$$

$$(a+1)(a-2)\left(\frac{3a}{a+1} + \frac{2}{a-2}\right) = (a+1)(a-2)(5)$$

$$(a+1)(a-2)\left(\frac{3a}{a+1}\right) + (a+1)(a-2)\left(\frac{2}{a-2}\right) = (a+1)(a-2)(5)$$

$$3a^2 - 6a + 2a + 2 = 5a^2 - 5a - 10$$

$$3a^2 - 4a + 2 = 5a^2 - 5a - 10$$

$$-4a + 2 = 2a^2 - 5a - 10$$

$$2 = 2a^2 - a - 10$$

$$0 = 2a^2 - a - 12 \qquad A = 2, \ B = -1, \ C = -12$$

$$a = \frac{-B \pm \sqrt{B^2 - 4AC}}{2A} = \frac{-(-1) \pm \sqrt{(-1)^2 - 4(2)(-12)}}{2(2)} = \frac{1 \pm \sqrt{1 + 96}}{4} = \frac{1 \pm \sqrt{97}}{4}$$

69. $\qquad 6x^2 - 3x - 4 = 0 \qquad\qquad A = 6, \ B = -3, \ C = -4$

$$x = \frac{-B \pm \sqrt{B^2 - 4AC}}{2A} = \frac{-(-3) \pm \sqrt{(-3)^2 - 4(6)(-4)}}{2(6)} = \frac{3 \pm \sqrt{9 + 96}}{12}$$

$$= \frac{3 \pm \sqrt{105}}{12}$$

$$x = \frac{3 + \sqrt{105}}{12} \cong \frac{3 + 10.247}{12} \cong 1.104 \ \text{ or } \ x = \frac{3 - \sqrt{105}}{12} \cong \frac{3 - 10.247}{12} \cong -0.604$$

73. The factoring and square root methods are easiest but do not always work. Completing the square and the quadratic always work, but are somewhat cumbersome. If the equation is in the form $(x + a)^2 = b$, use the square root method. If the equation is in the form $(x + a)(x + b) = 0$, then set each factor equal to zero and solve for x. Otherwise, get the equation into the standard form $Ax^2 + Bx + C = 0$. If $B = 0$, use the square root method. If $B \neq 0$, try the factoring method first. If the quadratic equation does not factor (easily or at all), use the quadratic formula.

74. (a) Since $B = -3$, we should get

$$x = \frac{-(-3) \pm \sqrt{9 - 4(1)}}{2} = \frac{3 \pm \sqrt{5}}{2}.$$

(b) Here, $B^2 - 4AC = (-5)^2 - 4(1)(-3) = 25 + 12 = 37$. So $x = \frac{5 \pm \sqrt{37}}{2}$.

(c) First, -3 should be 3, since $B = -3$ and the quadratic formula requires $-B$. Second, the formula requires 2A divide $-B$ as well as $\sqrt{B^2 - 4AC}$. Thus, we should get

$$x = \frac{3 \pm \sqrt{9 - 8}}{2} = \frac{3 \pm \sqrt{1}}{2} = \frac{3 \pm 1}{2}, \text{ so that}$$

$$x = \frac{3 + 1}{2} = \frac{4}{2} = 2, \text{ or } x = \frac{3 - 1}{2} = \frac{2}{2} = 1.$$

(d) The solution is correct up to the cancellation step, which is improperly done. Remember that it is wrong to cancel terms, as was done here. From

212

$$x = \frac{6 \pm 4\sqrt{3}}{2}, \text{ we get } x = \frac{2(3 \pm 2\sqrt{3})}{2} = 3 \pm 2\sqrt{3}.$$

75. (a) $(x-3)(x-4) = 0$

   $x^2 - 7x + 12 = 0$

   (b) $(x+4)(x-5) = 0$

   $x^2 - x - 20 = 0$

   (c) $\left(x - \dfrac{3}{5}\right)(x+2) = 0$

   $x^2 + \dfrac{7}{5}x - \dfrac{6}{5} = 0$ or $5x^2 + 7x - 6 = 0$

   (d) $(x-3i)(x+3i) = 0$

   $x^2 - 9i^2 = 0$

   $x^2 - 9(-1) = 0$

   $x^2 + 9 = 0$

Exercises 8.4

1. Let x = the positive number.

   $$x^2 - 5 = 5x + 1$$
   $$x^2 - 5x - 6 = 0$$
   $$(x-6)(x+1) = 0$$
   $$x - 6 = 0 \text{ or } x + 1 = 0$$
   $$x = 6 \text{ or } x = -1$$

   Eliminate x = –1, since x must be positive. Thus, the positive number is 6.

5. Let x = the number in question.

   $$(x+6)^2 = 169$$
   $$x + 6 = \pm\sqrt{169}$$
   $$x + 6 = \pm 13$$
   $$x + 6 = 13 \text{ or } x + 6 = -13$$
   $$x = 7 \text{ or } x = -19$$

   Thus, the numbers with the required property are 7 and –19.

7. Let x = the number in question.

   $$x + \frac{1}{x} = \frac{13}{6}$$

   $$6x\left(x + \frac{1}{x}\right) = 6x\left(\frac{13}{6}\right)$$

   $$6x(x) + 6x\left(\frac{1}{x}\right) = 6x\left(\frac{13}{6}\right)$$

$$6x^2 + 6 = 13x$$

$$6x^2 - 13x + 6 = 0$$

$$(3x - 2)(2x - 3) = 0$$

$$3x - 2 = 0 \text{ or } 2x - 3 = 0$$

$$3x = 2 \text{ or } 2x = 3$$

$$x = \frac{2}{3} \text{ or } x = \frac{3}{2}$$

Thus, the number is either $\frac{2}{3}$ or $\frac{3}{2}$.

11. $P = 10,000(-d^2 + 12d - 35)$

(a) When $d = 5$, $P = 10,000(-5^2 + 12(5) - 35) = 10,000(-25 + 60 - 35) = 10,000(0) = 0$

Thus, no profit is made by selling tickets for $5.

(b) When $P = 10,000$,

$$10,000 = 10,000(-d^2 + 12d - 35)$$

$$1 = -d^2 + 12d - 35$$

$$d^2 - 12d + 36 = 0$$

$$(d - 6)^2 = 0$$

$$d - 6 = 0$$

$$d = 6$$

Thus, a profit of $10,000 is made by selling tickets for $6.

13. $s = -16t^2 + 40$

(a) When $t = 1$, $s = -16(1)^2 + 40 = -16(1) + 40 = -16 + 40 = 24$.

Thus, the diver is 24 feet above the pool after the first second.

(b) When the diver hits the water, $s = 0$. So $0 = -16t^2 + 40$.

$$16t^2 = 40$$

$$t^2 = \frac{40}{16}$$

$$t^2 = \frac{5}{2}$$

$$t = \pm\sqrt{\frac{5}{2}} = \pm\frac{\sqrt{5}}{\sqrt{2}} = \pm\frac{\sqrt{5}\sqrt{2}}{\sqrt{2}\sqrt{2}} = \pm\frac{\sqrt{10}}{2}$$

Since t cannot be negative, we find $t = \frac{\sqrt{10}}{2}$. Thus, the diver hits the water after $\frac{\sqrt{10}}{2}$ seconds $\approx$ 1.58 seconds.

(c) To find the height of the diving board, let $t = 0$. Then

$$s = -16(0)^2 + 40 = -16(0) + 40 = 0 + 40 = 40.$$

Thus, the diving board is 40 feet above the pool.

15. Let w = width of the rectangle (in feet). Then w + 2 = length of the rectangle (in feet).

$$w(w + 2) = 80$$
$$w^2 + 2w = 80$$
$$w^2 + 2w - 80 = 0$$
$$(w + 10)(w - 8) = 0$$
$$w + 10 = 0 \text{ or } w - 8 = 0$$
$$w = -10 \text{ or } w = 8$$

Since w cannot be negative, eliminate w = –10. Thus, the width of the rectangle is 8 feet and the length of the rectangle is 8 + 2 = 10 feet.

21. Let x = one of the numbers. Then 3x + 2 = the other number.

$$x(3x + 2) = 85$$
$$3x^2 + 2x = 85$$
$$3x^2 + 2x - 85 = 0$$
$$(3x + 17)(x - 5) = 0$$
$$3x + 17 = 0 \text{ or } x - 5 = 0$$
$$3x = -17 \text{ or } x = 5$$
$$x = -\frac{17}{3} \text{ or } x = 5$$

When $x = -\frac{17}{3}, 3x + 2 = \cancel{3}\left(-\frac{17}{\cancel{3}}\right) + 2 = -17 + 2 = -15$. When x = 5, 3x + 2 = 3(5) + 2 = 17.

Thus, the numbers are either $-\frac{17}{3}$ and –15 or 5 and 17.

25. Let a = 3" and let b = 8" and use the Pythagorean Theorem to find c, the length of the hypotenuse.

$$a^2 + b^2 = c^2$$
$$3^2 + 8^2 = c^2$$
$$9 + 64 = c^2$$
$$73 = c^2$$
$$\pm\sqrt{73} = c$$

We eliminate the negative answer, since c is a length. Thus, the length of the hypotenuse is $\sqrt{73}$ inches $\cong 8.54$ inches.

27. Let a = 4" and let c = 7", and use the Pythagorean Theorem to find b, the length of the other leg.

$$a^2 + b^2 = c^2$$
$$4^2 + b^2 = 7^2$$
$$16 + b^2 = 49$$
$$b^2 = 33$$

$$b = \pm\sqrt{33}$$

We eliminate the negative answer, since b is a length. Thus, the other leg of the right triangle is $\sqrt{33}$ inches $\cong 5.74$ inches.

29.
$$a^2 + b^2 = c^2$$
$$8^2 + (x-2)^2 = (x+4)^2$$
$$64 + x^2 - 4x + 4 = x^2 + 8x + 16$$
$$x^2 - 4x + 68 = x^2 + 8x + 16$$
$$-4x + 68 = 8x + 16$$
$$52 = 12x$$
$$\frac{52}{12} = x$$
$$\frac{13}{3} = x$$

Thus, the legs of the right triangle are 8 (given) and $x - 2 = \frac{13}{3} - 2 = \frac{7}{3}$, and the hypotenuse is

$$x + 4 = \frac{13}{3} + 4 = \frac{25}{3}.$$

33. Let d = length of the diagonal of the square (in inches).

$$5^2 + 5^2 = d^2$$
$$25 + 25 = d^2$$
$$50 = d^2$$
$$\pm\sqrt{50} = d$$
$$\pm 5\sqrt{2} = d$$

Since d is a length, we can eliminate the negative answer. Thus, the length of the diagonal of the

square is $5\sqrt{2}$ inches $\cong 7.07$ inches.

37. Let x = height of the top of the ladder (in feet).

$$8^2 + x^2 = 30^2$$
$$64 + x^2 = 900$$
$$x^2 = 836$$
$$x = \pm\sqrt{836}$$
$$x = \pm 2\sqrt{209}$$

We eliminate the negative answer, since x represents a height. Thus, the top of the ladder is $2\sqrt{209}$ feet $\cong 28.9$ feet above the ground.

41. Let ℓ = length of the rectangle (in inches).

$$5^2 + \ell^2 = 12^2$$
$$25 + \ell^2 = 144$$
$$\ell^2 = 119$$
$$\ell = \pm\sqrt{119}$$

Since ℓ is a length, it cannot be negative. Therefore, $\ell = \sqrt{119}$. Thus, the area of the rectangle (which is the product of its width and its length) is $5\sqrt{119}$ square inches $\cong 54.54$ square inches.

43. Using the slope formula, the first line has slope $= \dfrac{n^2 - n}{3 - 1} = \dfrac{n^2 - n}{2}$ and the second has

slope $= \dfrac{6 - 0}{-5 - (-6)} = \dfrac{6}{1} = 6$. Since the lines are parallel, it must be true that $\dfrac{n^2 - n}{2} = 6$.

Then $n^2 - n = 12$

$$n^2 - n - 12 = 0$$
$$(n - 4)(n + 3) = 0$$
$$n - 4 = 0 \text{ or } n + 3 = 0$$
$$n = 4 \text{ or } \quad n = -3$$

47. Let w = width of the picture (in inches). Then 2w = length of the picture (in inches).

$$(w + 2)(2w + 2) = 60$$
$$2w^2 + 6w + 4 = 60$$
$$2w^2 + 6w - 56 = 0$$
$$w^2 + 3w - 28 = 0$$

$$w + 7 = 0 \text{ or } w - 4 = 0$$

$$w = -7 \text{ or } w = 4$$

Since w cannot be negative, eliminate $w = -7$. Thus, the width of the picture is 4 inches and the length of the picture is $2(4) = 8$ inches.

53. Let r = boat's rate in still water (in kph). Then r – 5 = boat's rate upstream (in kph), and r + 5 = boat's rate downstream (in kph).

$$\frac{20}{r - 5} + \frac{10}{r + 5} = \frac{4}{3}$$

$$3(r - 5)(r + 5)\left(\frac{20}{r - 5} + \frac{10}{r + 5}\right) = 3(r - 5)(r + 5)\left(\frac{4}{3}\right)$$

$$3(r - 5)(r + 5)\left(\frac{20}{r - 5}\right) + 3(r - 5)(r + 5)\left(\frac{10}{r + 5}\right) = 3(r - 5)(r + 5)\left(\frac{4}{3}\right)$$

$$60(r+5)+30(r-5)=4(r^2-25)$$
$$60r+300+30r-150=4r^2-100$$
$$90r+150=4r^2-100$$
$$0=4r^2-90r-250$$
$$0=2r^2-45r-125$$
$$0=(2r+5)(r-25)$$
$$0=2r+5 \text{ or } 0=r-25$$

$$-\frac{5}{2}=r \text{ or } 25=r$$

We eliminate the negative answer and conclude that the boat's rate in still water is 25 kph.

55. $125y^6-1=(5y^2)^3-1^3=(5y^2-1)((5y^2)^2+(5y^2)+1)=(5y^2-1)(25y^4+5y^2+1)$

57. $\dfrac{3}{2-i}=\dfrac{3}{2-i}\cdot\dfrac{2+i}{2+i}=\dfrac{3(2+i)}{4-i^2}=\dfrac{3(2+i)}{4-(-1)}=\dfrac{6+3i}{5}=\dfrac{6}{5}+\dfrac{3}{5}i$

59. $2x^{\frac{1}{4}}-1=9$
$$2x^{\frac{1}{4}}=10$$
$$x^{\frac{1}{4}}=5$$
$$(x^{\frac{1}{4}})^4=5^4$$
$$x=625$$

Exercises 8.5

1. $\sqrt{x}+3=2x$

$$\sqrt{x}=2x-3$$
$$\left(\sqrt{x}\right)^2=(2x-3)^2$$
$$x=4x^2-12x+9$$
$$0=4x^2-13x+9$$
$$0=(4x-9)(x-1)$$
$$0=4x-9 \text{ or } 0=x-1$$
$$\frac{9}{4}=x \text{ or } 1=x$$

CHECK $x=\dfrac{9}{4}$:

$$\sqrt{x}+3=2x$$
$$\sqrt{\frac{9}{4}}+3\overset{?}{=}2\left(\frac{9}{4}\right)$$
$$\frac{3}{2}+3\overset{?}{=}\frac{9}{2}$$
$$\frac{9}{2}\overset{\checkmark}{=}\frac{9}{2}$$

CHECK $x=1$:

$$\sqrt{x}+3=2x$$
$$\sqrt{1}+3\overset{?}{=}2(1)$$
$$1+3\overset{?}{=}2$$
$$4\neq 2$$

218

Thus, x = 1 is an extraneous root, and the only solution is $x = \dfrac{9}{4}$.

5. $\sqrt{5a-1} + 5 = a$

$\sqrt{5a-1} = a - 5$

$\left(\sqrt{5a-1}\right)^2 = (a-5)^2$

$5a - 1 = a^2 - 10a + 25$

$0 = a^2 - 15a + 26$

$0 = (a-2)(a-13)$

$0 = a - 2$ or $0 = a - 13$

$2 = a$ or $13 = a$

CHECK a = 2:

$\sqrt{5a-1} + 5 = a$

$\sqrt{5(2)-1} + 5 \overset{?}{=} 0$

$\sqrt{10-1} + 5 \overset{?}{=} 0$

$\sqrt{9} + 5 \overset{?}{=} 2$

$3 + 5 \overset{?}{=} 2$

$8 \neq 2$

CHECK a = 13:

$\sqrt{5a-1} + 5 = a$

$\sqrt{5(13)-1} + 5 \overset{?}{=} 13$

$\sqrt{65-1} + 5 \overset{?}{=} 13$

$\sqrt{64} + 5 \overset{?}{=} 13$

$8 + 5 \overset{?}{=} 13$

$13 \overset{\checkmark}{=} 13$

Thus, a = 2 is an extraneous root, and the only solution is a = 13.

9. $\sqrt{3x+1} + 3 = x$

$\sqrt{3x+1} = x - 3$

$\left(\sqrt{3x+1}\right)^2 = (x-3)^2$

$3x + 1 = x^2 - 6x + 9$

$0 = x^2 - 9x + 8$

$0 = (x-1)(x-8)$

$0 = x - 1$ or $0 = x - 8$

$1 = x$ or $8 = x$

CHECK x = 1:

$\sqrt{3x+1} + 3 = x$

$\sqrt{3(1)+1} + 3 \overset{?}{=} 1$

$\sqrt{3+1} + 3 \overset{?}{=} 1$

$\sqrt{4} + 3 \overset{?}{=} 1$

$2 + 3 \overset{?}{=} 1$

$5 \neq 1$

CHECK x = 8:

$\sqrt{3x+1} + 3 = x$

$\sqrt{3(8)+1} + 3 \overset{?}{=} 8$

$\sqrt{24+1} + 3 \overset{?}{=} 8$

$\sqrt{25} + 3 \overset{?}{=} 8$

$5 + 3 \overset{?}{=} 8$

$8 \overset{\checkmark}{=} 8$

Thus, x = 1 is an extraneous root, and the only solution is x = 8.

13. $\sqrt{y+3} = 1 + \sqrt{y}$

$\left(\sqrt{y+3}\right)^2 = \left(1+\sqrt{y}\right)^2$

$y + 3 = 1 + 2\sqrt{y} + y$

$3 = 1 + 2\sqrt{y}$

$2 = 2\sqrt{y}$

$1 = \sqrt{y}$

$1^2 = \left(\sqrt{y}\right)^2$

$1 = y$

CHECK y = 1:

$\sqrt{y+3} = 1 + \sqrt{y}$

$\sqrt{1+3} \overset{?}{=} 1 + \sqrt{1}$

$\sqrt{4} \overset{?}{=} 1 + \sqrt{1}$

$2 \overset{?}{=} 1 + 1$

$2 \overset{\checkmark}{=} 2$

Thus, the only solution is y = 1.

17. $\sqrt{7s+1} - 2\sqrt{s} = 2$

$\sqrt{7s+1} = 2 + 2\sqrt{s}$

$\left(\sqrt{7s+1}\right)^2 = \left(2+2\sqrt{s}\right)^2$

$7s + 1 = 4 + 8\sqrt{s} + 4s$

$3s - 3 = 8\sqrt{s}$

$(3s-3)^2 = \left(8\sqrt{s}\right)^2$

$9s^2 - 18s + 9 = 64s$

$9s^2 - 82s + 9 = 0$

$(9s - 1)(s - 9) = 0$

$s = \dfrac{1}{9}$ or $s = 9$

CHECK $s = \dfrac{1}{9}$:

$\sqrt{7s+1} - 2\sqrt{s} = 2$

$\sqrt{7\left(\frac{1}{9}\right)+1} - 2\sqrt{\frac{1}{9}} \overset{?}{=} 2$

$\sqrt{\frac{7}{9}+1} - 2\sqrt{\frac{1}{9}} \overset{?}{=} 2$

$\sqrt{\frac{16}{9}} - 2\sqrt{\frac{1}{9}} \overset{?}{=} 2$

$\frac{4}{3} - 2\left(\frac{1}{3}\right) \overset{?}{=} 2$

$\frac{4}{3} - \frac{2}{3} \overset{?}{=} 2$

$\frac{2}{3} \neq 2$

CHECK s = 9:

$\sqrt{7s+1} - 2\sqrt{s} = 2$

$\sqrt{7(9)+1} - 2\sqrt{9} \overset{?}{=} 2$

$\sqrt{64} - 2\sqrt{9} \overset{?}{=} 2$

$8 - 2(3) \overset{?}{=} 2$

$8 - 6 \overset{?}{=} 2$

$2 \overset{\checkmark}{=} 2$

Thus, $s = \dfrac{1}{9}$ is an extraneous root, and the only solution is s = 9.

19. $\sqrt{7-a} - \sqrt{3+a} = 2$

$\sqrt{7-a} = 2 + \sqrt{3-a}$

$\left(\sqrt{7-a}\right)^2 = \left(2+\sqrt{3+a}\right)^2$

$7 - a = 4 + 4\sqrt{3+a} + 3 + a$

$7 - a = 7 + 4\sqrt{3+a} + a$

CHECK a = 6:

$\sqrt{7-a} - \sqrt{3+a} = 2$

$\sqrt{7-6} - \sqrt{3+6} \overset{?}{=} 2$

$\sqrt{1} - \sqrt{9} \overset{?}{=} 2$

$1 - 3 \overset{?}{=} 2$

220

$$-2a = 4\sqrt{3+a}$$

$$-a = 2\sqrt{3+a}$$

$$(-a)^2 = \left(2\sqrt{3+a}\right)^2$$

$$a^2 = 4(3+a)$$

$$a^2 = 12 + 4a$$

$$a^2 - 4a - 12 = 0$$

$$(a-6)(a+2) = 0$$

$$a - 6 = 0 \text{ or } a + 2 = 0$$

$$a = 6 \text{ or } a = -2$$

$$-2 \neq 2$$

CHECK $a = -2$:

$$\sqrt{7-a} = \sqrt{3+a} = 2$$

$$\sqrt{7-(-2)} - \sqrt{3-2} \overset{?}{=} 2$$

$$\sqrt{9} - \sqrt{1} \overset{?}{=} 2$$

$$3 - 1 \overset{?}{=} 2$$

$$2 \overset{\surd}{=} 2$$

Thus, $a = 6$ is an extraneous root, and the only solution is $a = -2$.

21. $\sqrt{x} + a = b$

$$\sqrt{x} = b - a$$

$$\left(\sqrt{x}\right)^2 = (b-a)^2$$

$$x = (b-a)^2$$

25. $\sqrt{5x+b} = 6 + b$

$$\left(\sqrt{5x+b}\right)^2 = (6+b)^2$$

$$5x + b = 36 + 12b + b^2$$

$$5x = 36 + 11b + b^2$$

$$x = \frac{36 + 11b + b^2}{5}$$

29. $x^3 - 2x^2 - 15x = 0$

$$x(x^2 - 2x - 15) = 0$$

$$x(x-5)(x+3) = 0$$

$$x = 0 \text{ or } x - 5 = 0 \text{ or } x + 3 = 0$$

$$x = 0 \text{ or } x = 5 \text{ or } x = -3$$

33. $y^4 - 17y^2 + 16 = 0$

$$(y^2 - 1)(y^2 - 16) = 0$$

$$(y-1)(y+1)(y-4)(y+4) = 0$$

$$y - 1 = 0 \text{ or } y + 1 = 0 \text{ or } y - 4 = 0 \text{ or } y + 4 = 0$$

$$y = 1 \text{ or } y = -1 \text{ or } y = 4 \text{ or } y = -4$$

37. $b^4 + 112 = 23b^2$

$$b^4 - 23b^2 + 112 = 0$$

$$(b^2 - 16)(b^2 - 7) = 0$$

$$(b-4)(b+4)(b^2 - 7) = 0$$

$$b - 4 = 0 \text{ or } b + 4 = 0 \text{ or } b^2 - 7 = 0$$

$$b = 4 \text{ or } b = -4 \text{ or } b^2 = 7$$

$$b = 4 \text{ or } b = -4 \text{ or } b = \pm \sqrt{7}$$

41.
$$x^3 + x^2 - x - 1 = 0$$
$$x^2(x+1) - (x+1) = 0$$
$$(x^2 - 1)(x+1) = 0$$
$$(x-1)(x+1)(x+1) = 0$$
$$(x-1)(x+1)^2 = 0$$
$$x - 1 = 0 \text{ or } (x+1)^2 = 0$$
$$x = 1 \text{ or } x + 1 = 0$$
$$x = 1 \text{ or } x = -1$$

45.
$$x + x^{\frac{1}{2}} - 6 = 0$$

Let $u = x^{\frac{1}{2}}$ Then $u^2 = x$

$$u^2 + u - 6 = 0$$
$$(u+3)(u-2) = 0$$
$$u + 3 = 0 \text{ or } u - 2 = 0$$
$$u = -3 \text{ or } u = 2$$
$$x^{\frac{1}{2}} = -3 \text{ or } x^{\frac{1}{2}} = 2$$
$$\left(x^{\frac{1}{2}}\right)^2 = (-3)^2 \text{ or } \left(x^{\frac{1}{2}}\right)^2 = 2^2$$

$$x = 9 \text{ or } x = 4$$

CHECK $x = 9$:
$$x + x^{\frac{1}{2}} - 6 = 0$$
$$9 + 9^{\frac{1}{2}} - 6 \stackrel{?}{=} 0$$
$$9 + \sqrt{9} - 6 \stackrel{?}{=} 0$$
$$9 + 3 - 6 \stackrel{?}{=} 0$$
$$6 \neq 0$$

CHECK $x = 4$:
$$x + x^{\frac{1}{2}} - 6 = 0$$
$$4 + 4^{\frac{1}{2}} - 6 \stackrel{?}{=} 0$$
$$4 + \sqrt{4} - 6 \stackrel{?}{=} 0$$
$$4 + 2 - 6 \stackrel{?}{=} 0$$
$$0 \stackrel{\checkmark}{=} 0$$

Thus $x = 9$ is an extraneous root, and the only solution is $x = 4$.

49.
$$x^{\frac{1}{2}} + 8x^{\frac{1}{4}} + 7 = 0$$

Let $u = x^{\frac{1}{4}}$

Then $u^2 = \left(x^{\frac{1}{4}}\right)^2 = x^{\frac{1}{2}}$

$$u^2 + 8u + 7 = 0$$
$$(u+1)(u+7) = 0$$
$$u + 1 = 0 \text{ or } u + 7 = 0$$
$$u = -1 \text{ or } u = -7$$
$$x^{\frac{1}{4}} = -1 \text{ or } x^{\frac{1}{4}} = -7$$

CHECK $x = 1$:
$$x^{\frac{1}{2}} + 8x^{\frac{1}{4}} + 7 = 0$$
$$1^{\frac{1}{2}} + 8(1)^{\frac{1}{4}} + 7 \stackrel{?}{=} 0$$
$$1 + 8(1) + 7 \stackrel{?}{=} 0$$
$$1 + 8 + 7 \stackrel{?}{=} 0$$
$$16 \neq 0$$

CHECK $x = 2401$:
$$x^{\frac{1}{2}} + 8x^{\frac{1}{4}} + 7 = 0$$

$$\left(x^{\frac{1}{4}}\right)^4 = (-1)^4 \text{ or } \left(x^{\frac{1}{4}}\right)^4 = (-7)^4 \qquad (2401)^{\frac{1}{2}} + 8(2401)^{\frac{1}{4}} + 7 \overset{?}{=} 0$$

$$x = 1 \text{ or } x = 2401 \qquad \sqrt{2401} + 8\left(\sqrt[4]{2401}\right) + 7 \overset{?}{=} 0$$

$$49 + 8(7) + 7 \overset{?}{=} 0$$

$$49 + 56 + 7 \overset{?}{=} 0$$

$$112 \neq 0$$

Thus, both x = 1 and x = 2401 are extraneous roots, and the equation has no solution.

51. $x^{-2} - 5x^{-1} + 6 = 0$ 　　　　　CHECK $x = \dfrac{1}{2}$:

Let $u = x^{-1}$ 　　　　　　　　　　$x^{-2} - 5x^{-1} + 6 = 0$

Then $u^2 = (x^{-1})^2 = x^{-2}$ 　　　$\left(\dfrac{1}{2}\right)^{-2} - 5\left(\dfrac{1}{2}\right)^{-1} + 6 \overset{?}{=} 0$

$u^2 - 5u + 6 = 0$ 　　　　　　　　$4 - 5(2) + 6 \overset{?}{=} 0$

$(u - 2)(u - 3) = 0$ 　　　　　　　$4 - 10 + 6 \overset{?}{=} 0$

$u - 2 = 0 \text{ or } u - 3 = 0$ 　　　　　$0 \overset{\checkmark}{=} 0$

$u = 2 \text{ or } u = 3$

$x^{-1} = 2 \text{ or } x^{-1} = 3$ 　　　　　CHECK $x = \dfrac{1}{3}$:

$(x^{-1})^{-1} = 2^{-1} \text{ or } (x^{-1})^{-1} = 3^{-1}$ 　　$x^{-2} - 5x^{-1} + 6 = 0$

$x = \dfrac{1}{2} \text{ or } x = \dfrac{1}{3}$ 　　　　　$\left(\dfrac{1}{3}\right)^{-2} - 5\left(\dfrac{1}{3}\right)^{-1} + 6 \overset{?}{=} 0$

$$9 - 5(3) + 6 \overset{?}{=} 0$$

$$9 - 15 + 6 \overset{?}{=} 0$$

$$0 \overset{\checkmark}{=} 0$$

55. $x^{-4} - 13x^{-2} + 36 = 0$ 　　　　　CHECK $x = \pm\dfrac{1}{2}$:

Let $u = x^{-2}$ 　　　　　　　　　　$x^{-4} - 13x^{-2} + 36 = 0$

Then $u^2 = (x^{-2})^2 = x^{-4}$ 　　　$\left(\pm\dfrac{1}{2}\right)^{-4} - 13\left(\pm\dfrac{1}{2}\right)^{-2} + 36 \overset{?}{=} 0$

$u^2 - 13u + 36 = 0$ 　　　　　　$16 - 13(4) + 36 \overset{?}{=} 0$

$(u - 4)(u - 9) = 0$ 　　　　　　$16 - 52 + 36 \overset{?}{=} 0$

$u - 4 = 0 \text{ or } u - 9 = 0$ 　　　　$0 \overset{\checkmark}{=} 0$

$u = 4 \text{ or } u = 9$ 　　　　　　　CHECK $x = \pm\dfrac{1}{3}$:

$x^{-2} = 4 \text{ or } x^{-2} = 9$ 　　　　$x^{-4} - 13x^{-2} + 36 = 0$

$\dfrac{1}{x^2} = 4 \text{ or } \dfrac{1}{x^2} = 9$ 　　　　$\left(\pm\dfrac{1}{3}\right)^{-4} - 13\left(\pm\dfrac{1}{3}\right)^{-2} + 36 \overset{?}{=} 0$

$\dfrac{1}{4} = x^2 \text{ or } \dfrac{1}{9} = x^2$ 　　　　$81 - 13(9) + 36 \overset{?}{=} 0$

$\pm\dfrac{1}{2} = x \text{ or } \pm\dfrac{1}{3} = x$ 　　　　$81 - 117 + 36 \overset{?}{=} 0$

$$0 \overset{\checkmark}{=} 0$$

59. $\sqrt{x} - 4\sqrt[4]{x} = 5$

$x^{\frac{1}{2}} - 4x^{\frac{1}{4}} - 5 = 0$

Let $u = x^{\frac{1}{4}}$

Then $u^2 - 4u - 5 = 0$

$(u - 5)(u + 1) = 0$

$u - 5 = 0$ or $u + 1 = 0$

$u = 5$ or $u = -1$

$x^{\frac{1}{4}} = 5$ or $x^{\frac{1}{4}} = -1$

$\left(x^{\frac{1}{4}}\right)^4 = 5$ or $\left(x^{\frac{1}{4}}\right)^4 = (-1)^4$

$x = 625$ or $x = 1$

CHECK $x = 625$:

$\sqrt{x} - 4\sqrt[4]{x} = 5$

$\sqrt{625} - 4\sqrt[4]{625} \overset{?}{=} 5$

$25 - 4(5) \overset{?}{=} 5$

$25 - 20 \overset{?}{=} 5$

$5 \overset{\checkmark}{=} 5$

CHECK $x = 1$:

$\sqrt{x} - 4\sqrt[4]{x} = 5$

$\sqrt{1} - 4\sqrt[4]{1} \overset{?}{=} 5$

$1 - 4(1) \overset{?}{=} 5$

$-3 \neq 5$

Thus, $x = 1$ is an extraneous root, and the only solution is $x = 625$.

61. $(a + 4)^2 + 6(a + 4) + 9 = 0$

Let $u = a + 4$

$u^2 + 6u + 9 = 0$

$(u + 3)^2 = 0$

$u + 3 = 0$

$u = -3$

$a + 4 = -3$

$a = -7$

CHECK $a = -7$:

$(a + 4)^2 + 6(a + 4) + 9 = 0$

$(-7 + 4)^2 + 6(-7 + 4) + 9 \overset{?}{=} 0$

$(-3)^2 + 6(-3) + 9 \overset{?}{=} 0$

$9 - 18 + 9 \overset{?}{=} 0$

$0 \overset{\checkmark}{=} 0$

65. $(x^2 + x)^2 - 4 = 0$

$((x^2 + x) - 2)((x^2 + x) + 2) = 0$

$(x^2 + x - 2)(x^2 + x + 2) = 0$

$(x + 2)(x - 1)(x^2 + x + 2) = 0$

$x + 2 = 0$ or $x - 1 = 0$ or

$x = -2$ or $x = 1$ or

$x^2 + x + 2 = 0 \qquad A = 1, B = 1, C = 2$

$x = \dfrac{-B \pm \sqrt{B^2 - 4AC}}{2A}$

$x = \dfrac{-1 \pm \sqrt{1 - 4(1)(2)}}{2(1)} = \dfrac{-1 \pm \sqrt{1 - 8}}{2}$

$= \dfrac{-1 \pm \sqrt{-7}}{2} = \dfrac{-1 \pm \sqrt{7}\,i}{2}$

67. $\left(a - \dfrac{10}{a}\right)^2 - 12\left(a - \dfrac{10}{a}\right) + 27 = 0$

Let $u = a - \dfrac{10}{a}$

$u^2 - 12u + 27 = 0$

$(u - 3)(u - 9) = 0$

$u - 3 = 0$ or $u - 9 = 0$

$u = 3$ or $u = 9$

$a - \dfrac{10}{a} = 3$ or $a - \dfrac{10}{a} = 9$

$a\left(a - \dfrac{10}{a}\right) = a(3)$ or $a\left(a - \dfrac{10}{a}\right) = a(9)$

$a^2 - \not{a}\left(\dfrac{10}{\not{a}}\right) = 3a$ or $a^2 - \not{a}\left(\dfrac{10}{\not{a}}\right) = 9a$

$a^2 - 10 = 3a$ or $a^2 - 10 = 9a$

$a^2 - 3a - 10 = 0$ or $a^2 - 9a - 10 = 0$

$(a - 5)(a + 2) = 0$ or $(a - 10)(a + 1) = 0$

$a - 5 = 0$ or $a + 2 = 0$ or $a - 10 = 0$ or $a + 1 = 0$

$a = 5$ or $a = -2$ or $a = 10$ or $a = -1$

71. The square of a difference is in general not equal to the difference of the squares. Accordingly, it is not correct to say that $\left(\sqrt{3x - 2} - \sqrt{x}\right)^2$ is equal to $\left(\sqrt{3x - 2}\right)^2 - \left(\sqrt{x}\right)^2$ or $3x - 2 - x$.

73. $\dfrac{a + 2}{a - 1} + 5 = \dfrac{3}{a - 1}$

$(a - 1)\left(\dfrac{a + 2}{a - 1} + 5\right) = (a - 1)\left(\dfrac{3}{a - 1}\right)$

$a + 2 + 5(a - 1) = 3$

$a + 2 + 5a - 5 = 3$

$6a - 3 = 3$

$6a = 6$

$a = 1$

CHECK $a = 1$:

$\dfrac{a + 2}{a - 1} + 5 = \dfrac{3}{a - 1}$

$\dfrac{(1) + 2}{(1) - 1} + 5 \overset{?}{=} \dfrac{3}{(1) - 1}$

$\dfrac{3}{0} + 5 \neq \dfrac{3}{0}$

since we cannot divide by 0.

Thus, the equation has no solution.

75. $|5x + 1| \geq 9$

$5x + 1 \leq -9$ or $5x + 1 \geq 9$

$5x \leq -10$ or $5x \geq 8$

$x \leq -2$ or $x \geq \dfrac{8}{5}$

Exercises 8.6

3. $(x + 4)(x - 2) > 0$

    $x + 4 = 0 \rightarrow x = -4$

    $x - 2 = 0 \rightarrow x = 2$

    Cut points are –4 and 2. Intervals are x < –4, –4 < x < 2, and x > 2. For x < 4, let x = –5 be the test value. When x = –5, (x + 4)(x – 2) = (–5 + 4)(–5 – 2) is positive. For –4< x < 2, let x = 0 be the test value. When x = 0, (x + 4)(x – 2) = (0 + 4)(0 – 2) is negative. For x > 2, let x = 3 be the test value. When x = 3, (x + 4)(x – 2) = (3 + 4)(3 – 2) is positive. Thus, (x + 4)(x – 2) > 0 when x < –4 or x > 2.

7. $(x - 3)(2x - 1) \geq 0$

    $x - 3 = 0 \rightarrow x = 3$

    $2x - 1 = 0 \rightarrow x = \dfrac{1}{2}$

    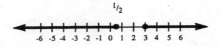

    Cut points are 3 and $\dfrac{1}{2}$. Intervals are $x < \dfrac{1}{2}$, $\dfrac{1}{2} < x < 3$, and x > 3. For $x < \dfrac{1}{2}$, let x = 0 be the test value. When x = 0, (x – 3)(2x – 1) = (0 – 3)(2(0) – 1) is positive. For $\dfrac{1}{2} < x < 3$, let x = 1 be the test value. When x = 1, (x – 3)(2x – 1) = (1 –3)(2(1) – 1) is negative. For x > 3, let x = 4 be the test value. When x = 4, (x – 3)(2x – 1) = (4 – 3)(2(4) – 1) is positive. Thus, (x – 3)(2x – 1) ≥ 0 when $x \leq \dfrac{1}{2}$ or x ≥ 3.

9. $a^2 - a - 20 < 0$

    $(a - 5)(a + 4) < 0$

    $a - 5 = 0 \rightarrow a = 5$

    $a + 4 = 0 \rightarrow a = -4$

    Cut points are 5 and –4. Intervals are a < –4, –4 < a < 5, and a > 5. For a < –4, let a = –5 be the test value. When a = –5, (a – 5)(a + 4) = (–5 – 5)(–5 + 4) is positive. For –4 < a < 5, let a = 0 be the test value. When a = 0, (a – 5)(a + 4) = (0 – 5)(0 + 4) is negative. For a >5, let a = 6 be the test value. When a = 6, (a – 5)(a + 4) = (6 – 5)(6 + 4) is positive. Thus, $a^2 - a - 20 < 0$ when

    –4 < a < 5.

13. $2a^2 - 9a \leq 5$

    $2a^2 - 9a - 5 \leq 0$

    $(2a + 1)(a - 5) \leq 0$

    $2a + 1 = 0 \rightarrow a = -\dfrac{1}{2}$

    $a - 5 = 0 \rightarrow a = 5$

Cut points are $-\dfrac{1}{2}$ and 5. Intervals are $a < -\dfrac{1}{2}, -\dfrac{1}{2} < a < 5,$ and $a > 5.$ For $a < -\dfrac{1}{2}$, let $a = -1$ be the test value. When $a = -1,$ $(2a + 1)(a - 5) = (2(-1) + 1)(-1 - 5)$ is positive. For $-\dfrac{1}{2} < a < 5$, let $a = 0$ be the test value. When $a = 0,$ $(2a + 1)(a - 5) = (2(0) + 1)(0 - 5)$ is negative. For $a > 5$, let $a = 6$ be the test value. When $a = 6,$ $(2a + 1)(a - 5) = (2(6) + 1)(6 - 5)$ is positive. Thus, $2a^2 - 9a \le 5$ when $-\dfrac{1}{2} \le a \le 5.$

17. $3x^2 \le 10 - 13x$

$3x^2 + 13x - 10 \le 0$

$(3x - 2)(x + 5) \le 0$

$3x - 2 = 0 \rightarrow x = \dfrac{2}{3}$

$x + 5 = 0 \rightarrow x = -5$

Cut points are $\dfrac{2}{3}$ and -5. Intervals are $x < -5,$ $-5 < x < \dfrac{2}{3},$ and $x > \dfrac{2}{3}.$ For $x < -5$, let $x = -6$ be the test value. When $x = -6,$ $(3x - 2)(x + 5) = (3(-6) - 2)(-6 + 5)$ is positive. For $-5 < x < \dfrac{2}{3}$, let $x = 0$ be the test value. When $x = 0,$ $(3x - 2)(x + 5) = (3(0) - 2)(0 + 5)$ is negative. For $x > \dfrac{2}{3}$, let $x = 1$ be the test value. When $x = 1,$ $(3x - 2)(x + 5) = (3(1) - 2)(1 + 5)$ is positive.

Thus, $3x^2 \le 10 - 13x$ when $-5 \le x \le \dfrac{2}{3}.$

21. $x^2 - 6x + 9 < 0$

$(x - 3)^2 < 0$

This inequality has no solution, since the square of a real number can never be negative.

25. $3y^2 \ge 5y + 2$

$3y^2 - 5y - 2 \ge 0$

$(3y + 1)(y - 2) \ge 0$

$3y + 1 = 0 \rightarrow y = -\dfrac{1}{3}$

$y - 2 = 0 \rightarrow y = 2$

Cut points are $-\dfrac{1}{3}$ and 2. Intervals are $y < -\dfrac{1}{3},$ $-\dfrac{1}{3} < y < 2,$ and $y > 2.$ For $y < -\dfrac{1}{3}$, let $y = -1$ be the test value. When $y = -1,$ $(3y + 1)(y - 2) = (3(-1) + 1)(-1 - 2)$ is positive. For $-\dfrac{1}{3} < y < 2$, let $y = 0$ be the test value. When $y = 0,$ $(3y + 1)(y - 2) = (3(0) + 1)(0 - 2)$ is negative. For $y > 2$, let $y = 3$ be the test value. When $y = 3,$ $(3y + 1)(y - 2) = (3(3) + 1)(3 - 2)$ is positive. Thus, $3y^2 \ge 5y + 2$ when $y \le -\dfrac{1}{3}$ or $y \ge 2.$

29. $\dfrac{x - 2}{x + 1} < 0$

$x - 2 = 0 \rightarrow x = 2$

$x + 1 = 0 \rightarrow x = -1$

Cut points are 2 and –1. Intervals are x < –1, –1 < x < 2, and x > 2. For x < –1, let x = –2 be the test value. When $x = -2$, $\dfrac{x-2}{x+1} = \dfrac{-2-2}{-2+1}$ is positive. For –1 < x < 2, let x = 0 be the test value.

When $x = 0$, $\dfrac{x-2}{x+1} = \dfrac{0-2}{0+1}$ is negative. For x > 2, let x = 3 be the test value. When x = 3,

$\dfrac{x-2}{x+1} = \dfrac{3-2}{3+1}$ is positive. Thus, $\dfrac{x-2}{x+1} < 0$ when –1 < x < 2.

35. $\dfrac{5}{y-4} > 0$

     $y - 4 = 0 \rightarrow y = 4$

Cut point is 4 (note that no cut point results from the numerator). Intervals are y < 4 and y > 4.
For y < 4, let y = 3 be the test value. When $y = 3$, $\dfrac{5}{y-4} = \dfrac{5}{3-4}$ is negative. For y > 4, let y = 5

be the test value. When $y = 5$, $\dfrac{5}{y-4} = \dfrac{5}{5-4}$ is positive. Thus, $\dfrac{5}{y-4} > 0$ when y > 4.

37. $\dfrac{3}{y-1} < 1$

     $\dfrac{3}{y-1} - 1 = 0$

     $\dfrac{3}{y-1} - \dfrac{y-1}{y-1} < 0$

     $\dfrac{3-(y-1)}{y-1} < 0$

     $\dfrac{4-y}{y-1} < 0$

     $4 - y = 0 \rightarrow y = 4$

     $y - 1 = 0 \rightarrow y = 1$

Cut points are 4 and 1. Intervals are y < 1, 1 < y < 4, and y > 4. For y < 1, let y = o be the test
value. When $y = 0$, $\dfrac{4-y}{y-1} = \dfrac{4-0}{0-1}$ is negative. For 1 < y < 4, let y = 2 be the test value. When

$y = 2$, $\dfrac{4-y}{y-1} = \dfrac{4-2}{2-1}$ is positive. For y > 4, let y = 5 be the test value. When $y = 5$, $\dfrac{4-1}{y-1} = \dfrac{4-5}{5-1}$

is negative. Thus, $\dfrac{3}{y-1} < 1$ when y < 1 or y > 4.

41.
$$\frac{2y+3}{y-1} \le 2$$

$$\frac{2y+3}{y-1} - 2 \le 0$$

$$\frac{2y+3}{y-1} - \frac{2(y-1)}{y-1} \le 0$$

$$\frac{2y+3-2(y-1)}{y-1} \le 0$$

$$\frac{5}{y-1} \le 0$$

$$y - 1 = 0 \rightarrow y = 1$$

Cut point is 1. Intervals are $y < 1$ and $y > 1$. For $y < 1$, let $y = 0$ be the test value. When $y = 0$, $\frac{5}{y-1} = \frac{5}{0-1}$ is negative. For $y > 1$, let $y = 2$ be the test value. When $y = 2$, $\frac{5}{y-1} = \frac{5}{2-1}$ is positive. Thus, $\frac{2y+3}{y-1} \le 2$ when $y < 1$. (Note that we exclude $y = 1$, since this value would lead to a zero denominator.)

45.
$$\frac{x}{x-1} \le \frac{3}{x-1}$$

$$\frac{x}{x-1} - \frac{3}{x-1} \le 0$$

$$\frac{x-3}{x-1} \le 0$$

$$x - 3 = 0 \rightarrow x = 3$$

$$x - 1 = 0 \rightarrow x = 1$$

Cut points are 3 and 1. Intervals are $x < 1$, $1 < x < 3$, and $x > 3$. For $x < 1$, let $x = 0$ be the test value. When $x = 0$, $\frac{x-3}{x-1} = \frac{0-3}{0-1}$ is positive. For $1 < x < 3$, let $x = 2$ be the test value. When $x = 2$, $\frac{x-3}{x-1} = \frac{2-3}{2-1}$ is negative. For $x > 3$, let $x = 4$ be the test value. When $x = 4$, $\frac{x-3}{x-1} = \frac{4-3}{4-1}$ is positive. Thus, $\frac{x}{x-1} \le \frac{3}{x-1}$ when $1 < x \le 3$.

49.
$$\frac{1}{x-2} + \frac{2}{x+3} \le \frac{3}{x+3}$$

$$\frac{1}{x-2} + \frac{2}{x+3} - \frac{3}{x+3} \le 0$$

$$\frac{1}{x-2} - \frac{1}{x+3} \le 0$$

$$\frac{(x+3)}{(x-2)(x+3)} - \frac{(x-2)}{(x-2)(x+3)} \le 0$$

$$\frac{(x+3)-(x-2)}{(x-2)(x+3)} \le 0$$

$$\frac{x+3-x+2}{(x-2)(x+3)} \le 0$$

$$\frac{5}{(x-2)(x+3)} \le 0$$

$$x - 2 = 0 \rightarrow x = 2$$

$$x + 3 = 0 \rightarrow x = -3$$

Cut points are 2 and –3. Intervals are $x < -3$, $-3 < x < 2$, and $x > 2$. For $x < -3$, let $x = -4$ be the test value. When $x = -4$, $\dfrac{5}{(x-2)(x+3)} = \dfrac{5}{(-4-2)(-4+3)}$ is positive. For $-3 < x < 2$, let $x = 0$ be the test value. When $x = 0$, $\dfrac{5}{(x-2)(x+3)} = \dfrac{5}{(0-2)(0+3)}$ is negative. For $x > 2$, let $x = 3$ be the test value. When $x = 3$, $\dfrac{5}{(x-2)(x+3)} = \dfrac{5}{(3-2)(3+3)}$ is positive. Thus,

$\dfrac{1}{x-2} + \dfrac{2}{x+3} \le \dfrac{3}{x+3}$ when $-3 < x < 2$.

51. Suppose there is a value of x such that $\dfrac{3}{x-2} = 0$. Then $(x-2)\left(\dfrac{3}{x-2}\right) = (x-2)(0)$, which implies that $3 = 0$, a contradiction. Therefore, it is impossible to find a solution to the given equation.

53.
$$\left(\frac{4x^{-4}y^{\frac{1}{2}}}{9x^{\frac{1}{2}}y^{-1}}\right)^{\frac{1}{2}} = \left(\frac{4x^{-\frac{9}{2}}y^{\frac{3}{2}}}{9}\right)^{\frac{1}{2}} = \frac{4^{\frac{1}{2}}(x^{-\frac{9}{2}})^{\frac{1}{2}}(y^{\frac{3}{2}})^{\frac{1}{2}}}{9^{\frac{1}{2}}} = \frac{2x^{-\frac{9}{4}}y^{\frac{3}{4}}}{3} = \frac{2y^{\frac{3}{4}}}{3x^{\frac{9}{4}}}$$

55.
$$\frac{3\sqrt{6}}{\sqrt{5}-\sqrt{2}} - \frac{6}{\sqrt{6}} = \frac{3\sqrt{6}}{\sqrt{5}-\sqrt{2}} \cdot \frac{\sqrt{5}+\sqrt{2}}{\sqrt{5}+\sqrt{2}} - \frac{6}{\sqrt{6}} \cdot \frac{\sqrt{6}}{\sqrt{6}} = \frac{3\sqrt{6}\left(\sqrt{5}+\sqrt{2}\right)}{5-2} - \frac{6\sqrt{6}}{6}$$

$$= \frac{3\sqrt{6}\left(\sqrt{5}+\sqrt{2}\right)}{3} - \sqrt{6} = \sqrt{30} + \sqrt{12} - \sqrt{6} = \sqrt{30} + 2\sqrt{3} - \sqrt{6}$$

Chapter 8 Review Exercises

1. $(x + 7)(x - 4) = 0$

   $x + 7 = 0$ or $x - 4 = 0$

   $x = -7$ or $x = 4$

3. $2y^2 - y - 1 = 0$

   $(2y + 1)(y - 1) = 0$

   $2y + 1 = 0$ or $y - 1 = 0$

   $y = -\dfrac{1}{2}$ or $y = 1$

5. $3x^2 - 17x = 28$

   $3x^2 - 17x - 28 = 0$

   $(3x + 4)(x - 7) = 0$

   $3x + 4 = 0$ or $x - 7 = 0$

   $x = -\dfrac{4}{3}$ or $x = 7$

7. $81 = a^2$

   $\pm\sqrt{81} = a$

   $\pm 9 = a$

9. $z^2 + 7 = 2$

   $z^2 = -5$

   $z = \pm\sqrt{-5} = \pm\sqrt{5}\, i$

11. $4x^2 + 36 = 24x$

    $4x^2 - 24x + 36 = 0$

    $x^2 - 6x + 9 = 0$

    $(x - 3)^2 = 0$

    $x - 3 = 0$

    $x = 3$

13. $(a + 7)(a + 3) = (3a + 1)(a + 1)$

    $a^2 + 10a + 21 = 3a^2 + 4a + 1$

    $0 = 2a^2 - 6a - 20$

    $0 = a^2 - 3a - 10$

    $0 = (a - 5)(a + 2)$

    $0 = a - 5$ or $0 = a + 2$

    $5 = a$ or $-2 = a$

15.
$$x - 2 = \frac{1}{x+2}$$

$$(x+2)(x-2) = (x+2)\left(\frac{1}{x+2}\right)$$

$$x^2 - 4 = 1$$

$$x^2 = 5$$

$$x = \pm\sqrt{5}$$

17.
$$\frac{2}{x-2} - \frac{5}{x+2} = 1$$

$$(x-2)(x+2)\left(\frac{2}{x-2} - \frac{5}{x+2}\right) = (x-2)(x+2)(1)$$

$$(x-2)(x+2)\left(\frac{2}{x-2}\right) - (x-2)(x+2)\left(\frac{5}{x+2}\right) = (x-2)(x+2)(1)$$

$$2(x+2) - 5(x-2) = (x-2)(x+2)$$

$$2x + 4 - 5x + 10 = x^2 - 4$$

$$-3x + 14 = x^2 - 4$$

$$0 = x^2 + 3x - 18$$

$$0 = (x+6)(x-3)$$

$$0 = x+6 \text{ or } 0 = x-3$$

$$-6 = x \text{ or } 3 = x$$

19.
$$x^2 + 2x - 4 = 0$$

$$\left(\frac{1}{2}(2)\right)^2 = 1^2 = 1$$

$$x^2 + 2x = 4$$
$$x^2 + 2x + 1 = 4 + 1$$

$$(x+1)^2 = 5$$

$$x + 1 = \pm\sqrt{5}$$

$$x = -1 \pm \sqrt{5}$$

21.
$$2y^2 + 4y - 3 = 0$$

$$2y^2 + 4y = 3$$

$$y^2 + 2y = \frac{3}{2}$$

$$\left(\frac{1}{2}(2)\right)^2 = 1^2 = 1$$

$$y^2 + 2y + 1 = \frac{3}{2} + 1$$

$$(y+1)^2 = \frac{5}{2}$$

$$y + 1 = \pm\sqrt{\frac{5}{2}}$$

$$y + 1 = \pm\frac{\sqrt{5}}{\sqrt{2}}$$

$$y + 1 = \pm\frac{\sqrt{5}\sqrt{2}}{\sqrt{2}\sqrt{2}}$$

$$y + 1 = \pm\frac{\sqrt{10}}{2}$$

$$y = -1 \pm \frac{\sqrt{10}}{2} = \frac{-2 \pm \sqrt{10}}{2}$$

23. $\quad 3a^2 + 6a - 5 = 0$

$$3a^2 + 6a = 5$$

$$a^2 + 2a = \frac{5}{3} \qquad\qquad \left(\frac{1}{2}(2)\right)^2 = 1^2 = 1$$

$$a^2 + 2a + 1 = \frac{5}{3} + 1$$

$$(a+1)^2 = \frac{8}{3}$$

$$a + 1 = \pm\sqrt{\frac{8}{3}}$$

$$a + 1 = \pm\frac{\sqrt{8}}{\sqrt{3}}$$

$$a + 1 = \pm\frac{\sqrt{8}\sqrt{3}}{\sqrt{3}\sqrt{3}}$$

$$a + 1 = \pm\frac{\sqrt{24}}{3} = \pm\frac{2\sqrt{6}}{3}$$

$$a = -1 \pm \frac{2\sqrt{6}}{3} = \frac{-3 \pm 2\sqrt{6}}{3}$$

25.
$$\frac{1}{a-5} + \frac{3}{a+2} = 4$$

$$(a-5)(a+2)\left(\frac{1}{a-5} + \frac{3}{a+2}\right) = (a-5)(a+2)(4)$$

$$(a-5)(a+2)\left(\frac{1}{a-5}\right) + (a-5)(a+2)\left(\frac{3}{a+2}\right) = (a-5)(a+2)(4)$$

$$a + 2 + 3(a - 5) = 4(a - 5)(a + 2)$$

$$a + 2 + 3a - 15 = 4(a^2 - 3a - 10)$$

$$4a - 13 = 4a^2 - 12a - 40$$

$$27 = 4a^2 - 16a$$

$$\frac{27}{4} = a^2 - 4a \qquad \left(\frac{1}{2}(-4)\right)^2 = (-2)^2 = 4$$

$$\frac{27}{4} + 4 = a^2 - 4a + 4$$

$$\frac{43}{4} = (a - 2)^2$$

$$\pm\sqrt{\frac{43}{4}} = a - 2$$

$$\pm\frac{\sqrt{43}}{\sqrt{4}} = a - 2$$

$$\pm\frac{\sqrt{43}}{2} = a - 2$$

$$2 \pm \frac{\sqrt{43}}{2} = a$$

$$\frac{4 \pm \sqrt{43}}{2} = a$$

27.
$$6a^2 - 13a = 5$$
$$6a^2 - 13a - 5 = 0$$
$$(3a + 1)(2a - 5) = 0$$
$$3a + 1 = 0 \ \text{ or } \ 2a - 5 = 0$$
$$a = -\frac{1}{3} \ \text{ or } \ a = \frac{5}{2}$$

29.
$$3a^2 - 7a = 6$$
$$3a^2 - 7a - 6 = 0$$
$$(3a + 2)(a - 3) = 0$$
$$3a + 2 = 0 \ \text{ or } \ a - 3 = 0$$
$$a = -\frac{2}{3} \ \text{ or } \ a = 3$$

31.
$$5a^2 - 3a = 3 - 3a + 2a^2$$
$$3a^2 = 3$$
$$a^2 = 1$$
$$a = \pm\sqrt{1} = \pm 1$$

33. $(x - 4)(x + 1) = x - 2$

$\quad x^2 - 3x - 4 = x - 2$

$\quad x^2 - 4x - 2 = 0$

$\quad A = 1, B = -4, C = -2$

$$x = \frac{-B \pm \sqrt{B^2 - 4AC}}{2A} = \frac{-(-4) \pm \sqrt{(-4)^2 - 4(1)(-2)}}{2(1)} = \frac{4 \pm \sqrt{16 + 8}}{2}$$

$$= \frac{4 \pm \sqrt{24}}{2} = \frac{4 \pm 2\sqrt{6}}{2} = \frac{\cancel{2}(2 \pm \sqrt{6})}{\cancel{2}} = 2 \pm \sqrt{6}$$

35. $(t + 3)(t - 4) = t(t + 2)$

$\quad t^2 - t - 12 = t^2 + 2t$

$\quad\quad -t - 12 = 2t$

$\quad\quad\quad -12 = 3t$

$\quad\quad\quad\; -4 = t$

37. $8x^2 = 12$

$\quad x^2 = \frac{12}{8} = \frac{3}{2}$

$$x = \pm\sqrt{\frac{3}{2}} = \pm\frac{\sqrt{3}}{\sqrt{2}} = \pm\frac{\sqrt{3}\sqrt{2}}{\sqrt{2}\sqrt{2}} = \pm\frac{\sqrt{6}}{2}$$

39. $3x^2 - 2x + 5 = 7x^2 - 2x + 5$

$\quad\quad\quad 0 = 4x^2$

$\quad\quad\quad 0 = x^2$

$\quad\quad\quad 0 = x$

41. $(x + 2)(x - 4) = 2x - 10$

$\quad x^2 - 2x - 8 = 2x - 10$

$\quad x^2 - 4x + 2 = 0$

$\quad A = 1, B = -4, C = 2$

$$x = \frac{-B \pm \sqrt{B^2 - 4AC}}{2A} = \frac{-(-4) \pm \sqrt{(-4)^2 - 4(1)(2)}}{2(1)} = \frac{4 \pm \sqrt{16 - 8}}{2}$$

$$= \frac{4 \pm \sqrt{8}}{2} = \frac{4 \pm 2\sqrt{2}}{2} = \frac{\cancel{2}(2 \pm \sqrt{2})}{\cancel{2}} = 2 \pm \sqrt{2}$$

43.

$$\frac{1}{z+2} = z - 4$$

$$(z+2)\left(\frac{1}{z+2}\right) = (z+2)(z-4)$$

$$1 = z^2 - 2z - 8$$

$$0 = z^2 - 2z - 9$$

$$A = 1, \ B = -2, \ C = -9$$

$$z = \frac{-B \pm \sqrt{B^2 - 4AC}}{2A} = \frac{-(-2) \pm \sqrt{(-2)^2 - 4(1)(-9)}}{2(1)} = \frac{2 \pm \sqrt{4 + 36}}{2}$$

$$= \frac{2 \pm \sqrt{40}}{2} = \frac{2 \pm 2\sqrt{10}}{2} = \frac{2\left(1 \pm \sqrt{10}\right)}{2} = 1 \pm \sqrt{10}$$

45.

$$\frac{1}{x+4} - \frac{3}{x+2} = 5$$

$$(x+4)(x+2)\left(\frac{1}{x+4} - \frac{3}{x+2}\right) = (x+4)(x+2)(5)$$

$$(x+4)(x+2)\left(\frac{1}{x+4}\right) - (x+4)(x+2)\left(\frac{3}{x+2}\right) = (x+4)(x+2)(5)$$

$$x + 2 - 3(x+4) = 5(x+4)(x+2)$$

$$x + 2 - 3x - 12 = 5(x^2 + 6x + 8)$$

$$-2x - 10 = 5x^2 + 30x + 40$$

$$0 = 5x^2 + 32x + 50$$

$$A = 5, \ B = 32, \ C = 50$$

$$z = \frac{-B \pm \sqrt{B^2 - 4AC}}{2A} = \frac{-32 \pm \sqrt{(32)^2 - 4(5)(50)}}{2(5)} = \frac{-32 \pm \sqrt{1024 - 1000}}{10}$$

$$= \frac{-32 \pm \sqrt{24}}{10} = \frac{-32 \pm 2\sqrt{6}}{10} = \frac{2\left(-16 \pm \sqrt{6}\right)}{10 \ \ 5} = \frac{-16 \pm \sqrt{6}}{5}$$

47.

$$\frac{3}{x-4} + \frac{2x}{x-5} = \frac{3}{x-5}$$

$$(x-4)(x-5)\left(\frac{3}{x-4} + \frac{2x}{x-5}\right) = (x-4)(x-5)\left(\frac{3}{x-5}\right)$$

$$(x-4)(x-5)\left(\frac{3}{x-4}\right) + (x-4)(x-5)\left(\frac{2x}{x-5}\right) = (x-4)(x-5)\left(\frac{3}{x-5}\right)$$

$$3(x-5) + 2x(x-4) = 3(x-4)$$

$$3x - 15 + 2x^2 - 8x = 3x - 12$$

$$2x^2 - 5x - 15 = 3x - 12$$

$$2x^2 - 8x - 3 = 0$$

$$A = 2, \ B = -8, \ C = -3$$

$$x = \frac{-B \pm \sqrt{B^2 - 4AC}}{2A} = \frac{-(-8) \pm \sqrt{(-8)^2 - 4(2)(-3)}}{2(2)} = \frac{8 \pm \sqrt{64 + 24}}{4}$$

$$= \frac{8 \pm \sqrt{88}}{4} = \frac{8 \pm 2\sqrt{22}}{4} = \frac{2\left(4 \pm \sqrt{22}\right)}{4 \ 2} = \frac{4 \pm \sqrt{22}}{2}$$

49.

$$A = \pi r^2 h, \quad r > 0$$

$$\frac{A}{\pi h} = r^2$$

$$\sqrt{\frac{A}{\pi h}} = r$$

$$\frac{\sqrt{A}\,\sqrt{\pi h}}{\sqrt{\pi h}\,\sqrt{\pi h}} = r$$

$$\frac{\sqrt{A\pi h}}{\pi h} = r$$

51.

$$2x^2 + xy - 3y^2 = 0$$

$$(2x + 3y)(x - y) = 0$$

$$2x + 3y = 0 \quad \text{or} \quad x - y = 0$$

$$x = -\frac{3y}{2} \quad \text{or} \quad x = y$$

53.

$$\sqrt{2a + 3} = a$$

$$\left(\sqrt{2a + 3}\right)^2 = a^2$$

$$2a + 3 = a^2$$

$$0 = a^2 - 2a - 3$$

$$0 = (a - 3)(a + 1)$$

$$0 = a - 3 \text{ or } 0 = a + 1$$

$$3 = a \text{ or } -1 = a$$

CHECK $a = 3$:

$$\sqrt{2a + 3} = a$$

$$\sqrt{2(3) + 3} \overset{?}{=} 3$$

$$\sqrt{6 + 3} \overset{?}{=} 3$$

$$\sqrt{9} \overset{?}{=} 3$$

$$3 \overset{\checkmark}{=} 3$$

CHECK $a = -1$:

$$\sqrt{2a + 3} = a$$

$$\sqrt{2(-1) + 3} \overset{?}{=} -1$$

$$\sqrt{-2 + 3} \overset{?}{=} -1$$

$$\sqrt{1} \overset{?}{=} -1$$

$$1 \neq -1$$

Thus, $a = -1$ is an extraneous root, and the only solution is $a = 3$.

55.

$$\sqrt{3a + 1} + 1 = a$$

$$\sqrt{3a + 1} = a - 1$$

$$\left(\sqrt{3a + 1}\right)^2 = (a - 1)^2$$

$$3a + 1 = a^2 - 2a + 1$$

$$0 = a^2 - 5a$$

$$0 = a(a - 5)$$

$$0 = a \text{ or } 0 = a - 5$$

$$0 = a \text{ or } 5 = a$$

CHECK $a = 0$:

$$\sqrt{3a + 1} + 1 = a$$

$$\sqrt{3(0) + 1} + 1 \overset{?}{=} 0$$

$$\sqrt{0 + 1} + 1 \overset{?}{=} 0$$

$$\sqrt{1} + 1 \overset{?}{=} 0$$

$$1 + 1 \overset{?}{=} 0$$

$$2 \neq 0$$

$$\sqrt{3a+1}+1 = a$$

$$\sqrt{3(5)+1}+1 \overset{?}{=} 5$$

$$\sqrt{15+1}+1 \overset{?}{=} 5$$

$$\sqrt{16}+1 \overset{?}{=} 5$$

$$4+1 \overset{?}{=} 5$$

$$5 \overset{\checkmark}{=} 5$$

Thus, a = 0 is an extraneous root, and the only solution is a = 5.

57. $$\sqrt{2x+1} - \sqrt{x-3} = 4$$

$$\sqrt{2x+1} = 4 + \sqrt{x-3}$$

$$\left(\sqrt{2x+1}\right)^2 = \left(4 + \sqrt{x-3}\right)^2$$

$$2x+1 = 16 + 8\sqrt{x-3} + x - 3$$

$$2x+1 = x + 13 + 8\sqrt{x-3}$$

$$x - 12 = 8\sqrt{x-3}$$

$$(x-12)^2 = \left(8\sqrt{x-3}\right)^2$$

$$x^2 - 24x + 144 = 64(x-3)$$

$$x^2 - 24x + 144 = 64x - 192$$

$$x^2 - 88x + 336 = 0$$

$$(x-4)(x-84) = 0$$

$$x - 4 = 0 \text{ or } x - 84 = 0$$

$$x = 4 \text{ or } x = 84$$

CHECK x = 4:

$$\sqrt{2x+1} - \sqrt{x-3} = 4$$

$$\sqrt{2(4)+1} - \sqrt{4-3} \overset{?}{=} 4$$

$$\sqrt{8+1} - \sqrt{4-3} \overset{?}{=} 4$$

$$\sqrt{9} - \sqrt{1} \overset{?}{=} 4$$

$$3 - 1 \overset{?}{=} 4$$

$$2 \neq 4$$

CHECK x = 84:

$$\sqrt{2x+1} - \sqrt{x-3} = 4$$

$$\sqrt{2(84)+1} - \sqrt{84-3} \overset{?}{=} 4$$

$$\sqrt{169} - \sqrt{81} \overset{?}{=} 4$$

$$13 - 9 \overset{?}{=} 4$$

$$4 \overset{\checkmark}{=} 4$$

Thus, x = 4 is an extraneous root, and the only solution is x = 84.

59. $$\sqrt{3x+4} - \sqrt{x-3} = 3$$

$$\sqrt{3x+4} = 3 + \sqrt{x-3}$$

$$\left(\sqrt{3x+4}\right)^2 = \left(3 + \sqrt{x-3}\right)^2$$

$$3x+4 = 9 + 6\sqrt{x-3} + x - 3$$

$$3x+4 = x + 6 + 6\sqrt{x-3}$$

$$2x - 2 = 6\sqrt{x-3}$$

$$x - 1 = 3\sqrt{x-3}$$

$$(x-1)^2 = \left(3\sqrt{x-3}\right)^2$$

$$x^2 - 2x + 1 = 9(x-3)$$

$$x^2 - 2x + 1 = 9x - 27$$

$$x^2 - 11x + 28 = 0$$

$$(x-4)(x-7) = 0$$

CHECK x = 4:

$$\sqrt{3x+4} - \sqrt{x-3} = 3$$

$$\sqrt{3(4)+4} - \sqrt{4-3} \overset{?}{=} 3$$

$$\sqrt{12+4} - \sqrt{4-3} \overset{?}{=} 3$$

$$\sqrt{16} - \sqrt{1} \overset{?}{=} 3$$

$$4 - 1 \overset{?}{=} 3$$

$$3 \overset{\checkmark}{=} 3$$

CHECK x = 7:

$$\sqrt{3x+4} - \sqrt{x-3} = 3$$

$$\sqrt{3(7)+4} - \sqrt{7-3} \overset{?}{=} 3$$

$$\sqrt{21+4} - \sqrt{7-3} \overset{?}{=} 3$$

$$x - 4 = 0 \text{ or } x - 7 = 0$$
$$x = 4 \text{ or } x = 7$$

$$\sqrt{25} - \sqrt{4} \overset{?}{=} 3$$
$$5 - 2 \overset{?}{=} 3$$
$$3 \overset{\checkmark}{=} 3$$

61.
$$\sqrt{3y + z} = x$$
$$\left(\sqrt{3y + z}\right)^2 = x^2$$
$$3y + z = x^2$$
$$3y = x^2 - z$$
$$y = \frac{x^2 - z}{3}$$

63.
$$\sqrt{3y} + z = x$$
$$\sqrt{3y} = x - z$$
$$\left(\sqrt{3y}\right)^2 = (x - z)^2$$
$$3y = (x - z)^2$$
$$y = \frac{(x - z)^2}{3}$$

65.
$$x^3 - 2x^2 - 15x = 0$$
$$x(x^2 - 2x - 15) = 0$$
$$x(x - 5)(x + 3) = 0$$
$$x = 0 \text{ or } x - 5 = 0 \text{ or } x + 3 = 0$$
$$x = 0 \text{ or } x = 5 \text{ or } x = -3$$

67.
$$4x^3 - 10x^2 - 6x = 0$$
$$2x(2x^2 - 5x - 3) = 0$$
$$2x(2x + 1)(x - 3) = 0$$
$$2x = 0 \text{ or } 2x + 1 = 0 \text{ or } x - 3 = 0$$
$$x = 0 \text{ or } x = -\frac{1}{2} \text{ or } x = 3$$

69.
$$a^4 - 17a^2 = -16$$
$$a^4 - 17a^2 + 16 = 0$$
$$(a^2 - 1)(a^2 - 16) = 0$$
$$(a - 1)(a + 1)(a - 4)(a + 4) = 0$$
$$a - 1 = 0 \text{ or } a + 1 = 0 \text{ or } a - 4 = 0 \text{ or } a + 4 = 0$$
$$a = 1 \text{ or } a = -1 \text{ or } a = 4 \text{ or } a = -4$$

71.
$$y^4 - 3y^2 = 4$$

$$y^4 - 3y^2 - 4 = 0$$
$$(y^2 - 4)(y^2 + 1) = 0$$
$$(y - 2)(y + 2)(y^2 + 1) = 0$$
$$y - 2 = 0 \quad \text{or} \quad y + 2 = 0 \quad \text{or} \quad y^2 + 1 = 0$$
$$y = 2 \quad \text{or} \quad y = -2 \quad \text{or} \quad y^2 = -1$$
$$y = 2 \quad \text{or} \quad y = -2 \quad \text{or} \quad y = \pm\sqrt{-1} = \pm i$$

73.
$$z^4 = 6z^2 - 5$$
$$z^4 - 6z^2 + 5 = 0$$
$$(z^2 - 1)(z^2 - 5) = 0$$
$$(z - 1)(z + 1)(z^2 - 5) = 0$$
$$z - 1 = 0 \quad \text{or} \quad z + 1 = 0 \quad \text{or} \quad z^2 - 5 = 0$$
$$z = 1 \quad \text{or} \quad z = -1 \quad \text{or} \quad z^2 = 5$$
$$z = 1 \quad \text{or} \quad z = -1 \quad \text{or} \quad z = \pm\sqrt{5}$$

75.
$$a^{\frac{1}{2}} - a^{\frac{1}{4}} - 6 = 0$$

Let $u = a^{\frac{1}{4}}$

Then $u^2 = \left(a^{\frac{1}{4}}\right)^2 = a^{\frac{1}{2}}$

$$u^2 - u - 6 = 0$$
$$(u - 3)(u + 2) = 0$$
$$u - 3 = 0 \text{ or } u + 2 = 0$$
$$u = 3 \text{ or } u = -2$$
$$a^{\frac{1}{4}} = 3 \text{ or } a^{\frac{1}{4}} = -2$$
$$\left(a^{\frac{1}{4}}\right)^4 = 3^4 \text{ or } \left(a^{\frac{1}{4}}\right)^4 = (-2)^4$$

$$a = 81 \text{ or } a = 16$$

CHECK a = 81:
$$a^{\frac{1}{2}} - a^{\frac{1}{4}} - 6 = 0$$
$$(81)^{\frac{1}{2}} - (81)^{\frac{1}{4}} - 6 \overset{?}{=} 0$$
$$9 - 3 - 6 \overset{?}{=} 0$$
$$0 \overset{\checkmark}{=} 0$$

CHECK a = 16:
$$a^{\frac{1}{2}} - a^{\frac{1}{4}} - 6 = 0$$
$$(16)^{\frac{1}{2}} - (16)^{\frac{1}{4}} - 6 \overset{?}{=} 0$$
$$4 - 2 - 6 \overset{?}{=} 0$$
$$-4 \neq 0$$

Thus, a = 16 is an extraneous root, and the only solution is a = 81.

77.
$$2x^{\frac{2}{3}} = 5x^{\frac{1}{3}} + 3$$

$$2x^{\frac{2}{3}} - 5x^{\frac{1}{3}} - 3 = 0$$

Let $u = x^{\frac{1}{3}}$

Then $u^2 = \left(x^{\frac{1}{3}}\right)^2 = x^{\frac{2}{3}}$

CHECK $x = -\frac{1}{8}$:
$$2x^{\frac{2}{3}} = 5x^{\frac{1}{3}} + 3$$
$$2\left(-\frac{1}{8}\right)^{\frac{2}{3}} \overset{?}{=} 5\left(-\frac{1}{8}\right)^{\frac{1}{3}} + 3$$
$$2\left(\frac{1}{4}\right) \overset{?}{=} 5\left(-\frac{1}{2}\right) + 3$$

240

$$2u^2 - 5u - 3 = 0$$

$$(2u + 1)(u - 3) = 0$$

$$2u + 1 = 0 \text{ or } u - 3 = 0$$

$$u = -\frac{1}{2} \text{ or } u = 3$$

$$x^{\frac{1}{3}} = -\frac{1}{2} \text{ or } x^{\frac{1}{3}} = 3$$

$$\left(x^{\frac{1}{3}}\right)^3 = \left(-\frac{1}{2}\right)^3 \text{ or } \left(x^{\frac{1}{3}}\right)^3 = 3^3$$

$$x = -\frac{1}{8} \text{ or } x = 27$$

$$\frac{1}{2} \overset{?}{=} -\frac{5}{2} + 3$$

$$\frac{1}{2} \overset{\checkmark}{=} \frac{1}{2}$$

CHECK x = 27:

$$2x^{\frac{2}{3}} = 5x^{\frac{1}{3}} + 3$$

$$2(27)^{\frac{2}{3}} \overset{?}{=} 5(27)^{\frac{1}{3}} + 3$$

$$2(9) \overset{?}{=} 5(3) + 3$$

$$18 \overset{?}{=} 15 + 3$$

$$18 \overset{\checkmark}{=} 18$$

79. $\sqrt{x} + 2\sqrt[4]{x} - 35 = 0$

$$x^{\frac{1}{2}} + 2x^{\frac{1}{4}} - 35 = 0$$

Let $u = x^{\frac{1}{4}}$

Then $u^2 = \left(x^{\frac{1}{4}}\right)^2 = x^{\frac{1}{2}}$

$$u^2 + 2u - 35 = 0$$

$$(u + 7)(u - 5) = 0$$

$$u = -7 \text{ or } u = 5$$

$$x^{\frac{1}{4}} = -7 \text{ or } x^{\frac{1}{4}} = 5$$

$$\left(x^{\frac{1}{4}}\right)^4 = (-7)^4 \text{ or } \left(x^{\frac{1}{4}}\right)^4 = 5^4$$

$$x = 2401 \text{ or } x = 625$$

CHECK x = 2401:

$$\sqrt{x} + 2\sqrt[4]{x} - 35 = 0$$

$$\sqrt{2401} + 2\sqrt[4]{2401} - 35 \overset{?}{=} 0$$

$$49 + 2(7) - 35 \overset{?}{=} 0$$

$$49 + 14 - 35 \overset{?}{=} 0$$

$$28 \neq 0$$

CHECK x = 625:

$$\sqrt{x} + 2\sqrt[4]{x} - 35 = 0$$

$$\sqrt{625} + 2\sqrt[4]{625} - 35 \overset{?}{=} 0$$

$$25 + 2(5) - 35 \overset{?}{=} 0$$

$$25 + 10 - 35 \overset{?}{=} 0$$

$$0 \overset{\checkmark}{=} 0$$

Thus, x = 2401 is an extraneous root, and the only solution is x = 625.

81. $3x^{-2} + x^{-1} - 2 = 0$

Let $u = x^{-1}$

Then $u^2 = (x^{-1})^2 = x^{-2}$

$$3u^2 + u - 2 = 0$$

CHECK $x = \frac{3}{2}$:

$$3x^{-2} + x^{-1} - 2 = 0$$

$$3\left(\frac{3}{2}\right)^{-2} + \left(\frac{3}{2}\right)^{-1} - 2 \overset{?}{=} 0$$

$$\overset{1}{\cancel{3}}\left(\frac{4}{\underset{3}{\cancel{9}}}\right) + \frac{2}{3} - 2 \overset{?}{=} 0$$

$$(3u - 2)(u + 1) = 0$$

$$3u - 2 = 0 \text{ or } u + 1 = 0$$

$$u = \frac{2}{3} \text{ or } u = -1$$

$$x^{-1} = \frac{2}{3} \text{ or } x^{-1} = -1$$

$$(x^{-1})^{-1} = \left(\frac{2}{3}\right)^{-1} \text{ or } (x^{-1})^{-1} = (-1)^{-1}$$

$$x = \frac{3}{2} \text{ or } x = -1$$

$$\frac{4}{3} + \frac{2}{3} - 2 \stackrel{?}{=} 0$$

$$\frac{6}{3} - 2 \stackrel{?}{=} 0$$

$$2 - 2 \stackrel{?}{=} 0$$

$$0 \stackrel{\checkmark}{=} 0$$

CHECK x = -1:

$$3x^{-2} + x^{-1} - 2 = 0$$

$$3(-1)^{-2} + (-1)^{-1} - 2 \stackrel{?}{=} 0$$

$$3(1) - 1 - 2 \stackrel{?}{=} 0$$

$$3 - 1 - 2 \stackrel{?}{=} 0$$

$$0 \stackrel{\checkmark}{=} 0$$

83. $(x - 2)(x + 1) > 0$

$$x - 2 = 0 \rightarrow x = 2$$
$$x + 1 = 0 \rightarrow x = -1$$

Cut points are 2 and –1. Intervals are $x < -1$, $-1 < x < 2$, and $x > 2$. For $x = -1$, let $x = -2$ be the test value. When $x = -2$, $(x - 2)(x + 1) = (-2 - 2)(-2 + 1)$ is positive. For $-1 < x < 2$, let $x = 0$ be the test value. When $x = 0$, $(x - 2)(x + 1) = (0 - 2)(0 + 1)$ is negative. For $x > 2$, let $x = 3$ be the test value. When $x = 3$, $(x - 2)(x + 1) = (3 - 2)(3 + 1)$ is positive. Thus, $(x - 2)(x + 1) > 0$ when $x < -1$ or $x > 2$.

85. $(3x + 1)(x - 2) \leq 0$

$$3x + 1 = 0 \rightarrow x = -\frac{1}{3}$$
$$x - 2 = 0 \rightarrow x = 2$$

Cut points are $-\frac{1}{3}$ and 2. Intervals are $x = -\frac{1}{3}$, $-\frac{1}{3} < x < 2$, and $x > 2$. For $x < -\frac{1}{3}$, let $x = -1$ be the test value. When $x = -1$, $(3x + 1)(x - 2) = (3(-1) + 1)(-1 - 2)$ is positive. For $-\frac{1}{3} < x < 2$, let $x = 0$ be the test value. When $x = 0$, $(3x + 1)(x - 2) = (3(0) + 1)(0 - 2)$ is negative. For $x > 2$, let $x = 3$ be the test value. When $x = 3$, $(3x + 1)(x - 2) = (3(3) + 1)(3 - 2)$ is positive. Thus, $(3x + 1)(x - 2) \leq 0$ when $-\frac{1}{3} \leq x \leq 2$.

87. $y^2 - 5y + 4 > 0$

$(y - 1)(y - 4) > 0$
$y - 1 = 0 \rightarrow y = 1$
$y - 4 = 0 \rightarrow y = 4$

Cut points are 1 and 4. Intervals are $y < 1$, $1 < y < 4$, and $y > 4$. For $y < 1$, let $y = 0$ be the test value. When $y = 0$, $(y - 1)(y - 4) = (0 - 1)(0 - 4)$ is positive. For $1 < y < 4$, let $y = 2$ be the test value. When y = 2, (y − 1)(y − 4) = (2 − 1)(2 − 4) is negative. For y > 4, let y = 5 be the test value. When y = 5, (y − 1)(y − 4) = (5 − 1)(5 − 4) is positive. Thus, $y^2 - 5y + 4 > 0$ when y < 1 or y > 4.

89. $a^2 < 81$

$a^2 - 81 < 0$

(a − 9)(a + 9) < 0

a − 9 = 0 → a = 9

a + 9 = 0 → a = −9

Cut points are 9 and −9. Intervals are a < −9, −9 < a < 9, and a > 9. For a < −9, let a = −10 be the test value. When a = −10, (a − 9)(a + 9) = (−10 − 9)(−10 + 9) is positive. For −9 < a < 9, let a = 0 be the test value. When a = 0, (a − 9)(a + 9) = (0 − 9)(0 + 9) is negative. For a > 9, let a = 10 be the test value. When a = 10, (a − 9)(a + 9) = (10 − 9)(10 + 9) is positive. Thus, $a^2 < 81$ when

−9 < a < 9.

91. $5s^2 - 18s \geq 8$

$5s^2 - 18s - 8 \geq 0$

(5s + 2)(s − 4) ≥ 0

$5s + 2 = 0 \rightarrow s = -\dfrac{2}{5}$

s − 4 = 0 → s = 4

Cut points are $-\dfrac{2}{5}$ and 4. Intervals are $s < -\dfrac{2}{5}$, $-\dfrac{2}{5} < s < 4$ and s > 4. For $s < -\dfrac{2}{5}$, let s = −1 be the test value. When s = −1, (5s + 2)(s − 4) = (5(−1) + 2)(−1 − 4) is positive. For $-\dfrac{2}{5} < s < 4$, let s = 0 be the test value. When s = 0, (5s + 2)(s − 4) = (5(0) + 2)(0 − 4) is negative. For s > 4 let s = 5 be the test value. When s = 5, (5s + 2)(s − 4) = (5(5) + 2)(5 − 4) is positive. Thus,

$5s^2 - 18s \geq 8$ when $s \leq -\dfrac{2}{5}$ or s ≥ 4.

93. $\dfrac{x-3}{x+2} < 0$

x − 3 = 0 → x = 3

x + 2 = 0 → x = −2

Cut points are 3 and −2. Intervals are x < −2, −2 < x < 3, and x > 3. For x < −2, let x = −3 be the test value. When x = −3, $\dfrac{x-3}{x+2} = \dfrac{-3-3}{-3+2}$ is positive. For −2 < x < 3, let x = 0 be the test value.

When x = 0, $\dfrac{x-3}{x+2} = \dfrac{0-3}{0+2}$ is negative. For x > 3, let x = 4 be the test value. When x = 4,

$\dfrac{x-3}{x+2} = \dfrac{4-3}{4+2}$ is positive. Thus, $\dfrac{x-3}{x+2} < 0$ when −2 < x < 3.

95. $\dfrac{x-3}{x+2} \geq 0$

x − 3 = 0 → x = 3

x + 2 = 0 → x = −2

Cut points are –2 and 3. Intervals are x < –2, –2 < x < 3, and x > 3. For x < –2, let x = –3 be the

test value. When x = –3, $\dfrac{x-3}{x+2} = \dfrac{-3-3}{-3+2}$ is positive. For –2 < x < 3, let x = 0 be the test value.

When x = 0, $\dfrac{x-3}{x+2} = \dfrac{0-3}{0+2}$ is negative. For x > 3, let x = 4 be the test value. When x = 4,

$\dfrac{x-3}{x+2} = \dfrac{4-3}{4+2}$ is positive. Thus, $\dfrac{x-3}{x+2} \geq 0$ when x < –2 or x ≥ 3. (Note that x ≠ –2 here.)

97.
$$\frac{2x+1}{x-3} < 2$$

$$\frac{2x+1}{x-3} - 2 < 0$$

$$\frac{2x+1}{x-3} - \frac{2(x-3)}{x-3} < 0$$

$$\frac{2x+1-2(x-3)}{x-3} < 0$$

$$\frac{2x+1-2x+6}{x-3} < 0$$

$$\frac{7}{x-3} < 0$$

$$x-3 = 0 \rightarrow x = 3$$

Cut point is 3. Intervals are x < 3 and x > 3. For x < 3, let x = 2 be the test value. When x = 2,

$\dfrac{7}{x-3} = \dfrac{7}{2-3}$ is negative. For x > 3, let x = 4 be the test value. When x = 4, $\dfrac{7}{x-3} = \dfrac{7}{4-3}$ is

positive. Thus, $\dfrac{2x+1}{x-3} < 2$ when x < 3.

99.
$$\frac{5}{x+4} \geq 4$$

$$\frac{5}{x+4} - 4 \geq 0$$

$$\frac{5}{x+4} - \frac{4(x+4)}{x+4} \geq 0$$

$$\frac{5-4(x+4)}{x+4} \geq 0$$

$$\frac{5-4x-16}{x+4} \geq 0$$

$$\frac{-4x-11}{x+4} \geq 0$$

$$-4x-11 = 0 \rightarrow x = -\frac{11}{4}$$

$$x+4 = 0 \rightarrow x = -4$$

Cut points are $-\dfrac{11}{4}$ and –4. Intervals are x < –4, $-4 < x < -\dfrac{11}{4}$, and $x > -\dfrac{11}{4}$. For x < –4, let

x = –5 be the test value. When x = –5, $\dfrac{-4x-11}{x+4} = \dfrac{-4(-5)-11}{-5+4}$ is negative. For $-4 < x < -\dfrac{11}{4}$,

let x = –3 be the test value. When x = –3, $\dfrac{-4x-11}{x+4} = \dfrac{-4(-3)-11}{-3+4}$ is positive. For $x > -\dfrac{11}{4}$, let

$x = -2$ be the test value. When $x = -2$, $\dfrac{-4x-11}{x+4} = \dfrac{-4(-2)-11}{-2+4}$ is negative. Thus, $\dfrac{5}{x+4} \geq 4$ when $-4 < x \leq -\dfrac{11}{4}$.

101. Let x = the number.

$$x^2 + 4 = 36$$
$$x^2 = 32$$
$$x = \pm\sqrt{32} = \pm 4\sqrt{2}$$

Thus, there are two numbers with the given property: $\pm 4\sqrt{2}$.

103. Let x = the number.

$$x + \frac{1}{x} = \frac{53}{14}$$
$$14x\left(x + \frac{1}{x}\right) = 14x\left(\frac{53}{14}\right)$$
$$14x(x) + 14x\left(\frac{1}{x}\right) = 14x\left(\frac{53}{14}\right)$$
$$14x^2 + 14 = 53x$$
$$14x^2 - 53x + 14 = 0$$
$$(7x - 2)(2x - 7) = 0$$
$$7x - 2 = 0 \ \text{ or } \ 2x - 7 = 0$$
$$x = \frac{2}{7} \ \text{ or } \ x = \frac{7}{2}$$

Thus, there are two numbers with the given property: $\dfrac{2}{7}$ and $\dfrac{7}{2}$.

105. Let w = width of the rectangle (in feet). Then $2w$ = length of the rectangle (in feet).

$$w(2w) = 50$$
$$2w^2 = 50$$
$$w^2 = 25$$
$$w = \pm 5$$

Since width cannot be negative, we eliminate the negative answer. Thus, the width of the rectangle is 5 feet and the length of the rectangle is $2(5) = 10$ feet.

107. Let w = width of the frame (in inches).

$$(5 + 2x)(8 + 2x) - 114 = (5)(8)$$
$$40 + 26x + 4x^2 - 114 = 40$$
$$4x^2 + 26x - 114 = 0$$
$$2x^2 + 13x - 57 = 0$$
$$(2x + 19)(x - 3) = 0$$

$$2x + 19 = 0 \quad \text{or} \quad x - 3 = 0$$

$$x = -\frac{19}{2} \quad \text{or} \quad x = 3$$

We eliminate the negative answer, since a width cannot be negative. Thus, the frame is 3 inches wide.

109. Let c = length of the hypotenuse (in inches).

$$5^2 + 15^2 = c^2$$

$$25 + 225 = c^2$$

$$250 = c^2$$

$$\pm\sqrt{250} = c$$

$$\pm 5\sqrt{10} = c$$

Since length cannot be negative, eliminate the negative answer. Thus, the length of the hypotenuse is $5\sqrt{10}$ inches $\cong$ 15.81 inches.

111. Let d = length of the diagonal (in inches).

$$5^2 + 4^2 = d^2$$

$$25 + 16 = d^2$$

$$41 = d^2$$

$$\pm\sqrt{41} = d$$

Eliminate the negative answer, since d is a length. Thus, the length of the diagonal is $\sqrt{41}$ inches $\cong$ 6.40 inches.

113. Let r = rate of the wind (in mph).

$$\frac{300}{200 - r} = \frac{300}{200 + r} = \frac{25}{8}$$

$$8(200 - r)(200 + r)\left(\frac{300}{200 - r} + \frac{300}{200 + r}\right) = 8(200 - r)(200 + r)\left(\frac{25}{8}\right)$$

$$8(200 - r)(200 + r)\left(\frac{300}{200 - r}\right) + 8(200 - r)(200 + r)\left(\frac{300}{200 + r}\right) = 8(200 - r)(200 + r)\left(\frac{25}{8}\right)$$

$$8(300)(200 + r) + 8(300)(200 - r) = 25(200 - r)(200 + r)$$

$$480000 + 2400r + 480000 - 2400r = 25(40000 - r^2)$$

$$960000 = 1000000 - 25r^2$$

$$-40000 = -25r^2$$

$$1600 = r^2$$

$$\pm\sqrt{1600} = r$$

$$\pm 40 = r$$

Since the rate of the wind cannot be negative, we eliminate the negative answer. Thus, the rate of the wind is 40 mph.

Chapter 8 Practice Test

1. (a)
$$(3z-1)(z-4) = z^2 - 8z + 7$$
$$3z^2 - 13z + 4 = z^2 - 8z + 7$$
$$2z^2 - 5z - 3 = 0$$
$$(2z+1)(z-3) = 0$$
$$2z+1 = 0 \text{ or } z-3 = 0$$
$$z = -\frac{1}{2} \text{ or } z = 3$$

(b)
$$3 + \frac{5}{x^2} = 4$$
$$\frac{5}{x^2} = 1$$
$$\cancel{x^2}\left(\frac{5}{\cancel{x^2}}\right) = x^2(1)$$
$$5 = x^2$$
$$\pm\sqrt{5} = x$$

3. $2x^2 - 3x + 5 = 0$

$A = 2, B = -3, C = 5$

$B^2 - 4AC = (-3)^2 - 4(2)(5) = 9 - 40 = -31$

Since the discriminant is < 0, the roots are not real.

5.
$$\sqrt{3x} = 2 + \sqrt{x+4}$$
$$\left(\sqrt{3x}\right)^2 = \left(2 + \sqrt{x+4}\right)^2$$
$$3x = 4 + 4\sqrt{x+4} + x + 4$$
$$3x = x + 8 + 4\sqrt{x+4}$$
$$2x - 8 = 4\sqrt{x+4}$$
$$x - 4 = 2\sqrt{x+4}$$
$$(x-4)^2 = \left(2\sqrt{x+4}\right)^2$$
$$x^2 - 8x + 16 = 4(x+4)$$
$$x^2 - 8x + 16 = 4x + 16$$
$$x^2 - 12x = 0$$
$$x(x-12) = 0$$
$$x = 0 \text{ or } x - 12 = 0$$
$$x = 0 \text{ or } x = 12$$

CHECK $x = 0$:
$$\sqrt{3x} = 2 + \sqrt{x+4}$$
$$\sqrt{3(0)} \overset{?}{=} 2 + \sqrt{0+4}$$
$$\sqrt{0} \overset{?}{=} 2 + \sqrt{4}$$
$$0 \overset{?}{=} 2 + 2$$
$$0 \neq 4$$

CHECK $x = 12$:
$$\sqrt{3x} = 2 + \sqrt{x+4}$$
$$\sqrt{3(12)} \overset{?}{=} 2 + \sqrt{12+4}$$
$$\sqrt{36} \overset{?}{=} 2 + \sqrt{16}$$
$$6 \overset{?}{=} 2 + 4$$
$$6 \overset{\checkmark}{=} 6$$

Thus, $x = 0$ is an extraneous root and the only solution is $x = 12$.

7.   (a) $x^2 - 5x \geq 36$

$x^2 - 5x - 36 \geq 0$

$(x - 9)(x + 4) \geq 0$

$x - 9 = 0 \rightarrow x = 9$

$x + 4 = 0 \rightarrow x = -4$

Cut points are 9 and –4. Intervals are $x < -4$, $-4 < x < 9$, $x > 9$. For $x < -4$, let $x = -5$ be the test value. When $x = -5$, $(x - 9)(x + 4) = (-5 - 9)(-5 + 4)$ is positive. For $-4 < x < 9$, let $x = 0$ be the test value. When $x = 0$, $(x - 9)(x + 4) = (0 - 9)(0 + 4)$ is negative. For $x > 9$, let $x = 10$ be the test value. When $x = 10$, $(x - 9)(x + 4) = (10 - 9)(10 + 4)$ is positive. Thus, $x^2 - 5x \geq 36$ when

$x \leq -4$ or $x \geq 9$.

(b) $\dfrac{3x - 2}{x - 6} < 0$

$3x - 2 = 0 \rightarrow x = \dfrac{2}{3}$

$x - 6 = 0 \rightarrow x = 6$

Cut points are $\dfrac{2}{3}$ and 6. Intervals are $x < \dfrac{2}{3}$, $\dfrac{2}{3} < x < 6$, and $x > 6$. For $x < \dfrac{2}{3}$, let $x = 0$ be the test value. When $x = 0$, $\dfrac{3x - 2}{x - 6} = \dfrac{3(0) - 2}{0 - 6}$ is positive. For $\dfrac{2}{3} < x < 6$, let $x = 1$ be the test value. When $x = 1$, $\dfrac{3x - 2}{x - 6} = \dfrac{3(1) - 2}{1 - 6}$ is negative. For $x > 6$, let $x = 7$ be the test value. When $x = 7$, $\dfrac{3x - 2}{x - 6} = \dfrac{3(7) - 2}{7 - 6}$ is positive. Thus, $\dfrac{3x - 2}{x - 6} < 0$ when $\dfrac{2}{3} < x < 6$.

9.   Let r = rate of the boat in still water (in mph).

$\dfrac{15}{r - 3} + \dfrac{12}{r + 3} = \dfrac{9}{4}$

$4(r - 3)(r + 3)\left(\dfrac{15}{r - 3} + \dfrac{12}{r + 3}\right) = 4(r - 3)(r + 3)\left(\dfrac{9}{4}\right)$

$4(r - 3)(r + 3)\left(\dfrac{15}{r - 3}\right) + 4(r - 3)(r + 3)\left(\dfrac{12}{r + 3}\right) = 4(r - 3)(r + 3)\left(\dfrac{9}{4}\right)$

$60(r + 3) + 48(r - 3) = 9(r - 3)(r + 3)$

$60r + 180 + 48r - 144 = 9(r^2 - 9)$

$108r + 36 = 9r^2 - 81$

$0 = 9r^2 - 108r - 117$

$0 = r^2 - 12r - 13$

$0 = (r - 13)(r + 1)$

$0 = r - 13$ or $0 = r + 1$

$13 = r$ or $-1 = r$

We eliminate $r = -1$, since the rate of the boat cannot be negative. Thus the boat's rate in still water is 13 mph.

CHAPTER 9
CONIC SECTIONS

Exercises 9.1

1. $d = \sqrt{(x_2 - x_1)^2 + (y_2 - y_1)^2} = \sqrt{(6-3)^2 + (9-5)^2} = \sqrt{3^2 + 4^2} = \sqrt{9+16} = \sqrt{25} = 5$

5. $d = \sqrt{(x_2 - x_1)^2 + (y_2 - y_1)^2} = \sqrt{(-6-6)^2 + (9-(-9))^2} = \sqrt{(-12)^2 + 18^2}$

    $= \sqrt{144 + 324} = \sqrt{468} = \sqrt{36 \cdot 13} = \sqrt{36}\sqrt{13} = 6\sqrt{13}$

7. $d = \sqrt{(x_2 - x_1)^2 + (y_2 - y_1)^2} = \sqrt{(-7-(-8))^2 + (-3-(-3))^2} = \sqrt{(-7+8)^2 + (-3+3)^2}$

    $= \sqrt{1^2 + 0^2} = \sqrt{1} = 1$

11. $d = \sqrt{(x_2 - x_1)^2 + (y_2 - y_1)^2} = \sqrt{(1.4-1.7)^2 + (0.8-1.2)^2} = \sqrt{(-0.3)^2 + (-0.4)^2}$

    $= \sqrt{0.09 + 0.16} = \sqrt{0.25} = 0.5$

13. $\left(\dfrac{x_1 + x_2}{2}, \dfrac{y_1 + y_2}{2}\right) = \left(\dfrac{0+0}{2}, \dfrac{5+7}{2}\right) = \left(\dfrac{0}{2}, \dfrac{12}{2}\right) = (0,6)$

17. $\left(\dfrac{x_1 + x_2}{2}, \dfrac{y_1 + y_2}{2}\right) = \left(\dfrac{-3+3}{2}, \dfrac{4+(-4)}{2}\right) = \left(\dfrac{0}{2}, \dfrac{0}{2}\right) = (0,0)$

19. $\left(\dfrac{x_1 + x_2}{2}, \dfrac{y_1 + y_2}{2}\right) = \left(\dfrac{\frac{2}{5}+\frac{1}{3}}{2}, \dfrac{\frac{3}{4}+2}{2}\right) = \left(\dfrac{\frac{11}{15}}{2}, \dfrac{\frac{11}{4}}{2}\right) = \left(\dfrac{11}{30}, \dfrac{11}{8}\right)$

21. $|PQ| = \sqrt{(8-5)^2 + (2-2)^2} = \sqrt{3^2 + 0^2} = \sqrt{9+0} = \sqrt{9} = 3$

    $|QR| = \sqrt{(8-8)^2 + (6-2)^2} = \sqrt{0^2 + 4^2} = \sqrt{0+16} = \sqrt{16} = 4$

    $|PR| = \sqrt{(8-5)^2 + (6-2)^2} = \sqrt{3^2 + 4^2} = \sqrt{9+16} = \sqrt{25} = 5$

    Since, $|PQ|^2 + |QR|^2 = |PR|^2$, the three points are the vertices of a right triangle.

25. Midpoint of PR: $\left(\dfrac{5+9}{2}, \dfrac{3+7}{2}\right) = (7,5)$

    Midpoint of QS: $\left(\dfrac{7+7}{2}, \dfrac{4+6}{2}\right) = (7,5)$

    Since the diagonals of quadrilateral PQRS have the same

    midpoint, they bisect each other.

    Therefore, the quadrilateral is a parallelogram.

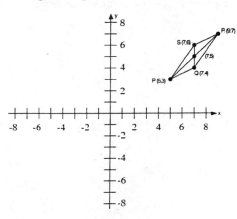

29.
$$(x-h)^2 + (y-k)^2 = r^2$$
$$(x-0)^2 + (y-0)^2 = 1^2$$
$$x^2 + y^2 = 1$$

33.
$$(x-h)^2 + (y-k)^2 = r^2$$
$$(x-2)^2 + (y-5)^2 = 6^2$$
$$(x-2)^2 + (y-5)^2 = 36$$

37.
$$(x-h)^2 + (y-k)^2 = r^2$$
$$(x-(-3))^2 + (y-(-2))^2 = 1^2$$
$$(x+3)^2 + (y+2)^2 = 1$$

39.
$$x^2 + y^2 = 16$$
$$(x-0)^2 + (y-0)^2 = 4^2$$
$(x-h)^2 + (y-k)^2 = r^2$, so h = 0, k = 0, and r = 4. The center is (0,0) and the radius is 4.

43.
$$(x-3)^2 + y^2 = 16$$
$$(x-3)^2 + (y-0)^2 = 4^2$$
$(x-h)^2 + (y-k)^2 = r^2$, so h = 3, k = 0, and r = 4. The center is (3,0) and the radius is 4.

47.
$$(x+1)^2 + (y-3)^2 = 25$$
$$(x+1)^2 + (y-3)^2 = 5^2$$
$(x-h)^2 + (y-k)^2 = r^2$, so h = −1, k = 3, and r = 5. The center is (-1,3) and the radius is 5.

51.
$$(x+7)^2 + (y+1)^2 = 2$$
$$(x+7)^2 + (y+1)^2 = \left(\sqrt{2}\right)^2$$
$$(x-h)^2 + (y-k)^2 = r^2$$
so h = −7, k = −1, and r = $\sqrt{2}$.
The center is $(-7,-1)$ and the radius is $\sqrt{2}$.

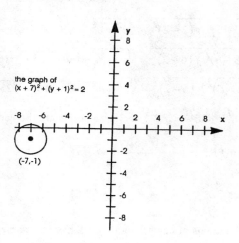

the graph of
$(x + 7)^2 + (y + 1)^2 = 2$

$(-7,-1)$

57. $x^2 + y^2 - 2x = 20 + 4y$

$\left(x^2 - 2x\right) + \left(y^2 - 4y\right) = 20$

$\left(x^2 - 2x + 1\right) + \left(y^2 - 4y + 4\right) = 20 + 1 + 4$

$(x - 1)^2 + (y - 2)^2 = 25$

Center $(1, 2)$; radius $= \sqrt{25} = 5$

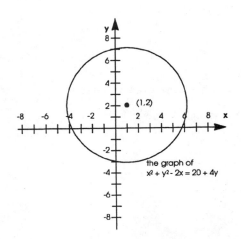

the graph of
$x^2 + y^2 - 2x = 20 + 4y$

61. $x^2 + y^2 = 2y - 6x - 2$

$\left(x^2 + 6x\right) + \left(y^2 - 2y\right) = -2$

$\left(x^2 + 6x + 9\right) + \left(y^2 - 2y + 1\right) = -2 + 9 + 1$

$(x + 3)^2 + (y - 1)^2 = 8$

Center $(-3, 1)$; radius $= \sqrt{8} = 2\sqrt{2}$

63. $2x^2 + 2y^2 - 4x + 4y = 22$

$x^2 + y^2 - 2x + 2y = 11$

$\left(x^2 - 2x\right) + \left(y^2 + 2y\right) = 11$

$\left(x^2 - 2x + 1\right) + \left(y^2 + 2y + 1\right) = 11 + 1 + 1$

$(x - 1)^2 + (y + 1)^2 = 13$

Center $(1, -1)$; radius $= \sqrt{13}$

67. By comparison with $(x - h)^2 + (y - k)^2 = r^2$

the graph of $x^2 + y^2 = 16$ is a circle with

center $(0,0)$ and radius $= 4$.

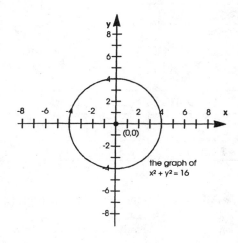

the graph of
$x^2 + y^2 = 16$

69. $x + y = 4$

$y = -x + 4$

By comparison with $y = mx + b$, the graph

of $x + y = 4$ is a straight line with

slope -1 and y-intercept 4.

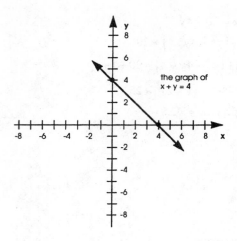

the graph of
$x + y = 4$

71. We can drop the absolute value in the equation in question, since $|a|^2 = a^2$ for every real number a.

72. Let P_1 have coordinates (x_1, y_1), let P_2 have coordinates (x_2, y_2), and let M have coordinates $\left(\dfrac{(x_1 + x_2)}{2}, \dfrac{(y_1 + y_2)}{2} \right)$. Then

$$|MP_1| = \sqrt{\left(x_1 - \frac{x_1 + x_2}{2} \right)^2 + \left(y_1 - \frac{y_1 + y_2}{2} \right)^2}$$

$$= \sqrt{\left(\frac{2x_1 - (x_1 + x_2)}{2} \right)^2 + \left(\frac{2y_1 - (y_1 + y_2)}{2} \right)^2} = \sqrt{\left(\frac{x_1 - x_2}{2} \right)^2 + \left(\frac{y_1 - y_2}{2} \right)^2}$$

$$= \sqrt{\left(\frac{x_1^2 - 2x_1x_2 + x_2^2}{4} \right) + \left(\frac{y_1^2 - 2y_1y_2 + y_2^2}{4} \right)} = \sqrt{\frac{x_1^2 - 2x_1x_2 + x_2^2 + y_1^2 - 2y_1y_2 + y_2^2}{4}}$$

$$|MP_2| = \sqrt{\left(x_2 - \frac{x_1 + x_2}{2} \right)^2 + \left(y_2 - \frac{y_1 + y_2}{2} \right)^2}$$

$$= \sqrt{\left(\frac{2x_2 - (x_1 + x_2)}{2} \right)^2 + \left(\frac{2y_2 - (y_1 + y_2)}{2} \right)^2} = \sqrt{\left(\frac{x_2 - x_1}{2} \right)^2 + \left(\frac{y_2 - y_1}{2} \right)^2}$$

$$= \sqrt{\left(\frac{x_2^2 - 2x_1x_2 + x_1^2}{4} \right) + \left(\frac{y_2^2 - 2y_1y_2 + y_1^2}{4} \right)} = \sqrt{\frac{x_1^2 - 2x_1x_2 + x_2^2 + y_1^2 - 2y_1y_2 + y_2^2}{4}}$$

Therefore, $|MP_1| = |MP_2|$, so M is the same distance from (x_1, y_1) as it is from (x_2, y_2).

73. (a) Since x^2 and y^2 are nonnegative and their sum is zero, it must be true that $x = 0$ and $y = 0$. That is, the only point whose coordinates will satisfy the equation is (0,0). Thus, the graph of $x^2 + y^2 = 0$ is a single point, located at the origin.

(b) Since the sum of two nonnegative quantities cannot be negative, it follows that the equation $x^2 + y^2 = -4$ has no real graph.

75. $\left(\dfrac{3x^2y^{-2}}{x^5y^{-8}}\right)^{-3} = \left(3x^{2-5}y^{-2-(-8)}\right)^{-3} = (3x^{-3}y^6)^{-3} = 3^{-3}\left(x^{-3}\right)^{-3}\left(y^6\right)^{-3}$

$$= 3^{-3}x^9y^{-18} = \dfrac{1}{3^3}\cdot x^9 \cdot \dfrac{1}{y^{18}} = \dfrac{x^9}{27y^{18}}$$

77. $2x + 5y = 8$

$\qquad 5y = -2x + 8$

$\qquad\quad y = -\dfrac{2}{5}x + \dfrac{8}{5}$

So $m = -\dfrac{2}{5}$

Exercises 9.2

1. $y = 2x^2$ $\qquad\qquad\qquad A = 2,\ B = 0,\ C = 0$

axis of symmetry: $x = -\dfrac{B}{2A} = -\dfrac{0}{2(2)} = -\dfrac{0}{4} = 0$

x-coordinate of vertex: 0

y-coordinate of vertex: $2(0)^2 = 2\cdot 0 = 0$

x-intercepts: $0 = 2x^2$, so $x = 0$

y-intercept: $y = 2(0)^2 = 0$

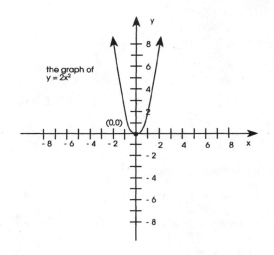

the graph of
$y = 2x^2$

5. $y = \dfrac{1}{2}x^2$ $\qquad\qquad\qquad A = \dfrac{1}{2},\ B = 0,\ C = 0$

axis of symmetry: $x = -\dfrac{B}{2A} = -\dfrac{0}{2\left(\dfrac{1}{2}\right)} = -\dfrac{0}{1} = 0$

x-coordinate of vertex: 0

y-coordinate of vertex: $\dfrac{1}{2}(0)^2 = \dfrac{1}{2}\cdot 0 = 0$

x-intercepts: $0 = \dfrac{1}{2}x^2$, so $x = 0$

y-intercept: $y = \dfrac{1}{2}(0)^2 = 0$

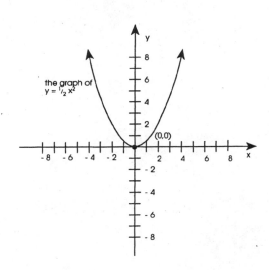

the graph of
$y = \frac{1}{2}x^2$

9. $y = x^2 - 4$ $\qquad$ A = 1, B = 0, C = −4

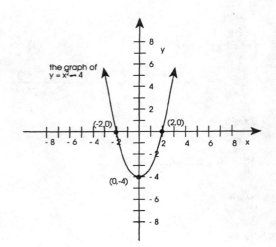

axis of symmetry: $x = -\dfrac{B}{2A} = -\dfrac{0}{2(1)} = -\dfrac{0}{2} = 0$

x-coordinate of vertex: 0

y-coordinate of vertex: $0^2 - 4 = 0 - 4 = -4$

x-intercepts: $0 = x^2 - 4$, so $x^2 = 4$

Thus, $x = \pm 2$.

y-intercept: $y = 0^2 - 4 = -4$

13. $y = 2x^2 - 8$ $\qquad$ A = 2, B = 0, C = −8

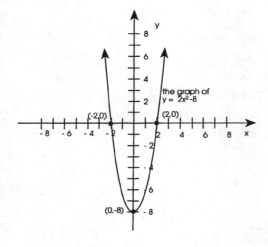

axis of symmetry: $x = -\dfrac{B}{2A} = -\dfrac{0}{2(2)} = -\dfrac{0}{4} = 0$

x-coordinate of vertex: 0

y-coordinate of vertex: $2(0)^2 - 8 = 0 - 8 = -8$

x-intercepts: $0 = 2x^2 - 8$, so $2x^2 = 8$ or $x^2 = 4$

Thus, $x = \pm 2$.

y-intercept: $y = 2(0)^2 - 8 = 0 - 8 = -8$

17. $y = -2x^2 - 8$ $A = -2, B = 0, C = -8$

axis of symmetry: $x = -\dfrac{B}{2A} = -\dfrac{0}{2(-2)} = \dfrac{-0}{-4} = 0$

x-coordinate of vertex: 0

y-coordinate of vertex: $-2(0)^2 - 8 = 0 - 8 = -8$

x-intercepts: $0 = -2x^2 - 8$, so $2x^2 = -8$ or $x^2 = -4$

Since this has no real solutions, there are

no x-intercepts.

y-intercept: $y = -2(0)^2 - 8 = 0 - 8 = -8$

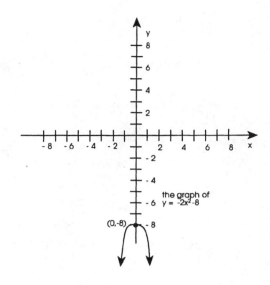

the graph of
$y = -2x^2 - 8$

(0,-8)

21. $y = 3x^2 + 9$ $A = 3, B = 0, C = 9$

axis of symmetry: $x = -\dfrac{B}{2A} = -\dfrac{0}{2(3)} = -\dfrac{0}{6} = 0$

x-coordinate of vertex: 0

y-coordinate of vertex: $3(0)^2 + 9 = 0 + 9 = 9$

x-intercepts: $0 = 3x^2 + 9$, so

$3x^2 = -9$ or $x^2 = -3$

Since this has no real solutions,

there are no x-intercepts.

y-intercept: $y = 3(0)^2 + 9 = 0 + 9 = 9$

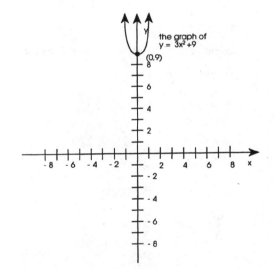

the graph of
$y = 3x^2 + 9$

(0,9)

25. $y = 2x^2 + 5x$ $A = 2, B = 5, C = 0$

axis of symmetry: $x = -\dfrac{B}{2A} = -\dfrac{5}{2(2)} = -\dfrac{5}{4}$

x-coordinate of vertex: $-\dfrac{5}{4}$

y-coordinate of vertex: $2\left(-\dfrac{5}{4}\right)^2 + 5\left(-\dfrac{5}{4}\right)$

$= 2\left(\dfrac{25}{16}\right) - \dfrac{25}{4} = \dfrac{25}{8} - \dfrac{50}{8} = -\dfrac{25}{8}$

x-intercepts: $0 = 2x^2 + 5x$

$\qquad 0 = x(2x + 5)$

$\qquad x = 0 \text{ or } 2x + 5 = 0$

$\qquad x = 0 \text{ or } x = -\dfrac{5}{2}$

y-intercept: $y = 2(0)^2 + 5(0) = 0 + 0 = 0$

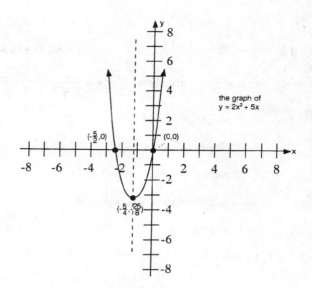

the graph of
$y = 2x^2 + 5x$

$\left(-\dfrac{5}{2},0\right)$ $(0,0)$ $\left(-\dfrac{5}{4}, -\dfrac{25}{8}\right)$

29. $\quad y = x^2 + 10x + 25 \qquad\qquad A = 1, B = 10, C = 25$

axis of symmetry: $x = -\dfrac{B}{2A} = -\dfrac{10}{2(1)} = -\dfrac{10}{2} = -5$

x-coordinate of vertex: -5

y-coordinate of vertex: $(-5)^2 + 10(-5) + 25$

$= 25 - 50 + 25 = 0$

x-intercepts: $0 = x^2 + 10x + 25$

$\qquad 0 = (x + 5)^2$

$\qquad 0 = x + 5$

$\qquad -5 = x$

y-intercept: $y = 0^2 + 10(0) + 25 = 0 + 0 + 25 = 25$

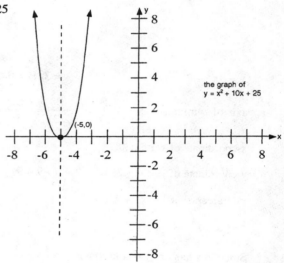

the graph of
$y = x^2 + 10x + 25$

$(-5,0)$

33. $\quad y = \dfrac{1}{3}x^2 - \dfrac{2}{3}x + \dfrac{2}{3} \qquad\qquad A = \dfrac{1}{3}, B = -\dfrac{2}{3}, C = \dfrac{2}{3}$

axis of symmetry: $x = -\dfrac{B}{2A} = \dfrac{-\left(-\dfrac{2}{3}\right)}{2\left(\dfrac{1}{3}\right)} = \dfrac{\dfrac{2}{3}}{\dfrac{2}{3}} = 1$

x-coordinate of vertex: 1

y-coordinate of vertex: $\dfrac{1}{3}(1)^2 - \dfrac{2}{3}(1) + \dfrac{2}{3}$

$= \dfrac{1}{3} - \dfrac{2}{3} + \dfrac{2}{3} = \dfrac{1}{3}$

x-intercepts: $0 = \dfrac{1}{3}x^2 - \dfrac{2}{3}x + \dfrac{2}{3}$

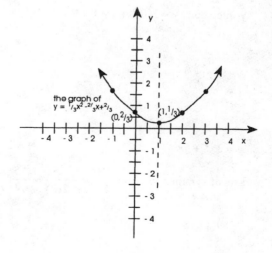

the graph of
$y = \frac{1}{3}x^2 - \frac{2}{3}x + \frac{2}{3}$

$\left(0, \frac{2}{3}\right)$ $\left(1, \frac{1}{3}\right)$

$$0 = x^2 - 2x + 2$$

Since $B^2 - 4AC = (-2)^2 - 4(1)(2)$

$= 4 - 8 = -4 < 0$, this equation has

no real solutions. Thus, there are

no x-intercepts.

y-intercept: $y = \frac{1}{3}(0)^2 - \frac{2}{3}(0) + \frac{2}{3} = 0 - 0 + \frac{2}{3} = \frac{2}{3}$

37. $y = -x^2 + 4x + 12$ $\qquad$ $A = -1, B = 4, C = 12$

axis of symmetry: $x = -\dfrac{B}{2A} = -\dfrac{4}{2(-1)} = \dfrac{-4}{-2} = 2$

x-coordinate of vertex: 2

y-coordinate of vertex: $-2^2 + 4(2) + 12$

$= -4 + 8 + 12 = 16$

x-intercepts: $0 = -x^2 + 4x + 12$

$x^2 - 4x - 12 = 0$

$(x - 6)(x + 2) = 0$

$x - 6 = 0$ or $x + 2 = 0$

$\qquad x = 6$ or $x = -2$

y-intercept: $y = -0^2 + 4(0) + 12 = 0 + 0 + 12 = 12$

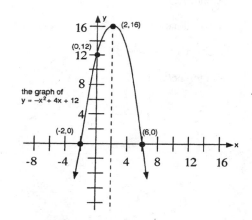

the graph of
$y = -x^2 + 4x + 12$

41. $x = y^2 - 2y - 35$ $\qquad$ $A = 1, B = -2, C = -35$

axis of symmetry: $y = -\dfrac{B}{2A} = -\dfrac{-2}{2(1)} = -\dfrac{-2}{2} = 1$

y-coordinate of vertex: 1

x-coordinate of vertex: $1^2 - 2(1) - 35$

$= 1 - 2 - 35 = -36$

y-intercepts: $0 = y^2 - 2y - 35$

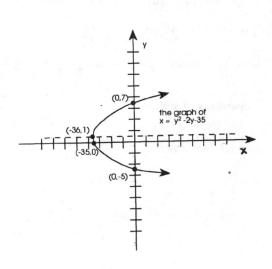

the graph of
$x = y^2 - 2y - 35$

$$0 = (y - 7)(y + 5)$$
$$0 = y - 7 \text{ or } 0 = y + 5$$
$$7 = y \text{ or } -5 = y$$

x-intercept: $x = 0^2 - 2(0) - 35 = 0 - 0 - 35 = -35$

45. $y + 4x = 2x^2 + 1$ $\hspace{2cm}$ A = 2, B = –4, C = 1

$y = 2x^2 - 4x + 1$

axis of symmetry: $x = -\dfrac{B}{2A} = -\dfrac{-4}{2(2)} = -\dfrac{-4}{4} = 1$

x-coordinate of vertex: 1

y-coordinate of vertex: $2(1)^2 - 4(1) + 1$

$= 2 - 4 + 1 = -1$

x-intercepts: $0 = 2x^2 - 4x + 1$

$$x = \frac{-B \pm \sqrt{B^2 - 4AC}}{2A}$$

$$= \frac{-(-4) \pm \sqrt{(-4)^2 - 4(2)(1)}}{2(2)}$$

$$= \frac{4 \pm \sqrt{16 - 8}}{4} = \frac{4 \pm \sqrt{8}}{4}$$

$$= \frac{4 \pm 2\sqrt{2}}{4} = \frac{2 \pm \sqrt{2}}{2}$$

y-intercept: $y = 2(0)^2 - 4(0) + 1 = 0 - 0 + 1 = 1$

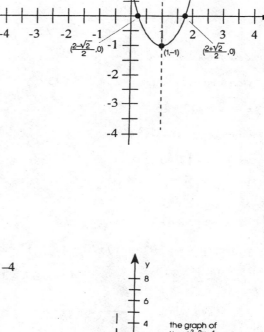

the graph of
$y + 4x = 2x^2 + 1$

$(0,1)$

$\left(\frac{2-\sqrt{2}}{2}, 0\right)$ $\hspace{1cm}$ $(1,-1)$ $\hspace{1cm}$ $\left(\frac{2+\sqrt{2}}{2}, 0\right)$

49. $y = -x^2 - 3x - 4$ $\hspace{2cm}$ A = –1, B = –3, C = –4

axis of symmetry: $x = -\dfrac{B}{2A} = -\dfrac{-3}{2(-1)} = -\dfrac{-3}{-2} = -\dfrac{3}{2}$

x-coordinate of vertex: $-\dfrac{3}{2}$

y-coordinate of vertex: $-\left(-\dfrac{3}{2}\right)^2 - 3\left(-\dfrac{3}{2}\right) - 4$

$= -\dfrac{9}{4} + \dfrac{9}{2} - 4 = \dfrac{9}{4} - 4 = \dfrac{9}{4} - \dfrac{16}{4} = -\dfrac{7}{4}$

x-intercepts: $0 = -x^2 - 3x - 4$

the graph of
$y = -x^2 - 3x - 4$

$(3/2, 7/4)$

$(0,-4)$

Since $B^2 - 4AC = (-3)^2 - 4(-1)(-4)$

$= 9 - 16 = -7 < 0$, this equation

has no real solutions. Thus, there

are no x-intercepts.

y-intercept: $y = -0^2 - 3(0) - 4 = 0 - 0 - 4 = -4$

53. $x = 2y^2 + y + 4$ $A = 2, B = 1, C = 4$

axis of symmetry: $y = -\dfrac{B}{2A} = -\dfrac{1}{2(2)} = -\dfrac{1}{4}$

y-coordinate of vertex: $-\dfrac{1}{4}$

x-coordinate of vertex: $2\left(-\dfrac{1}{4}\right)^2 + \left(-\dfrac{1}{4}\right) + 4$

$= 2\left(\dfrac{1}{16}\right) - \dfrac{1}{4} + 4 = \dfrac{1}{8} - \dfrac{1}{4} + 4$

$= -\dfrac{1}{8} + 4 = -\dfrac{1}{8} + \dfrac{32}{8} = \dfrac{31}{8}$

y-intercepts: $0 = 2y^2 + y + 4$

Since $B^2 - 4AC = 1^2 - 4(2)(4)$

$= 1 - 32 = -31 < 0$, this equation

has no real solutions. Thus, there

are no y-intercepts.

x-intercept: $x = 2(0)^2 + 0 + 4 = 0 + 0 + 4 = 4$

the graph of
$x = 2y^2 + y + 4$

57. $P = -x^2 + 112x - 535$ $A = -1, B = 112, C = -535$

$x = -\dfrac{B}{2A} = -\dfrac{112}{2(-1)} = -\dfrac{112}{-2} = 56$

Therefore, the number of cases of candy canes to be made daily in order to maximize the daily profit is 56.

Maximum profit $= -56^2 + 112(56) - 535 = -3136 + 6272 - 535 = \$2601$

61.   Let x = length of one of the two equal sides being fenced (in feet). Then 50 – 2x = length of the side opposite the house (in feet). If we let y = area of the garden (in square feet), then

$y = x(50 - 2x) = 50x - 2x^2 = -2x^2 + 50x$

$A = -2, B = 50, C = 0$

$x = -\dfrac{B}{2A} = -\dfrac{50}{2(-2)} = \dfrac{-50}{-4} = \dfrac{25}{2}$

So $x = \dfrac{25}{2}$ is the value of x that produces the maximum area. Therefore, the two equal sides to be fenced should measure $\dfrac{25}{2}$ feet each, while the side opposite the house should measure 25 feet.

62. (a) If A > 0, the parabola opens upward; if A < 0, the parabola opens downward.

(b) If A < 0, the parabola opens to the left; if A > 0, the parabola opens to the right.

63. If $B^2 - 4AC > 0$, the parabola crosses the x-axis twice; if $B^2 - 4AC < 0$, the parabola does not cross the x-axis at all; if $B^2 - 4AC = 0$, the parabola touches the x-axis in exactly one point.

64. $y = Ax^2 + Bx + C$. When $x = -\dfrac{B}{2A}$,

$$y = A\left(-\dfrac{B}{2A}\right)^2 + B\left(-\dfrac{B}{2A}\right) + C = A\left(\dfrac{B^2}{4A^2}\right) - \dfrac{B^2}{2A} + C = \dfrac{B^2}{4A} - \dfrac{B^2}{2A} + C$$

$$= \dfrac{B^2}{4A} - \dfrac{2B^2}{4A} + \dfrac{4AC}{4A} = \dfrac{B^2 - 2B^2 + 4AC}{4A} = \dfrac{4AC - B^2}{4A}$$

65. $d_1 = \sqrt{(x-0)^2 + (y-a)^2} = \sqrt{x^2 + (y-a)^2}$

$d_2 = \sqrt{(x-x)^2 + (y-(-a))^2} = \sqrt{(y+a)^2}$

$d_1 = d_2$

$\sqrt{x^2 + (y-a)^2} = \sqrt{(y+a)^2}$

$x^2 + (y-a)^2 = (y+a)^2$

$x^2 + y^2 - 2ay + a^2 = y^2 + 2ay + a^2$

$x^2 = 4ay$

If we solve this equation for y, we obtain $y = \left(\dfrac{1}{4a}\right)x^2$, which is an equation of the form

$y = Ax^2 + Bx + C$ with $A = \dfrac{1}{4a}$, $B = 0$, and $C = 0$.

Exercises 9.3

1. $a^2 = 16$, so $a = 4$; $b^2 = 9$, so $b = 3$. Vertices: (–4,0), (4,0), (0,–3), and (0,3).

3. $a^2 = 9$, so $a = 3$; $b^2 = 16$, so $b = 4$. Vertices: (–3,0), (3,0), (0,–4), and (0,4).

5. $a^2 = 25$, so $a = 5$; $b^2 = 36$, so $b = 6$. Vertices: (–5,0), (5,0), (0,–6), and (0,6).

9. $a^2 = 24$, so $a = \sqrt{24} = 2\sqrt{6}$; $b^2 = 20$, so $b = \sqrt{20} = 2\sqrt{5}$. Vertices: $(-2\sqrt{6},0)$, $(2\sqrt{6},0)$, $(0, -2\sqrt{5})$, and $(0, 2\sqrt{5})$.

13. $a^2 = 1$, so $a = 1$;

$b^2 = 16$, so $b = 4$.

Vertices: $(-1,0)$, $(1,0)$,

$(0,-4)$, and $(0,4)$

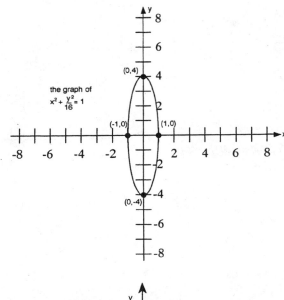

the graph of
$x^2 + \frac{y^2}{16} = 1$

17. $4x^2 + 25y^2 = 100$

$\dfrac{4x^2}{100} + \dfrac{25y^2}{100} = \dfrac{100}{100}$

$\dfrac{x^2}{25} + \dfrac{y^2}{4} = 1$

$a^2 = 25$, so $a = 5$;

$b^2 = 4$, so $b = 2$.

Vertices: $(-5,0)$, $(5,0)$,

$(0,-2)$, and $(0,2)$

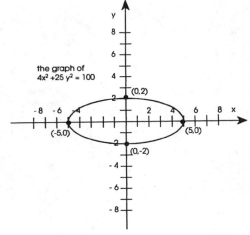

the graph of
$4x^2 + 25 y^2 = 100$

21. $8x^2 + 7y^2 = 56$

$\dfrac{8x^2}{56} + \dfrac{7y^2}{56} = \dfrac{56}{56}$

$\dfrac{x^2}{7} + \dfrac{y^2}{8} = 1$

$a^2 = 7$, so $a = \sqrt{7}$;

$b^2 = 8$, so $b = \sqrt{8} = 2\sqrt{2}$.

Vertices: $(-\sqrt{7}, 0)$, $(\sqrt{7}, 0)$,

$(0, -2\sqrt{2})$, and $(0, 2\sqrt{2})$

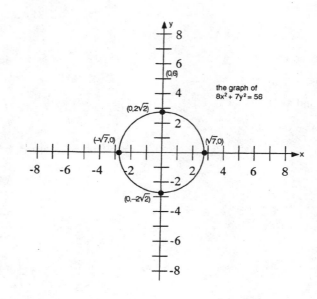

25. $4x^2 + y^2 = 1$

$\dfrac{x^2}{\dfrac{1}{4}} + \dfrac{y^2}{1} = 1$

$a^2 = \dfrac{1}{4}$, so $a = \dfrac{1}{2}$;

$b^2 = 1$, so $b = 1$.

Vertices: $\left(-\dfrac{1}{2}, 0\right)$, $\left(\dfrac{1}{2}, 0\right)$,

$(0, -1)$, and $(0, 1)$

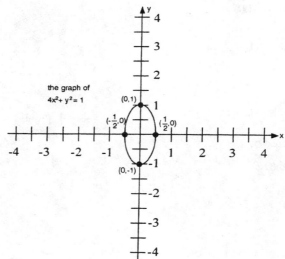

29. $\dfrac{x^2}{4} + \dfrac{y^2}{16} = 1$

ellipse; $a^2 = 4$, so $a = 2$; $b^2 = 16$, so $b = 4$. Vertices: $(-2, 0)$, $(2, 0)$, $(0, -4)$, and $(0, 4)$.

33. $\dfrac{x^2}{16} + \dfrac{y^2}{16} = 1$

$16\left(\dfrac{x^2}{16} + \dfrac{y^2}{16}\right) = 16(1)$

$x^2 + y^2 = 16$

circle: centered at $(0, 0)$ with radius $= 4$

35. $2x^2 + 4y^2 = 8$

$\dfrac{2x^2}{8} + \dfrac{4y^2}{8} = \dfrac{8}{8}$

$\dfrac{x^2}{4} + \dfrac{y^2}{2} = 1$

ellipse; $a^2 = 4$, so $a = 2$;

$b^2 = 2$, so $b = \sqrt{2}$.

Vertices: $(-2,0)$, $(2,0)$,

$(0, -\sqrt{2})$, and $(0, \sqrt{2})$

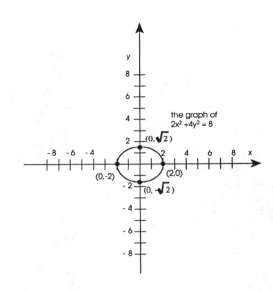

37. $2x + 4y = 8$

straight line;

x-intercept: 4

y-intercept: 2

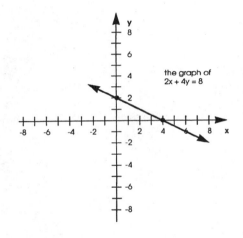

41. Place the base of the semiellipse along the

x-axis and place the center at the origin.

Since the base covers 50 feet, the horizontal

vertices are $(-25,0)$ and $(25,0)$. Since the highest

part of the semiellipse is 15 feet above its

center, the vertical vertex that is above the

x-axis is $(0,15)$.

So an equation of this semiellipse is $\dfrac{x^2}{25^2} + \dfrac{y^2}{15^2} = 1$,

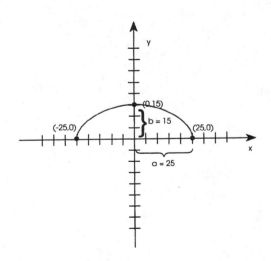

263

or $\dfrac{x^2}{625} + \dfrac{y^2}{225} = 1$, where $y \geq 0$.

Choosing $x = 11$, we find that $\dfrac{11^2}{625} + \dfrac{y^2}{225} = 1$,

or $\dfrac{121}{625} + \dfrac{y^2}{225} = 1$. Next, we solve for y:

$$\frac{y^2}{225} = 1 - \frac{121}{625} = \frac{625}{625} - \frac{121}{625} = \frac{504}{625}$$

$$y^2 = \overset{9}{\cancel{225}}\left(\frac{504}{\underset{25}{\cancel{625}}}\right) = \frac{4536}{25}$$

$$y = \sqrt{\frac{4536}{25}} = \frac{\sqrt{4536}}{\sqrt{25}} = \frac{18\sqrt{14}}{5} \approx 13.47$$

Since the truck is 14 feet high, it cannot pass

through the bridge staying right of the center line.

43. $\qquad \dfrac{x^2}{a^2} + \dfrac{y^2}{a^2} = 1$

$a^2 \left(\dfrac{x^2}{a^2} + \dfrac{y^2}{a^2} \right) = a^2(1)$

$x^2 + y^2 = a^2$

The graph of this equation is a circle centered at (0,0), with radius a.

44. $d_1 = |PF_1| = \sqrt{(x-(-s))^2 + (y-0)^2} = \sqrt{(x+s)^2 + y^2}$

$d_2 = |PF_2| = \sqrt{(x-s)^2 + (y-0)^2} = \sqrt{(x-s)^2 + y^2}$

$d_1 + d_2 = k$

$\sqrt{(x+s)^2 + y^2} + \sqrt{(x-s)^2 + y^2} = k$

$\sqrt{(x+s)^2 + y^2} = k - \sqrt{(x-s)^2 + y^2}$

$(x+s)^2 + y^2 = k^2 - 2k\sqrt{(x-s)^2 + y^2} + (x-s)^2 + y^2$

$2k\sqrt{(x-s)^2 + y^2} = k^2 + (x-s)^2 - (x+s)^2$

$2k\sqrt{(x-s)^2 + y^2} = k^2 - 4sx$

$4k^2((x-s)^2 + y^2) = k^4 - 8k^2sx + 16s^2x^2$

$4k^2(x^2 - 2sx + s^2 + y^2) = k^4 - 8k^2sx + 16s^2x^2$

$4k^2x^2 - 8k^2sx + 4k^2s^2 + 4k^2y^2 = k^4 - 8k^2sx + 16s^2x^2$

$4k^2x^2 - 16s^2x^2 + 4k^2y^2 = k^4 - 4k^2s^2$

$(4k^2 - 16s^2)x^2 + 4k^2y^2 = k^2(k^2 - 4s^2)$

$4(k^2 - 4s^2)x^2 + 4k^2y^2 = k^2(k^2 - 4s^2)$

$\dfrac{4(k^2 - 4s^2)}{k^2(k^2 - 4s^2)}x^2 + \dfrac{4k^2}{k^2(k^2 - 4s^2)}y^2 = \dfrac{k^2(k^2 - 4s^2)}{k^2(k^2 - 4s^2)}$

$\left(\dfrac{4}{k^2}\right)x^2 + \left(\dfrac{4}{k^2 - 4s^2}\right)y^2 = 1$

$\dfrac{x^2}{\left(\dfrac{k^2}{4}\right)} + \dfrac{y^2}{\left(\dfrac{k^2 - 4s^2}{4}\right)} = 1$

This is the standard form equation of an ellipse centered at the origin, with vertices $\left(-\dfrac{k}{2}, 0\right)$,

$\left(\dfrac{k}{2}, 0\right)$, $\left(0, \dfrac{-\sqrt{k^2 - 4s^2}}{2}\right)$, and $\left(0, \dfrac{\sqrt{k^2 - 4s^2}}{2}\right)$.

45. If the center of the ellipse is at (h,k), the generalized form of its standard equation is $\dfrac{(x-h)^2}{a^2} +$

$\dfrac{(y-k)^2}{b^2} = 1$. Note that if h = 0 and k = 0 (in other words, if the ellipse happens to be centered at

the origin), then this equation reduces to $\dfrac{x^2}{a^2} + \dfrac{y^2}{b^2} = 1$.

47.
$$(x-2)^2 = (x+1)^2 - 15$$
$$x^2 - 4x + 4 = x^2 + 2x + 1 - 15$$
$$x^2 - 4x + 4 = x^2 + 2x - 14$$
$$-4x + 4 = 2x - 14$$
$$4 = 6x - 14$$
$$18 = 6x$$
$$3 = x$$

49.   Let x = amount invested at 8%; y = amount invested at 10%

$$0.08x + 0.10y = 730 \xrightarrow{\text{multiply by 100}}$$ $$8x + 10y = 73,000 \xrightarrow{\text{multiply by 5}}$$

$$0.10x + 0.08y = 710 \xrightarrow{\text{multiply by 100}}$$ $$10x + 8y = 71,000 \xrightarrow{\text{multiply by -4}}$$

$$40x + 50y = 365,000$$
$$\underline{-40x - 32y = -284,000}$$
$$\text{Add: } 18y = 81,000$$
$$y = 4,500$$
$$8x + 10y = 73,000$$
$$8x + 10(4,500) = 73,000$$
$$8x + 45,000 = 73,000$$
$$8x = 28,000$$
$$x = 3,500$$

Thus, the total amount invested is $3,500 + $4,500 = $8,000

Exercises 9.4

1. $a^2 = 9$, so a = 3; $b^2 = 16$, so b = 4.   Vertices: $(-3,0)$ and $(3,0)$.

Asymptotes: $y = \pm \dfrac{b}{a} x = \pm \dfrac{4}{3} x$

5. $a^2 = 16$, so a = 4; $b^2 = 9$, so b = 3.   Vertices: $(0,-3)$ and $(0,3)$.

Asymptotes: $y = \pm \dfrac{b}{a} x = \pm \dfrac{3}{4} x$

9. $a^2 = 12$, so $a = \sqrt{12} = 2\sqrt{3}$; $b^2 = 4$, so b = 2.   Vertices: $(-2\sqrt{3},0)$ and $(2\sqrt{3},0)$.

Asymptotes: $y = \pm \dfrac{b}{a} x = \pm \dfrac{\cancel{2}}{\cancel{2}\sqrt{3}} x = \pm \dfrac{1}{\sqrt{3}} x = \pm \dfrac{\sqrt{3}}{3} x$

13. $9x^2 - 16y^2 = 144$

$$\frac{9x^2}{144} - \frac{16y^2}{144} = \frac{144}{144}$$

$$\frac{x^2}{16} - \frac{y^2}{9} = 1$$

$a^2 = 16$, so a = 4;

$b^2 = 9$, so b = 3.

Vertices: (–4,0) and (4,0)

Asymptotes: $y = \pm \dfrac{b}{a} x = \pm \dfrac{3}{4} x$

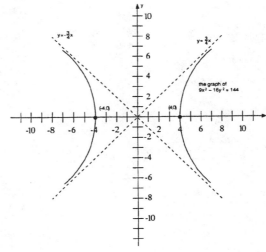

17. $x^2 - 2y^2 = 2$

$\dfrac{x^2}{2} - \dfrac{2y^2}{2} = \dfrac{2}{2}$

$\dfrac{x^2}{2} - y^2 = 1$

$a^2 = 2$, so $a = \sqrt{2}$;

$b^2 = 1$, so $b = 1$.

Vertices: $(-\sqrt{2}, 0)$ and $(\sqrt{2}, 0)$

Asymptotes: $y = \pm \dfrac{b}{a} x = \pm \dfrac{1}{\sqrt{2}} x = \pm \dfrac{\sqrt{2}}{2} x$

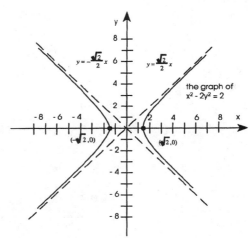

21. $12y^2 - 5x^2 = 60$

$\dfrac{12y^2}{60} - \dfrac{5x^2}{60} = \dfrac{60}{60}$

$\dfrac{y^2}{5} - \dfrac{x^2}{12} = 1$

$a^2 = 12$, so $a = \sqrt{12} = 2\sqrt{3}$;

$b^2 = 5$, so $b = \sqrt{5}$.

Vertices: $(0, -\sqrt{5})$ and $(0, \sqrt{5})$

Asymptotes: $y = \pm \dfrac{b}{a} x = \pm \dfrac{\sqrt{5}}{2\sqrt{3}} x = \pm \dfrac{\sqrt{15}}{6} x$

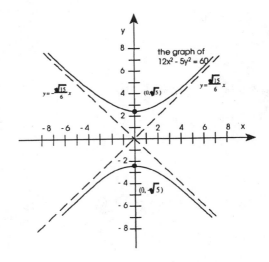

267

25. $16x^2 - 4y^2 = 1$

$$\frac{x^2}{\frac{1}{16}} - \frac{y^2}{\frac{1}{4}} = 1$$

$a^2 = \frac{1}{16}$, so $a = \frac{1}{4}$;

$b^2 = \frac{1}{4}$, so $b = \frac{1}{2}$.

Vertices: $\left(-\frac{1}{4}, 0\right)$ and $\left(\frac{1}{4}, 0\right)$

Asymptotes: $y = \pm \frac{b}{a} x = \pm \frac{\frac{1}{2}}{\frac{1}{4}} x = \pm 2x$

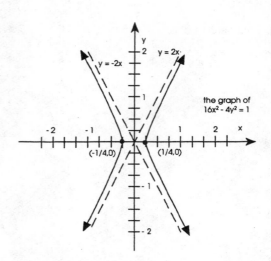

the graph of $16x^2 - 4y^2 = 1$

27. $\frac{x^2}{25} - \frac{y^2}{100} = 1$; hyperbola

$a^2 = 25$, so $a = 5$; $b^2 = 100$, so $b = 10$.

Vertices: $(-5,0)$ and $(5,0)$. Asymptotes: $y = \pm \frac{b}{a} x = \pm \frac{10}{5} x = \pm 2x$

29. $\frac{x^2}{100} + \frac{y^2}{25} = 1$; ellipse

$a^2 = 100$, so $a = 10$; $b^2 = 25$, so $b = 5$.

Vertices: $(-10,0)$, $(10,0)$, $(0,-5)$ and $(0,5)$

33. $8y^2 - 8x^2 = 16$

$$\frac{8y^2}{16} - \frac{8x^2}{16} = \frac{16}{16}$$

$$\frac{y^2}{2} - \frac{x^2}{2} = 1$$

$a^2 = 2$, so $a = \sqrt{2}$; $b^2 = 2$, so $b = \sqrt{2}$.

Vertices: $(0, -\sqrt{2})$ and $(0, \sqrt{2})$. Asymptotes: $y = \pm \frac{b}{a} x = \pm \frac{\sqrt{2}}{\sqrt{2}} x = \pm x$

37. $x^2 + y = 4$ parabola

$y = -x^2 + 4$ $A = -1, B = 0, C = 4$

axis of symmetry: $x = -\frac{B}{2A}$

$= -\frac{0}{2(-1)} = \frac{-0}{-2} = 0$

x–coordinate of vertex: 0

y–coordinate of vertex: $-0^2 + 4$

$= 0 + 4 = 4$

x–intercepts: $0 = -x^2 + 4$, so $x^2 = 4$.

Thus, $x = \pm 2$.

y–intercept: $y = -0^2 + 4 = 0 + 4 = 4$

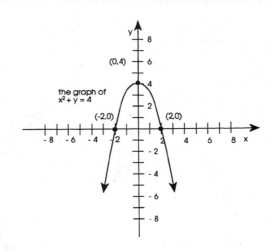

41. $4x^2 + y^2 = 16$; ellipse

$$\frac{4x^2}{16} + \frac{y^2}{16} = \frac{16}{16}$$

$$\frac{x^2}{4} + \frac{y^2}{16} = 1$$

$a^2 = 4$, so $a = 2$;

$b^2 = 16$, so $b = 4$.

Vertices: $(-2,0)$, $(2,0)$,

$(0,-4)$ and $(0,4)$

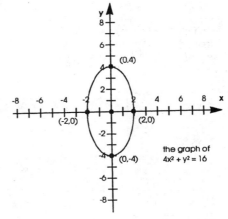

43. $4x^2 - y^2 = 16$; hyperbola

$$\frac{4x^2}{16} - \frac{y^2}{16} = \frac{16}{16}$$

$$\frac{x^2}{4} - \frac{y^2}{16} = 1$$

$a^2 = 4$, so $a = 2$;

$b^2 = 16$, so $b = 4$.

Vertices: $(-2,0)$ and $(2,0)$

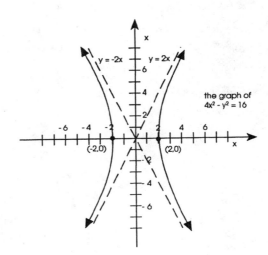

Asymptotes: $y = \pm \dfrac{b}{a} x$

$= \pm \dfrac{4}{2} = \pm 2x$

47. $d_1 = |PF_1| = \sqrt{(x-(-s))^2 + (y-0)^2} = \sqrt{(x+s)^2 + y^2}$

$d_2 = |PF_2| = \sqrt{(x-s)^2 + (y-0)^2} = \sqrt{(x-s)^2 + y^2}$

$|d_1 - d_2| = k$

$\left| \sqrt{(x+s)^2 + y^2} - \sqrt{(x-s)^2 + y^2} \right| = k$

$\left| \sqrt{(x+s)^2 + y^2} - \sqrt{(x-s)^2 + y^2} \right|^2 = k^2$

$\left(\sqrt{(x+s)^2 + y^2} - \sqrt{(x-s)^2 + y^2} \right)^2 = k^2$

$(x+s)^2 + y^2 - 2\sqrt{(x+s)^2 + y^2}\sqrt{(x-s)^2 + y^2} + (x-s)^2 + y^2 = k^2$

$x^2 + 2sx + s^2 + y^2 - 2\sqrt{(x+s)^2 + y^2}\sqrt{(x-s)^2 + y^2} + x^2 - 2sx + s^2 + y^2 = k^2$

$2x^2 + 2s^2 + 2y^2 - k^2 = 2\sqrt{(x+s)^2 + y^2}\sqrt{(x-s)^2 + y^2}$

$(2x^2 + 2s^2 + 2y^2 - k^2)^2 = 4((x+s)^2 + y^2)((x-s)^2 + y^2)$

$\cancel{4x^4} + 8s^2x^2 + \cancel{8x^2y^2} - 4k^2x^2 + \cancel{4s^4} + \cancel{8s^2y^2} - 4s^2k^2 + \cancel{4y^4} - 4k^2y^2 + k^4$

$= 4[\cancel{x^4} - 2s^2x^2 + \cancel{s^4} + \cancel{2x^2y^2} + \cancel{2s^2y^2} + \cancel{y^4}]$

$8s^2x^2 - 4k^2x^2 - 4s^2k^2 - 4k^2y^2 + k^4 = -8s^2x^2$

$16s^2x^2 - 4k^2x^2 - 4s^2k^2 - 4k^2y^2 + k^4 = 0$

$(16s^2 - 4k^2)x^2 - 4k^2y^2 = 4s^2k^2 - k^4$

$4(4s^2 - k^2)x^2 - 4k^2y^2 = k^2(4s^2 - k^2)$

$\dfrac{4\cancel{(4s^2 - k^2)}}{k^2\cancel{(4s^2 - k^2)}} x^2 - \dfrac{4\cancel{k^2}}{\cancel{k^2}(4s^2 - k^2)} y^2 = \dfrac{\cancel{k^2}\cancel{(4s^2 - k^2)}}{\cancel{k^2}\cancel{(4s^2 - k^2)}}$

$\dfrac{4}{k^2} x^2 - \dfrac{4}{4s^2 - k^2} y^2 = 1$

$\dfrac{x^2}{\left(\dfrac{k^2}{4}\right)} - \dfrac{y^2}{\left(\dfrac{4s^2 - k^2}{4}\right)} = 1$

270

This is the standard form equation of a hyperbola with $a = \dfrac{k}{2}$ and $b = \dfrac{\sqrt{4s^2 - k^2}}{2}$

48. (a)
$$\frac{x^2}{4} - \frac{y^2}{9} = 1$$
$$\frac{x^2}{4} - 1 = \frac{y^2}{9}$$
$$\frac{x^2 - 4}{4} = \frac{y^2}{9}$$
$$\frac{9}{4}(x^2 - 4) = y^2$$
$$\pm \frac{3}{2}\sqrt{x^2 - 4} = y$$

(b)

	$y = \pm \dfrac{3}{2}\sqrt{x^2 - 4}$	$y = \pm \dfrac{3}{2}x$
x = 4	±5.196	±6
x = 10	±14.697	±15
x = 20	±29.850	±30
x = 100	±149.970	±150
x = 200	±299.985	±300
x = 1000	±1499.997	±1500
x = 2000	±2999.998	±3000

49. The asymptotes of the hyperbola $xy = k$ are the coordinate axes, $x = 0$ and $y = 0$.

50. If the center of the hyperbola is at (h,k), the generalized form of its standard equation is either
$$\frac{(x - h)^2}{a^2} - \frac{(y - k)^2}{b^2} = 1 \text{ or } \frac{(y - k)^2}{b^2} - \frac{(x - h)^2}{a^2} = 1$$

51.
$$(2y - 3)(3y + 1) = 17$$
$$6y^2 - 7y - 3 = 17$$
$$6y^2 - 7y - 20 = 0$$
$$(2y - 5)(3y + 4) = 0$$
$$2y - 5 = 0 \text{ or } 3y + 4 = 0$$
$$2y = 5 \text{ or } 3y = -4$$
$$y = \frac{5}{2} \text{ or } y = -\frac{4}{3}$$

53. Let x = height of the ladder (in feet)
$$6^2 + x^2 = 40^2$$
$$36 + x^2 = 1600$$
$$x^2 = 1564$$
$$x = 2\sqrt{391}$$

Thus, the ladder is $2\sqrt{391}$ feet $\cong$ 39.53 feet high.

Exercises 9.5

1. circle

3. ellipse

5. parabola

7. straight line

9. hyperbola

11. hyperbola

13. circle

15. circle

17. hyperbola

19. parabola

21. Circle, center: (0,0)
    radius: 4

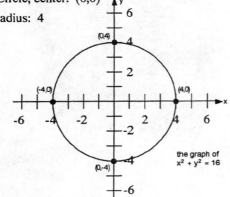

the graph of
$x^2 + y^2 = 16$

25. Hyperbola, vertices: (–5,0) and (5,0)
    asymptotes: $y = \pm \dfrac{1}{10} x$

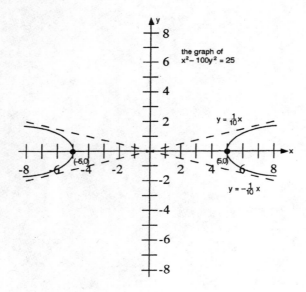

the graph of
$x^2 - 100y^2 = 25$

$y = \frac{1}{10} x$

$y = -\frac{1}{10} x$

272

27. Degenerate hyperbola (two intersecting

straight lines): $y = \pm \dfrac{4}{3} x$

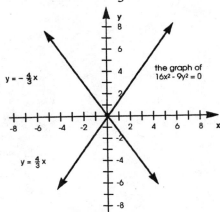

31. Parabola, axis of symmetry: $x = -1$

vertex: $(-1, 4)$

y-intercept: $(0, 6)$

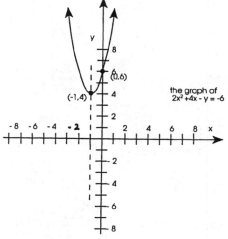

33. Ellipse, vertices: $(-2, 0)$, $(2, 0)$, $(0, -2\sqrt{2})$, and $(0, 2\sqrt{2})$

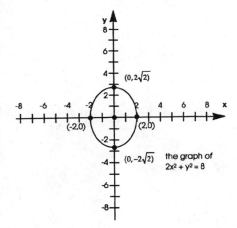

37. **(a)** If it is a parabola, put its equation into standard form ($y = Ax^2 + Bx + C$ or

$x = Ay^2 + By + C$). In the first case, the parabola opens up if $A > 0$, down if $A < 0$; in the second case, the parabola opens right if $A > 0$, left if $A < 0$.

(b) If it is a hyperbola, put its equation into standard form

$\left(\dfrac{(x-h)^2}{a^2} - \dfrac{(y-k)^2}{b^2} = 1 \text{ or } \dfrac{(y-k)^2}{b^2} - \dfrac{(x-h)^2}{a^2} = 1 \right)$. In the first case, the hyperbola opens left and right; in the

second case, the hyperbola opens up and down.

Chapter 9 Review Exercises

1. $|PQ| = \sqrt{(2-0)^2 + (6-0)^2} = \sqrt{2^2 + 6^2} = \sqrt{4 + 36} = \sqrt{40} = 2\sqrt{10}$

midpoint of PQ: $\left(\dfrac{0+2}{2}, \dfrac{0+6}{2}\right) = (1,3)$

3. $|PQ| = \sqrt{(-2-2)^2 + (5-5)^2} = \sqrt{(-4)^2 + 0^2} = \sqrt{16+0} = \sqrt{16} = 4$

   midpoint of PQ: $\left(\dfrac{2+(-2)}{2}, \dfrac{5+5}{2}\right) = (0,5)$

5. $|PQ| = \sqrt{(4-6)^2 + (-6-(-4))^2} = \sqrt{(-2)^2 + (-2)^2} = \sqrt{4+4} = \sqrt{8} = 2\sqrt{2}$

   midpoint of PQ: $\left(\dfrac{6+4}{2}, \dfrac{-4-6}{2}\right) = (5,-5)$

7. $|PQ| = \sqrt{(2-(-2))^2 + (5-(-5))^2} = \sqrt{4^2 + 10^2} = \sqrt{16+100} = \sqrt{116} = 2\sqrt{29}$

   midpoint of PQ: $\left(\dfrac{-2+2}{2}, \dfrac{-5+5}{2}\right) = (0,0)$

9. $|PQ| = \sqrt{(8-3)^2 + (-3-(-4))^2} = \sqrt{5^2 + 1^2} = \sqrt{25+1} = \sqrt{26}$

   midpoint of PQ: $\left(\dfrac{3+8}{2}, \dfrac{-4-3}{2}\right) = \left(\dfrac{11}{2}, -\dfrac{7}{2}\right)$

11. $|PQ| = \sqrt{(5-3)^2 + (9-6)^2} = \sqrt{2^2 + 3^2} = \sqrt{4+9} = \sqrt{13}$

    $|QR| = \sqrt{(8-5)^2 + (7-9)^2} = \sqrt{3^2 + (-2)^2} = \sqrt{9+4} = \sqrt{13}$

    $|PR| = \sqrt{(8-3)^2 + (7-6)^2} = \sqrt{5^2 + 1^2} = \sqrt{25+1} = \sqrt{26}$

    Since $|PQ|^2 + |QR|^2 = |PR|^2$, the Pythagorean theorem is satisfied, and P, Q and R are the vertices of a right triangle with right angle at Q.

13. $x^2 + y^2 = 100$

    $(x-0)^2 + (y-0)^2 = 10^2$

    Center: (0, 0); radius: 10

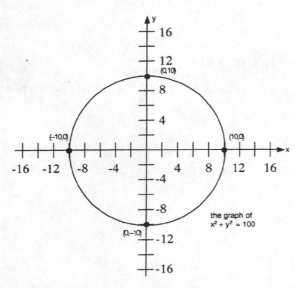

274

15. $x^2 + y^2 - 4x - 14y = -52$

$(x^2 + 4x + 4) + (y^2 - 14y + 49)$

$= -52 + 4 + 49$

$(x - 2)^2 + (y - 7)^2 = 1^2$

Center: (2, 7); radius: 1

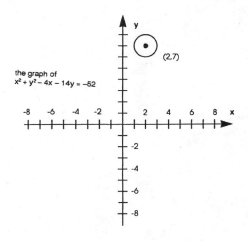

the graph of
$x^2 + y^2 - 4x - 14y = -52$

17. $x^2 + y^2 - 6x + 4y = 68$

$(x^2 - 6x + 9) + (y^2 + 4y + 4) = 68 + 9 + 4$

$(x - 3)^2 + (y + 2)^2 = 81$

$(x - 3)^2 + (y + 2)^2 = 9^2$

Center: (3, −2); radius: 9

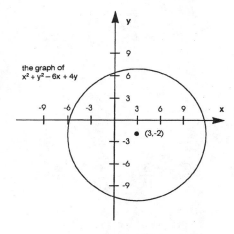

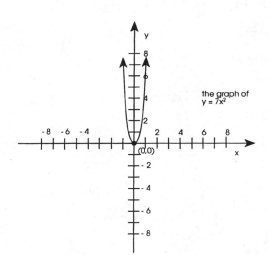

the graph of
$x^2 + y^2 - 6x + 4y$

19. $y = 7x^2$ A = 7, B = 0, C = 0

axis of symmetry: $x = -\dfrac{B}{2A} = -\dfrac{0}{2(7)} = -\dfrac{0}{14} = 0$

x-coordinate of vertex: 0

y-coordinate of vertex: $7(0)^2 = 7(0) = 0$

x-intercept: $0 = 7x^2$, so x = 0

y-intercept: $y = 7(0)^2 = 0$

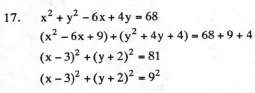

the graph of
$y = 7x^2$

21. $y = -7x^2 + 3$ $A = -7, B = 0, C = 3$

      axis of symmetry: $x = -\dfrac{B}{2A} = \dfrac{-0}{2(-7)} = \dfrac{-0}{-14} = 0$

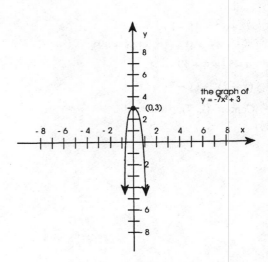

the graph of
$y = -7x^2 + 3$

      x-coordinate of vertex: 0

      y-coordinate of vertex: $-7(0)^2 + 3 = 0 + 3 = 3$

      x-intercepts: $0 = -7x^2 + 3$, so

      $7x^2 = 3$

      Then $x^2 = \dfrac{3}{7}$, so

      $x = \pm\sqrt{\dfrac{3}{7}} = \pm\dfrac{\sqrt{3}}{\sqrt{7}} = \pm\dfrac{\sqrt{21}}{7}$

      y-intercept: $y = -7(0)^2 + 3 = 0 + 3 = 3$

23. $y = x^2 - 6x$ $A = 1, B = -6, C = 0$

      axis of symmetry: $x = -\dfrac{B}{2A} = -\dfrac{-6}{2(1)} = -\dfrac{-6}{2} = 3$

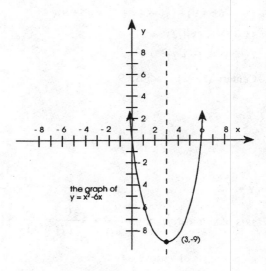

the graph of
$y = x^2 - 6x$

      x-coordinate of vertex: 3

      y-coordinate of vertex: $3^2 - 6(3) = 9 - 18 = -9$

      x-intercepts: $0 = x^2 - 6x = x(x - 6)$, so

      $x = 0$ or $x = 6$

      y-intercept: $y = 0^2 - 6(0) = 0 - 0 = 0$

25. $y = x^2 - 2x - 8$ $A = 1, B = -2, C = -8$

axis of symmetry: $x = -\dfrac{B}{2A} = -\dfrac{-2}{2(1)} = -\dfrac{-2}{2} = 1$

x-coordinate of vertex: 1

y-coordinate of vertex: $1^2 - 2(1) - 8$

$= 1 - 2 - 8 = -9$

x-intercepts: $0 = x^2 - 2x - 8$

$= (x - 4)(x + 2)$, so

$x = 4$ or $x = -2$

y-intercept: $y = 0^2 - 2(0) - 8$

$= 0 - 0 - 8 = -8$

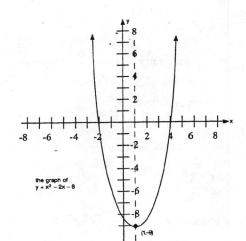

the graph of
$y = x^2 - 2x - 8$

(1,-9)

27. $x = y^2 - 2y - 8$ $A = 1, B = -2, C = -8$

axis of symmetry: $y = -\dfrac{B}{2A} = -\dfrac{-2}{2(1)} = -\dfrac{-2}{2} = 1$

y-coordinate of vertex: 1

x-coordinate of vertex: $1^2 - 2(1) - 8$

$= 1 - 2 - 8 = -9$

y-intercepts: $0 = y^2 - 2y - 8$

$= (y - 4)(y + 2)$, so

$y = 4$ or $y = -2$

x-intercept: $x = 0^2 - 2(0) - 8$

$= 0 - 0 - 8 = -8$

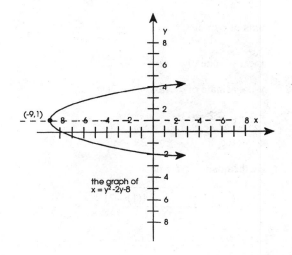

(-9,1)

the graph of
$x = y^2 - 2y - 8$

29. $y = x^2 - x - 12$ A = 1, B = –1, C = –12

axis of symmetry: $x = -\dfrac{B}{2A} = -\dfrac{-1}{2(1)} = -\dfrac{-1}{2} = \dfrac{1}{2}$

x-coordinate of vertex: $\dfrac{1}{2}$

y-coordinate of vertex: $\left(\dfrac{1}{2}\right)^2 - \dfrac{1}{2} - 12$

$= \dfrac{1}{4} - \dfrac{1}{2} - 12 = -\dfrac{49}{4}$

x-intercepts: $0 = x^2 - x - 12$

$= (x - 4)(x + 3)$, so

$x = 4$ or $x = -3$

y-intercept: $y = 0^2 - 0 - 12$

$= 0 - 0 - 12 = -12$

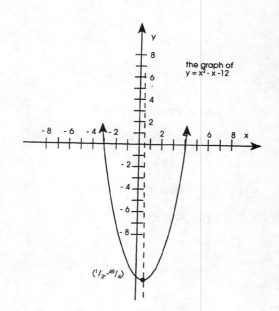

the graph of
$y = x^2 - x - 12$

$(^1/_2, -^{49}/_4)$

31. $y = x^2 - 2x - 2$ A = 1, B = –2, C = –2

axis of symmetry: $x = -\dfrac{B}{2A} = -\dfrac{-2}{2(1)} = -\dfrac{-2}{2} = 1$

x-coordinate of vertex: 1

y-coordinate of vertex: $1^2 - 2(1) - 2$

$= 1 - 2 - 2 = -3$

x-intercepts: $0 = x^2 - 2x - 2$

By the quadratic formula,

$x = \dfrac{-(-2) \pm \sqrt{(-2)^2 - 4(1)(-2)}}{2(1)}$

$= \dfrac{2 \pm \sqrt{4 + 8}}{2} = \dfrac{2 \pm \sqrt{12}}{2}$

$= \dfrac{2 \pm 2\sqrt{3}}{2} = 1 \pm \sqrt{3}$

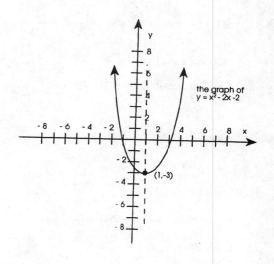

the graph of
$y = x^2 - 2x - 2$

$(1, -3)$

y-intercept: $y = 0^2 - 2(0) - 2 = 0 - 0 - 2 = -2$

33. $y = x^2 - 2x + 5$ $A = 1, B = -2, C = 5$

axis of symmetry: $x = -\dfrac{B}{2A} = -\dfrac{-2}{2(1)} = -\dfrac{-2}{2} = 1$

x-coordinate of vertex: 1

y-coordinate of vertex: $1^2 - 2(1) + 5$

$= 1 - 2 + 5 = 4$

x-intercepts: $0 = x^2 - 2x + 5$

Since $B^2 - 4AC = (-2)^2 - 4(1)(5)$

$= 4 - 20 = -16 < 0$, this equation

has no real solution. Thus, there

are no x-intercepts.

y-intercept: $y = 0^2 - 2(0) + 5 = 0 - 0 + 5 = 5$

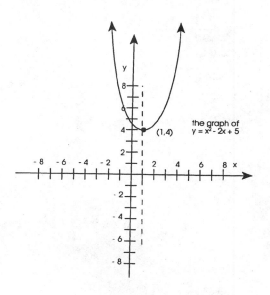

the graph of
$y = x^2 - 2x + 5$

(1,4)

35. $y = -x^2 + 2x - 5$ $A = -1, B = 2, C = -5$

axis of symmetry: $x = -\dfrac{B}{2A} = -\dfrac{2}{2(-1)} = -\dfrac{2}{-2} = 1$

x-coordinate of vertex: 1

y-coordinate of vertex: $-1^2 + 2(1) - 5$

$= -1 + 2 - 5 = -4$

x-intercepts: $0 = -x^2 + 2x - 5$

Since $B^2 - 4AC = 2^2 - 4(-1)(-5)$

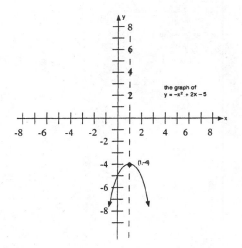

the graph of
$y = -x^2 + 2x - 5$

(1,-4)

279

= –16 < 0, this equation

has no real solution. Thus, there

are no x-intercepts.

y-intercept: $y = -0^2 + 2(0) - 5 = -0 + 0 - 5 = -5$

37. $144x^2 + 9y^2 = 1,296$

$$\frac{144x^2}{1,296} + \frac{9y^2}{1,296} = \frac{1,296}{1,296}$$

$$\frac{x^2}{9} + \frac{y^2}{144} = 1$$

$a^2 = 9$ so $a = 3$;

$b^2 = 144$, so $b = 12$.

Vertices: (–3,0), (3,0),

(0,–12) and (0,12)

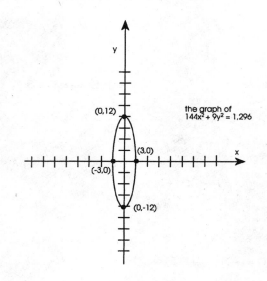

the graph of
$144x^2 + 9y^2 = 1,296$

39. $144x^2 + y^2 = 144$

$$\frac{144x^2}{144} + \frac{y^2}{144} = \frac{144}{144}$$

$$x^2 + \frac{y^2}{144} = 1$$

$a^2 = 1$ so $a = 1$;

$b^2 = 144$, so $b = 12$.

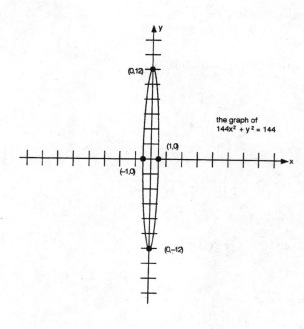

the graph of
$144x^2 + y^2 = 144$

Vertices: (−1,0), (1,0),

(0,−12) and (0,12)

41.　$3x^2 + y^2 = 24$

$\dfrac{3x^2}{24} + \dfrac{y^2}{24} = \dfrac{24}{24}$

$\dfrac{x^2}{8} + \dfrac{y^2}{24} = 1$

$a^2 = 8$, so $a = \sqrt{8} = 2\sqrt{2}$;

$b^2 = 24$, so $b = \sqrt{24} = 2\sqrt{6}$.

Vertices: $(-2\sqrt{2},0)$, $(2\sqrt{2},0)$,

$(0,-2\sqrt{6})$ and $(0,2\sqrt{6})$

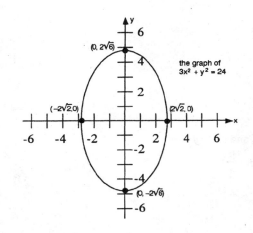

the graph of
$3x^2 + y^2 = 24$

43.　$144x^2 - 9y^2 = 1,296$

$\dfrac{144x^2}{1,296} - \dfrac{9y^2}{1,296} = \dfrac{1,296}{1,296}$

$\dfrac{x^2}{9} - \dfrac{y^2}{144} = 1$

$a^2 = 9$, so $a = 3$;

$b^2 = 144$, so $b = 12$.

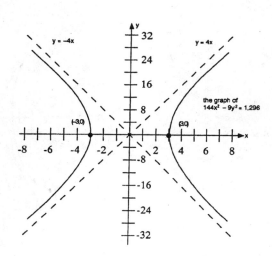

$y = -4x$　　$y = 4x$

the graph of
$144x^2 - 9y^2 = 1,296$

Vertices: $(-3,0)$ and $(3,0)$

Asymptotes: $y = \pm \dfrac{b}{a}x = \pm \dfrac{12}{3}x = \pm 4x$

45. $y^2 - 144x^2 = 144$

$\dfrac{y^2}{144} - \dfrac{144x^2}{144} = \dfrac{144}{144}$

$\dfrac{y^2}{144} - x^2 = 1$

$a^2 = 1$, so $a = 1$;

$b^2 = 144$, so $b = 12$.

Vertices: $(0,-12)$ and $(0,12)$

Asymptotes: $y = \pm \dfrac{b}{a}x = \pm 12x$

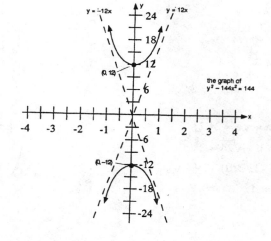

the graph of
$y^2 - 144x^2 = 144$

47. $3x^2 - y^2 = 24$

$\dfrac{3x^2}{24} - \dfrac{y^2}{24} = \dfrac{24}{24}$

$\dfrac{x^2}{8} - \dfrac{y^2}{24} = 1$

$a^2 = 8$, so $a = \sqrt{8} = 2\sqrt{2}$;

$b^2 = 24$, so $b = \sqrt{24} = 2\sqrt{6}$.

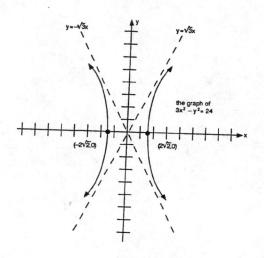

the graph of
$3x^2 - y^2 = 24$

282

Vertices: $(-2\sqrt{2}, 0)$ and $(2\sqrt{2}, 0)$

Asymptotes: $y = \pm\dfrac{b}{a} x = \pm\dfrac{2\sqrt{6}}{2\sqrt{2}} x = \pm\sqrt{3}\, x$

49. $4x^2 + y^2 = 16$

$\dfrac{4x^2}{16} + \dfrac{y^2}{16} = \dfrac{16}{16}$

$\dfrac{x^2}{4} + \dfrac{y^2}{16} = 1$; ellipse

$a^2 = 4$, so $a = 2$; $b^2 = 16$, so $b = 4$

Vertices: $(-2, 0)$, $(2, 0)$, $(0, -4)$ and $(0, 4)$

51. $x^2 - 10x - y = -25$

$y = x^2 - 10x + 25$; parabola

$A = 1$, $B = -10$, $C = 25$

axis of symmetry: $x = -\dfrac{B}{2A} = -\dfrac{-10}{2(1)} = -\dfrac{-10}{2} = 5$

x-coordinate of vertex: 5

y-coordinate of vertex: $5^2 - 10(5) + 25 = 25 - 50 + 25 = 0$

x-intercept: $0 = x^2 - 10x + 25 = (x - 5)^2$, so $x = 5$

y-intercept: $y = 0^2 - 10(0) + 25 = 0 - 0 + 25 = 25$

53. $x^2 + y^2 = 108$; circle; center: $(0,0)$; radius $= \sqrt{108} = 6\sqrt{3}$

55. $x + y = 9$; straight line; x-intercept: 9, y-intercept: 9

57. $25x^2 + 32y^2 = 800$

$\dfrac{25x^2}{800} + \dfrac{32y^2}{800} = \dfrac{800}{800}$

$\dfrac{x^2}{32} + \dfrac{y^2}{25} = 1$; ellipse

$a^2 = 32$, so $a = \sqrt{32} = 4\sqrt{2}$; $b^2 = 25$, so $b = 5$.

Vertices: $(-4\sqrt{2},0)$, $(4\sqrt{2},0)$, $(0,-5)$ and $(0,5)$

59.　$2x^2 - 3y^2 = 12$

$$\frac{2x^2}{12} - \frac{3y^2}{12} = \frac{12}{12}$$

$$\frac{x^2}{6} - \frac{y^2}{4} = 1; \text{ hyperbola}$$

$a^2 = 6$, so $a = \sqrt{6}$; $b^2 = 4$, so $b = 2$.

Vertices: $(-\sqrt{6},0)$ and $(\sqrt{6},0)$

Asymptotes: $\displaystyle y = \pm \frac{b}{a}x = \pm \frac{2}{\sqrt{6}}x = \pm \frac{\sqrt{6}}{3}x$

61.　$y^2 + 2y - x = 25$

$$x = y^2 + 2y - 25$$

Parabola; $A = 1$, $B = 2$, $C = -25$

axis of symmetry: $\displaystyle y = -\frac{B}{2A} = -\frac{2}{2(1)} = -\frac{2}{2} = -1$

y-coordinate of vertex: -1

x-coordinate of vertex: $(-1)^2 + 2(-1) - 25 = 1 - 2 - 25 = -26$

y-intercepts: $0 = y^2 + 2y - 25$. By the quadratic formula,

$$y = \frac{-2 \pm \sqrt{2^2 - 4(1)(-25)}}{2(1)} = \frac{-2 \pm \sqrt{104}}{2} = \frac{-2 \pm 2\sqrt{26}}{2} = -1 \pm \sqrt{26}$$

x-intercept: $x = 0^2 + 2(0) - 25 = 0 + 0 - 25 = -25$

63.　$$x^2 + y^2 - 14x - 6y = -50$$
$$(x^2 - 14x + 49) + (y^2 - 6y + 9) = -50 + 49 + 9$$
$$(x - 7)^2 + (y - 3)^2 = 8$$

Circle; center: $(7,3)$; radius: $\sqrt{8} = 2\sqrt{2}$

65.　$$x^2 + y^2 + 4x + 4y = -4$$
$$(x^2 + 4x + 4) + (y^2 + 4y + 4) = -4 + 4 + 4$$
$$(x + 2)^2 + (y + 2)^2 = 4$$

Circle; center: $(-2,-2)$; radius: 2

67.　$x^2 - y^2 = 16$

$$\frac{x^2}{16} - \frac{y^2}{16} = \frac{16}{16}$$

$$\frac{x^2}{16} - \frac{y^2}{16} = 1; \text{ hyperbola}$$

$a^2 = 16$, so $a = 4$; $b^2 = 16$, so $b = 4$.

Vertices: $(-4,0)$ and $(4,0)$

Asymptotes: $y = \pm \dfrac{b}{a}x = \pm \dfrac{4}{4}x = \pm x$

69. $x - y = 16$; straight line; x-intercept: 16, y-intercept: -16

71. $x^2 + y^2 = 0$; degenerate circle: the point $(0,0)$

73.
$$x^2 + y^2 + 6x - 4y = -15$$
$$(x^2 + 6x + 9) + (y^2 - 4y + 4) = -15 + 9 + 4$$
$$(x + 3)^2 + (y - 2)^2 = -2$$

No real graph

75. $-3x^2 - y = -2$
$$y = -3x^2 + 2$$

Parabola; $A = -3$, $B = 0$, $C = 2$

axis of symmetry: $x = -\dfrac{B}{2A} = -\dfrac{0}{2(-3)} = \dfrac{-0}{-6} = 0$

x-coordinate of vertex: 0

y-coordinate of vertex: $-3(0)^2 + 2$

$= 0 + 2 = 2$

x-intercepts: $0 = -3x^2 + 2$, so $3x^2 = 2$

Then $x = \dfrac{2}{3}$, so $x = \pm\sqrt{\dfrac{2}{3}} = \pm\dfrac{\sqrt{2}}{\sqrt{3}} = \pm\dfrac{\sqrt{6}}{3}$

y-intercept: $y = -3(0)^2 + 2 = 0 + 2 = 2$

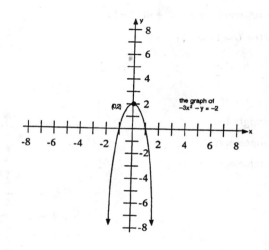

the graph of
$-3x^2 - y = -2$

77. $y^2 + 2y - x = 8$
$$x = y^2 + 2y - 8$$

Parabola; $A = 1$, $B = 2$, $C = -8$

axis of symmetry: $y = -\dfrac{B}{2A} = -\dfrac{2}{2(1)} = -\dfrac{2}{2} = -1$

y-coordinate of vertex: -1

x-coordinate of vertex: $(-1)^2 + 2(-1) - 8$

$= 1 - 2 - 8 = -9$

y-intercepts: $0 = y^2 + 2y - 8$

$= (y + 4)(y - 2)$, so $y = -4$ or $y = 2$

x-intercept: $x = 0^2 + 2(0) - 8$

$= 0 + 0 - 8 = -8$

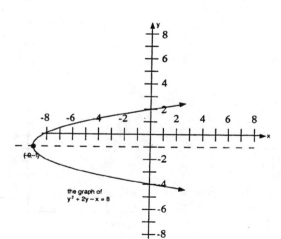

the graph of
$y^2 + 2y - x = 8$

79. $-x^2 + 4x - y = 11$

$$y = -x^2 + 4x - 11$$

Parabola; $A = -1$, $B = 4$, $C = -11$

285

axis of symmetry: $x = -\dfrac{B}{2A} = -\dfrac{4}{2(-1)} = \dfrac{-4}{-2} = 2$

x-coordinate of vertex: 2

y-coordinate of vertex: $-2^2 + 4(2) - 11$

$= -4 + 8 - 11 = -7$

x-intercepts: $0 = -x^2 + 4x - 11$

Since $B^2 - 4AC = (4)^2 - 4(-1)(-11)$

$= 16 - 44 = -28 < 0$, this equation

has no real solution. Thus, there

are no x-intercepts.

y-intercept: $y = -0^2 + 4(0) - 11$

$= 0 + 0 - 11 = -11$

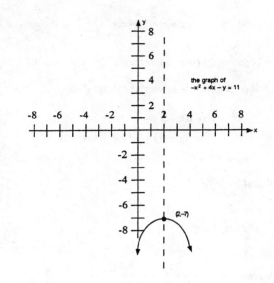

81.　$x + y = 36$

Straight line

x-intercept: 36

y-intercept: 36

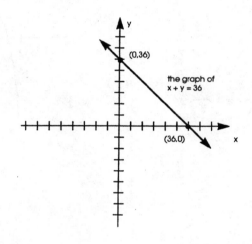

83.　$x^2 + y^2 = 36$

Circle

Center: $(0,0)$; radius $= 6$

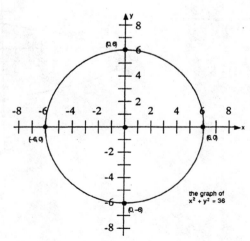

286

85. $x^2 - y^2 = 36$

$$\frac{x^2}{36} - \frac{y^2}{36} = \frac{36}{36}$$

$\dfrac{x^2}{36} - \dfrac{y^2}{36} = 1$; hyperbola

$a^2 = 36$, so $a = 6$;

$b^2 = 36$, so $b = 6$.

Vertices: $(-6,0)$ and $(6,0)$

Asymptotes: $y = \pm\dfrac{b}{a}x = \pm\dfrac{6}{6}x = \pm x$

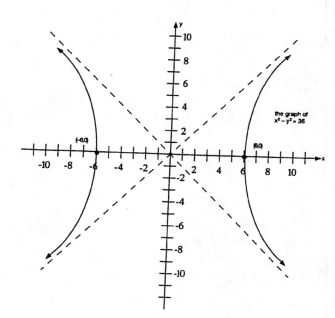

the graph of $x^2 - y^2 = 36$

87. $x^2 - y = 36$

$y = x^2 - 36$

Parabola; $A = 1$, $B = 0$, $C = -36$

axis of symmetry: $x = -\dfrac{B}{2A} = -\dfrac{0}{2(1)} = -\dfrac{0}{2} = 0$

x-coordinate of vertex: 0

y-coordinate of vertex: $0^2 - 36$

$= 0 - 36 = -36$

x-intercepts: $0 = x^2 - 36$, so

$x^2 = 36$. Thus, $x = \pm 6$.

y-intercept: $y = 0^2 - 36$

$= 0 - 36 = -36$

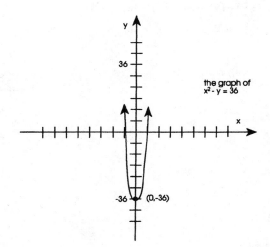

the graph of $x^2 - y = 36$

89. $x^2 - y^2 = 32$

$$\frac{x^2}{32} - \frac{y^2}{32} = \frac{32}{32}$$

$\dfrac{x^2}{32} - \dfrac{y^2}{32} = 1$; hyperbola

$a^2 = 32$ so $a = \sqrt{32} = 4\sqrt{2}$;

$b^2 = 32$ so $b = \sqrt{32} = 4\sqrt{2}$.

Vertices: $(-4\sqrt{2},0)$ and $(4\sqrt{2},0)$

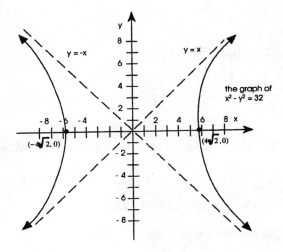

the graph of $x^2 - y^2 = 32$

Asymptotes: $y = \pm \dfrac{b}{a} x$

$$= \pm \dfrac{4\sqrt{2}}{4\sqrt{2}} x = \pm x$$

91.

$$x^2 + 6x + y^2 - 8y = -9$$
$$(x^2 + 6x + 9) + (y^2 - 8y + 16) = -9 + 9 + 16$$
$$(x + 3)^2 + (y - 4)^2 = 16$$

Circle

Center: $(-3, 4)$; radius: 4

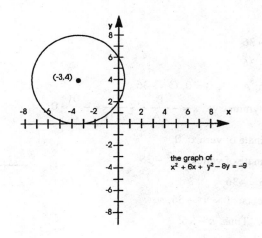

the graph of
$x^2 + 6x + y^2 - 8y = -9$

93. $C = -\dfrac{x^2}{10} + 100x - 24,000$

$$x = -\dfrac{B}{2A} = -\dfrac{100}{2\left(-\dfrac{1}{10}\right)} = \dfrac{-100}{-\dfrac{1}{5}} = 500$$

Thus, 500 widgets would produce the maximum cost. When $x = 500$,

$$C = -\dfrac{(500)^2}{10} + 100(500) - 24,000 = -\dfrac{250,000}{10} + 50,000 - 24,000$$
$$= -25,000 + 50,000 - 24,000 = 1,000$$

Thus, the maximum cost would be $1,000.

95.

$$S = -2t^2 + 5t + 5$$

(a)
$$t = -\frac{B}{2A} = -\frac{5}{2(-2)} = -\frac{5}{-4} = \frac{5}{4}. \text{ When } t = \frac{5}{4},$$

$$S = -2\left(\frac{5}{4}\right)^2 + 5\left(\frac{5}{4}\right) + 5 = -2\left(\frac{25}{16}\right) + \frac{25}{4} + 5 = -\frac{25}{8} + \frac{25}{4} + 5 = \frac{65}{8}$$

So the maximum height reached is $\frac{65}{8}$ feet.

(b)   When $t = 0$, $S = -2(0)^2 + 5(0) + 5 = 0 + 0 + 5 = 5$. Thus, the diving board is 5 feet high.

Chapter 9 Practice Test

1.   Distance between $(3,-2)$ and $(-2,3) = \sqrt{(-2-3)^2 + (3-(-2))^2} = \sqrt{(-5)^2 + 5^2} = \sqrt{25+25} = \sqrt{50} = 5\sqrt{2}$

3.   (a) $(x-3)^2 + (y+2)^2 = 12$

Center: $(3,-2)$; radius: $\sqrt{12} = 2\sqrt{3}$

(b) $x^2 + y^2 - 4x - 6y - 3 = 0$

$(x^2 - 4x + 4) + (y^2 - 6y + 9) = 3 + 4 + 9$

$(x-2)^2 + (y-3)^2 = 16$

Center: $(2,3)$; radius: 4

5. $8x^2 + 9y^2 = 72$

$\dfrac{8x^2}{72} + \dfrac{9y^2}{72} = \dfrac{72}{72}$

$\dfrac{x^2}{9} + \dfrac{y^2}{8} = 1$

$a^2 = 9$, so $a = 3$;

$b^2 = 8$, so $b = \sqrt{8} = 2\sqrt{2}$.

Vertices: $(-3,0)$, $(3,0)$,
$(0,-2\sqrt{2})$ and $(0,2\sqrt{2})$

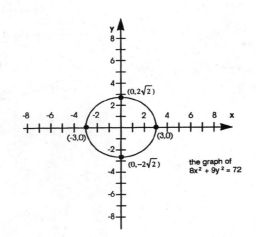

the graph of
$8x^2 + 9y^2 = 72$

7.   (a) $x^2 + 4y^2 = 64$

$\dfrac{x^2}{64} + \dfrac{4y^2}{64} = \dfrac{64}{64}$

$\dfrac{x^2}{64} + \dfrac{y^2}{16} = 1$; ellipse

$a^2 = 64$, so $a = 8$;

$b^2 = 16$, so $b = 4$.

Vertices: (–8,0), (8,0), (0,–4) and (0,4)

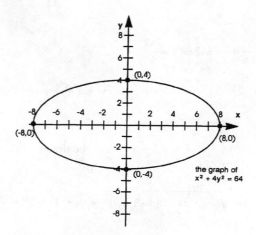

the graph of
$x^2 + 4y^2 = 64$

(b) $y^2 + 6y = x$

Parabola; A = 1, B = 6, C = 0

axis of symmetry: $y = -\dfrac{B}{2A} = -\dfrac{6}{2(1)} = -\dfrac{6}{2} = -3$

y-coordinate of vertex: –3

x-coordinate of vertex: $(-3)^2 + 6(-3)$

$= 9 - 18 = -9$

y-intercepts: $0 = y^2 + 6y$

$= y(y + 6)$, so $y = 0$ or $y = -6$

x-intercept: $x = 0^2 + 6(0)$

$= 0 + 0 = 0$

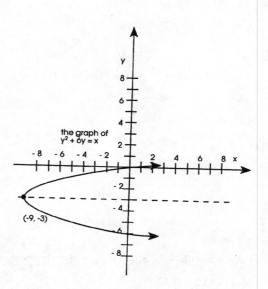

the graph of
$y^2 + 6y = x$

(-9, -3)

(c) $x^2 + y^2 - 2x + 4y - 3 = 0$

$(x^2 - 2x + 1) + (y^2 + 4y + 4) = 3 + 1 + 4$

$(x - 1)^2 + (y + 2)^2 = 8$

Circle

Center: (1,–2)

Radius: $\sqrt{8} = 2\sqrt{2}$

(1,-2)

the graph of
$x^2 + y^2 - 2x + 4y - 3 = 0$

290

Cumulative Review: Chapters 7–9

1. $\sqrt{16a^4b^8} = \sqrt{16}\sqrt{a^4}\sqrt{b^8} = 4a^2b^4$

3. $\left(3x\sqrt{2x^2y}\right)\left(2x\sqrt{8xy^3}\right) = (3x \cdot 2x)\sqrt{2x^2y}\sqrt{8xy^3} = 6x^2\sqrt{2x^2y \cdot 8xy^3} = 6x^2\sqrt{16x^3y^4}$

$$= 6x^2\sqrt{16x^2y^4 \cdot x} = 6x^2\sqrt{16x^2y^4}\sqrt{x} = 6x^2(4xy^2)\sqrt{x} = 24x^3y^2\sqrt{x}$$

5. $\dfrac{\sqrt{48}}{\sqrt{3}} = \sqrt{\dfrac{48}{3}} = \sqrt{16} = 4$

7. $\sqrt[8]{x^4} = (x^4)^{\frac{1}{8}} = x^{\frac{4}{8}} = x^{\frac{1}{2}} = \sqrt{x}$

9. $2\sqrt{5} - 3\sqrt{5} + 8\sqrt{5} = (2 - 3 + 8)\sqrt{5} = 7\sqrt{5}$

11. $3a\sqrt[3]{a^4} - a^2\sqrt[3]{a} = 3a\sqrt[3]{a^3 \cdot a} - a^2\sqrt[3]{a} = 3a\sqrt[3]{a^3}\sqrt[3]{a} - a^2\sqrt[3]{a} = 3a \cdot a\sqrt[3]{a} - a^2\sqrt[3]{a}$

$$= 3a^2\sqrt[3]{a} - a^2\sqrt[3]{a} = 2a^2\sqrt[3]{a}$$

13. $\sqrt{2}\left(\sqrt{2} - 1\right) + 2\sqrt{2} = \sqrt{2}\sqrt{2} - \sqrt{2} + 2\sqrt{2} = 2 - \sqrt{2} + 2\sqrt{2} = 2 + \sqrt{2}$

15. $\left(\sqrt{5} - \sqrt{3}\right)\left(\sqrt{5} + \sqrt{3}\right) = \sqrt{5}\sqrt{5} + \sqrt{5}\sqrt{3} - \sqrt{3}\sqrt{5} - \sqrt{3}\sqrt{3} = 5 + \sqrt{15} - \sqrt{15} - 3 = 2$

17. $\dfrac{5}{\sqrt{3} + \sqrt{2}} - \dfrac{3}{\sqrt{3}} = \dfrac{5}{\sqrt{3} + \sqrt{2}} \cdot \dfrac{\sqrt{3} - \sqrt{2}}{\sqrt{3} - \sqrt{2}} - \dfrac{3}{\sqrt{3}} \cdot \dfrac{\sqrt{3}}{\sqrt{3}}$

$$= \dfrac{5\left(\sqrt{3} - \sqrt{2}\right)}{\sqrt{3}\sqrt{3} - \sqrt{3}\sqrt{2} + \sqrt{2}\sqrt{3} - \sqrt{2}\sqrt{2}} - \dfrac{3\sqrt{3}}{3}$$

$$= \dfrac{5\left(\sqrt{3} - \sqrt{2}\right)}{3 - \sqrt{6} + \sqrt{6} - 2} - \sqrt{3} = \dfrac{5\left(\sqrt{3} - \sqrt{2}\right)}{1} - \sqrt{3}$$

$$= 5\sqrt{3} - 5\sqrt{2} - \sqrt{3} = 4\sqrt{3} - 5\sqrt{2}$$

19. $\sqrt{3x + 2} = 7$

   $3x + 2 = 49$

   $3x = 47$

   $x = \dfrac{47}{3}$

CHECK $x = \dfrac{47}{3}$:

$\sqrt{3x + 2} = 7$

$\sqrt{3\left(\dfrac{47}{3}\right) + 2} \overset{?}{=} 7$

$\sqrt{47 + 2} \overset{?}{=} 7$

$\sqrt{49} \overset{?}{=} 7$

$7 \overset{\checkmark}{=} 7$

21. $\sqrt[3]{x-1} = -2$

$\left(\sqrt[3]{x-1}\right)^3 = (-2)^3$

$x - 1 = -8$

$x = -7$

CHECK: $x = -7$:

$\sqrt[3]{x-1} = -2$

$\sqrt[3]{-7-1} \overset{?}{=} -2$

$\sqrt[3]{-8} \overset{?}{=} -2$

$-2 \overset{\checkmark}{=} -2$

23. $i^{35} = (i^4)^8 i^3 = 1^8 i^3 = 1(-i) = -i$

25. $(3 - 2i)(5 + i) = 3(5) + 3i - 2i(5) - 2i^2 = 15 + 3i - 10i - 2i^2 = 15 - 7i - 2(-1)$
$= 15 - 7i + 2 = 17 - 7i$

27. $a^2 - 2a - 15 = 0$

$(a - 5)(a + 3) = 0$

$a - 5 = 0$ or $a + 3 = 0$

$a = 5$ or $a = -3$

29. $2x^2 - 3x - 4 = 9 - 3(x - 2)$

$2x^2 - 3x - 4 = 9 - 3x + 6$

$2x^2 - 3x - 4 = -3x + 15$

$2x^2 - 4 = 15$

$2x^2 = 19$

$x^2 = \dfrac{19}{2}$

$x = \pm\sqrt{\dfrac{19}{2}} = \pm\dfrac{\sqrt{19}}{\sqrt{2}} = \pm\dfrac{\sqrt{19}\sqrt{2}}{\sqrt{2}\sqrt{2}} = \pm\dfrac{\sqrt{38}}{2}$

31. $y^2 + 6y - 1 = 0$ $\qquad\qquad \left(\dfrac{1}{2}(6)\right)^2 = 3^2 = 9$

$y^2 + 6y = 1$

$y^2 + 6y + 9 = 1 + 9$

$(y + 3)^2 = 10$

$y + 3 = \pm\sqrt{10}$

$y = -3 \pm \sqrt{10}$

33. $3a^2 - 2a - 2 = 0$

$a = \dfrac{-(-2) \pm \sqrt{(-2)^2 - 4(3)(-2)}}{2(3)}$

$a = \dfrac{2 \pm \sqrt{4 + 24}}{6} = \dfrac{2 \pm \sqrt{28}}{6} = \dfrac{2 \pm 2\sqrt{7}}{6} = \dfrac{1 \pm \sqrt{7}}{3}$

35.

$$\frac{1}{x+2} = x + 2$$

$$1 = (x+2)^2$$

$$\pm 1 = x + 2$$

$$-2 \pm 1 = x$$

$$-1 = x \text{ or } -3 = x$$

37.

$$\frac{2}{x+3} - \frac{3}{x} = -\frac{2}{3}$$

$$3x(x+3)\left(\frac{2}{x+3} - \frac{3}{x}\right) = 3x(x+3)\left(-\frac{2}{3}\right)$$

$$3x(x+3)\left(\frac{2}{x+3}\right) - 3x(x+3)\left(\frac{3}{x}\right) = 3x(x+3)\left(-\frac{2}{3}\right)$$

$$3x(2) - 3(x+3)(3) = x(x+3)(-2)$$

$$6x - 9(x+3) = -2x(x+3)$$

$$6x - 9x - 27 = -2x^2 - 6x$$

$$-3x - 27 = -2x^2 - 6x$$

$$2x^2 + 3x - 27 = 0$$

$$(2x+9)(x-3) = 0$$

$$2x + 9 = 0 \text{ or } x - 3 = 0$$

$$x = -\frac{9}{2} \text{ or } x = 3$$

39.

$$\frac{3}{x-3} + \frac{2x}{x+3} = \frac{5}{x-3}$$

$$(x-3)(x+3)\left(\frac{3}{x-3} + \frac{2x}{x+3}\right) = (x-3)(x+3)\left(\frac{5}{x-3}\right)$$

$$(x-3)(x+3)\left(\frac{3}{x-3}\right) + (x-3)(x+3)\left(\frac{2x}{x+3}\right) = (x-3)(x+3)\left(\frac{5}{x-3}\right)$$

$$3(x+3) + 2x(x-3) = 5(x+3)$$

$$3x + 9 + 2x^2 - 6x = 5x + 15$$

$$2x^2 - 3x + 9 = 5x + 15$$

$$2x^2 - 8x - 6 = 0$$

$$x^2 - 4x - 3 = 0$$

$$x = \frac{-(-4) \pm \sqrt{(-4)^2 - 4(1)(-3)}}{2(1)} = \frac{4 \pm \sqrt{16 + 12}}{2}$$

$$x = \frac{4 \pm \sqrt{28}}{2} = \frac{4 \pm 2\sqrt{7}}{2} = 2 \pm \sqrt{7}$$

41.

$$s = \frac{xy}{z^2}$$

$$z^2 s = \not{z^2} \left(\frac{xy}{\not{z^2}} \right)$$

$$z^2 s = xy$$

$$z^2 = \frac{xy}{s}$$

$$z = \pm \sqrt{\frac{xy}{s}} = \pm \frac{\sqrt{xy}}{\sqrt{s}} = \pm \frac{\sqrt{xy}\sqrt{s}}{\sqrt{s}\sqrt{s}} = \pm \frac{\sqrt{xys}}{s}$$

43.

$$\sqrt{5x-1} = x+1$$

$$\left(\sqrt{5x-1}\right)^2 = (x+1)^2$$

$$5x - 1 = x^2 + 2x + 1$$

$$0 = x^2 - 3x + 2$$

$$0 = (x-1)(x-2)$$

$$0 = x - 1 \text{ or } 0 = x - 2$$

$$1 = x \text{ or } 2 = x$$

CHECK x = 1:

$$\sqrt{5x-1} = x+1$$

$$\sqrt{5(1)-1} \overset{?}{=} 1+1$$

$$\sqrt{5-1} \overset{?}{=} 2$$

$$\sqrt{4} \overset{?}{=} 2$$

$$2 \overset{\checkmark}{=} 2$$

CHECK x = 2:

$$\sqrt{5x-1} = x+1$$

$$\sqrt{5(2)-1} \overset{?}{=} 2+1$$

$$\sqrt{10-1} \overset{?}{=} 3$$

$$\sqrt{9} \overset{?}{=} 3$$

$$3 \overset{\checkmark}{=} 3$$

45.

$$\sqrt{5x-1} = 2y$$

$$\left(\sqrt{5x-1}\right)^2 = (2y)^2$$

$$5x - 1 = 4y^2$$

$$5x = 4y^2 + 1$$

$$x = \frac{4y^2 + 1}{5}$$

47.

$$x^4 - 81 = 0$$

$$(x^2 - 9)(x^2 + 9) = 0$$

$$(x-3)(x+3)(x^2+9) = 0$$

$$x - 3 = 0 \text{ or } x + 3 = 0 \text{ or } x^2 + 9 = 0$$

$$x = 3 \text{ or } x = -3 \text{ or } x^2 = -9$$

$$x = 3 \text{ or } x = -3 \text{ or } x = \pm\sqrt{-9} = \pm 3i$$

49. $(x-5)(x-8) < 0$

Cut points: x = 5 and x = 8

Intervals: x < 5, 5 < x < 8, and x > 8

For x < 5, let x = 4 be the test value. When x = 4, $(x-5)(x-8) = (4-5)(4-8)$ is positive. For 5 < x < 8, let x = 6 be the test value. When x = 6, $(x - 5)(x - 8) = (6 - 5)(6 - 8)$ is negative. For x > 8, let x = 9 be the test value. When x = 9, $(x - 5)(x - 8) = (9 - 5)(9 - 8)$ is positive. Thus, $(x - 5)(x - 8) < 0$ when $5 < x < 8$.

51. $\dfrac{x - 2}{x + 3} > 0$

Cut points: x = 2 and x = -3

Intervals: $x < -3$, $-3 < x < 2$, and $x > 2$

For x < -3, let x = -4 be the test value. When x = -4, $\dfrac{x - 2}{x + 3} = \dfrac{-4 - 2}{-4 + 3}$ is positive. For -3 < x < 2, let x = 0 be the test value. When x = 0, $\dfrac{x - 2}{x + 3} = \dfrac{0 - 2}{0 + 3}$ is negative. For x > 2, let x = 3 be the test value. When x = 3, $\dfrac{x - 2}{x + 3} = \dfrac{3 - 2}{3 + 3}$ is positive. Thus, $\dfrac{x - 2}{x + 3} > 0$ when $x < -3$ or $x > 2$.

53. Let x = the number.

$$x - \frac{1}{x} = \frac{5}{6}$$

$$6x\left(x - \frac{1}{x}\right) = 6x\left(\frac{5}{6}\right)$$

$$6x(x) - 6x\left(\frac{1}{x}\right) = 6x\left(\frac{5}{6}\right)$$

$$6x^2 - 6 = 5x$$

$$6x^2 - 5x - 6 = 0$$

$$(3x + 2)(2x - 3) = 0$$

3x + 2 = 0 or 2x - 3 = 0

$$x = -\frac{2}{3} \text{ or } x = \frac{3}{2}$$

Thus, the number in question is either $-\dfrac{2}{3}$ or $\dfrac{3}{2}$.

CHECK $x = -\dfrac{2}{3}$:

$$-\frac{2}{3} - \left(-\frac{3}{2}\right) \stackrel{?}{=} \frac{5}{6}$$

$$-\frac{2}{3} + \frac{3}{2} \stackrel{?}{=} \frac{5}{6}$$

$$-\frac{4}{6} + \frac{9}{6} \stackrel{?}{=} \frac{5}{6}$$

$$\frac{5}{6} \stackrel{\checkmark}{=} \frac{5}{6}$$

CHECK $x = \dfrac{3}{2}$:

$$\frac{3}{2} - \frac{2}{3} \stackrel{?}{=} \frac{5}{6}$$

$$\frac{9}{6} - \frac{4}{6} \stackrel{?}{=} \frac{5}{6}$$

$$\frac{5}{6} \stackrel{\checkmark}{=} \frac{5}{6}$$

55. Let x = length of side of square (in inches).

$x^2 + x^2 = 28^2$ (by the Pythagorean theorem)

$$2x^2 = 784$$
$$x^2 = 392$$
$$x = \pm\sqrt{392} = \pm 14\sqrt{2}$$

Since x measures length, it cannot be negative. Thus, $x = 14\sqrt{2}$ inches $\cong 19.8$ inches.

57. $|PQ| = \sqrt{(5-2)^2 + (9-5)^2} = \sqrt{3^2 + 4^2} = \sqrt{9+16} = \sqrt{25} = 5$

       midpoint of PQ: $\left(\dfrac{2+5}{2}, \dfrac{5+9}{2}\right) = \left(\dfrac{7}{2}, 7\right)$

59. $|PQ| = \sqrt{(6-2)^2 + (3-(-4))^2} = \sqrt{4^2 + 7^2} = \sqrt{16+49} = \sqrt{65}$

       midpoint of PQ: $\left(\dfrac{2+6}{2}, \dfrac{-4+3}{2}\right) = \left(4, -\dfrac{1}{2}\right)$

61. $(x-2)^2 + (y-3)^2 = 25$

       $(x-2)^2 + (y-3)^2 = 5^2$

       $(x-h)^2 + (y-k)^2 = r^2$, so

       $h = 2$, $k = 3$, and $r = 5$

       Center: (2,3), radius: 5

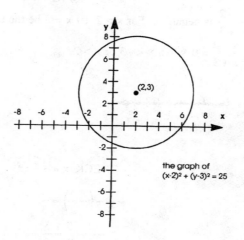

the graph of
$(x-2)^2 + (y-3)^2 = 25$

63. $x^2 + y^2 - 6x + 2y = -1$

       $(x^2 - 6x + 9) + (y^2 + 2y + 1) = -1 + 9 + 1$

       $(x-3)^2 + (y+1)^2 = 9$

       $(x-3)^2 + (y+1)^2 = 3^2$

       $(x-h)^2 + (y-k)^2 = r^2$, so

       $h = 3$, $k = -1$, and $r = 3$

       Center: (3,–1), radius: 3

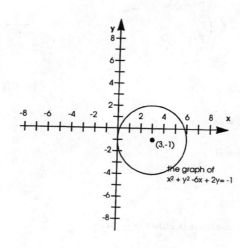

the graph of
$x^2 + y^2 - 6x + 2y = -1$

296

65. $y = 3x^2$

A = 3, B = 0, C = 0

axis of symmetry: $x = -\dfrac{B}{2A} = -\dfrac{0}{2(3)} = -\dfrac{0}{6} = 0$

x-coordinate of vertex: 0

y-coordinate of vertex: $3(0)^2 = 3 \cdot 0 = 0$

x-intercept: $0 = 3x^2$, so $x = 0$

y-intercept: $y = 3(0)^2 = 0$

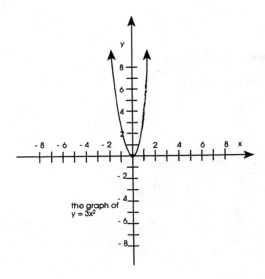

the graph of
$y = 3x^2$

67. $y = x^2 + 2x$

A = 1, B = 2, C = 0

axis of symmetry: $x = -\dfrac{B}{2A} = -\dfrac{2}{2(1)} = -\dfrac{2}{2} = -1$

x-coordinate of vertex: −1

y-coordinate of vertex: $(-1)^2 + 2(-1) = 1 - 2 = -1$

x-intercepts: $0 = x^2 + 2x = x(x + 2)$, so

x = 0 or x = −2

y-intercept: $y = 0^2 + 2(0) = 0 + 0 = 0$

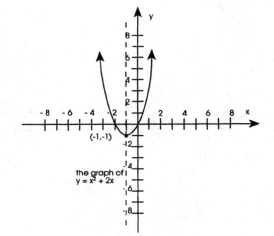

(-1,-1)

the graph of
$y = x^2 + 2x$

69. $y = -2x^2 - 4x + 6$

A = −2, B = −4, C = 6

axis of symmetry: $x = -\dfrac{B}{2A} = -\dfrac{-4}{2(-2)} = -\dfrac{-4}{-4} = -1$

x-coordinate of vertex: −1

y-coordinate of vertex: $-2(-1)^2 - 4(-1) + 6$

$= -2 + 4 + 6 = 8$

x-intercepts: $0 = -2x^2 - 4x + 6$

$= -2(x^2 + 2x - 3) = -2(x + 3)(x - 1)$, so

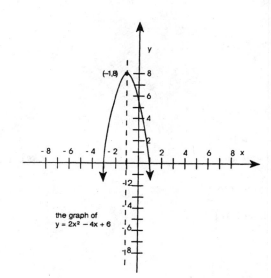

(-1,8)

the graph of
$y = 2x^2 - 4x + 6$

297

x = –3 or x = 1

y-intercept: $y = -2(0)^2 - 4(0) + 6$

$= 0 + 0 + 6 = 6$

71. $\dfrac{x^2}{16} + \dfrac{y^2}{5} = 1$
$a^2 = 16$, so a = 4;
$b^2 = 5$, so b = $\sqrt{5}$.
Vertices: (–4,0), (4,0),
(0, –$\sqrt{5}$) and (0, $\sqrt{5}$)

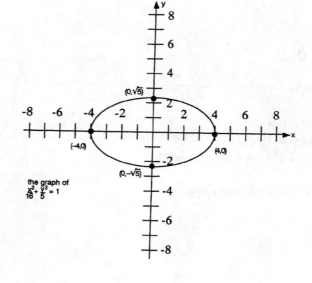

73. $9x^2 + y^2 = 1$

$\dfrac{x^2}{\frac{1}{9}} + y^2 = 1$

$a^2 = \dfrac{1}{9}$, so a = $\dfrac{1}{3}$;

$b^2 = 1$, so b = 1.

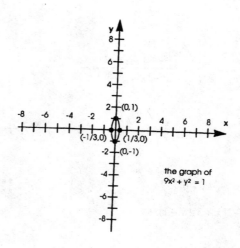

298

Vertices: $\left(-\frac{1}{3},0\right), \left(\frac{1}{3},0\right),$

$(0,-1)$ and $(0,1)$

75. $4x^2 - 49y^2 = 196$

$\dfrac{4x^2}{196} - \dfrac{49y^2}{196} = \dfrac{196}{196}$

$\dfrac{x^2}{49} - \dfrac{y^2}{4} = 1$

$a^2 = 49$, so $a = 7$;

$b^2 = 4$, so $b = 2$.

Vertices: $(-7,0)$ and $(7,0)$

Asymptotes: $y = \pm\dfrac{b}{a}x = \pm\dfrac{2}{7}x$

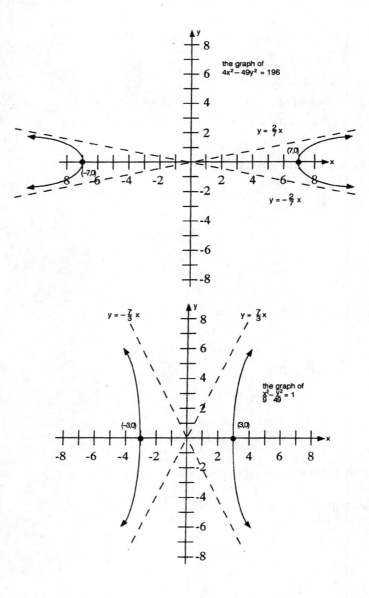

77. $\dfrac{x^2}{9} - \dfrac{y^2}{49} = 1$

Hyperbola

$a^2 = 9$, so $a = 3$;

$b^2 = 49$, so $b = 7$

Vertices: $(-3,0)$ and $(3,0)$

Asymptotes: $y = \pm\dfrac{b}{a}x = \pm\dfrac{7}{3}x$

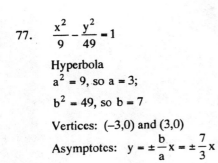

79. $\dfrac{x}{3} + \dfrac{y}{4} = 1$

Straight line

x-intercept: 3

y-intercept: 4

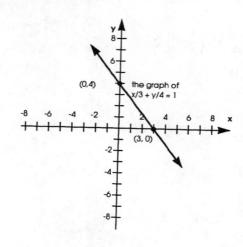

(0,4)

the graph of
x/3 + y/4 = 1

(3, 0)

81. $x^2 + y^2 + 4x - 6y = -12$

$(x^2 + 4x + 4) + (y^2 - 6y + 9) = -12 + 4 + 9$

$(x + 2)^2 + (y - 3)^2 = 1$

Circle

Center: $(-2, 3)$, radius: 1

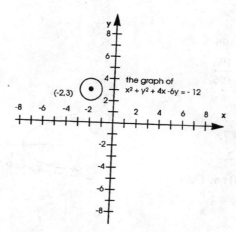

(-2,3)

the graph of
x² + y² + 4x -6y = - 12

83. $P = -200p^2 + 500p$

$A = -200, \ B = 500, \ C = 0$

$p = -\dfrac{B}{2A} = -\dfrac{500}{2(-200)} = \dfrac{-500}{-400} = \dfrac{5}{4}$

Thus, the ticket price that would produce the maximum profit is $\dfrac{5}{4}$ dollars, or $1.25. Then maximum profit

$= -200\left(\dfrac{5}{4}\right)^2 + 500\left(\dfrac{5}{4}\right) = \dfrac{-1250}{4} + \dfrac{2500}{4} = \dfrac{1250}{4} = \$312.50$

Cumulative Practice Test: Chapters 7–9

1. (a) $\sqrt{24x^2y^5} = \sqrt{4x^2y^4 \cdot 6y} = \sqrt{4x^2y^4}\sqrt{6y} = 2xy^2\sqrt{6y}$

   (b) $\left(a^2\sqrt{2ab^2}\right)\left(b\sqrt{4a^2b^3}\right) = (a^2b)\sqrt{2ab^2}\sqrt{4a^2b^3} = a^2b\sqrt{2ab^2 \cdot 4a^2b^3} = a^2b\sqrt{8a^3b^5}$

   $$= a^2b\sqrt{4a^2b^4 \cdot 2ab} = a^2b\sqrt{4a^2b^4}\sqrt{2ab} = a^2b(2ab^2)\sqrt{2ab} = 2a^3b^3\sqrt{2ab}$$

   (c) $\sqrt[3]{\dfrac{5}{a}} = \dfrac{\sqrt[3]{5}}{\sqrt[3]{a}} = \dfrac{\sqrt[3]{5}\sqrt[3]{a^2}}{\sqrt[3]{a}\sqrt[3]{a^2}} = \dfrac{\sqrt[3]{5a^2}}{\sqrt[3]{a^3}} = \dfrac{\sqrt[3]{5a^2}}{a}$

3. $\sqrt{2x} + 3 = 4$

   $\sqrt{2x} = 1$

   $2x = 1$

   $x = \dfrac{1}{2}$

   CHECK $x = \dfrac{1}{2}$:

   $\sqrt{2x} + 3 = 4$

   $\sqrt{2\left(\dfrac{1}{2}\right)} + 3 \overset{?}{=} 4$

   $\sqrt{1} + 3 \overset{?}{=} 4$

   $1 + 3 \overset{?}{=} 4$

   $4 \overset{\checkmark}{=} 4$

5. (a) $2x^2 + x = 3$

   $2x^2 + x - 3 = 0$

   $(2x + 3)(x - 1) = 0$

   $2x + 3 = 0$ or $x - 1 = 0$

   $x = -\dfrac{3}{2}$ or $x = 1$

   (b) $\dfrac{x}{3} = \dfrac{4}{x}$

   $\cancel{3}x\left(\dfrac{x}{\cancel{3}}\right) = 3\cancel{x}\left(\dfrac{4}{\cancel{x}}\right)$

   $x^2 = 12$

   $x = \pm\sqrt{12} = \pm 2\sqrt{3}$

   (c) $2x^2 + 4x - 3 = 0$ $A = 2, B = 4, C = -3$

   $$x = \frac{-B \pm \sqrt{B^2 - 4AC}}{2A} = \frac{-4 \pm \sqrt{4^2 - 4(2)(-3)}}{2(2)} = \frac{-4 \pm \sqrt{16 + 24}}{4}$$

   $$= \frac{-4 \pm \sqrt{40}}{4} = \frac{-4 \pm 2\sqrt{10}}{4} = \frac{-2 \pm \sqrt{10}}{2}$$

(d)
$$\frac{x}{x+1} + \frac{2}{x-3} = 4$$

$$(x+1)(x-3)\left(\frac{x}{x+1} + \frac{2}{x-3}\right) = (x+1)(x-3)(4)$$

$$\cancel{(x+1)}(x-3)\left(\frac{x}{\cancel{x+1}}\right) + (x+1)\cancel{(x-3)}\left(\frac{2}{\cancel{x-3}}\right) = (x+1)(x-3)(4)$$

$$x(x-3) + 2(x+1) = 4(x+1)(x-3)$$

$$x^2 - 3x + 2x + 2 = 4(x^2 - 2x - 3)$$

$$x^2 - x + 2 = 4x^2 - 8x - 12$$

$$0 = 3x^2 - 7x - 14$$

$$A = 3, \ B = -7, \ C = -14$$

$$x = \frac{-B \pm \sqrt{B^2 - 4AC}}{2A} = \frac{-(-7) \pm \sqrt{(-7)^2 - 4(3)(-14)}}{2(3)} = \frac{7 \pm \sqrt{49 + 168}}{6} = \frac{7 \pm \sqrt{217}}{6}$$

7.　(a)
$$\sqrt{2x+1} = 2\sqrt{x} - 1$$

$$\left(\sqrt{2x+1}\right)^2 = \left(2\sqrt{x} - 1\right)^2$$

$$2x + 1 = 4x - 4\sqrt{x} + 1$$

$$2x = 4x - 4\sqrt{x}$$

$$4\sqrt{x} = 2x$$

$$2\sqrt{x} = x$$

$$\left(2\sqrt{x}\right)^2 = x^2$$

$$4x = x^2$$

$$0 = x^2 - 4x$$

$$0 = x(x-4)$$

$$0 = x \ \text{or} \ 0 = x - 4$$

$$0 = x \ \text{or} \ 4 = x$$

CHECK x = 0:

$$\sqrt{2x+1} = 2\sqrt{x} - 1$$

$$\sqrt{2(0)+1} \overset{?}{=} 2\sqrt{0} - 1$$

$$\sqrt{0+1} \overset{?}{=} 2(0) - 1$$

$$\sqrt{1} \overset{?}{=} 0 - 1$$

$$1 \neq -1$$

CHECK x = 4:

$$\sqrt{2x+1} \overset{?}{=} 2\sqrt{x} - 1$$

$$\sqrt{2(4)+1} \overset{?}{=} 2\sqrt{4} - 1$$

$$\sqrt{8+1} \overset{?}{=} 2(2) - 1$$

$$\sqrt{9} \overset{?}{=} 4 - 1$$

$$3 \overset{\checkmark}{=} 3$$

Thus, x = 0 is an extraneous root, and the only solution is x = 4

(b)
$$x^3 + x^2 = 6x$$

$$x^3 + x^2 - 6x = 0$$

$$x(x^2 + x - 6) = 0$$

$$x(x-2)(x+3) = 0$$

$$x = 0 \ \text{or} \ x - 2 = 0 \ \text{or} \ x + 3 = 0$$

$$x = 0 \ \text{or} \ x = 2 \ \text{or} \ x = -3$$

9. Let r = the rate of the stream (in mph). Then Michael's rate downstream = 7 + r, and Michael's rate upstream = 7 − r.

$$\frac{4}{7+r} + \frac{4}{7-r} = \frac{7}{5}$$

$$5(7+r)(7-r)\left(\frac{4}{7+r} + \frac{4}{7-r}\right) = 5(7+r)(7-r)\left(\frac{7}{5}\right)$$

$$5(7+r)(7-r)\left(\frac{4}{7+r}\right) + 5(7+r)(7-r)\left(\frac{4}{7-r}\right) = 5(7+r)(7-r)\left(\frac{7}{5}\right)$$

$$20(7-r) + 20(7+r) = 7(7+r)(7-r)$$

$$140 - 20r + 140 + 20r = 7(49 - r^2)$$

$$280 = 7(49 - r^2)$$

$$40 = 49 - r^2$$

$$r^2 = 9$$

$$r = \pm 3$$

Since the rate of the stream cannot be negative, conclude that it is 3 mph.

11.
$$x^2 + y^2 - 4x + 4y = 4$$
$$(x^2 - 4x \quad) + (y^2 + 4y \quad) = 4$$
$$(x^2 - 4x + 4) + (y^2 + 4y + 4) = 4 + 4 + 4$$
$$(x - 2)^2 + (y + 2)^2 = 12$$
Center: (2, −2); radius: $\sqrt{12} = 2\sqrt{3}$

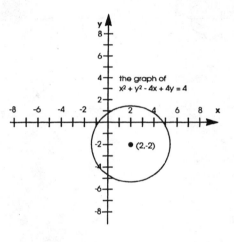

the graph of
$x^2 + y^2 - 4x + 4y = 4$

(2,-2)

13. (a) $\frac{x^2}{9} + y^2 = 1$; ellipse

$a^2 = 9$, so a = 3;

$b^2 = 1$, so b = 1.

Vertices: (−3,0), (3,0),

(0, −1) and (0,1)

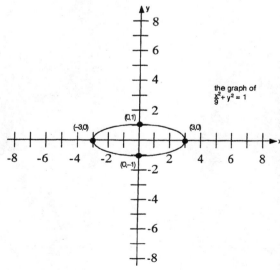

the graph of
$\frac{x^2}{9} + y^2 = 1$

(0,1)

(−3,0)   (3,0)

(0,−1)

(b) $\dfrac{x^2}{36} - \dfrac{y^2}{25} = 1$; hyperbola

$a^2 = 36$, so $a = 6$;

$b^2 = 25$, so $b = 5$.

Vertices: $(-6,0)$, $(6,0)$

Asymptotes: $y = \pm \dfrac{b}{a}x = \pm \dfrac{5}{6}x$

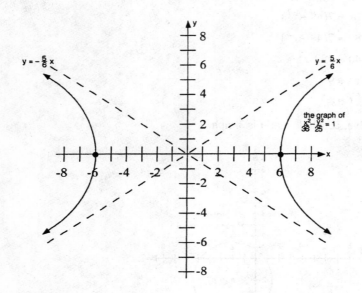

CHAPTER 10
MORE SYSTEMS OF EQUATIONS
AND SYSTEMS OF INEQUALITIES

Exercises 10.1

1. $\begin{cases} x + y + z = 9 \\ 2x - y + z = 9 \\ x - y + z = 3 \end{cases}$

Add equations (1) and (2), obtaining equation (4):

$x + y + z = 9$
$\underline{2x - y + z = 9}$
$3x + 2z = 18$

Add equations (1) and (3), obtaining equation (5):
$x + y + z = 9$
$\underline{x - y + z = 3}$
$2x + 2z = 12$

Multiply equation (5) by -1 and add the resulting equation to equation (4):
$3x + 2z = 18$
$\underline{-2x - 2z = -12}$
$x = 6$

Substitute into equation (4):
$3x + 2z = 18$
$3(6) + 2z = 18$
$18 + 2z = 18$
$2z = 0$
$z = 0$

Substitute into equation (l):
$x + y + z = 9$
$6 + y + 0 = 9$
$y + 6 = 9$
$y = 3$

Solution: $x = 6$, $y = 3$, $z = 0$

CHECK:

$x + y + z = 9$	$2x - y + z = 9$	$x - y + z = 3$
$6 + 3 + 0 \overset{?}{=} 9$	$2(6) - 3 + 0 \overset{?}{=} 9$	$6 - 3 + 0 \overset{?}{=} 3$
$9 + 0 \overset{?}{=} 9$	$12 - 3 \overset{?}{=} 9$	$3 + 0 \overset{?}{=} 3$
$9 \overset{\vee}{=} 9$	$9 \overset{\vee}{=} 9$	$3 \overset{\vee}{=} 3$

5. $\begin{cases} x + y - z = 1 \\ 2x + 2y + 2z = 0 \\ x - y + z = 3 \end{cases}$

Multiply equation (1) by 2 and add the resulting equation to equation (2), obtaining equation (4):

$$2x + 2y - 2z = 2$$
$$\underline{2x + 2y + 2z = 0}$$
$$4x + 4y = 2$$

Add equations (1) and (3), obtaining equation (5):

$$x + y - z = 1$$
$$\underline{x - y + z = 3}$$
$$2x = 4$$
$$\text{So } x = 2$$

Substitute into equation (4):

$$4x + 4y = 2$$
$$4(2) + 4y = 2$$
$$8 + 4y = 2$$
$$4y = -6$$
$$y = -\frac{3}{2}$$

Substitute into equation (1):

$$x + y - z = 1$$
$$2 + \left(-\frac{3}{2}\right) - z = 1$$
$$2 - \frac{3}{2} - z = 1$$
$$\frac{1}{2} - z = 1$$
$$-z = \frac{1}{2}$$
$$z = -\frac{1}{2}$$

$$\text{Solution: } x = 2, \ y = -\frac{3}{2}, \ z = -\frac{1}{2}$$

CHECK:

$$x + y - z = 1$$
$$2 + \left(-\frac{3}{2}\right) - \left(-\frac{1}{2}\right) \overset{?}{=} 1$$
$$2 - \frac{3}{2} + \frac{1}{2} \overset{?}{=} 1$$
$$\frac{1}{2} + \frac{1}{2} \overset{?}{=} 1$$
$$1 \overset{\checkmark}{=} 1$$

$$2x + 2y + 2z = 0$$
$$2(2) + 2\left(-\frac{3}{2}\right) + 2\left(-\frac{1}{2}\right) \overset{?}{=} 0$$
$$4 - 3 - 1 \overset{?}{=} 0$$
$$1 - 1 \overset{?}{=} 0$$
$$0 \overset{\checkmark}{=} 0$$

$$x - y + z = 3$$
$$2 - \left(-\frac{3}{2}\right) + \left(-\frac{1}{2}\right) \overset{?}{=} 3$$
$$2 + \frac{3}{2} - \frac{1}{2} \overset{?}{=} 3$$
$$2 + 1 \overset{?}{=} 3$$
$$3 \overset{\checkmark}{=} 3$$

7. $$\begin{cases} x + 2y + 3z = 1 \\ 3x + 6y + 9z = 3 \\ 4x + 8y + 12z = 4 \end{cases}$$

Note that if we were to multiply equation (1) by 3, we would get equation (2). Also, if we were to multiply equation (1) by 4, we would get equation (3). Thus, all three equations are equivalent, and the system has infinitely many solutions. Each is of the form (x,y,z), where $x + 2y + 3z = 1$.

11. $\begin{cases} 2a + b - 3c = -6 \\ 4a - 4b + 2c = 10 \\ 6a - 7b + c = 12 \end{cases}$

Multiply equation (1) by -2 and add the resulting equation to equation (2), obtaining equation (4):
$$\begin{array}{r} -4a - 2b + 6c = 12 \\ \underline{4a - 4b + 2c = 10} \\ -6b + 8c = 22 \end{array}$$

Multiply equation (1) by -3 and add the resulting equation to equation (3), obtaining equation (5):
$$\begin{array}{r} -6a - 3b + 9c = 18 \\ \underline{6a - 7b + c = 12} \\ -10b + 10c = 30 \end{array}$$

Multiply equation (4) by 5, equation (5) by -3, and add the resulting equations:
$$\begin{array}{r} -30b + 40c = 110 \\ \underline{30b - 30c = -90} \\ 10c = 20 \\ c = 2 \end{array}$$

Substitute into equation (4):
$$\begin{array}{r} -6b + 8c = 22 \\ -6b + 8(2) = 22 \\ -6b + 16 = 22 \\ -6b = 6 \\ b = -1 \end{array}$$

Substitute into equation (1):
$$\begin{array}{r} 2a + b - 3c = -6 \\ 2a + (-1) - 3(2) = -6 \\ 2a - 1 - 6 = -6 \\ 2a - 7 = -6 \\ 2a = 1 \\ a = \frac{1}{2} \end{array}$$

Solution: $a = \frac{1}{2}$, $b = -1$, $c = 2$

CHECK:

$$2a + b - 3c = -6$$
$$2\left(\frac{1}{2}\right) + (-1) - 3(2) \stackrel{?}{=} -6$$
$$1 - 1 - 6 \stackrel{?}{=} -6$$
$$0 - 6 \stackrel{?}{=} -6$$
$$-6 \stackrel{\checkmark}{=} -6$$

$$4a - 4b + 2c = 10$$
$$4\left(\frac{1}{2}\right) - 4(-1) + 2(2) \stackrel{?}{=} 10$$
$$2 + 4 + 4 \stackrel{?}{=} 10$$
$$6 + 4 \stackrel{?}{=} 10$$
$$10 \stackrel{\checkmark}{=} 10$$

$$6a - 7b + c = 12$$
$$6\left(\frac{1}{2}\right) - 7(-1) + 2 \stackrel{?}{=} 12$$
$$3 + 7 + 2 \stackrel{?}{=} 12$$
$$10 + 2 \stackrel{?}{=} 12$$
$$12 \stackrel{\checkmark}{=} 12$$

15. $\begin{cases} \dfrac{1}{2}s + \dfrac{1}{3}t + u = 3 \\ \dfrac{1}{3}s - \dfrac{1}{2}t - 2u = 1 \\ \dfrac{2}{3}s - \dfrac{1}{6}t + \dfrac{1}{2}u = 6 \end{cases}$

Multiply each equation by 6 in order to clear fractions:
$3s + 2t + 6u = 18$
$2s - 3t - 12u = 6$
$4s - t + 3u = 36$

Multiply equation (3) by 4 and add the resulting equation to equation (2), obtaining equation (4):
$2s - 3t - 12u = 6$
$\underline{16s - 4t + 12u = 144}$
$18s - 7t = 150$

Multiply equation (1) by 2 and add the resulting equation to equation (2), obtaining equation (5):
$6s + 4t + 12u = 36$
$\underline{2s - 3t - 12u = 6}$
$8s + t = 42$

Multiply equation (5) by 7 and add the resulting equation to equation (4):
$18s - 7t = 150$
$\underline{56s + 7t = 294}$
$74s = 444$
$s = 6$

Substitute into equation (5):
$8s + t = 42$
$8(6) + t = 42$
$48 + t = 42$
$t = -6$

Substitute into equation (3):

$4s - t + 3u = 36$
$4(6) - (-6) + 3u = 36$
$24 + 6 + 3u = 36$
$30 + 3u = 36$
$3u = 6$
$u = 2$

Solution: $s = 6$, $t = -6$, $u = 2$

CHECK:

$\dfrac{1}{2}s + \dfrac{1}{3}t + u = 3$ $\qquad$ $\dfrac{1}{3}s - \dfrac{1}{2}t - 2u = 1$ $\qquad$ $\dfrac{2}{3}s - \dfrac{1}{6}t + \dfrac{1}{2}u = 6$

$\dfrac{1}{2}(6) + \dfrac{1}{3}(-6) + 2 \overset{?}{=} 3$ $\qquad$ $\dfrac{1}{3}(6) - \dfrac{1}{2}(-6) - 2(2) \overset{?}{=} 1$ $\qquad$ $\dfrac{2}{3}(6) - \dfrac{1}{6}(-6) + \dfrac{1}{2}(2) \overset{?}{=} 6$

$3 - 2 + 2 \overset{?}{=} 3$ $\qquad\qquad$ $2 + 3 - 4 \overset{?}{=} 1$ $\qquad\qquad$ $4 + 1 + 1 \overset{?}{=} 6$

$1 + 2 \overset{?}{=} 3$ $\qquad\qquad\qquad$ $5 - 4 \overset{?}{=} 1$ $\qquad\qquad\qquad$ $5 + 1 \overset{?}{=} 6$

$3 \overset{\checkmark}{=} 3$ $\qquad\qquad\qquad$ $1 \overset{\checkmark}{=} 1$ $\qquad\qquad\qquad$ $6 \overset{\checkmark}{=} 6$

19. $\begin{cases} x + y = 0 \\ y + z = 0 \\ x + z = 2 \end{cases}$

Multiply equation (2) by -1 and add the resulting equation to equation (1), obtaining equation (4):

$$\begin{array}{l} x + y \quad\;\; = 0 \\ \underline{\;\; -y - z = 0} \\ x \qquad - z = 0 \end{array}$$

Add equations (3) and (4):

$$\begin{array}{l} x + z = 2 \\ \underline{x - z = 0} \\ \;\, 2x = 2 \\ \quad\;\, x = 1 \end{array}$$

Substitute into equation (4):

$$\begin{array}{l} x - z = 0 \\ 1 - z = 0 \\ \quad 1 = z \end{array}$$

Substitute into equation (1):

$$\begin{array}{l} x + y = 0 \\ 1 + y = 0 \\ \quad\; y = -1 \end{array}$$

Solution: $x = 1$, $y = -1$, $z = 1$

CHECK:

$x + y = 0$	$y + z = 0$	$x + z = 2$
$1 + (-1) \overset{?}{=} 0$	$-1 + 1 \overset{?}{=} 0$	$1 + 1 \overset{?}{=} 0$
$0 \overset{\checkmark}{=} 0$	$0 \overset{\checkmark}{=} 0$	$2 \overset{\checkmark}{=} 2$

23. $\begin{cases} 12a + 5b + 3c = 24,000 \\ 10a + 6b + 4c = 13,300 \\ 8a + 7b + 5c = 8,700 \end{cases}$

Multiply equation (1) by 4, equation (2) by -3, and add the resulting equations, obtaining equation (4):

$$\begin{array}{l} 48a + 20b + 12c = \quad 96,000 \\ \underline{-30a - 18b - 12c = -39,900} \\ \quad\; 18a + 2b = \quad 56,100 \end{array}$$

Multiply equation (1) by 5, equation (3) by -3, and add the resulting equations, obtaining equation (5):

$$\begin{array}{l} 60a + 25b + 15c = 120,000 \\ \underline{-24a - 21b - 15c = -26,100} \\ \quad\; 36a + 4b = \quad 93,900 \end{array}$$

Multiply equation (4) by -2 and add the resulting equation to equation (5):

$$\begin{array}{l} -36a - 4b = -112,200 \\ \underline{\;\; 36a + 4b = \quad 93,900} \\ \qquad\; 0 = -18,300, \text{ a contradiction} \end{array}$$

Thus, the original system is inconsistent. That is, it has no solutions.

27. Let d = # of dimes; q = # of quarters; h = # of half-dollars.

$d + q + h = 48$

$10d + 25q + 50h = 1,055$

$\qquad d = q + h - 2 \rightarrow d - q - h = -2$

Add equations (1) and (3):
$d + q + h = 48$
$\underline{d - q - h = -2}$
$\qquad 2d = 46$
$\qquad d = 23$

Substitute this value of d into both equations (1) and (2):

$23 + q + h = 48 \rightarrow \qquad\qquad\qquad q + h = 25$

$10(23) + 25q + 50h = 1,055 \rightarrow \qquad 25q + 50h = 825$

Multiply the first of these equations by −25 and add the resulting equation to the second:
$-25q - 25h = -625$
$\underline{25q + 50h = 825}$
$\qquad 25h = 200$
$\qquad h = 8$

Substitute: $q + h = 25$
$\qquad\quad q + 8 = 25$
$\qquad\qquad q = 17$

Conclude: The collection consists of 23 dimes, 17 quarters, and 8 half-dollars.

CHECK:

# of coins $= 23 + 17 + 8 \overset{\checkmark}{=} 48$

Value of coins $= 10(23) + 25(17) + 50(8) = 230 + 425 + 400$

$\qquad\qquad \overset{\checkmark}{=} 1,055$ (cents) or $10.55

# of quarters and half dollars combined $= 17 + 8 = 25$, and

# of dimes $= 23$, which is 2 fewer than 25.

31. Let x = # of orchestra tickets; y = # of mezzanine tickets; z = # of balcony tickets.

$x + y + z = 750$

$12x + 8y + 6z = 7,290$

$\qquad x = y + z + 100 \rightarrow x - y - z = 100$

Add equations (1) and (3):
$x + y + z = 750$
$\underline{x - y - z = 100}$
$\qquad 2x = 850$
$\qquad x = 425$

Substitute this value of x into both equations (1) and (2):
$425 + y + z = 750 \rightarrow y + z = 325$
$12(425) + 8y + 6z = 7,290 \rightarrow 8y + 6z = 2,190$

Multiply the first of these equations by –6 and add the resulting equation to the second:

$$-6y - 6z = -1,950$$
$$\underline{8y + 6z = 2,190}$$
$$2y = 240$$
$$y = 120$$

Substitute: $y + z = 325$
$120 + z = 325$
$z = 205$

Conclude: There are 425 orchestra tickets, 120 mezzanine tickets, and 205 balcony tickets.

CHECK:

# of tickets = $425 + 120 + 205 \overset{\checkmark}{=} 750$

Value of tickets = $12(425) + 8(120) + 6(205) = 5,100 + 960 + 1,230 \overset{\checkmark}{=} \$7,290$

# of mezzanine and balcony tickets combined = $120 + 205 = 325$,

and the # of orchestra seats = 425, which is 100 more than 325.

35. To find the x-intercept, set y = 0:
$$3x - 2y = 12$$
$$3x - 2(0) = 12$$
$$3x - 0 = 12$$
$$3x = 12$$
$$x = 4$$

Hence, the graph crosses the x-axis at (4,0).

To find the y-intercept, set x = 0:
$$3x - 2y = 12$$
$$3(0) - 2y = 12$$
$$0 - 2y = 12$$
$$-2y = 12$$
$$y = -6$$

Hence, the graph crosses the y-axis at (0,–6). To find a check point, choose x = 2:
$$3x - 2y = 12$$
$$3(2) - 2y = 12$$
$$6 - 2y = 12$$
$$-2y = 6$$
$$y = -3$$

Hence, the graph passes through (2,–3).

37.
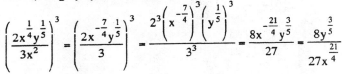
$$\left(\frac{2x^{\frac{1}{4}}y^{\frac{1}{5}}}{3x^2}\right)^3 = \left(\frac{2x^{-\frac{7}{4}}y^{\frac{1}{5}}}{3}\right)^3 = \frac{2^3\left(x^{-\frac{7}{4}}\right)^3\left(y^{\frac{1}{5}}\right)^3}{3^3} = \frac{8x^{-\frac{21}{4}}y^{\frac{3}{5}}}{27} = \frac{8y^{\frac{3}{5}}}{27x^{\frac{21}{4}}}$$

39. Let x = number of quarts of pure water to be added.

$$0x + 0.60(5) = 0.40(x + 5)$$
$$0.60(5) = 0.40(x + 5)$$
$$6(5) = 4(x + 5)$$
$$30 = 4x + 20$$
$$10 = 4x$$
$$\frac{5}{2} = x$$

Thus, $\frac{5}{2}$ or $2\frac{1}{2}$ quarts of pure water must be added.

Exercises 10.2

1. $\begin{bmatrix} 3 & -2 & | & 5 \\ 1 & -1 & | & 8 \end{bmatrix}$

3. $\begin{bmatrix} 1 & -2 & 3 & | & 4 \\ 0 & 1 & -1 & | & -3 \\ 2 & 3 & 0 & | & 8 \end{bmatrix}$

7. $\begin{bmatrix} 1 & -3 & | & 6 \\ 3 & 5 & | & -10 \end{bmatrix}$ $\qquad -3R_1 + R_2 \rightarrow R_2 \qquad \begin{bmatrix} 1 & -3 & | & 6 \\ 0 & 14 & | & -28 \end{bmatrix}$

$$\begin{cases} x - 3y = 6 \\ 14y = -28 \end{cases}$$

First solve for y:
14y = −28. Divide both sides of the equation by 14.
  y = −2

Then substitute −2 for y in the first equation and find x:
$$x - 3y = 6$$
$$x - 3(-2) = 6$$
$$x + 6 = 6$$
$$x = 0$$

Hence, the solution is (0,−2).

11. $\begin{bmatrix} 6 & 2 & | & 9 \\ 4 & -1 & | & -1 \end{bmatrix}$ $\qquad \begin{matrix} 2R_1 \rightarrow R_1 \\ -3R_2 \rightarrow R_2 \end{matrix} \qquad \begin{bmatrix} 12 & 4 & | & 18 \\ -12 & 3 & | & 3 \end{bmatrix}$

$\qquad\qquad\qquad R_1 + R_2 \rightarrow R_2 \qquad \begin{bmatrix} 12 & 4 & | & 18 \\ 0 & 7 & | & 21 \end{bmatrix}$

$$\begin{cases} 12x + 4y = 18 \\ 7y = 21 \end{cases}$$

First solve for y:
7y = 21. Divide both sides of the equation by 7.
  y = 3

Then substitute 3 for y in the first equation and find x:

$12x + 4y = 18$

$12x + 4(3) = 18$

$12x + 12 = 18$

$12x = 6$

$x = \dfrac{1}{2}$

Hence, the solution is $\left(\dfrac{1}{2}, 3\right)$.

15. $\begin{bmatrix} 1 & 3 & 1 & | & 8 \\ 1 & 2 & 1 & | & 7 \\ 1 & -2 & 2 & | & 6 \end{bmatrix}$ $\quad \begin{array}{l} -R_1 + R_2 \rightarrow R_2 \\ -R_1 + R_3 \rightarrow R_3 \end{array}$ $\quad \begin{bmatrix} 1 & 3 & 1 & | & 8 \\ 0 & -1 & 0 & | & -1 \\ 0 & -5 & 1 & | & -2 \end{bmatrix}$

$\quad -5R_2 + R_3 \rightarrow R_3 \quad \begin{bmatrix} 1 & 3 & 1 & | & 8 \\ 0 & -1 & 0 & | & -1 \\ 0 & 0 & 1 & | & 3 \end{bmatrix}$

$\begin{cases} x + 3y + z = 8 \\ \quad - y \quad\quad = -1 \\ \quad\quad\quad z = 3 \end{cases}$

The third equation gives z = 3. Multiply the second equation by –1 to find y = 1. Then substitute 3 for z and 1 for y in the first equation and find x:

$x + 3y + z = 8$

$x + 3(1) + (3) = 8$

$x + 3 + 3 = 8$

$x + 6 = 8$

$x = 2$

Hence, the solution is (2,1,3).

21. $\begin{bmatrix} 4 & -1 & 2 & | & 6 \\ 2 & 3 & -1 & | & 4 \\ 2 & -2 & 1 & | & 0 \end{bmatrix}$ $\quad R_1 \leftrightarrow R_3 \quad \begin{bmatrix} 2 & -2 & 1 & | & 0 \\ 2 & 3 & -1 & | & 4 \\ 4 & -1 & 2 & | & 6 \end{bmatrix}$

$\quad \begin{array}{l} -R_1 + R_2 \rightarrow R_2 \\ -2R_1 + R_3 \rightarrow R_3 \end{array}$ $\quad \begin{bmatrix} 2 & -2 & 1 & | & 0 \\ 0 & 5 & -2 & | & 4 \\ 0 & 3 & 0 & | & 6 \end{bmatrix}$

$\quad R_2 \leftrightarrow R_3 \quad \begin{bmatrix} 2 & -2 & 1 & | & 0 \\ 0 & 3 & 0 & | & 6 \\ 0 & 5 & -2 & | & 4 \end{bmatrix}$

$\quad \dfrac{1}{3}R_2 \rightarrow R_2 \quad \begin{bmatrix} 2 & -2 & 1 & | & 0 \\ 0 & 1 & 0 & | & 2 \\ 0 & 5 & -2 & | & 4 \end{bmatrix}$

$\quad -5R_2 + R_3 \rightarrow R_3 \quad \begin{bmatrix} 2 & -2 & 1 & | & 0 \\ 0 & 1 & 0 & | & 2 \\ 0 & 0 & -2 & | & -6 \end{bmatrix}$

$$\begin{cases} 2x - 2y + z = 0 \\ \qquad y \quad = 2 \\ \qquad\quad -2z = -6 \end{cases}$$

First solve the third equation for z:

$-2z = -6$. Divide both sides of the equation by -2.

$z = 3$

The second equation gives $y = 2$. Then substitute 3 for z and 2 for y in the first equation and find x:

$$2x - 2y + z = 0$$
$$2x - 2(2) + (3) = 0$$
$$2x - 4 + 3 = 0$$
$$2x - 1 = 0$$
$$2x = 1$$
$$x = \frac{1}{2}$$

Hence, the solution is $\left(\dfrac{1}{2}, 2, 3\right)$.

25.
$$\begin{bmatrix} 1 & -2 & 1 & -1 & 2 \\ 1 & 2 & 2 & 1 & 0 \\ 2 & -2 & 1 & -1 & 3 \\ 2 & 0 & -2 & 1 & 5 \end{bmatrix}$$

$-R_1 + R_2 \rightarrow R_2$
$-2R_1 + R_3 \rightarrow R_3$
$-2R_1 + R_4 \rightarrow R_4$

$$\begin{bmatrix} 1 & -2 & 1 & -1 & 2 \\ 0 & 4 & 1 & 2 & -2 \\ 0 & 2 & -1 & 1 & -1 \\ 0 & 4 & -4 & 3 & 1 \end{bmatrix}$$

$R_2 \leftrightarrow R_3$

$$\begin{bmatrix} 1 & -2 & 1 & -1 & 2 \\ 0 & 2 & -1 & 1 & -1 \\ 0 & 4 & 1 & 2 & -2 \\ 0 & 4 & -4 & 3 & 1 \end{bmatrix}$$

$-2R_2 + R_3 \rightarrow R_3$
$-2R_2 + R_4 \rightarrow R_4$

$$\begin{bmatrix} 1 & -2 & 1 & -1 & 2 \\ 0 & 2 & -1 & 1 & -1 \\ 0 & 0 & 3 & 0 & 0 \\ 0 & 0 & -2 & 1 & 3 \end{bmatrix}$$

$\dfrac{1}{3} R_3 \rightarrow R_3$

$$\begin{bmatrix} 1 & -2 & 1 & -1 & 2 \\ 0 & 2 & -1 & 1 & -1 \\ 0 & 0 & 1 & 0 & 0 \\ 0 & 0 & -2 & 1 & 3 \end{bmatrix}$$

$2R_3 + R_4 \rightarrow R_4$

$$\begin{bmatrix} 1 & -2 & 1 & -1 & 2 \\ 0 & 2 & -1 & 1 & -1 \\ 0 & 0 & 1 & 0 & 0 \\ 0 & 0 & 0 & 1 & 3 \end{bmatrix}$$

$$\begin{cases} w - 2x + y - z = 2 \\ \quad 2x - y + z = -1 \\ \qquad\quad y \quad = 0 \\ \qquad\qquad z = 3 \end{cases}$$

The third and fourth equations give y = 0 and z = 3. Substitute 3 for z and 0 for y in the second equation and find x:

$$2x - y + z = -1$$
$$2x - (0) + (3) = -1$$
$$2x - 0 + 3 = -1$$
$$2x + 3 = -1$$
$$2x = -4$$
$$x = -2$$

Finally, substitute 3 for z, 0 for y, and –2 for x in the first equation and find w:

$$w - 2x + y - z = 2$$
$$w - 2(-2) + (0) + (3) = 2$$
$$w + 4 + 0 - 3 = 2$$
$$w + 1 = 2$$
$$w = 1$$

Hence, the solution is (1, –2, 0, 3).

27. $\begin{bmatrix} 2 & 3 & | & 5 \\ 1 & 4 & | & 0 \end{bmatrix}$

$R_1 \leftrightarrow R_2$ $\begin{bmatrix} 1 & 4 & | & 0 \\ 2 & 3 & | & 5 \end{bmatrix}$

$-2R_1 + R_2 \rightarrow R_2$ $\begin{bmatrix} 1 & 4 & | & 0 \\ 0 & -5 & | & 5 \end{bmatrix}$

$-\dfrac{1}{5}R_2 \rightarrow R_2$ $\begin{bmatrix} 1 & 4 & | & 0 \\ 0 & 1 & | & -1 \end{bmatrix}$

$-4R_2 + R_1 \rightarrow R_1$ $\begin{bmatrix} 1 & 0 & | & 4 \\ 0 & 1 & | & -1 \end{bmatrix}$

29. $\begin{bmatrix} 1 & 1 & 2 & | & 1 \\ 2 & 4 & 2 & | & 6 \\ 3 & 1 & 2 & | & 5 \end{bmatrix}$

$\begin{matrix} -2R_1 + R_2 \rightarrow R_2 \\ -3R_1 + R_3 \rightarrow R_3 \end{matrix}$ $\begin{bmatrix} 1 & 1 & 2 & | & 1 \\ 0 & 2 & -2 & | & 4 \\ 0 & -2 & -4 & | & 2 \end{bmatrix}$

$\dfrac{1}{2}R_2 \rightarrow R_2$ $\begin{bmatrix} 1 & 1 & 2 & | & 1 \\ 0 & 1 & -1 & | & 2 \\ 0 & -2 & -4 & | & 2 \end{bmatrix}$

$\begin{matrix} -R_2 + R_1 \rightarrow R_1 \\ 2R_2 + R_3 \rightarrow R_3 \end{matrix}$ $\begin{bmatrix} 1 & 0 & 3 & | & -1 \\ 0 & 1 & -1 & | & 2 \\ 0 & 0 & -6 & | & 6 \end{bmatrix}$

$-\dfrac{1}{6}R_3 \rightarrow R_3$ $\begin{bmatrix} 1 & 0 & 3 & | & -1 \\ 0 & 1 & -1 & | & 2 \\ 0 & 0 & 1 & | & -1 \end{bmatrix}$

$\begin{matrix} -3R_3 + R_1 \rightarrow R_1 \\ R_3 + R_2 \rightarrow R_2 \end{matrix}$ $\begin{bmatrix} 1 & 0 & 0 & | & 2 \\ 0 & 1 & 0 & | & 1 \\ 0 & 0 & 1 & | & -1 \end{bmatrix}$

33. $\begin{bmatrix} 1 & 2 & 1 & | & 8 \\ 1 & 4 & -1 & | & 12 \\ 1 & -2 & 1 & | & -4 \end{bmatrix}$

$\begin{matrix} -R_1 + R_2 \rightarrow R_2 \\ -R_1 + R_3 \rightarrow R_3 \end{matrix}$ $\begin{bmatrix} 1 & 2 & 1 & | & 8 \\ 0 & 2 & -2 & | & 4 \\ 0 & -4 & 0 & | & -12 \end{bmatrix}$

$$R_2 \leftrightarrow R_3 \qquad \begin{bmatrix} 1 & 2 & 1 & | & 8 \\ 0 & -4 & 0 & | & -12 \\ 0 & 2 & -2 & | & 4 \end{bmatrix}$$

$$-\frac{1}{4}R_2 \rightarrow R_2 \qquad \begin{bmatrix} 1 & 2 & 1 & | & 8 \\ 0 & 1 & 0 & | & 3 \\ 0 & 2 & -2 & | & 4 \end{bmatrix}$$

$$\begin{matrix} -2R_2 + R_1 \rightarrow R_1 \\ -2R_2 + R_3 \rightarrow R_3 \end{matrix} \qquad \begin{bmatrix} 1 & 0 & 1 & | & 2 \\ 0 & 1 & 0 & | & 3 \\ 0 & 0 & -2 & | & -2 \end{bmatrix}$$

$$-\frac{1}{2}R_3 \rightarrow R_3 \qquad \begin{bmatrix} 1 & 0 & 1 & | & 2 \\ 0 & 1 & 0 & | & 3 \\ 0 & 0 & 1 & | & 1 \end{bmatrix}$$

$$-R_3 + R_1 \rightarrow R_1 \qquad \begin{bmatrix} 1 & 0 & 0 & | & 1 \\ 0 & 1 & 0 & | & 3 \\ 0 & 0 & 1 & | & 1 \end{bmatrix}$$

So $x = 1$, $y = 3$, $z = 1$.

35. $\begin{bmatrix} 1 & -2 & 0 & | & -4 \\ 0 & 2 & 2 & | & 4 \\ 1 & 0 & 1 & | & 1 \end{bmatrix}$ $\qquad -R_1 + R_3 \rightarrow R_3 \qquad \begin{bmatrix} 1 & -2 & 0 & | & -4 \\ 0 & 2 & 2 & | & 4 \\ 0 & 2 & 1 & | & 5 \end{bmatrix}$

$$\frac{1}{2}R_2 \rightarrow R_2 \qquad \begin{bmatrix} 1 & -2 & 0 & | & -4 \\ 0 & 1 & 1 & | & 2 \\ 0 & 2 & 1 & | & 5 \end{bmatrix}$$

$$\begin{matrix} 2R_2 + R_1 \rightarrow R_1 \\ -2R_2 + R_3 \rightarrow R_3 \end{matrix} \qquad \begin{bmatrix} 1 & 0 & 2 & | & 0 \\ 0 & 1 & 1 & | & 2 \\ 0 & 0 & -1 & | & 1 \end{bmatrix}$$

$$-R_3 \rightarrow R_3 \qquad \begin{bmatrix} 1 & 0 & 2 & | & 0 \\ 0 & 1 & 1 & | & 2 \\ 0 & 0 & 1 & | & -1 \end{bmatrix}$$

$$\begin{matrix} -2R_3 + R_1 \rightarrow R_1 \\ -R_3 + R_2 \rightarrow R_2 \end{matrix} \qquad \begin{bmatrix} 1 & 0 & 0 & | & 2 \\ 0 & 1 & 0 & | & 3 \\ 0 & 0 & 1 & | & -1 \end{bmatrix}$$

So $x = 2$, $y = 3$, $z = -1$.

Exercises 10.3

1. $\begin{vmatrix} 1 & 2 \\ 3 & 4 \end{vmatrix} = 1(4) - 2(3) = 4 - 6 = -2$

5. $\begin{vmatrix} -3 & -1 \\ 2 & -2 \end{vmatrix} = -3(-2) - (-1)(2) = 6 + 2 = 8$

9. $\begin{vmatrix} 4 & 6 \\ 6 & 9 \end{vmatrix} = 4(9) - 6(6) = 36 - 36 = 0$

13. $\begin{vmatrix} 1 & 2 & 3 \\ 2 & 4 & 6 \\ 1 & 1 & 1 \end{vmatrix} = 1 \cdot \begin{vmatrix} 4 & 6 \\ 1 & 1 \end{vmatrix} - 2 \cdot \begin{vmatrix} 2 & 6 \\ 1 & 1 \end{vmatrix} + 3 \cdot \begin{vmatrix} 2 & 4 \\ 1 & 1 \end{vmatrix}$

$\qquad = 1(4 - 6) - 2(2 - 6) + 3(2 - 4) = 1(-2) - 2(-4) + 3(-2)$

$\qquad = -2 + 8 - 6 = 0$

15. $\begin{vmatrix} -3 & 0 & 4 \\ 5 & 2 & -3 \\ 7 & 0 & 6 \end{vmatrix} = -0 \cdot \begin{vmatrix} 5 & -3 \\ 7 & 6 \end{vmatrix} + 2 \cdot \begin{vmatrix} -3 & 4 \\ 7 & 6 \end{vmatrix} - 0 \cdot \begin{vmatrix} -3 & 4 \\ 5 & -3 \end{vmatrix}$ (expansion down second column)

$\qquad = 0 + 2(-18 - 28) + 0 = 0 + 2(-46) + 0 = -92$

19. $\begin{vmatrix} 2x & 4 \\ 3 & 5 \end{vmatrix} = 18$

$$2x(5) - 4(3) = 18$$
$$10x - 12 = 18$$
$$10x = 30$$
$$x = 3$$

23. $\begin{vmatrix} x & 3 \\ 2 & x+2 \end{vmatrix} = 9$

$$x(x + 2) - 3(2) = 9$$
$$x^2 + 2x - 6 = 9$$
$$x^2 + 2x - 15 = 0$$
$$(x + 5)(x - 3) = 0$$
$$x + 5 = 0 \text{ or } x - 3 = 0$$
$$x = -5 \text{ or } x = 3$$

27. $\begin{cases} 2x - 4y = 5 \\ -3x + 6y = 7 \end{cases}$

$D = \begin{vmatrix} 2 & -4 \\ -3 & 6 \end{vmatrix} = 2(6) - (-4)(-3) = 12 - 12 = 0$

This means that the system does not have a unique solution.

Since $D_x = \begin{vmatrix} 5 & -4 \\ 7 & 6 \end{vmatrix} = 5(6) - (-4)(7) = 30 + 28 = 58 \neq 0$, the system is inconsistent and has no solutions.

31. $\begin{cases} 2x - 7y = -4 \to 2x - 7y = -4 \\ 3y - 4x = 8 \to -4x + 3y = 8 \end{cases}$

$D = \begin{vmatrix} 2 & -7 \\ -4 & 3 \end{vmatrix} = 2(3) - (-7)(-4) = 6 - 28 = -22$

$D_x = \begin{vmatrix} -4 & -7 \\ 8 & 3 \end{vmatrix} = -4(3) - (-7)(8) = -12 + 56 = 44$

$D_y = \begin{vmatrix} 2 & -4 \\ -4 & 8 \end{vmatrix} = 2(8) - (-4)(-4) = 16 - 16 = 0$

Then $x = \dfrac{D_x}{D} = \dfrac{44}{-22} = -2$ and $y = \dfrac{D_y}{D} = \dfrac{0}{-22} = 0$

CHECK: $2x - 7y = -4$ $3y - 4x = 8$

$2(-2) - 7(0) \overset{?}{=} -4$ $3(0) - 4(-2) \overset{?}{=} 8$

$-4 - 0 \overset{?}{=} -4$ $0 + 8 \overset{?}{=} 8$

$-4 \overset{\checkmark}{=} -4$ $8 \overset{\checkmark}{=} 8$

35. $\begin{cases} 2s - 9t = 4 \\ 3s + 5t = 6 \end{cases}$

$D = \begin{vmatrix} 2 & -9 \\ 3 & 5 \end{vmatrix} = 2(5) - (-9)(3) = 10 + 27 = 37$

$D_s = \begin{vmatrix} 4 & -9 \\ 6 & 5 \end{vmatrix} = 4(5) - (-9)(6) = 20 + 54 = 74$

$D_t = \begin{vmatrix} 2 & 4 \\ 3 & 6 \end{vmatrix} = 2(6) - 4(3) = 12 - 12 = 0$

Then $s = \dfrac{D_s}{D} = \dfrac{74}{37} = 2$ and $t = \dfrac{D_t}{D} = \dfrac{0}{37} = 0$

CHECK: $2s - 9t = 4$ $3s + 5t = 6$

$2(2) - 9(0) \overset{?}{=} 4$ $3(2) + 5(0) \overset{?}{=} 6$

$4 - 0 \overset{?}{=} 4$ $6 + 0 \overset{?}{=} 6$

$4 \overset{\checkmark}{=} 4$ $6 \overset{\checkmark}{=} 6$

39. $\begin{cases} \dfrac{1}{2}x - 12y = 6 \\ \dfrac{1}{3}x - 8y = 6 \end{cases}$

$D = \begin{vmatrix} \dfrac{1}{2} & -12 \\ \dfrac{1}{3} & -8 \end{vmatrix} = \dfrac{1}{2}(-8) - (-12)\left(\dfrac{1}{3}\right) = -4 + 4 = 0$

This means that the system does not have a unique solution.

Since $D_x = \begin{vmatrix} 6 & -12 \\ 6 & -8 \end{vmatrix} = 6(-8) - (-12)(6) = -48 + 72 = 24 \neq 0$, the system is inconsistent and has no solutions.

43. $\begin{cases} 3x + 4y + 2z = 1 \\ 2x + 3y + z = 1 \\ 6x + y + 5z = 1 \end{cases}$

$$D = \begin{vmatrix} 3 & 4 & 2 \\ 2 & 3 & 1 \\ 6 & 1 & 5 \end{vmatrix} = 3\begin{vmatrix} 3 & 1 \\ 1 & 5 \end{vmatrix} - 4\begin{vmatrix} 2 & 1 \\ 6 & 5 \end{vmatrix} + 2\begin{vmatrix} 2 & 3 \\ 6 & 1 \end{vmatrix}$$

$$= 3(15-1) - 4(10-6) + 2(2-18)$$

$$= 3(14) - 4(4) + 2(-16) = 42 - 16 - 32 = -6$$

$$D_x = \begin{vmatrix} 1 & 4 & 2 \\ 1 & 3 & 1 \\ 1 & 1 & 5 \end{vmatrix} = 1\begin{vmatrix} 3 & 1 \\ 1 & 5 \end{vmatrix} - 4\begin{vmatrix} 1 & 1 \\ 1 & 5 \end{vmatrix} + 2\begin{vmatrix} 1 & 3 \\ 1 & 1 \end{vmatrix}$$

$$= 1(15-1) - 4(5-1) + 2(1-3)$$

$$= 1(14) - 4(4) + 2(-2) = 14 - 16 - 4 = -6$$

$$D_y = \begin{vmatrix} 3 & 1 & 2 \\ 2 & 1 & 1 \\ 6 & 1 & 5 \end{vmatrix} = 3\begin{vmatrix} 1 & 1 \\ 1 & 5 \end{vmatrix} - 1\begin{vmatrix} 2 & 1 \\ 6 & 5 \end{vmatrix} + 2\begin{vmatrix} 2 & 1 \\ 6 & 1 \end{vmatrix}$$

$$= 3(5-1) - 1(10-6) + 2(2-6)$$

$$= 3(4) - 1(4) + 2(-4) = 12 - 4 - 8 = 0$$

$$D_z = \begin{vmatrix} 3 & 4 & 1 \\ 2 & 3 & 1 \\ 6 & 1 & 1 \end{vmatrix} = 3\begin{vmatrix} 3 & 1 \\ 1 & 1 \end{vmatrix} - 4\begin{vmatrix} 2 & 1 \\ 6 & 1 \end{vmatrix} + 1\begin{vmatrix} 2 & 3 \\ 6 & 1 \end{vmatrix}$$

$$= 3(3-1) - 4(2-6) + 1(2-18)$$

$$= 3(2) - 4(-4) + 1(-16) = 6 + 16 - 16 = 6$$

Then $x = \dfrac{D_x}{D} = \dfrac{-6}{-6} = 1$, $y = \dfrac{D_y}{D} = \dfrac{0}{-6} = 0$, and $z = \dfrac{D_z}{D} = \dfrac{6}{-6} = -1$

CHECK: $3x + 4y + 2z = 1$ $\qquad\qquad\qquad$ $2x + 3y + z = 1$

$3(1) + 4(0) + 2(-1) \overset{?}{=} 1$ $\qquad\qquad$ $2(1) + 3(0) + (-1) \overset{?}{=} 1$

$3 + 0 - 2 \overset{?}{=} 1$ $\qquad\qquad\qquad\qquad$ $2 + 0 - 1 \overset{?}{=} 1$

$1 \overset{\checkmark}{=} 1$ $\qquad\qquad\qquad\qquad\qquad\qquad$ $1 \overset{\checkmark}{=} 1$

$6x + y + 5z = 1$

$6(1) + (0) + 5(-1) \overset{?}{=} 1$

$6 + 0 - 5 \overset{?}{=} 1$

$1 \overset{\checkmark}{=} 1$

47. $\begin{cases} 3x - y = 4 \\ 2y + z = 6 \\ 3x + 4z = 14 \end{cases}$

$$D = \begin{vmatrix} 3 & -1 & 0 \\ 0 & 2 & 1 \\ 3 & 0 & 4 \end{vmatrix} = 3\begin{vmatrix} 2 & 1 \\ 0 & 4 \end{vmatrix} - (-1)\begin{vmatrix} 0 & 1 \\ 3 & 4 \end{vmatrix} + 0\begin{vmatrix} 0 & 2 \\ 3 & 0 \end{vmatrix}$$

$$= 3(8-0) + 1(0-3) + 0(0-6)$$

$$= 3(8) + 1(-3) + 0 = 24 - 3 = 21$$

$$D_x = \begin{vmatrix} 4 & -1 & 0 \\ 6 & 2 & 1 \\ 14 & 0 & 4 \end{vmatrix} = 4\begin{vmatrix} 2 & 1 \\ 0 & 4 \end{vmatrix} - (-1)\begin{vmatrix} 6 & 1 \\ 14 & 4 \end{vmatrix} + 0\begin{vmatrix} 6 & 2 \\ 14 & 0 \end{vmatrix}$$

$$= 4(8-0) + 1(24-14) + 0(0-28)$$

$$= 4(8) + 1(10) + 0 = 32 + 10 = 42$$

$$D_y = \begin{vmatrix} 3 & 4 & 0 \\ 0 & 6 & 1 \\ 3 & 14 & 4 \end{vmatrix} = 3\begin{vmatrix} 6 & 1 \\ 14 & 4 \end{vmatrix} - 4\begin{vmatrix} 0 & 1 \\ 3 & 4 \end{vmatrix} + 0\begin{vmatrix} 0 & 6 \\ 3 & 14 \end{vmatrix}$$

$$= 3(24-14) - 4(0-3) + 0(0-18)$$

$$= 3(10) - 4(-3) + 0 = 30 + 12 = 42$$

$$D_z = \begin{vmatrix} 3 & -1 & 4 \\ 0 & 2 & 6 \\ 3 & 0 & 14 \end{vmatrix} = 3\begin{vmatrix} -1 & 4 \\ 2 & 6 \end{vmatrix} - 0\begin{vmatrix} 3 & 4 \\ 0 & 6 \end{vmatrix} + 14\begin{vmatrix} 3 & -1 \\ 0 & 2 \end{vmatrix} \quad \text{(expansion across third row)}$$

$$= 3(-6-8) - 0(18-0) + 14(6-0)$$

$$= 3(-14) + 0 + 14(6) = -42 + 84 = 42$$

Then $x = \dfrac{D_x}{D} = \dfrac{42}{21} = 2$, $y = \dfrac{D_y}{D} = \dfrac{42}{21} = 2$, and $z = \dfrac{D_z}{D} = \dfrac{42}{21} = 2$.

CHECK:
$$3x - y = 4 \qquad 2y + z = 6 \qquad 3x + 4z = 14$$
$$3(2) - 2 \overset{?}{=} 4 \qquad 2(2) + 2 \overset{?}{=} 6 \qquad 3(2) + 4(2) \overset{?}{=} 14$$
$$6 - 2 \overset{?}{=} 4 \qquad 4 + 2 \overset{?}{=} 6 \qquad 6 + 8 \overset{?}{=} 14$$
$$4 \overset{\checkmark}{=} 4 \qquad 6 \overset{\checkmark}{=} 6 \qquad 14 \overset{\checkmark}{=} 14$$

51.
$$x = y + z + 2 \rightarrow x - y - z = 2$$
$$y = x - z + 3 \rightarrow -x + y + z = 3$$
$$z = x + y + 4 \rightarrow -x - y + z = 4$$

$$D = \begin{vmatrix} 1 & -1 & -1 \\ -1 & 1 & 1 \\ -1 & -1 & 1 \end{vmatrix} = 1\begin{vmatrix} 1 & 1 \\ -1 & 1 \end{vmatrix} - (-1)\begin{vmatrix} -1 & 1 \\ -1 & 1 \end{vmatrix} + (-1)\begin{vmatrix} -1 & 1 \\ -1 & -1 \end{vmatrix}$$

$$= 1(1-(-1)) + 1(-1+1) - 1(1-(-1))$$

$$= 1(2) + 1(0) - 1(2) = 2 + 0 - 2 = 0$$

This means that the system does not have a unique solution.

$$\text{Since } D_x = \begin{vmatrix} 2 & -1 & -1 \\ 3 & 1 & 1 \\ 4 & -1 & 1 \end{vmatrix} = 2\begin{vmatrix} 1 & 1 \\ -1 & 1 \end{vmatrix} - (-1)\begin{vmatrix} 3 & 1 \\ 4 & 1 \end{vmatrix} + (-1)\begin{vmatrix} 3 & 1 \\ 4 & -1 \end{vmatrix}$$

$$= 2(1-(-1)) + 1(3-4) - 1(-3-4)$$

$$= 2(2) + 1(-1) - 1(-7) = 4 - 1 + 7 = 10 \neq 0,$$

the system is inconsistent and has no solutions.

55. $\begin{vmatrix} 68 & 85 \\ 920 & 743 \end{vmatrix} = (68)(743) - (85)(920) = 50{,}524 - 78{,}200 = -27{,}676$

57. $5x - 3y = -0.181$
$0.2x + 4y = 0.2482$

$D = \begin{vmatrix} 5 & \pm 3 \\ 0.2 & 4 \end{vmatrix} = (5)(4) - (-3)(0.2) = 20 + 0.6 = 20.6$

$D_x = \begin{vmatrix} \pm 0.181 & \pm 3 \\ 0.2482 & 4 \end{vmatrix} = (-0.181)(4) - (-3)(0.2482) = -0.724 + 0.7446 = 0.0206$

$D_y = \begin{vmatrix} 5 & \pm 0.181 \\ 0.2 & 0.2482 \end{vmatrix} = (5)(0.2482) - (-0.181)(0.2) = 1.241 + 0.0362 = 1.2772$

Then $x = \dfrac{D_x}{D} = \dfrac{0.0206}{20.6} = .001$ and $y = \dfrac{D_y}{D} = \dfrac{1.2772}{20.6} = .062$

61. $\begin{vmatrix} a_1 & b_1 & c_1 \\ a_2 & b_2 & c_2 \\ a_2 & b_2 & c_2 \end{vmatrix} = a_1 \begin{vmatrix} b_2 & c_2 \\ b_3 & c_3 \end{vmatrix} - b_1 \begin{vmatrix} a_2 & c_2 \\ a_3 & c_3 \end{vmatrix} + c_1 \begin{vmatrix} a_2 & b_2 \\ a_3 & b_3 \end{vmatrix}$

$= a_1(b_2 c_3 - c_2 b_3) - b_1(a_2 c_3 - c_2 a_3) + c_1(a_2 b_3 - b_2 a_3)$

$= a_1 b_2 c_3 - a_1 b_3 c_2 - a_2 b_1 c_3 + a_3 b_1 c_2 + a_2 b_3 c_1 - a_3 b_2 c_1$

$= a_1(b_2 c_3 - b_3 c_2) - a_2(b_1 c_3 - b_3 c_1) + a_3(b_1 c_2 - b_2 c_1)$

$= a_1 \begin{vmatrix} b_2 & c_2 \\ b_3 & c_3 \end{vmatrix} - a_2 \begin{vmatrix} b_1 & c_1 \\ b_3 & c_3 \end{vmatrix} + a_3 \begin{vmatrix} b_1 & c_1 \\ b_2 & c_2 \end{vmatrix}$

which is the expansion down the first column.

63. $\dfrac{2}{x - 4} = \dfrac{1}{x + 2} + 1$

$(x - 4)(x + 2)\left(\dfrac{2}{x - 4}\right) = (x - 4)(x + 2)\left(\dfrac{1}{x + 2} + 1\right)$

$2(x + 2) = x - 4 + (x - 4)(x + 2)$

$2x + 4 = x - 4 + x^2 - 2x - 8$

$2x + 4 = x^2 - x - 12$

$4 = x^2 - 3x - 12$

$0 = x^2 - 3x - 16 \qquad A = 1,\ B = -3,\ C = -16$

$x = \dfrac{-B \pm \sqrt{B^2 - 4AC}}{2A} = \dfrac{-(-3) \pm \sqrt{(-3)^2 - 4(1)(-16)}}{2(1)} = \dfrac{3 \pm \sqrt{9 + 64}}{2} = \dfrac{3 \pm \sqrt{73}}{2}$

65. boundary: dotted line

x-intercept: 6

y-intercept: −4

test point: (0,0)

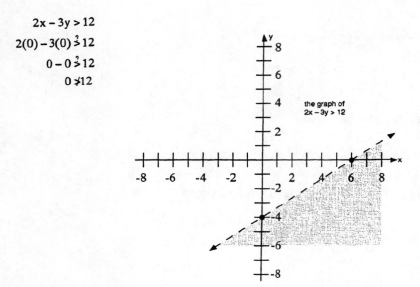

$$2x - 3y > 12$$
$$2(0) - 3(0) \overset{?}{>} 12$$
$$0 - 0 \overset{?}{>} 12$$
$$0 \not> 12$$

the graph of
$2x - 3y > 12$

Exercises 10.4

1. $\begin{cases} x^2 + y^2 = 10 \\ \underline{x^2 - y^2 = 8} \end{cases}$

Add: $2x^2 = 18$
$\quad\quad x^2 = 9$
$\quad\quad\quad x = \pm 3$

CHECK (3,1):

$x^2 + y^2 = 8$
$3^2 + 1^2 \overset{?}{=} 10$
$9 + 1 \overset{?}{=} 10$
$10 \overset{\checkmark}{=} 10$

$x^2 - y^2 = 8x$
$3^2 - 1^2 \overset{?}{=} 8$
$9 - 1 \overset{?}{=} 8$
$8 \overset{\checkmark}{=} 8$

Substitute: $x^2 + y^2 = 10$
$\quad\quad\quad\quad 9 + y^2 = 10$
$\quad\quad\quad\quad\quad y^2 = 1$
$\quad\quad\quad\quad\quad\quad y = \pm 1$

CHECK (3,−1):

$x^2 + y^2 = 10$
$3^2 + (-1)^2 \overset{?}{=} 10$
$9 + 1 \overset{?}{=} 10$
$10 \overset{\checkmark}{=} 10$

$x^2 - y^2 = 8$
$3^2 - (-1)^2 \overset{?}{=} 8$
$9 - 1 \overset{?}{=} 8$
$8 \overset{\checkmark}{=} 8$

Solutions: (3,1), (3,−1), (−3,1), (−3,−1)

CHECK (−3,1):

$x^2 + y^2 = 10$
$(-3)^2 + 1^2 \overset{?}{=} 10$
$9 + 1 \overset{?}{=} 10$
$10 \overset{\checkmark}{=} 10$

$x^2 - y^2 = 8$
$(-3)^2 - 1^2 \overset{?}{=} 8$
$9 - 1 \overset{?}{=} 8$
$8 \overset{\checkmark}{=} 8$

CHECK (–3,–1):

$$x^2 + y^2 = 10 \qquad\qquad x^2 - y^2 = 8$$
$$(-3)^2 + (-1)^2 \overset{?}{=} 10 \qquad (-3)^2 - (-1)^2 \overset{?}{=} 8$$
$$9 + 1 \overset{?}{=} 10 \qquad\qquad 9 - 1 \overset{?}{=} 8$$
$$10 \overset{\checkmark}{=} 10 \qquad\qquad 8 \overset{\checkmark}{=} 8$$

5. $\begin{cases} 16x^2 - 4y^2 = 64 \xrightarrow{\text{as is}} \\ x^2 + y^2 = 9 \xrightarrow{\text{multiply by 4}} \end{cases}$

$$16x^2 - 4y^2 = 64$$
$$\underline{4x^2 + 4y^2 = 36}$$

Add: $20x^2 = 100$

$$x^2 = 5$$
$$x = \pm\sqrt{5}$$

Substitute: $x^2 + y^2 = 9$
$$5 + y^2 = 9$$
$$y^2 = 4$$
$$y = \pm 2$$

CHECK $(\sqrt{5}, 2)$:

$$16x^2 - 4y^2 = 64 \qquad\qquad x^2 + y^2 = 9$$
$$16\left(\sqrt{5}\right)^2 - 4(2)^2 \overset{?}{=} 64 \qquad \left(\sqrt{5}\right)^2 + 2^2 \overset{?}{=} 9$$
$$16(5) - 4(4) \overset{?}{=} 64 \qquad\qquad 5 + 4 \overset{?}{=} 9$$
$$80 - 16 \overset{?}{=} 64 \qquad\qquad 9 \overset{\checkmark}{=} 9$$
$$64 \overset{\checkmark}{=} 64$$

Solutions: $\left(\sqrt{5}, 2\right), \left(\sqrt{5}, -2\right),$
$\left(-\sqrt{5}, 2\right)$ and $\left(-\sqrt{5}, -2\right)$

CHECK $\left(\sqrt{5}, -2\right)$:

$$16x^2 - 4y^2 = 64 \qquad\qquad x^2 + y^2 = 9$$
$$16\left(\sqrt{5}\right)^2 - 4(-2)^2 \overset{?}{=} 64 \qquad \left(\sqrt{5}\right)^2 + (-2)^2 \overset{?}{=} 9$$
$$16(5) - 4(4) \overset{?}{=} 64 \qquad\qquad 5 + 4 \overset{?}{=} 9$$
$$80 - 16 \overset{?}{=} 64 \qquad\qquad 9 \overset{\checkmark}{=} 9$$
$$64 \overset{\checkmark}{=} 64$$

CHECK $\left(-\sqrt{5}, 2\right)$:

$$16x^2 - 4y^2 = 64 \qquad\qquad x^2 + y^2 = 9$$
$$16\left(-\sqrt{5}\right)^2 - 4(2)^2 \overset{?}{=} 64 \qquad \left(-\sqrt{5}\right)^2 + 2^2 \overset{?}{=} 9$$
$$16(5) - 4(4) \overset{?}{=} 64 \qquad\qquad 5 + 4 \overset{?}{=} 9$$
$$80 - 16 \overset{?}{=} 64 \qquad\qquad 9 \overset{\checkmark}{=} 9$$
$$64 \overset{\checkmark}{=} 64$$

CHECK $\left(-\sqrt{5}, -2\right)$:

$$16x^2 - 4y^2 = 64$$
$$16\left(-\sqrt{5}\right)^2 - 4(-2)^2 \overset{?}{=} 64$$
$$16(5) - 4(4) \overset{?}{=} 64$$
$$80 - 16 \overset{?}{=} 64$$
$$64 \overset{\checkmark}{=} 64$$

$$x^2 + y^2 = 9$$
$$\left(-\sqrt{5}\right)^2 + (-2)^2 \overset{?}{=} 9$$
$$5 + 4 \overset{?}{=} 9$$
$$9 \overset{\checkmark}{=} 9$$

7. $\begin{cases} x^2 + y^2 = 9 \\ 2x - y = 3 \end{cases}$

From the second equation, $y = 2x - 3$. Substitute this result into the first equation:

$$x^2 + (2x - 3)^2 = 9$$

CHECK (0,3):

$$x^2 + y^2 = 9$$
$$0^2 + (-3)^2 \overset{?}{=} 9$$
$$0 + 9 \overset{?}{=} 9$$
$$9 \overset{\checkmark}{=} 9$$

$$2x - y = 3$$
$$2(0) - (-3) \overset{?}{=} 3$$
$$0 + 3 \overset{?}{=} 3$$
$$3 \overset{\checkmark}{=} 3$$

$$x^2 + 4x^2 - 12x + 9 = 9$$
$$5x^2 - 12x = 0$$
$$x(5x - 12) = 0$$
$$x = 0 \text{ or } 5x - 12 = 0$$
$$x = 0 \text{ or } x = \frac{12}{5}$$

CHECK $(\frac{12}{5}, \frac{9}{5})$:

$$x^2 + y^2 = 9$$
$$\left(\frac{12}{5}\right)^2 + \left(\frac{9}{5}\right)^2 \overset{?}{=} 9$$
$$\frac{144}{25} + \frac{81}{25} \overset{?}{=} 9$$
$$\frac{225}{25} \overset{?}{=} 9$$
$$9 \overset{\checkmark}{=} 9$$

$$2x - y = 3$$
$$2\left(\frac{12}{5}\right) - \frac{9}{5} \overset{?}{=} 3$$
$$\frac{24}{5} - \frac{9}{5} \overset{?}{=} 3$$
$$\frac{15}{5} \overset{?}{=} 3$$
$$3 \overset{\checkmark}{=} 3$$

When $x = 0$, $y = 2(0) - 3 = -3$.

When $x = \frac{12}{5}$, $y = 2\left(\frac{12}{5}\right) - 3 = \frac{24}{5} - 3 = \frac{9}{5}$.

Solutions: $(0, -3)$ and $(\frac{12}{5}, \frac{9}{5})$

13. $\begin{cases} y = 1 - x^2 \\ x + y = 2 \end{cases}$

Substitute the result of the first equation into the second:

$$x + (1 - x^2) = 2$$
$$x + 1 - x^2 = 2$$
$$-x^2 + x + 1 = 2$$
$$-x^2 + x - 1 = 0$$
$$x^2 - x + 1 = 0$$

$A = 1, B = -1, C = 1$

So $B^2 - 4AC = (-1)^2 - 4(1)(1) = 1 - 4 = -3 < 0$, which means that the equation has no real solutions. Therefore, the system has no real solutions.

15. $\begin{cases} y = x^2 - 6x \\ y = x - 12 \end{cases}$

Substitute the result of the
first equation into the second:

$x^2 - 6x = x - 12$

$x^2 - 7x + 12 = 0$

$(x - 3)(x - 4) = 0$

$x - 3 = 0 \quad \text{or} \quad x - 4 = 0$

$x = 3 \quad \text{or} \quad x = 4$

When $x = 3$, $y = 3 - 12 = -9$.

When $x = 4$, $y = 4 - 12 = -8$.

Solutions: $(3, -9)$ and $(4, -8)$

CHECK $(3, -9)$:

$y = x^2 - 6x$

$-9 \overset{?}{=} 3^2 - 6(3)$

$-9 \overset{?}{=} 9 - 18$

$-9 \overset{\checkmark}{=} -9$

$y = x - 12$

$-9 \overset{?}{=} 3 - 12$

$-9 \overset{\checkmark}{=} -9$

CHECK $(4, -8)$:

$y = x^2 - 6x$

$-8 \overset{?}{=} 4^2 - 6(4)$

$-8 \overset{?}{=} 16 - 24$

$-8 \overset{\checkmark}{=} -8$

$y = x - 12$

$-8 \overset{?}{=} 4 - 12$

$-8 \overset{\checkmark}{=} -8$

19. $\begin{cases} x^2 + y^2 - 25 = 0 \\ x + y - 7 = 0 \end{cases}$

From the second equation,
$y = 7 - x$. Substitute this

result into the first equation:

$x^2 + (7 - x)^2 - 25 = 0$

$x^2 + 49 - 14x + x^2 - 25 = 0$

$2x^2 - 14x + 24 = 0$

$x^2 - 7x + 12 = 0$

$(x - 3)(x - 4) = 0$

$x - 3 = 0 \quad \text{or} \quad x - 4 = 0$

$x = 3 \quad \text{or} \quad x = 4$

When $x = 3$, $y = 7 - 3 = 4$.

When $x = 4$, $y = 7 - 4 = 3$.

Solutions: $(3, 4)$ and $(4, 3)$

CHECK $(3, 4)$:

$x^2 + y^2 - 25 = 0$

$3^2 + 4^2 - 25 \overset{?}{=} 0$

$9 + 16 - 25 \overset{?}{=} 0$

$25 - 25 \overset{?}{=} 0$

$0 \overset{\checkmark}{=} 0$

$x + y - 7 = 0$

$3 + 4 - 7 \overset{?}{=} 0$

$7 - 7 \overset{?}{=} 0$

$0 \overset{\checkmark}{=} 0$

CHECK $(4, 3)$:

$x^2 + y^2 - 25 = 0$

$4^2 + 3^2 - 25 \overset{?}{=} 0$

$16 + 9 - 25 \overset{?}{=} 0$

$25 - 25 \overset{?}{=} 0$

$0 \overset{\checkmark}{=} 0$

$x + y - 7 = 0$

$4 + 3 - 7 \overset{?}{=} 0$

$7 - 7 \overset{?}{=} 0$

$0 \overset{\checkmark}{=} 0$

23. $\begin{cases} 9x^2 + y^2 = 9 \\ 3x + y = 3 \end{cases}$

From the second equation,
$y = 3 - 3x$. Substitute this
result into the first equation:

$9x^2 + (3 - 3x)^2 = 9$

$9x^2 + 9 - 18x + 9x^2 = 9$

$18x^2 - 18x + 9 = 9$

$18x^2 - 18x = 0$

$18x(x - 1) = 0$

CHECK $(0, 3)$:

$9x^2 + y^2 = 9$

$9(0)^2 + 3^2 \overset{?}{=} 9$

$0 + 9 \overset{?}{=} 9$

$9 \overset{\checkmark}{=} 9$

$3x + y = 3$

$3(0) + 3 \overset{?}{=} 3$

$0 + 3 \overset{?}{=} 3$

$3 \overset{\checkmark}{=} 3$

CHECK $(1, 0)$:

$9x^2 + y^2 = 9$

$9(1)^2 + 0^2 \overset{?}{=} 9$

$9 + 0 \overset{?}{=} 9$

$9 \overset{\checkmark}{=} 9$

$3x + y = 3$

$3(1) + 0 \overset{?}{=} 3$

$3 + 0 \overset{?}{=} 3$

$3 \overset{\checkmark}{=} 3$

$18x = 0$ or $x - 1 = 0$

$x = 0$ or $x = 1$

When $x = 0$, $y = 3 - 3(0) = 3 - 0 = 3$. When $x = 1$, $y = 3 - 3(1) = 3 - 3 = 0$.

Solutions: $(0,3)$ and $(1,0)$

27. $\begin{cases} x = y^2 - 3 \\ x + y = 9 \end{cases}$

Substitute the result of the first equation into the second:

$y^2 - 3 + y = 9$

$y^2 + y - 3 = 9$

$y^2 + y - 12 = 0$

$(y + 4)(y - 3) = 0$

$y + 4 = 0$ or $y - 3 = 0$

$y = -4$ or $y = 3$

When $y = -4$, $x = (-4)^2 - 3 = 16 - 3 = 13$.

When $y = 3$, $x = 3^2 - 3 = 9 - 3 = 6$.

Solutions: $(6,3)$ and $(13,-4)$

CHECK $(6,3)$:

$x = y^2 - 3$

$6 \overset{?}{=} 3^2 - 3$

$6 \overset{?}{=} 9 - 3$

$6 \overset{\checkmark}{=} 6$

$x + y = 9$

$6 + 3 \overset{?}{=} 9$

$9 \overset{\checkmark}{=} 9$

CHECK $(13,-4)$:

$x = y^2 - 3$

$13 \overset{?}{=} (-4)^2 - 3$

$13 \overset{?}{=} 16 - 3$

$13 \overset{\checkmark}{=} 13$

$x + y = 9$

$13 + (-4) \overset{?}{=} 9$

$9 \overset{\checkmark}{=} 9$

31. $\begin{cases} x^2 + y^2 = 10 \\ xy = 4 \end{cases}$

From the second equation,

$y = \dfrac{4}{x}$. Substitute this result into the first equation:

$x^2 + \left(\dfrac{4}{x}\right)^2 = 10$

$x^2 + \dfrac{16}{x^2} = 10$

$x^4 + 16 = 10x^2$

$x^4 - 10x^2 + 16 = 0$

$(x^2 - 2)(x^2 - 8) = 0$

$x^2 - 2 = 0$ or $x^2 - 8 = 0$

$x^2 = 2$ or $x^2 = 8$

CHECK $(\sqrt{2}, 2\sqrt{2})$:

$x^2 + y^2 = 10$

$(\sqrt{2})^2 + (2\sqrt{2})^2 \overset{?}{=} 10$

$2 + 8 \overset{?}{=} 10$

$10 \overset{\checkmark}{=} 10$

$xy = 4$

$(\sqrt{2})(2\sqrt{2}) \overset{?}{=} 4$

$2 \cdot 2 \overset{?}{=} 4$

$4 \overset{\checkmark}{=} 4$

CHECK $(-\sqrt{2}, -2\sqrt{2})$:

$x^2 + y^2 = 10$

$(-\sqrt{2})^2 + (-2\sqrt{2})^2 \overset{?}{=} 10$

$2 + 8 \overset{?}{=} 10$

$10 \overset{\checkmark}{=} 10$

$xy = 4$

$(-\sqrt{2})(-2\sqrt{2}) \overset{?}{=} 4$

$2 \cdot 2 \overset{?}{=} 4$

$4 \overset{\checkmark}{=} 4$

$x = \pm\sqrt{2}$ or $x = \pm\sqrt{8} = \pm2\sqrt{2}$

When $x = \sqrt{2}, y = \dfrac{4}{\sqrt{2}} = \dfrac{4\sqrt{2}}{2} = 2\sqrt{2}$.

When $x = -\sqrt{2}, y = \dfrac{4}{-\sqrt{2}} = \dfrac{-4\sqrt{2}}{2} = -2\sqrt{2}$.

When $x = 2\sqrt{2}, y = \dfrac{4}{2\sqrt{2}} = \dfrac{4\sqrt{2}}{4} = \sqrt{2}$.

When $x = -2\sqrt{2}, y = \dfrac{4}{-2\sqrt{2}} = -\dfrac{4\sqrt{2}}{4} = -\sqrt{2}$.

Solutions: $(\sqrt{2}, 2\sqrt{2}), (-\sqrt{2}, -2\sqrt{2}),$
$(2\sqrt{2}, \sqrt{2})$ and $(-2\sqrt{2}, -\sqrt{2})$

CHECK $(2\sqrt{2}, \sqrt{2})$:

$x^2 + y^2 = 10$	$xy = 4$
$(2\sqrt{2})^2 + (\sqrt{2})^2 \overset{?}{=} 10$	$(2\sqrt{2})(\sqrt{2}) \overset{?}{=} 4$
$8 + 2 \overset{?}{=} 10$	$2 \cdot 2 \overset{?}{=} 4$
$10 \overset{\checkmark}{=} 10$	$4 \overset{\checkmark}{=} 4$

CHECK $\left(-2\sqrt{2}, -\sqrt{2}\right)$:

$x^2 + y^2 = 10$	$xy = 4$
$(-2\sqrt{2})^2 + (-\sqrt{2})^2 \overset{?}{=} 10$	$(-2\sqrt{2})(-\sqrt{2}) \overset{?}{=} 4$
$8 + 2 \overset{?}{=} 10$	$2 \cdot 2 \overset{?}{=} 4$
$10 \overset{\checkmark}{=} 10$	$4 \overset{\checkmark}{=} 4$

35. $\begin{cases} x^2 + y^2 = 25 \\ x^2 - xy + y^2 = 13 \end{cases}$

Subtract the second equation
from the first

$\begin{array}{l} x^2 + y^2 = 25 \\ \underline{x^2 - xy + y^2 = 13} \\ xy = 12 \end{array}$

From this equation, find
that $y = \dfrac{12}{x}$. Substitute
this result into the first
equation:

$x^2 + \left(\dfrac{12}{x}\right)^2 = 25$

$x^2 + \dfrac{144}{x^2} = 25$

$x^4 + 144 = 25x^2$

$x^4 - 25x^2 + 144 = 0$

$(x^2 - 9)(x^2 - 16) = 0$

CHECK $(3, 4)$:

$x^2 + y^2 = 25$	$x^2 - xy + y^2 = 13$
$3^2 + 4^2 \overset{?}{=} 25$	$3^2 - 3(4) + 4^2 \overset{?}{=} 13$
$9 + 16 \overset{?}{=} 25$	$9 - 12 + 16 \overset{?}{=} 13$
$25 \overset{\checkmark}{=} 25$	$13 \overset{\checkmark}{=} 13$

CHECK $(-3, -4)$:

$x^2 + y^2 = 25$	$x^2 - xy + y^2 = 13$
$(-3)^2 + (-4)^2 \overset{?}{=} 25$	$(-3)^2 - (-3)(-4) + (-4)^2 \overset{?}{=} 13$
$9 + 16 \overset{?}{=} 25$	$9 - 12 + 16 \overset{?}{=} 13$
$25 \overset{\checkmark}{=} 25$	$13 \overset{\checkmark}{=} 13$

CHECK $(4, 3)$:

$x^2 + y^2 = 25$	$x^2 - xy + y^2 = 13$
$4^2 + 3^2 \overset{?}{=} 25$	$4^2 - 4(3) + 3^2 \overset{?}{=} 13$
$16 + 9 \overset{?}{=} 25$	$16 - 12 + 9 \overset{?}{=} 13$
$25 \overset{\checkmark}{=} 25$	$13 \overset{\checkmark}{=} 13$

$(x - 3)(x + 3)(x - 4)(x + 4) = 0$

CHECK $(-4, -3)$:

$$x^2 + y^2 = 25 \qquad\qquad x^2 - xy + y^2 = 13$$

$x - 3 = 0$ or $x + 3 = 0$ or

$x - 4 = 0$ or $x + 4 = 0$

$x = 3$ or $x = -3$ or $x = 4$ or $x = -4$

$$(-4)^2 + (-3)^2 \overset{?}{=} 25 \qquad (-4)^2 - (-4)(-3) + (-3)^2 \overset{?}{=} 13$$

$$16 + 9 \overset{?}{=} 25 \qquad\qquad 16 - 12 + 9 \overset{?}{=} 13$$

$$25 \overset{\checkmark}{=} 25 \qquad\qquad 13 \overset{\checkmark}{=} 13$$

When $x = 3$, $y = \dfrac{12}{3} = 4$. When $x = -3$, $y = \dfrac{12}{-3} = -4$.

When $x = 4$, $y = \dfrac{12}{4} = 3$. When $x = -4$, $y = \dfrac{12}{-4} = -3$.

Solutions: $(3, 4)$, $(-3, -4)$, $(4, 3)$ and $(-4, -3)$

37.
$$\begin{cases} x^2 + y^2 = 29 \\ x - y = 3 \end{cases}$$

CHECK $(-2, -5)$:

$$x^2 + y^2 = 29 \qquad\qquad x - y = 3$$

$$(-2)^2 + (-5)^2 \overset{?}{=} 29 \qquad -2 - (-5) \overset{?}{=} 3$$

$$4 + 25 \overset{?}{=} 29 \qquad\qquad -2 + 5 \overset{?}{=} 3$$

$$29 \overset{\checkmark}{=} 29 \qquad\qquad 3 \overset{\checkmark}{=} 3$$

From the second equation,
$x = y + 3$. Substitute this
result into the first equation:

$(y + 3)^2 + y^2 = 29$

$y^2 + 6y + 9 + y^2 = 29$

$2y^2 + 6y + 9 = 29$

$2y^2 + 6y - 20 = 0$

$y^2 + 3y - 10 = 0$

$(y + 5)(y - 2) = 0$

CHECK $(5, 2)$:

$$x^2 + y^2 = 29 \qquad\qquad x - y = 3$$

$$5^2 + 2^2 \overset{?}{=} 29 \qquad\qquad 5 - 2 \overset{?}{=} 3$$

$$25 + 4 \overset{?}{=} 29 \qquad\qquad 3 \overset{\checkmark}{=} 3$$

$$29 \overset{\checkmark}{=} 29$$

$y + 5 = 0$ or $y - 2 = 0$

$y = -5$ or $y = 2$

When $y = -5$, $x = -5 + 3 = -2$.

When $y = 2$, $x = 2 + 3 = 5$.

Solutions: $(-2, -5)$ and $(5, 2)$

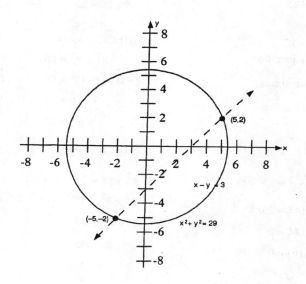

39.
$$\begin{cases} x^2 + y^2 = 29 \\ x^2 - y^2 = 3 \end{cases}$$

Add: $2x^2 = 32$

$x^2 = 16$

$x = \pm 4$

Substitute: $x^2 + y^2 = 29$

$16 + y^2 = 29$

$y^2 = 13$

$y = \pm\sqrt{13}$

Solutions: $(4, \sqrt{13}), (4, -\sqrt{13}),$

$(-4, \sqrt{13})$ and $(-4, -\sqrt{13})$

CHECK $\left(4, \sqrt{13}\right)$:

$x^2 + y^2 = 29$	$x^2 - y^2 = 3$
$4^2 + \left(\sqrt{13}\right)^2 \overset{?}{=} 29$	$4^2 - \left(\sqrt{13}\right)^2 \overset{?}{=} 3$
$16 + 13 \overset{?}{=} 29$	$16 - 13 \overset{?}{=} 3$
$29 \overset{\checkmark}{=} 29$	$3 \overset{\checkmark}{=} 3$

CHECK $(4, -\sqrt{13})$:

$x^2 + y^2 = 29$	$x^2 - y^2 = 3$
$4^2 + \left(-\sqrt{13}\right)^2 \overset{?}{=} 29$	$4^2 - \left(-\sqrt{13}\right)^2 \overset{?}{=} 3$
$16 + 13 \overset{?}{=} 29$	$16 - 13 \overset{?}{=} 3$
$29 \overset{\checkmark}{=} 29$	$3 \overset{\checkmark}{=} 3$

CHECK $(-4, \sqrt{13})$:

$x^2 + y^2 = 29$	$x^2 - y^2 = 3$
$(-4)^2 + \left(\sqrt{13}\right)^2 \overset{?}{=} 29$	$(-4)^2 - \left(-\sqrt{13}\right)^2 \overset{?}{=} 3$
$16 + 13 \overset{?}{=} 29$	$16 - 13 \overset{?}{=} 3$
$29 \overset{\checkmark}{=} 29$	$3 \overset{\checkmark}{=} 3$

CHECK $(-4, -\sqrt{13})$:

$x^2 + y^2 = 29$	$x^2 - y^2 = 3$
$(-4)^2 + \left(-\sqrt{13}\right)^2 \overset{?}{=} 29$	$(-4)^2 - \left(-\sqrt{13}\right)^2 \overset{?}{=} 3$
$16 + 13 \overset{?}{=} 29$	$16 - 13 \overset{?}{=} 3$
$29 \overset{\checkmark}{=} 29$	$3 \overset{\checkmark}{=} 3$

45. $2x^2 + y^2 = 50 \xrightarrow{\text{as is}}$

$x^2 + 2y^2 = 25 \xrightarrow{\text{multiply by } -2}$

$2x^2 + y^2 = 50$

$\underline{-2x^2 - 4y^2 = -50}$

Add: $-3y^2 = 0$

$y^2 = 0$

$y = 0$

Substitute:

$2x^2 + y^2 = 50$

$2x^2 + 0^2 = 50$

$2x^2 = 50$

$x^2 = 25$

$x = \pm 5$

Solutions: $(5,0)$ and $(-5,0)$

CHECK $(5,0)$:

$2x^2 + y^2 = 50$

$2(5)^2 + 0^2 \overset{?}{=} 50$

$2(25) + 0 \overset{?}{=} 50$

$50 + 0 \overset{?}{=} 50$

$50 \overset{\checkmark}{=} 50$

$x^2 + 2y^2 = 25$

$5^2 + 2(0)^2 \overset{?}{=} 25$

$25 + 2(0) \overset{?}{=} 25$

$25 + 0 \overset{?}{=} 25$

$25 \overset{\checkmark}{=} 25$

CHECK $(-5,0)$:

$2x^2 + y^2 = 50$

$2(-5)^2 + 0^2 \overset{?}{=} 50$

$2(25) + 0 \overset{?}{=} 50$

$50 + 0 \overset{?}{=} 50$

$50 \overset{\checkmark}{=} 50$

$x^2 + y^2 = 25$

$(-5)^2 + 2(0)^2 \overset{?}{=} 25$

$25 + 2(0) \overset{?}{=} 25$

$25 + 0 \overset{?}{=} 25$

$25 \overset{\checkmark}{=} 25$

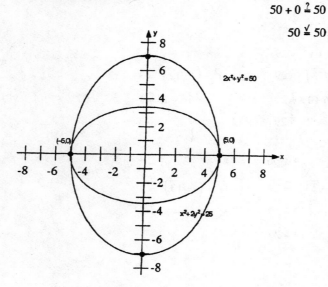

47. Let L = length of the rectangle (in inches); W = width of the rectangle (in inches).

$LW = 210$

$2L + 2W = 59$

From the first equation,

$L = \dfrac{210}{W}$. Substitute this

result into the second equation:

CHECK: $(17.5)(12) \overset{?}{=} 210$

$210 \overset{\checkmark}{=} 210$

$2(17.5) + 2(12) \overset{?}{=} 59$

$35 + 24 \overset{?}{=} 59$

$59 \overset{\checkmark}{=} 59$

$$2\left(\frac{210}{W}\right) + 2W = 59$$

$$\frac{420}{W} + 2W = 59$$

$$420 + 2W^2 = 59W$$

$$2W^2 - 59W + 420 = 0$$

$$(W - 12)(2W - 35) = 0$$

$$W - 12 = 0 \text{ or } 2W - 35 = 0.$$

$$W = 12 \text{ or } W = \frac{35}{2}.$$

When $W = 12$, $L = \frac{210}{12} = \frac{35}{2}$. When $W = \frac{35}{2}$, $L = \frac{210}{\frac{35}{2}} = 12$.

Thus, the dimensions of the rectangle are 12 inches and $\frac{35}{2} = 17.5$ inches.

49. Let r = Maria's rate going (in mph); t = Maria's time going (in hours).
$$rt = 60$$

$$(r + 12)\left(t - \frac{1}{4}\right) = 60 \rightarrow rt + 12t - \frac{1}{4}r - 3 = 60$$

Substitute the result of the first equation into the second:

$$60 + 12t - \frac{1}{4}r - 3 = 60$$

$$12t - \frac{1}{4}r - 3 = 0$$

$$12t - 3 = \frac{1}{4}r$$

$$48t - 12 = r$$

Substitute this result into the first equation:

$$(48t - 12)t = 60$$

$$48t^2 - 12t = 60 \qquad\qquad \text{CHECK: } 48\left(\frac{5}{4}\right) \overset{\checkmark}{=} 60$$

$$48t^2 - 12t - 60 = 0$$

$$4t^2 - t - 5 = 0 \qquad\qquad\qquad 60(1) \overset{\checkmark}{=} 60$$

$$(4t - 5)(t + 1) = 0$$

$$4t - 5 = 0 \text{ or } t + 1 = 0$$

$$t = \frac{5}{4} \text{ or } t = -1$$

Since t cannot be negative, we reject $t = -1$. Thus, $t = \frac{5}{4}$. Then $r = 48\left(\frac{5}{4}\right) - 12 = 60 - 12 = 48$.

Conclude: Maria's rate going is 48 mph, and she travels for $\frac{5}{4} = 1.25$ hours. Maria's rate returning is $48 + 12 = 60$ mph, and she travels for $\frac{5}{4} - \frac{1}{4} = 1$ hour.

51. It is incorrect ot say that the square of $x + y$ is equal to $x^2 + y^2$. In fact, $(x + y)^2 = x^2 + 2xy + y^2$.

Exercises 10.5

1. $x + y \leq 6$

     boundary: solid

     x-intercept: 6

     y-intercept: 6

     test point: (0,0)

     $0 + 0 \overset{?}{\leq} 6$

        $0 \overset{\checkmark}{\leq} 6$

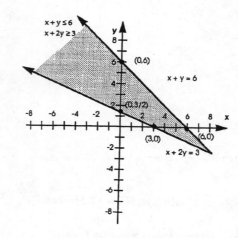

     $x + 2y \geq 3$

     boundary: solid

     x-intercept: 3

     y-intercept: $\dfrac{3}{2}$

     test point: (0,0)

     $0 + 2(0) \overset{?}{\geq} 3$

       $0 + 0 \overset{?}{\geq} 3$

         $0 \not\geq 3$

5. $x - y \geq 2$

     boundary: solid

     x-intercept: 2

     y-intercept: -2

     test point: (0,0)

     $0 - 0 \overset{?}{\geq} 2$

        $0 \not\geq 2$

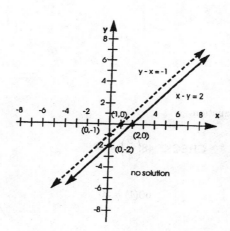

     $y - x > -1$

     boundary: dotted

     x-intercept: 1

     y-intercept: -1

     test point: (0,0)

     $0 - 0 \overset{?}{\geq} -1$

        $0 \overset{\checkmark}{\geq} -1$

9. $3x + 2y \leq 12$

     boundary: solid

     x-intercept: 4

     y-intercept: 6

test point: (0,0)

$3(0) + 2(0) \overset{?}{\leq} 12$

$0 + 0 \overset{?}{\leq} 12$

$0 \overset{\checkmark}{\leq} 12$

$y \leq x$

boundary: solid

x-intercept: 0

y-intercept: 0

second boundary point: (1,1)

test point: (1,0)

$0 \overset{?}{\leq} 1$

$0 \overset{\checkmark}{\leq} 1$

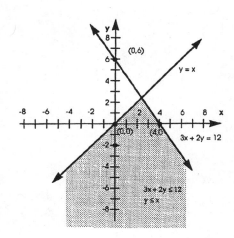

13. $5x - 3y \leq 15$

boundary: solid

x-intercept: 3

y-intercept: −5

test point: (0,0)

$5(0) - 3(0) \overset{?}{\leq} 15$

$0 - 0 \overset{?}{\leq} 15$

$0 \overset{\checkmark}{\leq} 15$

$x < 3$

boundary: dotted

vertical line, 3 units to the right of the y − axis

test point: (0,0)

$0 \overset{?}{<} 3$

$0 \overset{\checkmark}{<} 3$

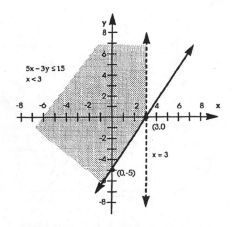

17. $x - 2y < 10$

boundary: dotted

x-intercept: 10

y-intercept: −5

test point: (0,0)

$0 - 2(0) \overset{?}{<} 10$

$0 - 0 \overset{?}{<} 10$

$0 \overset{\checkmark}{<} 10$

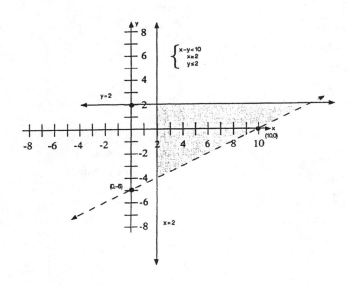

x ≥ 2

boundary: solid

vertical line, 2 units to

to the right of the y – axis

test point: (0,0)

$0 \overset{?}{\geq} 2$

$0 \not\geq 2$

y ≤ 2

boundary: solid

horizontal line, 2 units above

the x – axis

test point: (0,0)

$0 \overset{?}{\leq} 2$

$0 \overset{\checkmark}{\leq} 2$

21. x + 3y ≥ 6

boundary: solid

x-intercept: 6

y-intercept: 2

test point: (0,0)

$0 + 3(0) \overset{?}{\geq} 6$

$0 + 0 \overset{?}{\geq} 6$

$0 \not\geq 6$

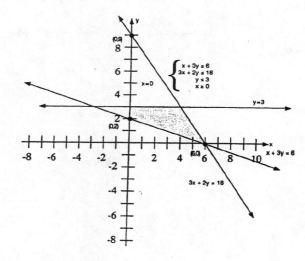

3x + 2y ≤ 18

boundary: solid

x-intercept: 6

y-intercept: 9

test point: (0,0)

$3(0) + 2(0) \overset{?}{\leq} 18$

$0 + 0 \overset{?}{\leq} 18$

$0 \overset{\checkmark}{\leq} 18$

y ≤ 3

boundary: solid

horizontal line, 3 units

above the x - axis

test point: (0, 0)

x ≥ 0

boundary: solid

the y - axis

test point: (1, 0)

$$0 \overset{?}{\leqq} 3 \qquad\qquad\qquad 1 \overset{?}{\geqq} 0$$

$$0 \overset{\checkmark}{\leqq} 3 \qquad\qquad\qquad 1 \overset{\checkmark}{\geqq} 0$$

27. Let x = # of ounces of Brand X; y = # of ounces of Brand Y

$10x + 13y \geq 21$

$0.33x + 0.37y \leq 1$

$x \geq 0$

$y \geq 0$

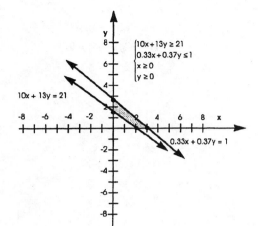

Chapter 10 Review Exercises

1.
$$\begin{cases} x + 2y - z = 2 \\ 2x + 3y + 4z = 9 \\ 3x + y - 2z = 2 \end{cases}$$

Multiply equation (1) by 4 and add the resulting equation to equation (2), obtaining equation (4):

$4x + 8y - 4z = 8$

$\underline{2x + 3y + 4z = 9}$

$\quad 6x + 11y = 17$

Multipy equation (3) by 2 and add the resulting equation to equation (2), obtaining equation (5):

$6x + 2y - 4z = 4$

$\underline{2x + 3y + 4z = 9}$

$\quad 8x + 5y = 13$

Multiply equation (4) by 4, equation (5) by −3, and add the resulting equations:

$24x + 44y = 68$

$\underline{-24x - 15y = -39}$

$\quad\quad 29y = 29$

$\quad\quad\quad y = 1$

Substitute into equation (4):

$$6x + 11y = 17$$
$$6x + 11(1) = 17$$
$$6x + 11 = 17$$
$$6x = 6$$
$$x = 1$$

Substitute into equation (1):

$$x + 2y - z = 2$$
$$1 + 2(1) - z = 2$$
$$1 + 2 - z = 2$$
$$3 - z = 2$$

$$-z = -1$$
$$z = 1$$

Solution: $x = 1$, $y = 1$, $z = 1$

CHECK: $x + 2y - z = 2$
$$1 + 2(1) - 1 \overset{?}{=} 2$$
$$1 + 2 - 1 \overset{?}{=} 2$$
$$3 - 1 \overset{?}{=} 2$$
$$2 \overset{\checkmark}{=} 2$$

$$2x + 3y + 4z = 9$$
$$2(1) + 3(1) + 4(1) \overset{?}{=} 9$$
$$2 + 3 + 4 \overset{?}{=} 9$$
$$5 + 4 \overset{?}{=} 9$$
$$9 \overset{\checkmark}{=} 9$$

$$3x + y - 2z = 2$$
$$3(1) + 1 - 2(1) \overset{?}{=} 2$$
$$3 + 1 - 2 \overset{?}{=} 2$$
$$4 - 2 \overset{?}{=} 2$$
$$2 \overset{\checkmark}{=} 2$$

3.
$$\begin{cases} 2x \quad\;\; - 3z = 7 \\ 3x + 4y \quad\quad = -11 \\ \quad\;\; 3y - 2z = 0 \end{cases}$$

Multiply equation (1) by 2, equation (3) by -3, and add the resulting equations, obtaining equation (4):

$$4x \quad\;\; - 6z = 14$$
$$-9y + 6z = 0$$
$$4x - 9y \quad\;\; = 14$$

Multiply equation (2) by 4, equation (4) by -3, and add the resulting equations:

$$12x + 16y = -44$$
$$\underline{-12x + 27y = -42}$$
$$43y = -86$$
$$y = -2$$

Substitute into equation (2):

$$3x + 4y = -11$$
$$3x + 4(-2) = -11$$
$$3x - 8 = -11$$
$$3x = -3$$
$$x = -1$$

Substitute into equation (3):

$$3y - 2z = 0$$
$$3(-2) - 2z = 0$$
$$-6 - 2z = 0$$
$$-2z = 6$$
$$z = -3$$

Solution: $x = -1, y = -2, z = -3$

CHECK: $2x - 3z = 7$
$$2(-1) - 3(-3) \overset{?}{=} 7$$
$$-2 + 9 \overset{?}{=} 7$$
$$7 \overset{\checkmark}{=} 7$$

$$3x + 4y = -11$$
$$3(-1) + 4(-2) \overset{?}{=} -11$$
$$-3 - 8 \overset{?}{=} -11$$
$$-11 \overset{\checkmark}{=} -11$$

$$3y - 2z = 0$$
$$3(-2) - 2(-3) \overset{?}{=} 0$$
$$-6 + 6 \overset{?}{=} 0$$
$$0 \overset{\checkmark}{=} 0$$

5. $\begin{cases} 3x + y - z = 2 \\ 3x + y - z = 5 \\ x - y + z = 1 \end{cases}$

Observe that the left-hand sides of equations (1) and (2) are identical. This would require that the right-hand sides of these equations must be equal. In other words, $2 = 5$, which is obviously false. Therefore, the system has no solution.

7. $\begin{bmatrix} 1 & 2 & | & 6 \\ 2 & 6 & | & 8 \end{bmatrix}$

$-2R_1 + R_2 \rightarrow R_2$ $\begin{bmatrix} 1 & 2 & | & 6 \\ 0 & 2 & | & -4 \end{bmatrix}$

$\frac{1}{2}R_2 \rightarrow R_2$ $\begin{bmatrix} 1 & 2 & | & 6 \\ 0 & 1 & | & -2 \end{bmatrix}$

$-2R_2 + R_1 \rightarrow R_1$ $\begin{bmatrix} 1 & 0 & | & 10 \\ 0 & 1 & | & -2 \end{bmatrix}$

9. $\begin{bmatrix} 2 & 2 & 6 & | & 4 \\ 2 & 5 & 9 & | & -2 \\ 1 & 2 & 3 & | & -1 \end{bmatrix}$

$\frac{1}{2}R_1 \rightarrow R_1$ $\begin{bmatrix} 1 & 1 & 3 & | & 2 \\ 2 & 5 & 9 & | & -2 \\ 1 & 2 & 3 & | & -1 \end{bmatrix}$

$-2R_1 + R_2 \rightarrow R_2$
$-R_1 + R_3 \rightarrow R_3$ $\begin{bmatrix} 1 & 1 & 3 & | & 2 \\ 0 & 3 & 3 & | & -6 \\ 0 & 1 & 0 & | & -3 \end{bmatrix}$

$R_2 \leftrightarrow R_3$ $\begin{bmatrix} 1 & 1 & 3 & | & 2 \\ 0 & 1 & 0 & | & -3 \\ 0 & 3 & 3 & | & -6 \end{bmatrix}$

$-R_2 + R_1 \rightarrow R_1$
$-3R_2 + R_3 \rightarrow R_3$ $\begin{bmatrix} 1 & 0 & 3 & | & 5 \\ 0 & 1 & 0 & | & -3 \\ 0 & 0 & 3 & | & 3 \end{bmatrix}$

$\frac{1}{3}R_3 \rightarrow R_3$ $\begin{bmatrix} 1 & 0 & 3 & | & 5 \\ 0 & 1 & 0 & | & -3 \\ 0 & 0 & 1 & | & 1 \end{bmatrix}$

$-3R_3 + R_1 \rightarrow R_1$ $\begin{bmatrix} 1 & 0 & 0 & | & 2 \\ 0 & 1 & 0 & | & -3 \\ 0 & 0 & 1 & | & 1 \end{bmatrix}$

11. $\begin{bmatrix} 1 & -2 & | & 1 \\ 2 & 3 & | & 9 \end{bmatrix}$

$-2R_1 + R_2 \rightarrow R_2$ $\begin{bmatrix} 1 & -2 & | & 1 \\ 0 & 7 & | & 7 \end{bmatrix}$

$\frac{1}{7}R_2 \rightarrow R_2$ $\begin{bmatrix} 1 & -2 & | & 1 \\ 0 & 1 & | & 1 \end{bmatrix}$

$\begin{cases} x - 2y = 1 \\ \quad\;\; y = 1 \end{cases}$

The second equation gives $y = 1$. Then substitute 1 for y in the first equation and find x:

$$x - 2y = 1$$
$$x - 2(1) = 1$$
$$x - 2 = 1$$
$$x = 3$$

Hence, the solution is (3,1).

13. $\begin{bmatrix} 1 & 2 & -1 & | & 5 \\ 1 & -1 & 1 & | & 0 \\ 1 & 1 & 2 & | & 1 \end{bmatrix}$ $\quad \begin{array}{l} -R_1 + R_2 \rightarrow R_2 \\ -R_1 + R_3 \rightarrow R_3 \end{array}$ $\quad \begin{bmatrix} 1 & 2 & -1 & | & 5 \\ 0 & -3 & 2 & | & -5 \\ 0 & -1 & 3 & | & -4 \end{bmatrix}$

$\quad\quad\quad\quad\quad\quad\quad\quad\quad R_2 \leftrightarrow R_3 \quad\quad\quad \begin{bmatrix} 1 & 2 & -1 & | & 5 \\ 0 & -1 & 3 & | & -4 \\ 0 & -3 & 2 & | & -5 \end{bmatrix}$

$\quad\quad\quad\quad\quad\quad\quad -3R_2 + R_3 \rightarrow R_3 \quad \begin{bmatrix} 1 & 2 & -1 & | & 5 \\ 0 & -1 & 3 & | & -4 \\ 0 & 0 & -7 & | & 7 \end{bmatrix}$

$$\begin{cases} x + 2y - z = 5 \\ \quad - y + 3z = -4 \\ \quad\quad\quad -7z = 7 \end{cases}$$

First solve the third equation for z:

−7z = 7. Divide both sides of the equation by −7.

z = −1

Then substitute −1 for z in the second equation and find y:

$$-y + 3z = -4$$
$$-y + 3(-1) = -4$$
$$-y - 3 = -4$$
$$-y = -1$$
$$y = 1$$

Finally, substitute −1 for z and 1 for y in the first equation and find x:

$$x + 2y - z = 5$$
$$x + 2(1) - (-1) = 5$$
$$x + 2 + 1 = 5$$
$$x + 3 = 5$$
$$x = 2$$

Hence, the solution is (2, 1, −1).

15. $\begin{bmatrix} 1 & -3 & | & -1 \\ 2 & 1 & | & 5 \end{bmatrix}$ $\quad\quad -2R_1 + R_2 \rightarrow R_2 \quad \begin{bmatrix} 1 & -3 & | & -1 \\ 0 & 7 & | & 7 \end{bmatrix}$

$\quad\quad\quad\quad\quad\quad\quad\quad \frac{1}{7}R_2 \rightarrow R_2 \quad\quad \begin{bmatrix} 1 & -3 & | & -1 \\ 0 & 1 & | & 1 \end{bmatrix}$

$$3R_2 + R_1 \to R_1 \qquad \begin{bmatrix} 1 & 0 & | & 2 \\ 0 & 1 & | & 1 \end{bmatrix}$$

So $x = 2$, $y = 1$.

17. $\begin{vmatrix} 2 & 3 \\ 4 & -1 \end{vmatrix} = 2(-1) - 3(4) = -2 - 12 = -4$

19. $\begin{vmatrix} 1 & 3 & -2 \\ 2 & 4 & 3 \\ 5 & -1 & -3 \end{vmatrix} = 1\begin{vmatrix} 4 & 3 \\ -1 & -3 \end{vmatrix} - 3\begin{vmatrix} 2 & 3 \\ 5 & -3 \end{vmatrix} + (-2)\begin{vmatrix} 2 & 4 \\ 5 & -1 \end{vmatrix}$

$$= 1(-12 - (-3)) - 3(-6 - 15) - 2(-2 - 20)$$

$$= 1(-9) - 3(-21) - 2(-22) = -9 + 63 + 44 = 98$$

21. $\begin{cases} 3x + 7y = 4 \\ 5x + 4y = 2 \end{cases}$

$D = \begin{vmatrix} 3 & 7 \\ 5 & 4 \end{vmatrix} = 3(4) - 7(5) = 12 - 35 = -23$

$D_x = \begin{vmatrix} 4 & 7 \\ 2 & 4 \end{vmatrix} = 4(4) - 7(2) = 16 - 14 = 2$

$D_y = \begin{vmatrix} 3 & 4 \\ 5 & 2 \end{vmatrix} = 3(2) - 4(5) = 6 - 20 = -14$

Then $x = \dfrac{D_x}{D} = \dfrac{2}{-23} = -\dfrac{2}{23}$ and $y = \dfrac{D_y}{D} = \dfrac{-14}{-23} = \dfrac{14}{23}$

CHECK: $3x + 7y = 4$ $\qquad\qquad\qquad\qquad\qquad$ $5x + 4y = 2$

$3\left(-\dfrac{2}{23}\right) + 7\left(\dfrac{14}{23}\right) \overset{?}{=} 4$ $\qquad\qquad$ $5\left(-\dfrac{2}{23}\right) + 4\left(\dfrac{14}{23}\right) \overset{?}{=} 2$

$\qquad -\dfrac{6}{23} + \dfrac{98}{23} \overset{?}{=} 4$ $\qquad\qquad\qquad$ $-\dfrac{10}{23} + \dfrac{56}{23} \overset{?}{=} 2$

$\qquad\qquad \dfrac{92}{23} \overset{?}{=} 4$ $\qquad\qquad\qquad\qquad\quad$ $\dfrac{46}{23} \overset{?}{=} 2$

$\qquad\qquad 4 \overset{\checkmark}{=} 4$ $\qquad\qquad\qquad\qquad\qquad$ $2 \overset{\checkmark}{=} 2$

23. $\begin{cases} 4a + 2b = 7 \\ 6a + 3b = 9 \end{cases}$

$D = \begin{vmatrix} 4 & 2 \\ 6 & 3 \end{vmatrix} = 4(3) - 2(6) = 12 - 12 = 0$

This means that the system does not have a unique solution.

Since $D_a = \begin{vmatrix} 7 & 2 \\ 9 & 3 \end{vmatrix} = 7(3) - 2(9) = 21 - 18 = 3 \neq 0$, the system is inconsistent and has no solutions.

25. $\begin{cases} 3s + 4t + u = 3 \\ 5s - 3t + 6u = 2 \\ 4s - 5t - 5u = 1 \end{cases}$

$$D = \begin{vmatrix} 3 & 4 & 1 \\ 5 & -3 & 6 \\ 4 & -5 & -5 \end{vmatrix} = 3\begin{vmatrix} -3 & 6 \\ -5 & -5 \end{vmatrix} - 4\begin{vmatrix} 5 & 6 \\ 4 & -5 \end{vmatrix} + 1\begin{vmatrix} 5 & -3 \\ 4 & -5 \end{vmatrix}$$

$$= 3(15 - (-30)) - 4(-25 - 24) + 1(-25 - (-12))$$

$$= 3(45) - 4(-49) + 1(-13) = 135 + 196 - 13 = 318$$

$$D_s = \begin{vmatrix} 3 & 4 & 1 \\ 2 & -3 & 6 \\ 1 & -5 & -5 \end{vmatrix} = 3\begin{vmatrix} -3 & 6 \\ -5 & -5 \end{vmatrix} - 4\begin{vmatrix} 2 & 6 \\ 1 & -5 \end{vmatrix} + 1\begin{vmatrix} 2 & -3 \\ 1 & -5 \end{vmatrix}$$

$$= 3(15 - (-30)) - 4(-10 - 6) + 1(-10 - (-3))$$

$$= 3(45) - 4(-16) + 1(-7) = 135 + 64 - 7 = 192$$

$$D_t = \begin{vmatrix} 3 & 3 & 1 \\ 5 & 2 & 6 \\ 4 & 1 & -5 \end{vmatrix} = 3\begin{vmatrix} 2 & 6 \\ 1 & -5 \end{vmatrix} - 3\begin{vmatrix} 5 & 6 \\ 4 & -5 \end{vmatrix} + 1\begin{vmatrix} 5 & 2 \\ 4 & 1 \end{vmatrix}$$

$$= 3(-10 - 6) - 3(-25 - 24) + 1(5 - 8)$$

$$= 3(-16) - 3(-49) + 1(-3) = -48 + 147 - 3 = 96$$

$$D_u = \begin{vmatrix} 3 & 4 & 3 \\ 5 & -3 & 2 \\ 4 & -5 & 1 \end{vmatrix} = 3\begin{vmatrix} -3 & 2 \\ -5 & 1 \end{vmatrix} - 4\begin{vmatrix} 5 & 2 \\ 4 & 1 \end{vmatrix} + 3\begin{vmatrix} 5 & -3 \\ 4 & -5 \end{vmatrix}$$

$$= 3(-3 - (-10)) - 4(5 - 8) + 3(-25 - (-12))$$

$$= 3(7) - 4(-3) + 3(-13) = 21 + 12 - 39 = -6$$

Then $s = \dfrac{D_s}{D} = \dfrac{192}{318} = \dfrac{32}{53}$, $t = \dfrac{D_t}{D} = \dfrac{96}{318} = \dfrac{16}{53}$, and $u = \dfrac{D_u}{D} = \dfrac{-6}{318} = -\dfrac{1}{53}$

CHECK: $3s + 4t + u = 3$ $\qquad\qquad\qquad\qquad$ $5s - 3t + 6u = 2$

$3\left(\dfrac{32}{53}\right) + 4\left(\dfrac{16}{53}\right) - \dfrac{1}{53} \overset{?}{=} 3$ $\qquad$ $5\left(\dfrac{32}{53}\right) - 3\left(\dfrac{16}{53}\right) + 6\left(-\dfrac{1}{53}\right) \overset{?}{=} 2$

$\dfrac{96}{53} + \dfrac{64}{53} - \dfrac{1}{53} \overset{?}{=} 3$ $\qquad\qquad$ $\dfrac{160}{53} - \dfrac{48}{53} - \dfrac{6}{53} \overset{?}{=} 2$

$\dfrac{159}{53} \overset{?}{=} 3$ $\qquad\qquad\qquad\qquad\qquad$ $\dfrac{106}{53} \overset{?}{=} 2$

$3 \overset{\checkmark}{=} 3$ $\qquad\qquad\qquad\qquad\qquad\qquad$ $2 \overset{\checkmark}{=} 2$

$4s - 5t - 5u = 1$

$4\left(\dfrac{32}{53}\right) - 5\left(\dfrac{16}{53}\right) - 5\left(-\dfrac{1}{53}\right) \overset{?}{=} 1$

$\dfrac{128}{53} - \dfrac{80}{53} + \dfrac{5}{53} \overset{?}{=} 1$

$\dfrac{53}{53} \overset{?}{=} 1$

$1 \overset{\checkmark}{=} 1$

27. $\begin{cases} x^2 + y^2 = 8 \\ x + y = 4 \end{cases}$

From the second equation,
$y = 4 - x$. Substitute
this result into the
first equation:

$$x^2 + (4-x)^2 = 8$$

$$x^2 + 16 - 8x + x^2 = 8$$
$$2x^2 - 8x + 16 = 8$$
$$2x^2 - 8x + 8 = 0$$
$$x^2 - 4x + 4 = 0$$
$$(x-2)^2 = 0$$
$$x - 2 = 0$$
$$x = 2$$

When $x = 2$, $y = 4 - 2 = 2$

Solution: (2,2)

29. $\begin{cases} 2x^2 - y^2 = 14 \\ 2x - 3y = -8 \end{cases}$

From the second equation,
$2x - 3y - 8$. So

$x = \dfrac{3y-8}{2} = \dfrac{3}{2}y - 4.$

Substitute this result
into the first equation:

$$2\left(\frac{3}{2}y - 4\right)^2 - y^2 = 14$$

$$2\left(\frac{9}{4}y^2 - 12y + 16\right) - y^2 = 14$$

$$\frac{9}{2}y^2 - 24y + 32 - y^2 = 14$$

$$\frac{7}{2}y^2 - 24y + 32 = 14$$

$$\frac{7}{2}y^2 - 24y + 18 = 0$$

$$7y^2 - 48y + 36 = 0$$

$$(7y - 6)(y - 6) = 0$$

$7y - 6 = 0$ or $y - 6 = 0$

$y = \dfrac{6}{7}$ or $y = 6$

CHECK (2,2):

$$x^2 + y^2 = 8$$
$$2^2 + 2^2 \overset{?}{=} 8$$
$$4 + 4 \overset{?}{=} 8$$
$$8 \overset{\checkmark}{=} 8$$

$$x + y = 4$$
$$2 + 2 \overset{?}{=} 4$$
$$4 \overset{\checkmark}{=} 4$$

CHECK $\left(-\dfrac{19}{7}, \dfrac{6}{7}\right)$:

$$2x^2 - y^2 = 14$$

$$2\left(-\frac{19}{7}\right)^2 - \left(\frac{6}{7}\right)^2 \overset{?}{=} 14$$

$$2\left(\frac{361}{49}\right) - \frac{36}{49} \overset{?}{=} 14$$

$$\frac{722}{49} - \frac{36}{49} \overset{?}{=} 14$$

$$\frac{686}{49} \overset{?}{=} 14$$

$$14 \overset{\checkmark}{=} 14$$

$$2x - 3y = -8$$

$$2\left(-\frac{19}{7}\right) - 3\left(\frac{6}{7}\right) \overset{?}{=} -8$$

$$-\frac{38}{7} - \frac{18}{7} \overset{?}{=} -8$$

$$-\frac{56}{7} \overset{?}{=} -8$$

$$-8 \overset{\checkmark}{=} -8$$

When $y = \frac{6}{7}$, $x = \frac{3}{2}\left(\frac{6}{7}\right) - 4 = \frac{-19}{7}$

When $y = 6$, $x = \frac{3}{2}(6) - 4 = 5$

Solutions: $\left(\frac{-19}{7}, \frac{6}{7}\right)$ and $(5,6)$

CHECK $(5,6)$:

$2x^2 - y^2 = 14$

$2(5)^2 - 6^2 \overset{?}{=} 14$

$2(25) - 36 \overset{?}{=} 14$

$50 - 36 \overset{?}{=} 14$

$14 \overset{\checkmark}{=} 14$

$2x - 3y = -8$

$2(5) - 3(6) \overset{?}{=} -8$

$10 - 18 \overset{?}{=} -8$

$-8 \overset{\checkmark}{=} -8$

31.
$$\begin{cases} x^2 + y^2 = 4 & \xrightarrow{\text{multiply by } -3} \\ 2x^2 + 3y^2 = 18 & \xrightarrow{\text{as is}} \end{cases}$$

$-3x^2 - 3y^2 = -12$

$\underline{2x^2 + 3y^2 = 18}$

Add: $\quad -x^2 = 6$

$\qquad\quad x^2 = -6$

Since there is no real number whose square is negative, the system has no real solutions.

33. $\quad x + y \le 2$
boundary: solid
x-intercept: 2
y-intercept: 2
test point: $(0,0)$

$0 + 0 \overset{?}{\le} 2$

$0 \overset{\checkmark}{\le} 2$

$2x - 3y \le 6$
boundary: solid
x-intercept: 3
y-intercept: -2
test point: (0.0)

$2(0) - 3(0) \overset{?}{\le} 6$

$0 - 0 \overset{?}{\le} 6$

$0 \overset{\checkmark}{\le} 6$

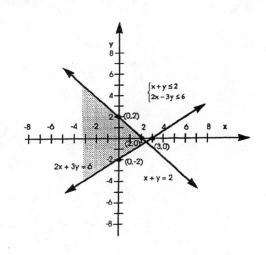

35. $\quad 2x + 3y \le 12$
boundary: solid
x-intercept: 6
y-intercept: 4
test point: $(0,0)$

342

$$2(0) + 3(0) \overset{?}{\leq} 12$$
$$0 + 0 \overset{?}{\leq} 12$$
$$0 \overset{\checkmark}{\leq} 12$$

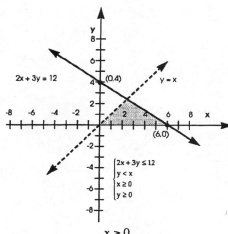

$y < x$

boundary: dotted

x-intercept: 0

y-intercept: 0

second boundary point: (1,1)

test point: (1,0)

$0 \overset{?}{<} 1$

$0 \overset{\checkmark}{<} 1$

$x \geq 0$

boundary: solid

y-axis

test point: (1,0)

$1 \overset{?}{\geq} 0$

$1 \overset{\checkmark}{\geq} 0$

$y \geq 0$

boundary: solid

x-axis

test point: (0,1)

$1 \overset{?}{\geq} 0$

$1 \overset{\checkmark}{\geq} 0$

37. Let x = amount invested in the bank account

y = amount invested in the corporate bond

z = amount invested in the municipal bond

$$\begin{array}{rcl} x + y + z = 20,000 & \rightarrow & x + y + z = 20,000 \\ y + z = x + 3,000 & \rightarrow & -x + y + z = 3,000 \\ 0.077x + 0.086y + 0.098z = 1,719.10 & \rightarrow & 77x + 86y + 98z = 1,719,100 \end{array}$$

Subtract equation (2) from equation (1), obtaining equation (4):

$2x = 17,000$

$x = 8,500$

Substitute this value of x into both equations (1) and (3):

$$8,500 + y + z = 20,000 \rightarrow y + z = 11,500$$
$$77(8,500) + 86y + 98z = 1,719,100 \rightarrow 86y + 98z = 1,064,600$$

Multiply the first of these equations by −86 and add the resulting equation to the second:

$$-86y - 86z = -989,000$$
$$\underline{86y + 98z = 1,064,600}$$
Add: $12z = 75,600$
$$z = 6,300$$

Then from $y + z = 11{,}500$, it follows that
$$y + 6{,}300 = 11{,}500$$
$$y = 5{,}200$$

CHECK: $\$8{,}500 + \$5{,}200 + \$6{,}300 \overset{\checkmark}{=} \$20{,}000$
$\$5{,}200 + \$6{,}300 = \$11{,}500$, which is
$\$3{,}000$ more than $\$8{,}500$
7.7% of $\$8{,}500 = (0.077)(8{,}500) = \$654.50$
8.6% of $\$5{,}200 = (0.086)(5{,}200) = \$447.20$
$\underline{9.8\% \text{ of } \$6{,}300 = (0.098)(6{,}300) = \$617.40}$

$$\text{Total} \overset{\checkmark}{=} \$1{,}719.10$$

39.   Let $x =$ width of the rectangle (in feet)

$y =$ length of the rectangle (in feet)

$$2x + 2y = 41$$
$$xy = 100$$

From the first equation, $2y = 41 - 2x$, so $y = \dfrac{41}{2} - x$.

Substitute this result into the second equation:

$$x\left(\frac{41}{2} - x\right) = 100$$

$$\frac{41}{2}x - x^2 = 100$$

$$41x - 2x^2 = 200$$

$$0 = 2x^2 - 41x + 200$$

$$0 = (2x - 25)(x - 8)$$

$0 = 2x - 25$ or $0 = x - 8$

$12.5 = x$ or $8 = x$

When $x = 12.5$, $y = \dfrac{41}{2} - 12.5 = 20.5 - 12.5 = 8$

When $x = 8$, $y = \dfrac{41}{2} - 8 = \dfrac{25}{2} = 12.5$

Conclude: The dimensions of the rectangle are 12.5 ft. and 8 ft.

CHECK: $2(12.5) + 2(8) = 25 + 16 \overset{\checkmark}{=} 41$
$(12.5)(8) \overset{\checkmark}{=} 100.$

Chapter 10 Practice Test

1. $\begin{cases} 2x - 3y + 4z = 2 \\ 3x + 2y - \ z = 10 \\ 2x - 4y + 3z = 3 \end{cases}$

Multiply equation (2) by 4 and add the resulting equation to equation (1), obtaining equation (4):

$$12x + 8y - 4z = 40$$
$$\underline{2x - 3y + 4z = \ \ 2}$$
$$14x + 5y = 42$$

Multiply equation (2) by 3 and add the resulting equation to equation (3), obtaining equation (5):

$$9x + 6y - 3z = 30$$
$$\underline{2x - 4y + 3z = \ \ 3}$$
$$11x + 2y = 33$$

Multiply equation (4) by –2, equation (5) by 5, and add the resulting equations:

$$-28x - 10y = -84$$
$$\underline{55x + 10y = 165}$$
$$27x = 81$$
$$x = 3$$

Substitute this value into (4):

$$14(3) + 5y = 42$$
$$42 + 5y = 42$$
$$5y = 0$$
$$y = 0$$

Finally, substitute these values into (1):

$$2(3) - 3(0) + 4z = 2$$
$$6 - 0 + 4z = 2$$
$$6 + 4z = 2$$
$$4z = -4$$
$$z = -1$$

Solution: $x = 3$, $y = 0$, $z = -1$

CHECK: $2x - 3y + 4z = 2$

$$3x + 2y - z = 10$$
$$3(3) + 2(0) - (-1) \overset{?}{=} 10$$
$$9 + 0 + 1 \overset{?}{=} 10$$
$$10 \overset{\checkmark}{=} 10$$

$$2x - 4y + 3z = 3$$
$$2(3) - 4(0) + 3(-1) \overset{?}{=} 3$$
$$6 - 0 - 3 \overset{?}{=} 3$$
$$3 \overset{\checkmark}{=} 3$$

3. (a) $\begin{vmatrix} 2 & 3 \\ -1 & 4 \end{vmatrix} = 2(4) - 3(-1) = 8 + 3 = 11$

(b) $\begin{vmatrix} 5 & 0 & 2 \\ 2 & 3 & 1 \\ 1 & 1 & 2 \end{vmatrix} = 5 \cdot \begin{vmatrix} 3 & 1 \\ 1 & 2 \end{vmatrix} - 0 \cdot \begin{vmatrix} 2 & 1 \\ 1 & 2 \end{vmatrix} + 2 \cdot \begin{vmatrix} 2 & 3 \\ 1 & 1 \end{vmatrix}$

$$= 5(6 - 1) - 0 + 2(2 - 3)$$
$$= 5(5) - 0 + 2(-1)$$
$$= 25 - 0 - 2 = 23$$

5.   Let x = # of $5 bills

          2y = # of $10 bills

           y = # of $20 bills

$$x + 2y + y = 40$$
$$5x + 10(2y) + 20y = 500$$

$$x + 3y = 40$$
$$5x + 40y = 500$$

Multiply the first equation by –5 and add the resulting equation to the second equation:

$$-5x - 15y = -200$$
$$\underline{5x + 40y = 500}$$
$$25y = 300$$
$$y = 12$$

Substitute: $x + 3y = 40$

$$x + 3(12) = 40$$
$$x + 36 = 40$$
$$x = 4$$

Conclude: The customer received 4 $5 bills, 24 $10 bills, and 12 $20 bills.

    CHECK: $4 + 24 + 12 \overset{\checkmark}{=} 40$

      4 $5 bills = 4($5) = $20

  24 $10 bills = 24($10) = $240

  $\underline{12\ \$20\ bills = 12(\$20) = \$240}$

             Total value $\overset{\checkmark}{=}$ $500

7. $y < 4$

    boundary: dotted

    horizontal line, 4 units above x-axis

    test point: (0,0)

    $0 \overset{?}{<} 4$

    $0 \overset{\checkmark}{<} 4$

    $x \le 2$

    boundary: solid

    vertical line, 2 units to the right of the y-axis

    test point: (0,0)

    $0 \overset{?}{\le} 2$

    $0 \overset{\checkmark}{\le} 2$

    $x + y > 3$

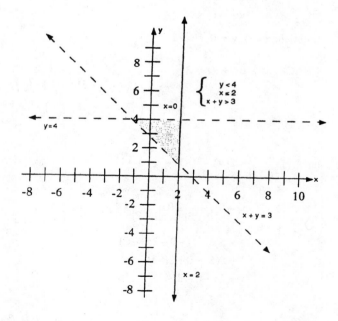

boundary: dotted

x-intercept: 3

y-intercept: 3

test point: (0,0)

$0 + 0 \overset{?}{>} 3$

$\quad 0 \not> 3$

CHAPTER 11
FUNCTIONS

Exercises 11.1

1. {(3,9), (3,7), (8,2), (7,2)}

5. domain = {3,4,5}; range = {2,3}

9. domain = $\{x|x \neq 0\}$

13. domain = $\left\{x \mid x \neq \dfrac{-3}{2}\right\}$

17. domain = $\{x|x \geq 4\}$

21. domain = $\left\{x \mid x \leq \dfrac{5}{4}\right\}$

23. domain = $\{x|x \geq 0\}$

25. domain = $\{x|x > 3\}$

27. domain = $\{x|{-3} \leq x \leq 3\}$
    range   = $\{y|{-3} \leq y \leq 3\}$

29. domain = $\{x|x \text{ is real}\}$
    range   = $\{y|y \geq -4\}$

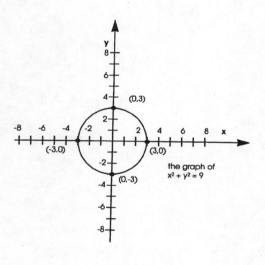

the graph of
$x^2 + y^2 = 9$

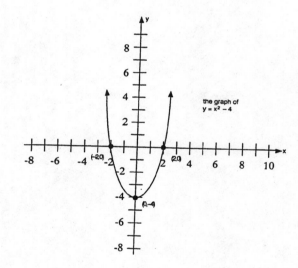

the graph of
$y = x^2 - 4$

31. domain = $\{x|x \geq -4\}$

range = $\{y|y \text{ is real}\}$

37. domain = $\{x|x \leq -2 \text{ or } x \geq 2\}$

range = $\{y|y \text{ is real}\}$

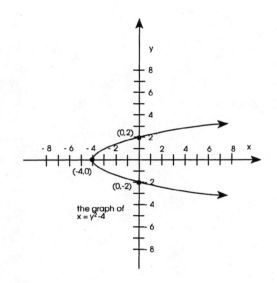

the graph of $x = y^2 - 4$

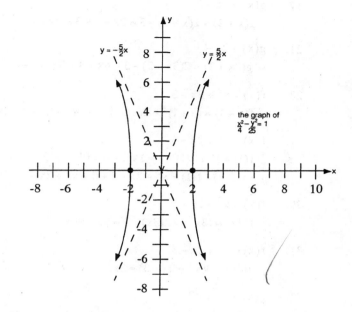

the graph of $\frac{x^2}{4} - \frac{y^2}{25} = 1$

39. function

43. function

45. not a function

49. function

53. function

59. function

63. function

65. not a function

67. (a) yes

(b) domain = $\{r|\$200 \leq r \leq \$400\}$; range = $\{P|\$35,000 \leq P \leq \$50,000\}$

(c) A rent of $300 per unit generates the maximum profit, which is $50,000.

68. A function is a rule that assigns to each element of one set a <u>unique</u> element of a second set.

Exercises 11.2

3. $g(x) = 3x^2 - x + 1$

$g(2) = 3(2)^2 - 2 + 1 = 3(4) - 2 + 1 = 12 - 2 + 1 = 11$

5. $g(x) = 3x^2 - x + 1$

$g(-2) = 3(-2)^2 - (-2) + 1 = 3(4) + 2 + 1 = 12 + 2 + 1 = 15$

9. $h(x) = \sqrt{x + 5}$

$h(-3) = \sqrt{-3 + 5} = \sqrt{2}$

11. $h(x) = \sqrt{x + 5}$

$h(a) = \sqrt{a + 5}$

13. $f(x) = x^2 + 2$

$f(x+1) = (x+1)^2 + 2 = x^2 + 2x + 1 + 2 = x^2 + 2x + 3$

17. $g(x) = 2x - 3$

$g(x+2) = 2(x+2) - 3 = 2x + 4 - 3 = 2x + 1$

21. $g(x) = 2x - 3$

$g(3x+2) = 2(3x+2) - 3 = 6x + 4 - 3 = 6x + 1$

23. $f(x) = x^2 + 2x - 3$

$f(x+1) = (x+1)^2 + 2(x+1) - 3 = x^2 + 2x + 1 + 2x + 2 - 3 = x^2 + 4x$

27. $f(x) = x^2 + 2x - 3$

$f(1) = 1^2 + 2(1) - 3 = 1 + 2 - 3 = 0$

So $f(x) + f(1) = x^2 + 2x - 3 + 0 = x^2 + 2x - 3$

29. $f(x) = x^2 + 2x - 3$

$f(3x) = (3x)^2 + 2(3x) - 3 = 9x^2 + 6x - 3$

31. $f(x) = x^2 + 2x - 3$

$3f(x) = 3(x^2 + 2x - 3) = 3x^2 + 6x - 9$

33. $g(x) = \sqrt{3x + 2}$

$g(3x+1) = \sqrt{3(3x+1) + 2} = \sqrt{9x + 3 + 2} = \sqrt{9x + 5}$

37. $g(x) = \sqrt{3x + 2}$

$g(x+h) = \sqrt{3(x+h) + 2} = \sqrt{3x + 3h + 2}$

39. $g(x) = \sqrt{3x + 2}$

$g(h) = \sqrt{3h + 2}$

So $g(x) + g(h) = \sqrt{3x + 2} + \sqrt{3h + 2}$

41. $g(x) = \sqrt{3x + 2}$

$g(x) + h = \sqrt{3x + 2} + h$

47. $f(x) = x^2 + 3x - 4$

$f(x+2) = (x+2)^2 + 3(x+2) - 4 = x^2 + 4x + 4 + 3x + 6 - 4 = x^2 + 7x + 6$

$f(x+2) - f(x) = x^2 + 7x + 6 - (x^2 + 3x - 4)$

$\qquad = x^2 + 7x + 6 - x^2 - 3x + 4 = 4x + 10$

So $\dfrac{f(x+2) - f(x)}{2} = \dfrac{4x + 10}{2} = 2x + 5$

49. $f(x) = x^2 + 3x - 4$

$f(x+h) = (x+h)^2 + 3(x+h) - 4 = x^2 + 2hx + h^2 + 3x + 3h - 4$

$f(x+h) - f(x) = x^2 + 2hx + h^2 + 3x + 3h - 4 - (x^2 + 3x - 4)$

$\qquad = x^2 + 2hx + h^2 + 3x + 3h - 4 - x^2 - 3x + 4 = 2hx + h^2 + 3h$

So $\dfrac{f(x+h) - f(x)}{h} = \dfrac{2hx + h^2 + 3h}{h} = \dfrac{\cancel{h}(2x + h + 3)}{\cancel{h}} = 2x + h + 3$

51. $g(x) = \sqrt{x+1}$

$g(3) = \sqrt{3+1} = \sqrt{4} = 2$

$f(x) = x^2 - 4$

$f(g(3)) = f(2) = 2^2 - 4 = 4 - 4 = 0$

55. $f(x) = x^2 - 4$, $g(x) = \sqrt{x+1}$

$f(g(x)) = f\left(\sqrt{x+1}\right) = \left(\sqrt{x+1}\right)^2 - 4 = x + 1 - 4 = x - 3$

57. $f(x) = x^2 - 4$, $g(x) = \sqrt{x+1}$

$g(f(x)) = g(x^2 - 4) = \sqrt{(x^2 - 4) + 1} = \sqrt{x^2 - 3}$

59. $h(x) = \dfrac{1}{x}$

$h(3) = \dfrac{1}{3}$

$g(x) = \sqrt{x+1}$

$g(h(3)) = g\left(\dfrac{1}{3}\right) = \sqrt{\dfrac{1}{3} + 1} = \sqrt{\dfrac{4}{3}} = \dfrac{\sqrt{4}}{\sqrt{3}} = \dfrac{2}{\sqrt{3}} = \dfrac{2\sqrt{3}}{\sqrt{3}\sqrt{3}} = \dfrac{2\sqrt{3}}{3}$

63. $g(x) = \sqrt{x+1}$, $h(x) = \dfrac{1}{x}$

$g(h(x)) = g\left(\dfrac{1}{x}\right) = \sqrt{\dfrac{1}{x} + 1} = \sqrt{\dfrac{1+x}{x}} = \dfrac{\sqrt{1+x}}{\sqrt{x}} = \dfrac{\sqrt{1+x}\sqrt{x}}{\sqrt{x}\sqrt{x}} = \dfrac{\sqrt{x+x^2}}{x}$

65. $f(x) = x^2 - 2x - 3$, $g(x) = x - 1$

$f(0) = 0^2 - 2(0) - 3 = 0 - 0 - 3 = -3$

$g(0) = 0 - 1 = -1$

So $(f + g)(0) = f(0) + g(0) = -3 - 1 = -4$

69. $f(x) = x^2 - 2x - 3$, $g(x) = x - 1$

$f(3) = 3^2 - 2(3) - 3 = 9 - 6 - 3 = 0$

$g(3) = 3 - 1 = 2$

So $\left(\dfrac{f}{g}\right)(3) = \dfrac{f(3)}{g(3)} = \dfrac{0}{2} = 0$

71. $f(x) = x^2 - 2x - 3$, $g(x) = x - 1$

$f(3) = 3^2 - 2(3) - 3 = 9 - 6 - 3 = 0$

$g(3) = 3 - 1 = 2$

Since $\left(\dfrac{g}{f}\right)(3) = \dfrac{g(3)}{f(3)}$ and $f(3) = 0$, $\left(\dfrac{g}{f}\right)(3)$ is undefined. That is, 3 is not in the domain of $\dfrac{g}{f}$.

75. (a) It is false to say that $f(x + 3) = f(x) + 3$.

(b) It is false to say that $f(3x) = 3f(x)$.

(c) It is false to say that $f(x + 3) = f(x) \cdot (x + 3)$.

77. $m = \dfrac{y_2 - y_1}{x_2 - x_1} = \dfrac{7 - 3}{6 - (-2)} = \dfrac{4}{8} = \dfrac{1}{2}$

$$y - y_1 = m(x - x_1)$$

$$y - 7 = \frac{1}{2}(x - 6)$$

$$y - 7 = \frac{1}{2}x - 3$$

$$y = \frac{1}{2}x + 4$$

79. $\begin{cases} 2x + 3y - z = 4 \\ 3x - y + z = -2 \\ x + y + 2z = -1 \end{cases}$

Add equations (1) and (2), obtaining equation (4):

$$\begin{array}{r} 2x + 3y - z = 4 \\ \underline{3x - y + z = -2} \\ 5x + 2y \phantom{{}- z} = 2 \end{array}$$

Multiply equation (1) by 2 and add the resulting equation to equation (3), obtaining equation (5):

$$\begin{array}{r} 4x + 6y - 2z = 8 \\ \underline{x + y + 2z = -1} \\ 5x + 7y = 7 \end{array}$$

Multiply equation (4) by −1 and add the resulting equation to equation (5):

$$\begin{array}{r} -5x - 2y = -2 \\ \underline{5x + 7y = 7} \\ 5y = 5 \\ y = 1 \end{array}$$

Substitute into equation (4): $\quad 5x + 2y = 2$

$$5x + 2(1) = 2$$
$$5x + 2 = 2$$
$$5x = 0$$
$$x = 0$$

Substitute into equation (2): $\quad 3x - y + z = -2$

$$3(0) - (1) + z = -2$$
$$0 - 1 + z = -2$$
$$-1 + z = -2$$
$$z = -1$$

Solution: $\quad x = 0, \; y = 1, \; z = -1$

Exercises 11.3

1. linear function

5. polynomial function

7. quadratic function

9. square root function

11. absolute value function

17. $f(x) = |4x - 1|$

$f(-2) = |4(-2) - 1| = |-8 - 1| = |-9| = 9$

$f(-1) = |4(-1) - 1| = |-4 - 1| = |-5| = 5$

$f(0) = |4(0) - 1| = |0 - 1| = |-1| = 1$

$f(1) = |4(1) - 1| = |4 - 1| = |3| = 3$

$f(2) = |4(2) - 1| = |8 - 1| = |7| = 7$

19. $f(x) = |4x| - 1$

$f(-2) = |4(-2)| - 1 = |-8| - 1 = 8 - 1 = 7$

$f(-1) = |4(-1)| - 1 = |-4| - 1 = 4 - 1 = 3$

$f(0) = |4(0)| - 1 = |0| - 1 = 0 - 1 = -1$

$f(1) = |4(1)| - 1 = |4| - 1 = 4 - 1 = 3$

$f(2) = |4(2)| - 1 = |8| - 1 = 8 - 1 = 7$

25. linear function

graph: straight line

slope −3, y-intercept 4

27. quadratic function

graph: parabola, opens upward

axis of symmetry: x = 0

vertex: (0,−9)

x-intercepts: (3,0) and (−3,0)

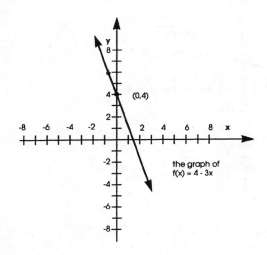

the graph of
f(x) = 4 - 3x

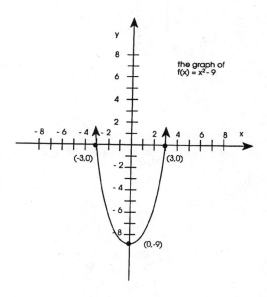

the graph of
f(x) = x² - 9

31. polynomial function

table of values:

x	f(x)
–2	–16
–1	–2
0	0
1	2
2	16

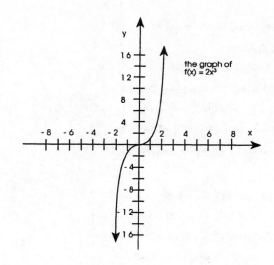

the graph of
$f(x) = 2x^3$

35. square root function

table of values:

x	f(x)
–5	0
–4	1
–1	2
4	3

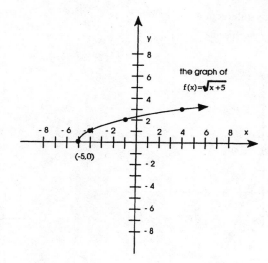

the graph of
$f(x) = \sqrt{x + 5}$

(-5,0)

37. square root function

table of values:

x	f(x)
0	5
1	6
4	7
9	8

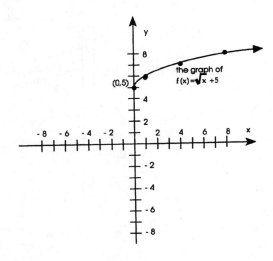

the graph of
$f(x) = \sqrt{x} + 5$

(0,5)

41. quadratic function

graph: parabola, opens upward

axis of symmetry: $x = 2$

vertex: $(2, -3)$

x-intercepts: $(2 + \sqrt{3}, 0)$ and $(2 - \sqrt{3}, 0)$

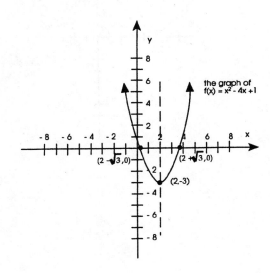

the graph of
$f(x) = x^2 - 4x + 1$

$(2 - \sqrt{3}, 0)$

$(2 + \sqrt{3}, 0)$

$(2, -3)$

355

45. quadratic function

graph: parabola, opens downward

axis of symmetry: $x = -1$

vertex: $(-1,9)$

x-intercepts: $(2,0)$ and $(-4,0)$

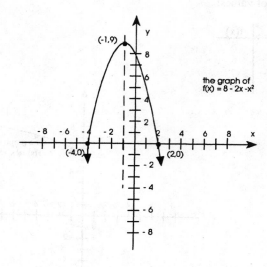

the graph of
$f(x) = 8 - 2x - x^2$

51. absolute value function

table of values:

x	f(x)
−7	2
−6	1
−5	0
−4	1
−3	2

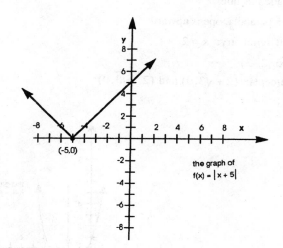

the graph of
$f(x) = |x + 5|$

53. absolute value function

table of values:

x	f(x)
−2	7
−1	6
0	5
1	6
2	7

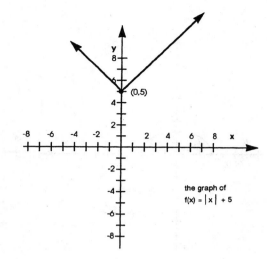

(0,5)

the graph of
$f(x) = |x| + 5$

57. absolute value function

table of values:

x	f(x)
3	2
4	1
5	0
6	1
7	2

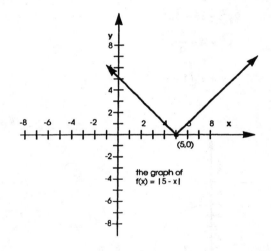

(5,0)

the graph of
$f(x) = |5 - x|$

59. (a)

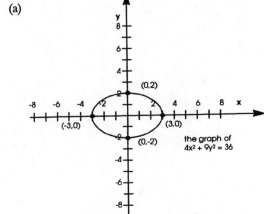

(0,2)

(-3,0)   (3,0)

(0,-2)   the graph of
$4x^2 + 9y^2 = 36$

357

(b)

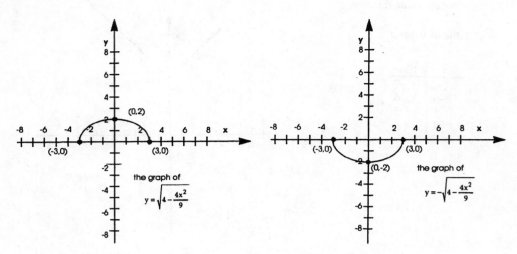

(c) When the two graphs of (b) are drawn on the same set of axes, we get the graph of (a).

(d) $4x^2 + 9y^2 = 36$

$9y^2 = 36 - 4x^2$

$y^2 = 4 - \dfrac{4}{9} x^2$

$y = \pm\sqrt{4 - \dfrac{4}{9} x^2}$

60. (a)

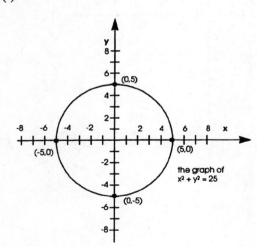

(b) $\quad x^2 + y^2 = 25$

$$y^2 = 25 - x^2$$

$$y = \pm\sqrt{25 - x^2}$$

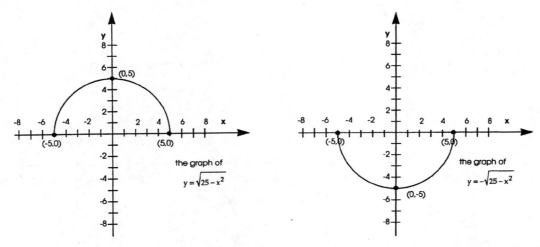

the graph of

$$y = \sqrt{25 - x^2}$$

the graph of

$$y = -\sqrt{25 - x^2}$$

(c) $\qquad$ When the two graphs of (b) are drawn on the same set of axes, we get the graph of (a).

61. (a)

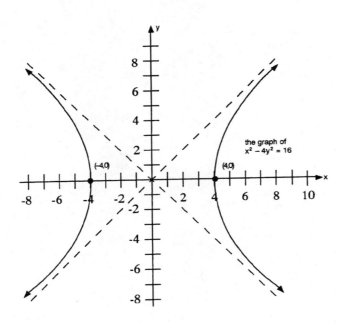

the graph of
$x^2 - 4y^2 = 16$

(b) $$x^2 - 4y^2 = 16$$
$$4y^2 = x^2 - 16$$
$$y^2 = \frac{1}{4}x^2 - 4$$
$$y = \pm\sqrt{\frac{1}{4}x^2 - 4}$$

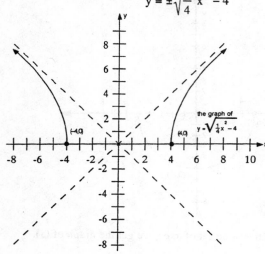

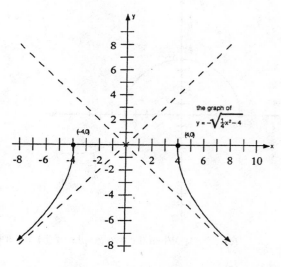

(c)   When the two graphs of (b) are drawn on the same set of axes, we get the graph of (a).

63. $$\frac{3}{2x-4} = \frac{x-2}{x+1}$$

$$(2x-4)(x+1)\left(\frac{3}{2x-4}\right) = (2x-4)(x+1)\left(\frac{x-2}{x+1}\right)$$

$$3(x+1) = (2x-4)(x-2)$$

$$3x+3 = 2x^2 - 8x + 8$$

$$3 = 2x^2 - 11x + 8$$

$$0 = 2x^2 - 11x + 5$$

$$0 = (2x-1)(x-5)$$

$$0 = 2x - 1 \text{ or } 0 = x - 5$$

$$1 = 2x \text{ or } 5 = x$$

$$\frac{1}{2} = x$$

65. $$(x-h)^2 + (y-k)^2 = r^2$$
$$(x-3)^2 + (y-(-4))^2 = 5^2$$
$$(x-3)^2 + (y+4)^2 = 25$$

Exercises 11.4

1.   {(3,1), (5,2)}

5.   {(5,3), (−5,2), (5,2), (6,2)}

7. $y = 3x - 4$

Switch x and y: $x = 3y - 4$

Solve for y: $x + 4 = 3y$

$$\frac{x + 4}{3} = y$$

13. $y = x^2 + 4$

Switch x and y: $x = y^2 + 4$

Solve for y: $x - 4 = y^2$

$$\pm\sqrt{x - 4} = y$$

17. one-to-one function

21. function

25. one-to-one function

29. neither

31. one-to-one function

35. Domain: $\{6, 2, -3\}$

Inverse: $\{(-3,6), (-4,2), (6, -3)\}$

Domain of inverse: $\{-3, -4, 6\}$

39. Domain of f(x): R (all real numbers)

Let $y = f(x)$, so that $y = 3x + 4$

Switch x and y: $x = 3y + 4$

Solve for y: $x - 4 = 3y$

$$\frac{x - 4}{3} = y$$

So $f^{-1}(x) = \dfrac{x - 4}{3}$

domain of $f^{-1}(x)$: R (all real numbers)

45. Domain of f(x): R (all real numbers)

Let $y = f(x)$, so that $y = x^2 + 2$

Switch x and y: $x = y^2 + 2$

Solve for y: $x - 2 = y^2$

$$\pm\sqrt{x - 2} = y$$

Since $x = 3$ gives $y = \pm\sqrt{3 - 2} = \pm\sqrt{1} = \pm1$, the inverse of f is not a function.

51. Domain of g(x): $\{x|x \neq -3\}$

Let $y = g(x)$, so that $y = \dfrac{2}{x+3}$

Switch x and y: $x = \dfrac{2}{y+3}$

Solve for y: $x(y+3) = 2$

$$xy + 3x = 2$$
$$xy = 2 - 3x$$
$$y = \dfrac{2-3x}{x}$$

So $g^{-1}(x) = \dfrac{2-3x}{x}$

Domain of $g^{-1}(x)$: $\{x|x \neq 0\}$

55. Domain of h(x): $\{x|x \neq 1\}$

Let $y = h(x)$, so that $y = \dfrac{x+2}{x-1}$

Switch x and y: $x = \dfrac{y+2}{y-1}$

Solve for y: $x(y-1) = y+2$

$$xy - x = y + 2$$
$$xy - y = x + 2$$
$$y(x-1) = x+2$$
$$y = \dfrac{x+2}{x-1}$$

So $h^{-1}(x) = \dfrac{x+2}{x-1}$

Domain of $h^{-1}(x)$: $\{x|x \neq 1\}$

(Note that this function is equal to its inverse!)

59. The graph of the inverse of the given function is identical to the graph of the function. This occurs because the given graph is its own reflection in the line $y = x$.

61. (a) Let $y = f(x)$, so that $y = 2x - 3$

Switch x and y: $x = 2y - 3$

Solve for y: $x + 3 = 2y$

$$\dfrac{x+3}{2} = y$$

So $f^{-1}(x) = \dfrac{x+3}{2}$

(b) $\quad f^{-1}(2) = \dfrac{2+3}{2} = \dfrac{5}{2}$

Thus $f\left[f^{-1}(2)\right] = f\left(\dfrac{5}{2}\right) = 2\left(\dfrac{5}{2}\right) - 3 = 5 - 3 = 2.$

$f(2) = 2(2) - 3 = 4 - 3 = 1$

Thus $f^{-1}[f(2)] = f^{-1}(1) = \dfrac{1+3}{2} = \dfrac{4}{2} = 2.$

(c) $\quad f\left[f^{-1}(x)\right] = f\left(\dfrac{x+3}{2}\right) = 2\left(\dfrac{x+3}{2}\right) - 3 = x + 3 - 3 = x$

$\quad f^{-1}[f(x)] = f^{-1}(2x - 3) = \dfrac{(2x-3)+3}{2} = \dfrac{2x}{2} = x$

62. (a) The graph of the relation $x^2 + y^2 = 25$ is a circle centered at the origin, with radius 5.

(b) Switch x and y, obtaining $y^2 + x^2 = 25$ or $x^2 + y^2 = 25$. This is the same as the original

equation, so the graph of the inverse relation is identical to the graph of the relation itself.

(c) We should expect the graph of the inverse relation to be the same as the graph of the relation

whenever the graph of the relation is its own reflection in the line $y = x$. This is true in the

case of any circle centered at the origin.

63. $\quad (5 - 3i)(5 - 2i) = 25 - 10i - 15i + 6i^2 = 25 - 25i + 6(-1)$
$\qquad\qquad\qquad\qquad\quad = 25 - 25i - 6 = 19 - 25i$

65. $\quad x^2 - 4x < 5$

$\qquad x^2 - 4x - 5 < 0$

$\qquad (x - 5)(x + 1) < 0$

$\qquad x - 5 = 0 \rightarrow x = 5$

$\qquad x + 1 = 0 \rightarrow x = -1$

Cut points are 5 and −1. Intervals are $x < -1, -1 < x < 5$, and $x > 5$. For $x < -1$, let $x = -2$ be the
test value. When $x = -2$, $(x - 5)(x + 1) = (-2 - 5)(-2 + 1)$ is positive. For $-1 < x < 5$, let $x = 0$ be
the test value. When $x = 0$, $(x - 5)(x + 1) = (0 - 5)(0 + 1)$ is negative. For $x > 5$, let $x = 6$ be the
test value. When $x = 6$, $(x - 5)(x + 1) = (6 - 5)(6 + 1)$ is positive. Thus, $x^2 - 4x < 5$ when

$-1 < x < 5.$

67. $\quad 2x - 3y < 6$

boundary: dotted

x-intercept: 3

y-intercept: −2

test point: (0,0)

$\qquad 2(0) - 3(0) \overset{?}{<} 6$

$\qquad\qquad 0 - 0 \overset{?}{<} 6$

$\qquad\qquad\quad 0 \overset{\checkmark}{<} 6$

$x + 3y > 3$

boundary: dotted

x-intercept: 3

y-intercept: 1

test point: (0,0)

$0 + 3(0) \overset{?}{>} 3$

$0 + 0 \overset{?}{>} 3$

$0 \not> 3$

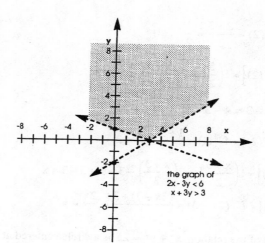

the graph of
$2x - 3y < 6$
$x + 3y > 3$

Exercises 11.5

1. $y = kx$ $\qquad$ $y = 2x$

    $8 = k(4)$ $\qquad$ $y = 2(3)$

    $2 = k$ $\qquad$ $y = 6$

5. $y = kx$ $\qquad$ $y = \dfrac{22}{3} x$

    $22 = k(3)$ $\qquad$ $5 = \dfrac{22}{3} x$

    $\dfrac{22}{3} = k$ $\qquad$ $\dfrac{15}{22} = x$

9. $r = ks^4$ $\qquad$ $r = \dfrac{3}{4} s^4$

    $12 = k(2)^4$ $\qquad$ $r = \dfrac{3}{4} (3)^4$

    $12 = k(16)$ $\qquad$ $r = \dfrac{3}{4} (81)$

    $\dfrac{3}{4} = k$ $\qquad$ $r = \dfrac{243}{4}$

13. $y = \dfrac{k}{x}$ $\qquad$ $y = \dfrac{252}{x}$

    $21 = \dfrac{k}{12}$ $\qquad$ $9 = \dfrac{252}{x}$

    $252 = k$ $\qquad$ $9x = 252$

    $\qquad$ $\qquad$ $x = 28$

17. $a = \dfrac{k}{b^3}$ $a = \dfrac{48}{b^3}$

       $6 = \dfrac{k}{2^3}$ $a = \dfrac{48}{(16)^3}$

       $6 = \dfrac{k}{8}$ $a = \dfrac{3}{256}$

       $48 = k$

21. $z = kxy$ $z = \dfrac{3}{2}\,xy$

       $12 = k(2)(4)$ $z = \dfrac{3}{2}(5)(2)$

       $12 = 8k$ $z = \dfrac{3}{2}(10)$

       $\dfrac{3}{2} = k$ $z = 15$

25. $a = kcd$ $a = \dfrac{5}{2}\,cd$

       $20 = k(2)(4)$ $25 = \dfrac{5}{2}(8)d$

       $20 = 8k$ $25 = 20d$

       $\dfrac{5}{2} = k$ $\dfrac{5}{4} = d$

29. $z = \dfrac{kx}{y}$ $z = \dfrac{\left(\frac{32}{3}\right)x}{y}$

       $16 = \dfrac{k(3)}{2}$ $z = \dfrac{\left(\frac{32}{3}\right)5}{3}$

       $32 = 3k$ $z = \dfrac{\frac{160}{3}}{3}$

       $\dfrac{32}{3} = k$ $z = \dfrac{160}{9}$

33. $z = \dfrac{kx}{y^2}$ $z = \dfrac{32x}{y^2}$

       $32 = \dfrac{k(4)}{2^2}$ $z = \dfrac{32(3)}{3^2}$

       $32 = \dfrac{4k}{4}$ $z = \dfrac{32}{3}$

       $32 = k$

37. $V = kr^3$ $V = \dfrac{4\pi}{3}\,r^3$

       $36\pi = k(3)^3$ $V = \dfrac{4\pi}{3}(4)^3$

$$36\pi = k(27) \qquad V = \frac{4\pi}{3}(64)$$

$$\frac{4\pi}{3} = k \qquad V = \frac{256\pi}{3}$$

The volume in question is $\frac{256\pi}{3}$ cm^3.

41. $E = \dfrac{k}{d^2}$ $\qquad$ $E = \dfrac{400}{d^2}$

$25 = \dfrac{k}{4^2}$ $\qquad$ $E = \dfrac{400}{8^2}$

$25 = \dfrac{k}{16}$ $\qquad$ $E = \dfrac{400}{64}$

$400 = k$ $\qquad$ $E = \dfrac{25}{4}$

When the distance is 8 feet, the illumination is $\frac{25}{4}$ footcandles.

45. $R = \dfrac{k\ell}{d^2}$ $\qquad$ $R = \dfrac{0.000015\ell}{d^2}$

$12 = \dfrac{k(80)}{(0.01)^2}$ $\qquad$ $R = \dfrac{0.000015(100)}{(0.02)^2}$

$12 = \dfrac{80k}{0.0001}$ $\qquad$ $R = \dfrac{0.0015}{0.0004}$

$0.0012 = 80k$ $\qquad$ $R = \dfrac{15}{4}$

$0.000015 = k$

The resistance in question is $\frac{15}{4}$ ohms.

47. $$\frac{\dfrac{2x}{y} + 7 + \dfrac{5y}{x}}{\dfrac{3x}{y} + 2 - \dfrac{y}{x}} = \frac{xy\left(\dfrac{2x}{y} + 7 + \dfrac{5y}{x}\right)}{xy\left(\dfrac{3x}{y} + 2 - \dfrac{y}{x}\right)} = \frac{2x^2 + 7xy + 5y^2}{3x^2 + 2xy - y^2}$$

$$= \frac{(2x + 5y)(x + y)}{(3x - y)(x + y)} = \frac{2x + 5y}{3x - y}$$

49. $$\sqrt[3]{5x^2y^5} \; \sqrt[3]{50x^5y^4} = \sqrt[3]{\left(5x^2y^5\right)\left(50x^5y^4\right)} = \sqrt[3]{250x^7y^9} = \sqrt[3]{\left(125x^6y^9\right)(2x)}$$

$$= \sqrt[3]{125x^6y^9} \; \sqrt[3]{2x} = 5x^2y^3 \; \sqrt[3]{2x}$$

Chapter 11 Review Exercises

1. Domain: {3,2}

Range: {4,5}

3. Domain: R (all real numbers)

Range: $\{y|y \geq -1\}$

(Graph is a parabola that opens upward
and has vertex $(0,-1)$.)

5. Domain: $\{x|-6 \leq x \leq 6\}$

   Range: $\{y|-6 \leq y \leq 6\}$

   (Graph is a circle centered at
   the origin, with radius 6.)

7. Domain: R (all real numbers)

9. Since $4 - x \geq 0$, it follows that $4 \geq x$ or $x \leq 4$. Domain: $\{x|x \leq 4\}$

11. Since $x + 2 \neq 0$, it follows that $x \neq -2$. Domain: $\{x|x \neq -2\}$

13. function

15. not a function (because both $(4,2)$ and $(4,7)$ are in the set that defines the relation)

17. function

19. function

21. function

23. not a function

25. $f(x) = 3x + 5$

   $f(-1) = 3(-1) + 5 = -3 + 5 = 2$

   $f(0) = 3(0) + 5 = 0 + 5 = 5$

   $f(1) = 3(1) + 5 = 3 + 5 = 8$

   $f(2) = 3(2) + 5 = 6 + 5 = 11$

27. $f(x) = 2x^2 - 3x + 2$

   $f(-1) = 2(-1)^2 - 3(-1) + 2 = 2(1) - 3(-1) + 2 = 2 + 3 + 2 = 7$

   $f(0) = 2(0)^2 - 3(0) + 2 = 2(0) - 3(0) + 2 = 0 - 0 + 2 = 2$

   $f(1) = 2(1)^2 - 3(1) + 2 = 2(1) - 3(1) + 2 = 2 - 3 + 2 = 1$

   $f(2) = 2(2)^2 - 3(2) + 2 = 2(4) - 3(2) + 2 = 8 - 6 + 2 = 4$

29. $h(x) = \sqrt{x - 5}$

   $h(6) = \sqrt{6 - 5} = \sqrt{1} = 1$

   $h(5) = \sqrt{5 - 5} = \sqrt{0} = 0$

   $h(4)$ cannot be computed, since 4 is not in the domain of $h(x)$.

31. $h(x) = \dfrac{x - 1}{x + 3}$

   $h(1) = \dfrac{1 - 1}{1 + 3} = \dfrac{0}{4} = 0$

   $h(3) = \dfrac{3 - 1}{3 + 3} = \dfrac{2}{6} = \dfrac{1}{3}$

   $h(-3)$ cannot be computed, since -3 is not in the domain of $h(x)$.

33. $f(x) = 2x^2 + 4x - 1$

$\quad f(a) = 2a^2 + 4a - 1$

$\quad f(z) = 2z^2 + 4z - 1$

35. $f(x) = 5x + 2$

$\quad f(x + 2) = 5(x + 2) + 2 = 5x + 10 + 2 = 5x + 12$

37. $f(x) = 5x + 2$

$\quad f(x) + 2 = 5x + 2 + 2 = 5x + 4$

39. $f(x) = 5x + 2$

$\quad f(2) = 5(2) + 2 = 10 + 2 = 12$

$\quad f(x) + f(2) = 5x + 2 + 12 = 5x + 14$

41. $g(x) = \sqrt{3 - 2x}$

$\quad g(x + 2) = \sqrt{3 - 2(x + 2)} = \sqrt{3 - 2x - 4} = \sqrt{-1 - 2x}$

43. $g(x) = \sqrt{3 - 2x}$

$\quad g(2x) = \sqrt{3 - 2(2x)} = \sqrt{3 - 4x}$

45. $g(x) = \sqrt{3 - 2x}$

$\quad 2g(x) = 2\sqrt{3 - 2x}$

47. $f(x) = 5x + 2$

$\quad f(x + h) = 5(x + h) + 2 = 5x + 5h + 2$

$\quad f(x + h) - f(x) = 5x + 5h + 2 - (5x + 2) = 5x + 5h + 2 - 5x - 2 = 5h$

49. $f(x) = 5x + 2, \ g(x) = \sqrt{3 - 2x}$

$\quad f(g(x)) = f\left(\sqrt{3 - 2x}\right) = 5\sqrt{3 - 2x} + 2$

51. linear function

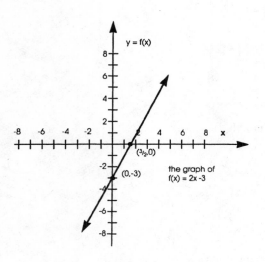

53. square root function

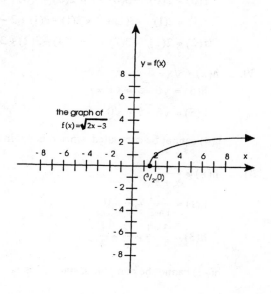

55. quadratic function

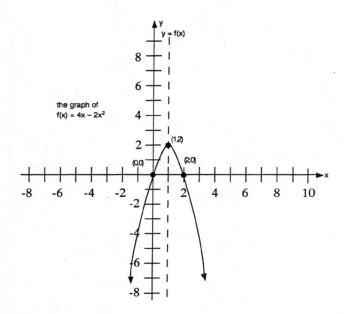

the graph of
$f(x) = 4x - 2x^2$

57. absolute value function

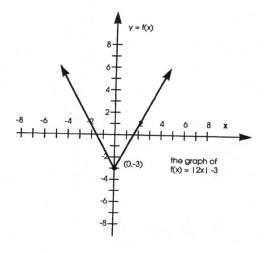

the graph of
$f(x) = |2x| - 3$

59. $\{(-1,0), (0,2), (-1,3)\}$

61. $y = x^2 - 3$

Switch x and y: $x = y^2 - 3$

Solve for y: $x + 3 = y^2$

$\pm\sqrt{x + 3} = y$

63. function

65. one-to-one function

67. one-to-one function

69. not a function

71. $\{(3,2), (4,3)\}$

73. $y = 3x + 8$

Switch x and y: $x = 3y + 8$

Solve for y: $x - 8 = 3y$

$\dfrac{x - 8}{3} = y$

75. $y = x^3$

Switch x and y: $x = y^3$

Solve for y: $\sqrt[3]{x} = y$

77. $y = \dfrac{3}{x + 1}$

Switch x and y: $x = \dfrac{3}{y + 1}$

Solve for y: $x(y+1) = 3$
$$xy + x = 3$$
$$xy = 3 - x$$
$$y = \frac{3 - x}{x}$$

79.

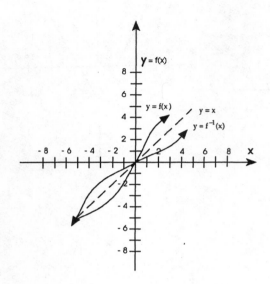

81. $x = ky$ $x = \frac{8}{3} y$

$8 = k(3)$ $x = \frac{8}{3} (2)$

$\frac{8}{3} = k$ $x = \frac{16}{3}$

83. $r = \frac{k}{s^2}$ $r = \frac{32}{s^2}$

$8 = \frac{k}{2^2}$ $4 = \frac{32}{s^2}$

$8 = \frac{k}{4}$ $4s^2 = 32$

$32 = k$ $s^2 = 8$

$s = \pm\sqrt{8} = \pm 2\sqrt{2}$

85. $z = kxy$ $z = \frac{8}{7} xy$

$16 = k(7)(2)$ $z = \frac{8}{7} (3)(4)$

$16 = 14k$ $z = \frac{96}{7}$

$\frac{8}{7} = k$

87. $V = kr^3$ $V = \dfrac{4\pi}{3} r^3$

$\dfrac{500\pi}{3} = k(5)^3$ $V = \dfrac{4\pi}{3}(6)^3$

$\dfrac{500\pi}{3} = 125k$ $V = \dfrac{4\pi}{3}(216)$

$\dfrac{4\pi}{3} = k$ $V = 288\pi$

When the radius of the sphere is 6 inches, its volume is 288π in^3.

89. $V = \dfrac{kT}{P}$ $V = \dfrac{120T}{P}$

$80 = \dfrac{k(20)}{30}$ $V = \dfrac{120(10)}{20}$

$80 = \dfrac{2}{3}k$ $V = \dfrac{1,200}{20}$

$120 = k$ $V = 60$

The volume in question is 60m^3.

Chapter 11 Practice Test

1.   (a)   Domain: $\{2,3,4\}$

               Range: $\{-3,5,6\}$

     (b)   Domain: $\{x|-1 \le x \le 1\}$; $\{y|-1 \le y \le 1\}$

       (Graph is a circle centered at the origin, with radius 1)

3.   (a)   not a function   (b)   function

     (c)   not a function   (d)   function

5.

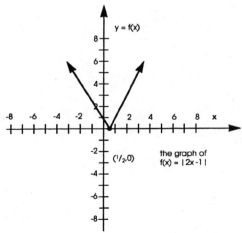

7.   The given function is not one-to-one, since the horizontal line test is violated.

9.
$$y = \frac{k}{x^2} \qquad y = \frac{96}{x^2}$$

$$24 = \frac{k}{2^2} \qquad y = \frac{96}{6^2}$$

$$24 = \frac{k}{4} \qquad y = \frac{96}{36}$$

$$96 = k \qquad y = \frac{8}{3}$$

CHAPTER 12
EXPONENTS AND
LOGARITHMIC FUNCTIONS

Exercises 12.1

1.

x	-3	-2	-1	0	1	2	3
$y = 2^x$	$\frac{1}{8}$	$\frac{1}{4}$	$\frac{1}{2}$	1	2	4	8

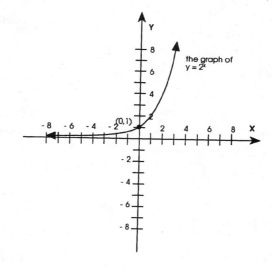

the graph of $y = 2^x$

5.

x	-3	-2	-1	0	1	2	3
$y = 3^{-x}$	27	9	3	1	$\frac{1}{3}$	$\frac{1}{9}$	$\frac{1}{27}$

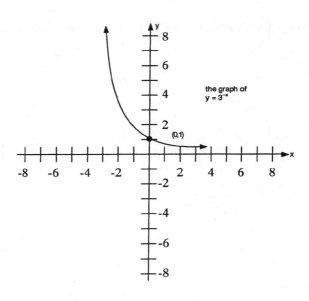

the graph of $y = 3^{-x}$

11. $2^x = 2^{3x-2}$ CHECK   x = 1:

           $x = 3x - 2$ $2^1 \overset{?}{=} 2^{3(1)-2}$

         $-2x = -2$ $2^1 \overset{?}{=} 2^{3-2}$

            $x = 1$ $2^1 \overset{\checkmark}{=} 2^1$

17. $8^x = 4^{x+1}$ CHECK   x = 2:

      $(2^3)^x = (2^2)^{x+1}$ $8^2 \overset{?}{=} 4^{2+1}$

         $2^{3x} = 2^{2x+2}$ $8^2 \overset{?}{=} 4^3$

          $3x = 2x + 2$ $64 \overset{\checkmark}{=} 64$

            $x = 2$

21. $4^{\sqrt{x}} = 2^{x-3}$ CHECK   x = 1:

       $(2^2)^{\sqrt{x}} = 2^{x-3}$ $4^{\sqrt{1}} \overset{?}{=} 2^{1-3}$

        $2^{2\sqrt{x}} = 2^{x-3}$ $4^1 \overset{?}{=} 2^{-2}$

        $2\sqrt{x} = x - 3$ $4 \neq \dfrac{1}{4}$

      $\left(2\sqrt{x}\right)^2 = (x-3)^2$

      $4x = x^2 - 6x + 9$ CHECK   x = 9:

      $0 = x^2 - 10x + 9$ $4^{\sqrt{9}} \overset{?}{=} 2^{9-3}$

      $0 = (x-1)(x-9)$ $4^3 \overset{?}{=} 2^6$

      $0 = x - 1$ or $0 = x - 9$ $64 \overset{\checkmark}{=} 64$

      $1 = x$ or $9 = x$

Since x = 1 does not check, the only solution of the given equation is x = 9.

23.

$$16^{x^2-1} = 8^{x-1}$$

$$(2^4)^{x^2-1} = (2^3)^{x-1}$$

$$2^{4x^2-4} = 2^{3x-3}$$

$$4x^2 - 4 = 3x - 3$$

$$4x^2 - 3x - 1 = 0$$

$$(4x+1)(x-1) = 0$$

$$4x+1 = 0 \text{ or } x-1 = 0$$

$$x = -\frac{1}{4} \text{ or } x = 1$$

CHECK $x = -\frac{1}{4}$:

$$16^{(-\frac{1}{4})^2-1} \stackrel{?}{=} 8^{-\frac{1}{4}-1}$$

$$16^{\frac{1}{16}-1} \stackrel{?}{=} 8^{-\frac{1}{4}-1}$$

$$16^{-\frac{15}{16}} \stackrel{?}{=} 8^{-\frac{5}{4}}$$

$$(2^4)^{-\frac{15}{16}} \stackrel{?}{=} (2^3)^{-\frac{5}{4}}$$

$$2^{-\frac{15}{4}} \stackrel{\checkmark}{=} 2^{-\frac{15}{4}}$$

CHECK $x = 1$:

$$16^{(1)^2-1} \stackrel{?}{=} 8^{1-1}$$

$$16^{1-1} \stackrel{?}{=} 8^{1-1}$$

$$16^0 \stackrel{?}{=} 8^0$$

$$1 \stackrel{\checkmark}{=} 1$$

25. Since the number of bacteria doubles every 10 hours, there will be $2(2,500) = 5,000$ bacteria after 10 hours, $(2(2(25)) =$ $2^2(2,500) = 10,000$ bacteria after 20 hours, and $2^5(2,500) = 80,000$ bacteria

after 50 hours. After t hours, the number of bacteria will be $2^{\frac{t}{10}}(2,500)$.

29. After 2 hours, the number of bacteria present is $\frac{1}{2}\left(\frac{1}{2}(10,000)\right) = \frac{1}{2^2}(10,000) = 2,500$; after 3 hours, the number of

bacteria present is $\frac{1}{2^3}(10,000) = 1,250$. After t hours, the number of bacteria present is

$\frac{1}{2^t}(10,000)$ or $\left(\frac{1}{2}\right)^t(10,000)$.

31.

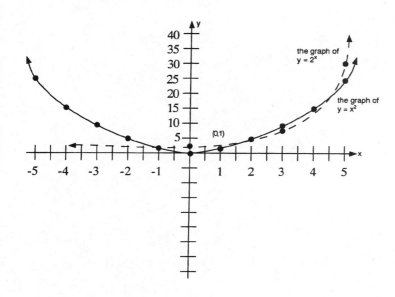

In both cases, y increases as x increases, where $x \geq 0$.

The following chart shows how much each function changes per unit change in x:

when x changes from	the change in x^2	the change in 2^x
0 to 1	1	1
1 to 2	3	2
2 to 3	5	4
3 to 4	7	8
4 to 5	9	16

When $x > 4$, the graph of $y = 2^x$ rises much faster than the graph of $y = x^2$.

Exercises 12.2

1. $7^2 = 49$

7. $\log_{10} 1{,}000 = 3$

11. $81^{\frac{1}{2}} = 9$

17. $\log_{25} 5 = \frac{1}{2}$

21. $\log_8 = 0$

27. $8^{-\frac{1}{3}} = \frac{1}{2}$

33. $7^0 = 1$

37. $\log_6 \sqrt{6} = \frac{1}{2}$

39. $\log_2 8 = \log_2 (2^3) = 3$

45. $\log_5 \frac{1}{125} = \log_5 (5^{-3}) = -3$

49. Let $\log_8 4 = t$. Then translating this into exponential form, we get $8^t = 4$. So $(2^3)^t = 2^2$ or $2^{3t} = 2^2$.

This means that $3t = 2$, so that $t = \frac{2}{3}$.

53. Let $\log_4 \frac{1}{8} = t$. Then the exponential form of this equation is $4^t = \frac{1}{8}$. So $(2^2)^t = 2^{-3}$, or $2^{2t} = 2^{-3}$.

This means that $2t = -3$, so that $t = -\frac{3}{2}$.

59. $\log_3 243 = \log_3(3^5) = 5$. Then $\log_5 (\log_3 243) = \log_5 5 = 1$.

63. $5^{\log_5 7} = 7$

67. $y = \log_{10} 1{,}000$

$10^y = 1{,}000$

$10^y = 10^3$

$y = 3$

71. $\log_6 x = -2$

$6^{-2} = x$

$\frac{1}{36} = x$

77. $\log_b \dfrac{1}{8} = -3$

$$b^{-3} = \dfrac{1}{8}$$
$$\dfrac{1}{b^3} = \dfrac{1}{8}$$
$$b^3 = 8$$
$$b = 2$$

81.

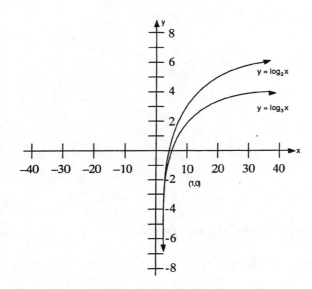

In exercises 83–86, we refer to the graph of $y = \log_3 x$ that appears in Figure 12.4(a) of the text.

83. The graph of $y = \log_3 (x - 1)$ is obtained from the graph of $y = \log_3 x$ by moving each point one unit to the right.

84. The graph of $y = \log_3 (x + 1)$ is obtained from the graph of $y = \log_3 x$ by moving each point one unit to the left.

85. The graph of $y = \log_3 x - 1$ is obtained from the graph of $y = \log_3 x$ by moving each point one unit down.

86. The graph of $y = \log_3 x + 1$ is obtained from the graph of $y = \log_3 x$ by moving each point one unit up.

87.
$$\left(\frac{x^{\frac{1}{3}}y^{\frac{1}{4}}}{xy^2}\right)^{12} = \left(x^{\frac{1}{3}-1}y^{\frac{1}{4}-2}\right)^{12} = \left(x^{-\frac{2}{3}}y^{-\frac{7}{4}}\right)^{12} = \left(x^{-\frac{2}{3}}\right)^{12}\left(y^{-\frac{7}{4}}\right)^{12}$$

$$= x^{-8}y^{-21} = \frac{1}{x^8}\cdot\frac{1}{y^{21}} = \frac{1}{x^8y^{21}}$$

89. $\quad g[f(x)] = g\left[\sqrt{x-3}\right] = 5\left(\sqrt{x-3}\right)^2 - 2 = 5(x-3) - 2 = 5x - 15 - 2 = 5x - 17$

Exercises 12.3

1. $\quad \log_5 xyz = \log_5 x + \log_5 y + \log_5 z$

7. $\quad \log_b a^{\frac{2}{3}} = \frac{2}{3}\log_b a$

13. $\quad \log_b x^2 y^3 = \log_b x^2 + \log_b y^3 = 2\log_b x + 3\log_b y$

17. $\quad \log_b \sqrt{xy} = \log_b (xy)^{\frac{1}{2}} = \frac{1}{2}\log_b (xy) = \frac{1}{2}(\log_b x + \log_b y)$

$$= \frac{1}{2}\log_b x + \frac{1}{2}\log_b y$$

21. $\quad \log_b (xy + z^2)$ cannot be simplified.

25. $\quad \log_6 \sqrt{\frac{6m^2n}{p^5q}} = \log_6\left(\frac{6m^2n}{p^5q}\right)^{\frac{1}{2}} = \frac{1}{2}\log_6\left(\frac{6m^2n}{p^5q}\right) = \frac{1}{2}(\log_6(6m^2n) - \log_6(p^5q))$

$$= \frac{1}{2}(\log_6 6 + \log_6 m^2 + \log_6 n - \log_6 p^5 - \log_6 q)$$

$$= \frac{1}{2}(1 + 2\log_6 m + \log_6 n - 5\log_6 p - \log_6 q)$$

$$= \frac{1}{2} + \log_6 m + \frac{1}{2}\log_6 n - \frac{5}{2}\log_6 p - \frac{1}{2}\log_6 q$$

29. $\quad 2\log_b m - 3\log_b n = \log_b m^2 - \log_b n^3 = \log_b\left(\frac{m^2}{n^3}\right)$

33. $\quad \frac{1}{3}\log_b x + \frac{1}{4}\log_b y - \frac{1}{5}\log_b z = \log_b x^{\frac{1}{3}} + \log_b y^{\frac{1}{4}} - \log_b z^{\frac{1}{5}}$

$$= \log_b x^{\frac{1}{3}}y^{\frac{1}{4}} - \log_b z^{\frac{1}{5}} = \log_b\left(\frac{x^{\frac{1}{3}}y^{\frac{1}{4}}}{z^{\frac{1}{5}}}\right)$$

37. $\quad 2\log_b x - (\log_b y + 3\log_b z) = 2\log_b x - (\log_b y + \log_b z^3)$

$$= 2\log_b x - (\log_b yz^3) = \log_b x^2 - \log_b yz^3 = \log_b\left(\frac{x^2}{yz^3}\right)$$

41. $\quad \log_b 10 = \log_b 2\cdot 5 = \log_b 2 + \log_b 5 = 1.2 + 2.1 = 3.3$

45. $\quad \log_b \frac{1}{3} = \log_b 1 - \log_b 3 = 0 - 1.42 = -1.42$

51. $\log_b \sqrt{20} = \log_b 20^{\frac{1}{2}} = \frac{1}{2} \log_b 20 = \frac{1}{2} \log_b 2^2 \cdot 5 = \frac{1}{2}(\log_b 2^2 + \log_b 5)$

$= \frac{1}{2}(2 \log_b 2 + \log_b 5) = \log_b 2 + \frac{1}{2} \log_b 5 = 1.2 + \frac{1}{2}(2.1)$

$= 1.2 + 1.05 = 2.25$

55. $\log_b \dfrac{x^3 y^2}{z} = \log_b x^3 y^2 - \log_b z = \log_b x^3 + \log_b y^2 - \log_b z$

$= 3 \log_b x + 2 \log_b y - \log_b z = 3A + 2B - C$

59. (a) $\log_b (x^3 + y^4) \neq \log_b x^3 + \log_b y^4$ 　　　(b) $\dfrac{\log_b x^3}{\log_b y^2} \neq \log_b x^3 - \log_b y^2$

60. Proof of the Quotient Rule:

Rule 4 for exponents says that $\dfrac{b^m}{b^n} = b^{m-n}$. Let $u = b^m$ and $v = b^n$. Writing these statements in

logarithmic form, we get $\log_b u = m$ and $\log_b v = n$. Then $\dfrac{u}{v} = \dfrac{b^m}{b^n} = b^{m-n}$, which is equivalent to

$\log_b \left(\dfrac{u}{v}\right) = m - n$. Substitution gives $\log_b \left(\dfrac{u}{v}\right) = \log_b u - \log_b v$.

Proof of the Power Rule:

Rule 2 for exponents says that $(b^m)^r = b^{mr}$ or b^{rm}. Let $u = b^m$ so that $u^r = (b^m)^r = b^{rm}$. Writing

these statements in logarithmic form, we get $\log_b u = m$ and $\log_b (u^r) = rm$. Substitution gives

$\log_b (u^r) = r \log_b u$.

Exercises 12.4

1. $\log 584 = 2.7664$

5. $\log 280,000 = 5.4472$

7. $\log 0.0000553 = -4.2573$

11. antilog $2.8420 = 695.0$

15. antilog $4.1875 = 1,540$

17. $\ln 0.941 = -0.0608$

19. antilog $0.941 = 8.730$

23. antilog $4.85 = 70,794.6$

25. $\ln 0.0045 = -5.4037$

27. $e^{4.5} = 90.02$

31. $\log_5 x = \dfrac{\log x}{\log 5}$

33. $\log_7 8 = \dfrac{\log 8}{\log 7}$

37. $\log_5 87 = \dfrac{\log 87}{\log 5} = \dfrac{1.9395}{0.6990} = 2.7747$

39. $\log_4 265 = \dfrac{\log 265}{\log 4} = \dfrac{2.4233}{0.6021} = 4.0247$

43. $\log_7 52 = \dfrac{\log 52}{\log 7} = \dfrac{1.7160}{0.8451} = 2.0305$

45. (a) $y = \log_a x$ (b) $a^y = x$ (c) $\log_b(a^y) = \log_b x$ (d) $y \log_b a = \log_b x$

     (e) $y = \dfrac{\log_b x}{\log_b a}$

     Comparing (a) and (e), $\log_a x = \dfrac{\log_b x}{\log_b a}$

47. $|5x - 2| > 3$

     $5x - 2 < -3$ or $5x - 2 > 3$

     $5x < -1$ or $5x > 5$

     $x < -\dfrac{1}{5}$ or $x > 1$

49.

$$\dfrac{3}{x-2} + 1 = \dfrac{10}{x}$$

$$x(x-2)\left(\dfrac{3}{x-2} + 1\right) = x(x-2)\left(\dfrac{10}{x}\right)$$

$$3x + x(x-2) = 10(x-2)$$

$$3x + x^2 - 2x = 10x - 20$$

$$x^2 + x = 10x - 20$$

$$x^2 - 9x = -20$$

$$x^2 - 9x + 20 = 0$$

$$(x-4)(x-5) = 0$$

$$x - 4 = 0 \text{ or } x - 5 = 0$$

$$x = 4 \text{ or } x = 5$$

CHECK $x = 4$:

$$\dfrac{3}{x-2} + 1 = \dfrac{10}{x}$$

$$\dfrac{3}{4-2} + 1 \overset{?}{=} \dfrac{10}{4}$$

$$\dfrac{3}{2} + 1 \overset{?}{=} \dfrac{5}{2}$$

$$\dfrac{5}{2} \overset{\checkmark}{=} \dfrac{5}{2}$$

CHECK $x = 5$:

$$\dfrac{3}{x-2} + 1 = \dfrac{10}{x}$$

$$\dfrac{3}{5-2} + 1 \overset{?}{=} \dfrac{10}{5}$$

$$\dfrac{3}{3} + 1 \overset{?}{=} 2$$

$$1 + 1 \overset{?}{=} 2$$

$$2 \overset{\checkmark}{=} 2$$

Exercises 12.5

1. $\log_3 5 + \log_3 x = 2$

     $\log_3 5x = 2$

       $5x = 3^2$

       $5x = 9$

         $x = \dfrac{9}{5}$

5.	$2 \log_5 x = \log_5 36$

$\log_5 x^2 = \log_5 36$

$x^2 = 36$

$x = \pm 6$

We reject $x = -6$, since we cannot take the logarithm of a negative number. Thus, the only solution is

$x = 6$.

9.	$\log_2 a + \log_2 (a + 2) = 3$

$\log_2 a(a + 2) = 3$

$a(a + 2) = 2^3$

$a^2 + 2a = 8$

$a^2 + 2a - 8 = 0$

$(a + 4)(a - 2) = 0$

$a + 4 = 0 \text{ or } a - 2 = 0$

$a = -4 \text{ or } a = 2$

We reject $a = -4$, since we cannot take the logarithm of a negative number. Thus, the only solution is

$a = 2$.

13.	$\log_3 x - \log_3 (x + 3) = 5$

$\log_3 \left(\dfrac{x}{x + 3} \right) = 5$

$\dfrac{x}{x + 3} = 3^5$

$\dfrac{x}{x + 3} = 243$

$x = 243(x + 3)$

$x = 243x + 729$

$-242x = 729$

$x = -\dfrac{729}{242}$

We reject this value, since we cannot take the logarithm of a negative number. Thus, the given

equation has no solution.

17.	$\log_p x - \log_p 2 = \log_p 7$

$\log_p \left(\dfrac{x}{2} \right) = \log_p 7$

$\dfrac{x}{2} = 7$

$x = 14$

21. $\log_3 x - \log_3 (x - 2) = \log_3 4$

$$\log_3 \left(\frac{x}{x-2} \right) = \log_3 4$$

$$\frac{x}{x-2} = 4$$

$$x = 4(x-2)$$

$$x = 4x - 8$$

$$-3x = -8$$

$$x = \frac{8}{3}$$

27. $\frac{1}{2} \log_3 x = \log_3 (x-6)$

$$\log_3 x^{\frac{1}{2}} = \log_3 (x-6)$$

$$x^{\frac{1}{2}} = x - 6$$

$$(x^{\frac{1}{2}})^2 = (x-6)^2$$

$$x = x^2 - 12x + 36$$

$$0 = x^2 - 13x + 36$$

$$0 = (x-4)(x-9)$$

$0 = x - 4$ or $0 = x - 9$

$4 = x$ or $9 = x$

We reject $x = 4$, since this value makes $x - 6$ negative, and we cannot take the logarithm of a negative number. Thus, the only solution is $x = 9$.

33. $2^{x+1} = 6$

$\log 2^{x+1} = \log 6$

$(x + 1) \log 2 = \log 6$

$x + 1 = \dfrac{\log 6}{\log 2}$

$$x = \frac{\log 6}{\log 2} - 1 = \frac{\log 6 - \log 2}{\log 2} = \frac{\log \left(\frac{6}{2} \right)}{\log 2} = \frac{\log 3}{\log 2} = \frac{0.4771}{0.3010} = 1.585$$

37. $7^{y+1} = 3y$

$\log 7^{y+1} = \log 3^y$

$(y + 1) \log 7 = y \log 3$

$y \log 7 + \log 7 = y \log 3$

$\log 7 = y \log 3 - y \log 7$

$\log 7 = y(\log 3 - \log 7)$

$$y = \frac{\log 7}{\log 3 - \log 7} = \frac{0.8451}{0.4881 - 0.8451} = -2.296$$

41.
$$8^{3x-2} = 9^{x+2}$$
$$\log 8^{3x-2} = \log 9^{x+2}$$
$$(3x-2)\log 8 = (x+2)\log 9$$
$$3x \log 8 - 2 \log 8 = x \log 9 + 2 \log 9$$
$$3x \log 8 - x \log 9 = 2 \log 8 + 2 \log 9$$
$$x(3 \log 8 - \log 9) = 2 \log 8 + 2 \log 9$$
$$x = \frac{2 \log 8 + 2 \log 9}{3 \log 8 - \log 9} = \frac{2(0.9031) + 2(0.9542)}{3(0.9031) - (0.9542)} = 2.116$$

45.
$$2^y \, 5^y = 3$$
$$\log 2^y \, 5^y = \log 3$$
$$\log 2^y + \log 5^y = \log 3$$
$$y \log 2 + y \log 5 = \log 3$$
$$y(\log 2 + \log 5) = \log 3$$
$$y = \frac{\log 3}{\log 2 + \log 5} = \frac{\log 3}{\log(2 \cdot 5)} = \frac{\log 3}{\log 10} = \log 3 = 0.4771$$

49. Let x = the first odd integer; $x + 2$ = the second odd integer; $x + 4$ = the third odd integer.
$$x + (x+2) + (x+4) = -27$$
$$3x + 6 = -27$$
$$3x = -33$$
$$x = -11$$
$$\text{Then } x + 2 = -11 + 2 = -9$$
$$x + 4 = -11 + 4 = -7$$
Thus, the three consecutive odd integers are -11, -9, and -7.

51. Let x = the number.
$$x + 2\left(\frac{1}{x}\right) = \frac{11}{3}$$
$$3x\left(x + \frac{2}{x}\right) = 3x\left(\frac{11}{3}\right)$$
$$3x^2 + 6 = 11x$$
$$3x^2 - 11x + 6 = 0$$
$$(x-3)(3x-2) = 0$$
$$x - 3 = 0 \text{ or } 3x - 2 = 0$$
$$x = 3 \text{ or } 3x = 2$$
$$x = \frac{2}{3}$$
Thus, there are two numbers with the given property: 3 and $\frac{2}{3}$.

53. Let s = distance that the object falls (in feet); t = time that the object falls (in seconds)

$$s = kt^2$$
$$64 = k(2)^2$$
$$64 = 4k$$
$$16 = k$$

So $s = 16t^2$. When $t = 5$, $s = 16(5)^2 = 400$. Thus, the object falls 400 ft. in 5 seconds.

Exercises 12.6

1. $A = P\left(1 + \dfrac{r}{n}\right)^{nt}$

Substitute $P = 8,000$, $r = 0.06$, $n = 2$, and $t = 5$

$A = 8,000\left(1 + \dfrac{0.06}{2}\right)^{2(5)} = 8,000(1 + 0.03)^{10}$

$= 8,000(1.03)^{10} = \$10,751.36$

3. $A = P\left(1 + \dfrac{r}{n}\right)^{nt}$

Substitute $P = 8,000$, $r = 0.06$, $n = 4$, and $t = 5$

$A = 8,000\left(1 + \dfrac{0.06}{4}\right)^{4(5)} = 8,000(1 + 0.015)^{20}$

$= 8,000(1.015)^{20} = \$10,774.80$

7. $A = P\left(1 + \dfrac{r}{n}\right)^{nt}$

Substitute $A = 2P$, $r = 0.062$, and $n = 12$

$2P = P\left(1 + \dfrac{0.062}{12}\right)^{12t}$

$2 = (1 + 0.0052)^{12t}$

$2 = (1.0052)^{12t}$

$\log 2 = \log(1.0052)^{12t} = 12t \log 1.0052$

So $t = \dfrac{\log 2}{12 \log 1.0052} = \dfrac{0.3010}{0.0270} = 11.15$

It would take 11.15 years for the money to double.

11. $A = Pe^{rt}$

Substitute $P = 5,000$, $A = 20,000$, and $r = 0.073$

$$20,000 = 5,000e^{0.073t}$$
$$4 = e^{0.073t}$$
$$\ln 4 = 0.073t$$
$$t = \frac{\ln 4}{0.073} = \frac{1.3863}{0.073} = 18.99$$

It would take 18.99 years.

13. $P = P_o e^{rt}$

Substitute $P_O = 2,000$, $r = 0.08$, and $t = 15$

$P = 2,000e^{(0.08)(15)} = 2,000e^{1.2}$
   $= 2,000(3.320) = 6,640$

The population of Rabbittville in 1995 will be 6,640.

17. $A = 10,000e^{0.0542t}$ $A = 10,000e^{0.122t}$

$100,000 = 10,000e^{0.0542t}$ $100,000 = 10,000e^{0.122t}$

       $10 = e^{0.0542t}$ $10 = e^{0.122t}$

   $\ln 10 = 0.0542t$ $\ln 10 = 0.122t$

         $t = \frac{\ln 10}{0.0542} = \frac{2.3026}{0.0542}$ $t = \frac{\ln 10}{0.122} = \frac{2.3026}{0.122}$

           $= 42.48$ $= 18.87$

21. $A = A_o e^{rt}$

Substitute $A_O = 1,000$, $A = 3,000$, and $t = 5$

$3,000 = 1,000e^{r(5)}$

     $3 = e^{5r}$

   $\ln 3 = 5r$

     $r = \frac{\ln 3}{5} = \frac{1.0986}{5} = 0.2197$

So $A = A_o e^{0.2197t}$

Now substitute $A_O = 1,000$ and $A = 50,000$

$50,000 = 1,000e^{0.2197t}$

     $50 = e^{0.2197t}$

   $\ln 50 = 0.2197t$

     $t = \frac{\ln 50}{0.2197} = \frac{3.9120}{0.2197} = 17.81$

It takes 17.81 hours for the bacteria colony to grow to 50,000.

23. $A = A_o e^{-0.0004t}$

   $\frac{1}{2}A_o = A_o e^{-0.0004t}$

   $\frac{1}{2} = e^{-0.0004t}$

   $\ln\left(\frac{1}{2}\right) = -0.0004t$

   $-\ln 2 = -0.0004t$

   $t = \frac{-\ln 2}{-0.0004} = \frac{-0.6931}{-0.0004} = 1,732.75$

So the half-life of radium is 1,732.75.

27.   Assuming $I_0 = 1$, $R = \log I$

   $R_1 = \log I_1$ $R_2 = \log I_2$
   $3.6 = \log I_1$ $7.2 = \log I_2$
   $I_1 = $ antilog 3.6 $I_2 = $ antilog 7.2
   $= 3,981$ $= 15,848,900$

So the earthquake with $R_2 = 7.2$ is almost 4,000 times as intense as the earthquake with $R_1 = 3.6$.

33. $N = 10 \log I + 160$

   $200 = 10 \log I + 160$

   $40 = 10 \log I$

$\log I = 4$

   $I = $ antilog $4 = 10,000$

So 10,000 watts/cm^2 is the intensity of sound at 200 decibels.

Chapter 12 Review Exercises

1.

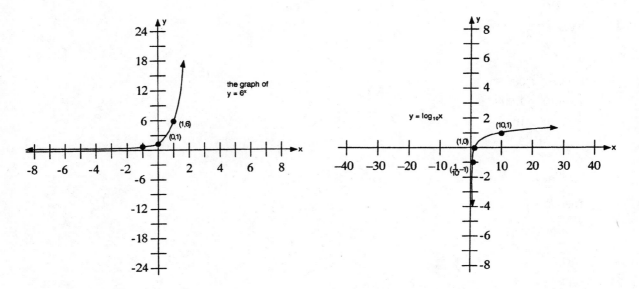

the graph of
$y = 6^x$

3.

$y = \log_{10}x$

5.

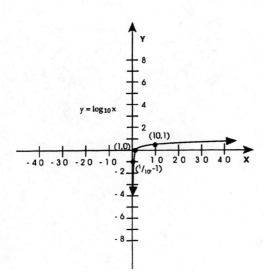

$y = \log_{10}x$

7. $3^4 = 81$

9. $\log_4 \dfrac{1}{64} = -3$

11. $8^{\frac{2}{3}} = 4$

13. $\log_{25} 5 = \dfrac{1}{2}$

15. $7^{\frac{1}{2}} = \sqrt{7}$

17. $6^0 = 1$

19. Let $\log_{10} 1{,}000 = t$
    Then $10^t = 1{,}000$
    $\qquad 10^t = 10^3$
    $\qquad\quad t = 3$

21. Let $\log_3 \dfrac{1}{9} = t$
    Then $3^t = \dfrac{1}{9}$
    $\qquad 3^t = 3^{-2}$
    $\qquad\ t = -2$

23. Let $\log_2 \dfrac{1}{4} = t$
    Then $2^t = \dfrac{1}{4}$
    $\qquad 2^t = 2^{-2}$
    $\qquad\ t = -2$

25. Let $\log_{\frac{1}{3}} 9 = t$
    Then $\left(\dfrac{1}{3}\right)^t = 9$
    $\qquad 3^{-t} = 3^2$
    $\qquad\ -t = 2$
    $\qquad\quad t = -2$

27. Let $\log_3 \dfrac{1}{9} = t$
    Then $3^t = \dfrac{1}{9}$
    $\qquad 3^t = 3^{-2}$
    $\qquad\ t = -2$

29. Let $\log_b \sqrt{b} = t$
    Then $b^t = \sqrt{b}$
    $\qquad b^t = b^{\frac{1}{2}}$
    $\qquad\ t = \dfrac{1}{2}$

31. Let $\log_{16} 32 = t$

Then $16^t = 32$

$(2^4)^t = 2^5$

$2^{4t} = 2^5$

$4t = 5$

$t = \dfrac{5}{4}$

33. $\log_b x^3 y^7 = \log_b x^3 + \log_b y^7 = 3 \log_b x + 7 \log_b y$

35. $\log_b \dfrac{u^2 v^5}{w^3} = \log_b (u^2 v^5) - \log_b w^3 = \log_b u^2 + \log_b v^5 - \log_b w^3$

$= 2 \log_b u + 5 \log_b v - 3 \log_b w$

37. $\log_b \sqrt[3]{xy} = \log_b (xy)^{\frac{1}{3}} = \dfrac{1}{3} \log_b (xy) = \dfrac{1}{3} (\log_b x + \log_b y)$

$= \dfrac{1}{3} \log_b x + \dfrac{1}{3} \log_b y$

39. $\log_b (x^3 + y^4)$ cannot be simplified

41. $\log_b \sqrt[4]{\dfrac{x^6 y^2}{z^2}} = \log_b \left(\dfrac{x^6 y^2}{z^2} \right)^{\frac{1}{4}} = \dfrac{1}{4} \log_b \left(\dfrac{x^6 y^2}{z^2} \right)$

$= \dfrac{1}{4} \left[\log_b (x^6 y^2) - \log_b z^2 \right] = \dfrac{1}{4} \left[\log_b x^6 + \log_b y^2 - \log_b z^2 \right]$

$= \dfrac{1}{4} \left[6 \log_b x + 2 \log_b y - 2 \log_b z \right]$

$= \dfrac{3}{2} \log_b x + \dfrac{1}{2} \log_b y - \dfrac{1}{2} \log_b z$

43. $\log_b 28 = \log_b (2^2 \cdot 7) = \log_b 2^2 + \log_b 7 = 2 \log_b 2 + \log_b 7$

$= 2(1.1) + 1.32 = 2.2 + 1.32 = 3.52$

45. $9^x = \dfrac{1}{81}$

$9^x = 9^{-2}$

$x = -2$

47. $16^x = 32$

$(2^4)^x = 2^5$

$2^{4x} = 2^5$

$4x = 5$

$x = \dfrac{5}{4}$

49.
$$5^{x+1} = 3$$
$$\log 5^{x+1} = \log 3$$
$$(x+1) \log 5 = \log 3$$
$$x \log 5 + \log 5 = \log 3$$
$$x \log 5 = \log 3 - \log 5$$
$$x = \frac{\log 3 - \log 5}{\log 5} = -0.3174$$

51.
$$\log (x+10) - \log (x+1) = 1$$
$$\log \left(\frac{x+10}{x+1} \right) = 1$$
$$10^1 = \frac{x+10}{x+1}$$
$$10 = \frac{x+10}{x+1}$$
$$10(x+1) = x+10$$
$$10x+10 = x+10$$
$$10x = x$$
$$9x = 0$$
$$x = 0$$

53.
$$\log_2 (t+1) + \log_2 (t-1) = 3$$
$$\log_2 ((t+1)(t-1)) = 3$$
$$2^3 = (t+1)(t-1)$$
$$8 = t^2 - 1$$
$$9 = t^2$$
$$\pm 3 = t$$

Reject $t = -3$, since this leads to logarithms of negative numbers. Thus, $t = 3$.

55.
$$\log_b 3x + \log_b (x+2) = \log_b 9$$
$$\log_b (3x(x+2)) = \log_b 9$$
$$3x(x+2) = 9$$
$$x(x+2) = 3$$
$$x^2 + 2x = 3$$
$$x^2 + 2x - 3 = 0$$
$$(x+3)(x-1) = 0$$

$x + 3 = 0$ or $x - 1 = 0$

$x = -3$ or $x = 1$

Reject $x = -3$, since this leads to logarithms of negative numbers. Thus, $x = 1$.

57. $\log 783 = 2.8938$

59. antilog $(-3) = 0.001$

61. $\log 0.00499 = -2.3019$

63. $\ln 0.0063 = -5.0672$

65. $e^{7.8} = 2,440.6$

67. $\log_5 73 = \dfrac{\log 73}{\log 5}$

$= \dfrac{1.8633}{0.6990} = 2.6657$

69. $\log_{12} 764 = \dfrac{\log 764}{\log 12}$

$= \dfrac{2.8831}{1.0792} = 2.6715$

71. $\log_{0.2} 190 = \dfrac{\log 190}{\log 0.2}$

$= \dfrac{2.2788}{-0.6990} = -3.2601$

73. $A = P\left(1 + \dfrac{r}{n}\right)^{nt}$

Substitute $P = 6,000$, $r = 0.082$, $n = 2$, and $t = 8$.

$A = 6,000\left(1 + \dfrac{.082}{2}\right)^{2(8)} = 6,000(1 + 0.041)^{16} = 6,000(1.041)^{16} = \$11,412.$

75. $A = A_o e^{-0.045t}$

$A_o = 100$ and $A = 25$

$25 = 100e^{-0.045t}$

$\dfrac{1}{4} = e^{-0.045t}$

$\ln \dfrac{1}{4} = -0.045t$

$-\ln 4 = -0.045t$

$t = \dfrac{-\ln 4}{-0.045} = \dfrac{-1.3863}{-0.045}$

$= 30.81$ years

77. $pH = -\log\left[H_3O^+\right] = -\log(6.21 \times 10^{-9}) = -(\log 6.21 + (-9))$

$= -\log 6.21 + 9 = -0.7931 + 9 = 8.2069$

1.

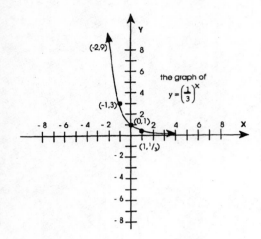

3. (a) $2^4 = 16$

(b) $9^{-\frac{1}{2}} = \frac{1}{3}$

5. (a) Let $\log_3 \frac{1}{3} = t$

Then $3^t = \frac{1}{3}$

$3^t = 3^{-1}$

$t = -1$

(b) Let $\log_{81} 9 = t$

Then $81^t = 9$

$(9^2)^t = 9$

$9^{2t} = 9^1$

$2t = 1$

$t = \frac{1}{2}$

(c)   Let $\log_8 32 = t$

Then $8^t = 32$

$(2^3)^t = 2^5$

$2^{3t} = 2^5$

$3t = 5$

$t = \dfrac{5}{3}$

7.   (a) $\log 27,900 = 4.4456$

(b) $\text{antilog}(-2.4) = 0.004$

(c) $\ln(0.004) = -5.5215$

(d) $e^{-0.02} = 0.9802$

9.   (a) $3^x = \dfrac{1}{81}$

$3^x = 3^{-4}$

$x = -4$

(b) $\log_2(x+4) - \log_2(x-2) = 4$

$\log_2\left(\dfrac{x+4}{x-2}\right) = 4$

$\dfrac{x+4}{x-2} = 2^4$

$\dfrac{x+4}{x-2} = 16$

$x + 4 = 16(x-2)$

$x + 4 = 16x - 32$

$4 = 15x - 32$

$36 = 15x$

$\dfrac{12}{5} = x$

(c) $4^{5x} = 32^{3x-4}$

$(2^2)^{5x} = (2^5)^{3x-4}$

$2^{10x} = 2^{15x-20}$

$10x = 15x - 20$

$-5x = -20$

$x = 4$

(d) $\log 5x - \log(x - 5) = 1$

$$\log\left(\frac{5x}{x-5}\right) = 1$$

$$\frac{5x}{x-5} = 10^1$$

$$\frac{5x}{x-5} = 10$$

$$5x = 10(x - 5)$$

$$5x = 10x - 50$$

$$-5x = -50$$

$$x = 10$$

(e) $$9^{x+3} = 5$$

$$\log(9^{x+3}) = \log 5$$

$$(x + 3)\log 9 = \log 5$$

$$x \log 9 + 3 \log 9 = \log 5$$

$$x \log 9 = \log 5 - 3 \log 9$$

$$x = \frac{\log 5 - 3 \log 9}{9}$$

$$x = \frac{0.6990 - 2.8627}{0.9542}$$

$$x = -2.2676$$

11. $A = A_o e^{-0.04t}$

Substitute $A_o = 100$ and $A = 25$

$$25 = 100e^{-0.04t}$$

$$\frac{1}{4} = e^{-0.04t}$$

$$\ln\left(\frac{1}{4}\right) = -0.04t$$

$$-\ln 4 = -0.04t$$

$$t = \frac{-\ln 4}{-0.04} = \frac{-1.3863}{-0.04} = 34.66 \text{ years}$$

Cumulative Review: Chapters 10–12

1. $$\begin{cases} x + y + z = 6 \\ 2x + y - 2z = 6 \\ 3x - y + 3z = 10 \end{cases}$$

Add equations (1) and (3), obtaining equation (4):
$$\begin{aligned} x + y + z &= 6 \\ \underline{3x - y + 3z} &= \underline{10} \\ 4x + 4z &= 16 \end{aligned}$$

Add equations (2) and (3), obtaining equations (5):

$$2x + y - 2z = 6$$
$$\underline{3x - y + 3z = 10}$$
$$5x + z = 16$$

Multiply equation (5) by −4 and add the resulting equation to equation (4):

$$4x + 4z = 16$$
$$\underline{-20x - 4z = -64}$$
$$-16x = -48$$
$$x = 3$$

Substitute into equation (5):

$$5x + z = 16$$
$$5(3) + z = 16$$
$$15 + z = 16$$
$$z = 1$$

Substitute into equation (1):

$$x + y + z = 6$$
$$3 + y + 1 = 6$$
$$y + 4 = 6$$
$$y = 2$$

Solution : $x = 3,\ y = 2,\ z = 1$

CHECK:

$x + y + z = 6$	$2x + y - 2z = 6$	$3x - y + 3z = 10$
$3 + 2 + 1 \overset{?}{=} 6$	$2(3) + 2 - 2(1) \overset{?}{=} 6$	$3(3) - 2 + 3(1) \overset{?}{=} 10$
$5 + 1 \overset{?}{=} 6$	$6 + 2 - 2 \overset{?}{=} 6$	$9 - 2 + 3 \overset{?}{=} 10$
$6 \overset{\checkmark}{=} 6$	$8 - 2 \overset{?}{=} 6$	$7 + 3 \overset{?}{=} 10$
	$6 \overset{\checkmark}{=} 6$	$10 \overset{\checkmark}{=} 10$

3. $\begin{cases} 3x - 4y + 5z = 1 \\ 2x - y + 3z = 2 \\ x - 2y + z = 3 \end{cases}$

Multiply equation (3) by −3 and add the resulting equation to equation (1), obtaining equation (4):

$$3x - 4y + 5z = 1$$
$$\underline{-3x + 6y - 3z = -9}$$
$$2y + 2z = -8$$

Multiply equation (3) by −2 and add the resulting equation to equation (2), obtaining equation (5):

$$2x - y + 3z = 2$$
$$\underline{-2x + 4y - 2z = -6}$$
$$3y + z = -4$$

Multiply equation (5) by −2 and add the resulting equation to equation (4):

$$2y + 2z = -8$$
$$\underline{-6y - 2z = 8}$$
$$-4y = 0$$
$$y = 0$$

Substitute into equation (5):

$$3y + z = -4$$
$$3(0) + z = -4$$
$$0 + z = -4$$
$$z = -4$$

Substitute into equation (3):
$$x - 2y + z = 3$$
$$x - 2(0) + (-4) = 3$$
$$x - 0 - 4 = 3$$
$$x - 4 = 3$$
$$x = 7$$
Solution: $x = 7$, $y = 0$, $z = -4$

CHECK:

$$3x - 4y + 5z = 1 \qquad\qquad 2x - y + 3z = 2 \qquad\qquad x - 2y + z = 3$$
$$3(7) - 4(0) + 5(-4) \overset{?}{=} 1 \qquad 2(7) - 0 + 3(-4) \overset{?}{=} 2 \qquad 7 - 2(0) + (-4) \overset{?}{=} 3$$
$$21 - 0 - 20 \overset{?}{=} 1 \qquad\qquad 14 - 0 - 12 \overset{?}{=} 2 \qquad\qquad 7 - 0 - 4 \overset{?}{=} 3$$
$$1 \overset{\checkmark}{=} 1 \qquad\qquad\qquad 2 \overset{\checkmark}{=} 2 \qquad\qquad\qquad 3 \overset{\checkmark}{=} 3$$

5. $\begin{cases} x - 2y + 3z = 4 \\ \dfrac{3}{2}x - 3y + \dfrac{9}{2}z = 6 \\ -3x + 6y - 9z = -12 \end{cases}$

Note that if we were to multiply equation (1) by $\dfrac{3}{2}$, we would get equation 2. Also, if we were

to multiply equation (1) by -3, we would get equation (3). Thus, all three equations are equivalent

and the system has infinitely many solutions. Each is of the form (x, y, z), where $x - 2y + 3z = 4$.

7. $\begin{vmatrix} 3 & 2 \\ 4 & 5 \end{vmatrix} = (3)(5) - (2)(4) = 15 - 8 = 7$

9. $\begin{vmatrix} 1 & 2 \\ 2 & 4 \end{vmatrix} = (1)(4) - (2)(2) = 4 - 4 = 0$

11. $\begin{vmatrix} 4 & -1 & 2 \\ 2 & 1 & 0 \\ -1 & 2 & -3 \end{vmatrix} = 4\begin{vmatrix} 1 & 0 \\ 2 & -3 \end{vmatrix} - (-1)\begin{vmatrix} 2 & 0 \\ -1 & -3 \end{vmatrix} + 2\begin{vmatrix} 2 & 1 \\ -1 & 2 \end{vmatrix}$

$= 4(-3 - 0) + (-6 - 0) + 2(4 - (-1)) = 4(-3) + (-6) + 2(5) = -8$

13. $\begin{vmatrix} 5 & 4 & 3 \\ 2 & 0 & 0 \\ 3 & 1 & 1 \end{vmatrix} = -2\begin{vmatrix} 4 & 3 \\ 1 & 1 \end{vmatrix} + 0\begin{vmatrix} 5 & 3 \\ 3 & 1 \end{vmatrix} - 0\begin{vmatrix} 5 & 4 \\ 3 & 1 \end{vmatrix}$

$= -2(4 - 3) + 0 - 0 = -2(1) = -2$

15. $\begin{bmatrix} 1 & 2 & | & 0 \\ 2 & -3 & | & 7 \end{bmatrix} \quad -2R_1 + R_2 \rightarrow R_2 \quad \begin{bmatrix} 1 & 2 & | & 0 \\ 0 & -7 & | & 7 \end{bmatrix}$

$\begin{cases} x + 2y = 0 \\ -7y = 7 \end{cases}$

First solve for y:
$-7y = 7$. Divide both sides of the equation by -7.
$y = -1$

Then substitute -1 for y in the first equation and find x:

$x + 2y = 0$

$x + 2(-1) = 0$

$x - 2 = 0$

$x = 2$

Hence, the solution is $(2, -1)$.

17. $\begin{cases} 3x + 7y = 2 \\ 10x + 5y = 11 \end{cases}$

$D = \begin{vmatrix} 3 & 7 \\ 10 & 5 \end{vmatrix} = (3)(5) - (7)(10) = 15 - 70 = -55$

$D_x = \begin{vmatrix} 2 & 7 \\ 11 & 5 \end{vmatrix} = (2)(5) - (7)(11) = 10 - 77 = -67$

$D_y = \begin{vmatrix} 3 & 2 \\ 10 & 11 \end{vmatrix} = (3)(11) - (2)(10) = 33 - 20 = 13$

Then $x = \dfrac{D_x}{D} = \dfrac{-67}{-55} = \dfrac{67}{55}$ and $y = \dfrac{D_y}{D} = \dfrac{13}{-55} = -\dfrac{13}{55}$

CHECK:

$3x + 7y = 2$ $10x + 5y = 11$

$3\left(\dfrac{67}{55}\right) + 7\left(-\dfrac{13}{55}\right) \overset{?}{=} 2$ $10\left(\dfrac{67}{55}\right) + 5\left(-\dfrac{13}{55}\right) \overset{?}{=} 11$

$\dfrac{201}{55} - \dfrac{91}{55} \overset{?}{=} 2$ $\dfrac{670}{55} - \dfrac{65}{55} \overset{?}{=} 11$

$\dfrac{110}{55} \overset{?}{=} 2$ $\dfrac{605}{55} \overset{?}{=} 11$

$2 \overset{\checkmark}{=} 2$ $11 \overset{\checkmark}{=} 11$

19. $\begin{cases} 2x + 3y + 4z = 3 \\ 5x + 2y - 3z = 2 \\ 3x - 7y + 5z = -7 \end{cases}$

$D = \begin{vmatrix} 2 & 3 & 4 \\ 5 & 2 & -3 \\ 3 & -7 & 5 \end{vmatrix} = 2\begin{vmatrix} 2 & -3 \\ -7 & 5 \end{vmatrix} - 3\begin{vmatrix} 5 & -3 \\ 3 & 5 \end{vmatrix} + 4\begin{vmatrix} 5 & 2 \\ 3 & -7 \end{vmatrix}$

$= 2(10 - 21) - 3(25 - (-9)) + 4(-35 - 6)$

$= 2(-11) - 3(34) + 4(-41) = -22 - 102 - 164 = -288$

$$D_x = \begin{vmatrix} 3 & 3 & 4 \\ 2 & 2 & -3 \\ -7 & -7 & 5 \end{vmatrix} = 3\begin{vmatrix} 2 & -3 \\ -7 & 5 \end{vmatrix} - 3\begin{vmatrix} 2 & -3 \\ -7 & 5 \end{vmatrix} + 4\begin{vmatrix} 2 & 2 \\ -7 & -7 \end{vmatrix}$$

$$= 3(10 - 21) - 3(10 - 21) + 4(-14 - (-14))$$

$$= 3(-11) - 3(-11) + 4(0) = -33 + 33 + 0 = 0$$

$$D_y = \begin{vmatrix} 2 & 3 & 4 \\ 5 & 2 & -3 \\ 3 & -7 & 5 \end{vmatrix} = -288, \text{ since it is identical to D.}$$

$$D_z = \begin{vmatrix} 2 & 3 & 3 \\ 5 & 2 & 2 \\ 3 & -7 & -7 \end{vmatrix} = 2\begin{vmatrix} 2 & 2 \\ -7 & -7 \end{vmatrix} - 3\begin{vmatrix} 5 & 2 \\ 3 & -7 \end{vmatrix} + 3\begin{vmatrix} 5 & 2 \\ 3 & -7 \end{vmatrix}$$

$$= 2(-14 - (-14)) - 3(-35 - 6) + 3(-35 - 6)$$

$$= 2(0) - 3(-41) + 3(-41) = 0 + 123 - 123 = 0$$

Then $x = \dfrac{D_x}{D} = \dfrac{0}{-288} = 0$, $\quad y = \dfrac{D_y}{D} = \dfrac{-288}{-288} = 1$, and $\quad z = \dfrac{D_z}{D} = \dfrac{0}{-288} = 0$

CHECK:

$2x + 3y + 4z = 3$	$5x + 2y - 3z = 2$	$3x - 7y + 5z = -7$
$2(0) + 3(1) + 4(0) \overset{?}{=} 3$	$5(0) + 2(1) - 3(0) \overset{?}{=} 2$	$3(0) - 7(1) + 5(0) \overset{?}{=} -7$
$0 + 3 + 0 \overset{?}{=} 3$	$0 + 2 - 0 \overset{?}{=} 0$	$0 - 7 + 0 \overset{?}{=} -7$
$3 \overset{\checkmark}{=} 3$	$2 \overset{\checkmark}{=} 2$	$-7 \overset{\checkmark}{=} -7$

21. $\begin{cases} x + y = 1 \\ x^2 + y^2 = 5 \end{cases}$

From the first equation, $y = 1 - x$. Substitute this result into the second equation:

$$x^2 + (1 - x)^2 = 5$$

$$x^2 + 1 - 2x + x^2 = 5$$

$$2x^2 - 2x + 1 = 5$$

$$2x^2 - 2x - 4 = 0$$

$$x^2 - x - 2 = 0$$

$$(x - 2)(x + 1) = 0$$

$$x - 2 = 0 \text{ or } x + 1 = 0$$

$$x = 2 \text{ or } x = -1$$

When $x = 2$, $y = 1 - 2 = -1$. When $x = -1$, $y = 1 - (-1) = 2$.

Solutions: $(2, -1)$ and $(-1, 2)$

CHECK $(2, -1)$:

$x + y = 1$	$x^2 + y^2 = 5$
	$2^2 + (-1)^2 \overset{?}{=} 5$
$2 + (-1) \overset{?}{=} 1$	$4 + 1 \overset{?}{=} 5$
$1 \overset{\checkmark}{=} 1$	$5 \overset{\checkmark}{=} 5$

CHECK (−1, 2):

$$x + y = 1 \qquad x^2 + y^2 = 5$$

$$\qquad\qquad (-1)^2 + 2^2 \overset{?}{=} 5$$

$$-1 + 2 \overset{?}{=} 1 \qquad\qquad 1 + 4 \overset{?}{=} 5$$

$$1 \overset{\checkmark}{=} 1 \qquad\qquad 5 \overset{\checkmark}{=} 5$$

23. $\begin{cases} y - x^2 = 4 \\ x^2 + y = 1 \end{cases}$

From the first equation, $y = x^2 + 4$. Substitute this result into the second equation:

$$x^2 + y = 1$$

$$x^2 + x^2 + 4 = 1$$

$$2x^2 + 4 = 1$$

$$2x^2 = -3$$

$$x^2 = -\frac{3}{2}$$

This equation cannot be solved, since the square of any real number is never negative. Therefore, the system is inconsistent.

25. $\begin{cases} x^2 + y^2 = 10 \xrightarrow{\text{multiply by 4}} \\ 3x^2 - 4y^2 = 23 \xrightarrow{\text{as is}} \end{cases}$

$$4x^2 + 4y^2 = 40$$

$$\underline{3x^2 - 4y^2 = 23}$$

Add $\quad 7x^2 = 63$

$$x^2 = 9$$

$$x = \pm 3$$

Substitute:

$$x^2 + y^2 = 10$$

$$9 + y^2 = 10$$

$$y^2 = 1$$

$$y = \pm 1$$

Solutions: (3,1), (−3,1), (3,−1), and (−3,−1)

CHECK (3,1):

$$3x^2 - 4y^2 = 23$$

$$x^2 + y^2 = 10 \qquad\qquad 3(3)^2 - 4(1)^2 \overset{?}{=} 23$$

$$3^2 + 1^2 \overset{?}{=} 10 \qquad\qquad 3(9) - 4(1) \overset{?}{=} 23$$

$$9 + 1 \overset{?}{=} 10 \qquad\qquad 27 - 4 \overset{?}{=} 23$$

$$10 \overset{\checkmark}{=} 10 \qquad\qquad 23 \overset{\checkmark}{=} 23$$

CHECK (–3, 1):

$$x^2 + y^2 = 10$$
$$(-3)^2 + 1^2 \overset{?}{=} 10$$
$$9 + 1 \overset{?}{=} 10$$
$$10 \overset{\checkmark}{=} 10$$

$$3x^2 - 4y^2 = 23$$
$$3(-3)^2 - 4(1)^2 \overset{?}{=} 23$$
$$3(9) - 4(1) \overset{?}{=} 23$$
$$27 - 4 \overset{?}{=} 23$$
$$23 \overset{\checkmark}{=} 23$$

CHECK (3, –1):

$$x^2 + y^2 = 10$$
$$3^2 + (-1)^2 \overset{?}{=} 10$$
$$9 + 1 \overset{?}{=} 10$$
$$10 \overset{\checkmark}{=} 10$$

$$3x^2 - 4y^2 = 23$$
$$3(3)^2 - 4(-1)^2 \overset{?}{=} 23$$
$$3(9) - 4(1) \overset{?}{=} 23$$
$$27 - 4 \overset{?}{=} 23$$
$$23 \overset{\checkmark}{=} 23$$

CHECK (–3, –1):

$$x^2 + y^2 = 10$$
$$(-3)^2 + (-1)^2 \overset{?}{=} 10$$
$$9 + 1 \overset{?}{=} 10$$
$$10 \overset{\checkmark}{=} 10$$

$$3x^2 - 4y^2 = 23$$
$$3(-3)^2 - 4(-1)^2 \overset{?}{=} 23$$
$$3(9) - 4(1) \overset{?}{=} 23$$
$$27 - 4 \overset{?}{=} 23$$
$$23 \overset{\checkmark}{=} 23$$

27. $\begin{cases} x - y^2 = 3 \\ 3x - 2y = 9 \end{cases}$

From the first equation, $x = y^2 + 3$. Substitute this result into the second equation:

$$3(y^2 + 3) - 2y = 9$$
$$3y^2 + 9 - 2y = 9$$
$$3y^2 - 2y = 0$$
$$y(3y - 2) = 0$$
$$y = 0 \text{ or } 3y - 2 = 0$$
$$y = 0 \text{ or } y = \frac{2}{3}$$

When $y = 0$, $x = 0^2 + 3 = 3$. When $y = \frac{2}{3}$, $x = \left(\frac{2}{3}\right)^2 + 3 = \frac{31}{9}$.

Solutions: $(3, 0)$ and $\left(\frac{31}{9}, \frac{2}{3}\right)$

CHECK (3, 0):

$$x - y^2 = 3$$
$$3 - 0^2 \overset{?}{=} 3$$
$$3 - 0 \overset{?}{=} 3$$
$$3 \overset{\checkmark}{=} 3$$

$$3x - 2y = 9$$
$$3(3) - 2(0) \overset{?}{=} 9$$
$$9 - 0 \overset{?}{=} 9$$
$$9 \overset{\checkmark}{=} 9$$

CHECK $\left(\frac{31}{9}, \frac{2}{3}\right)$:

$x - y^2 = 3$

$\frac{31}{9} - \left(\frac{2}{3}\right)^2 \overset{?}{=} 3$

$\frac{31}{9} - \frac{4}{9} \overset{?}{=} 3$

$\frac{27}{9} \overset{?}{=} 3$

$3 \overset{\checkmark}{=} 3$

$3x - 2y = 9$

$3\left(\frac{31}{9}\right) - 2\left(\frac{2}{3}\right) \overset{?}{=} 9$

$\frac{31}{3} - \frac{4}{3} \overset{?}{=} 9$

$\frac{27}{3} \overset{?}{=} 9$

$9 \overset{\checkmark}{=} 9$

29. $x + y \leq 6$

boundary: solid

x-intercept: 6

y-intercept: 6

test point: (0,0)

$0 + 0 \overset{?}{\leq} 6$

$0 \overset{\checkmark}{\leq} 6$

$2x + y \geq 4$

boundary: solid

x-intercept: 2

y-intercept: 4

test point: (0,0)

$2(0) + 0 \overset{?}{\geq} 4$

$0 + 0 \overset{?}{\geq} 4$

$0 \not\geq 4$

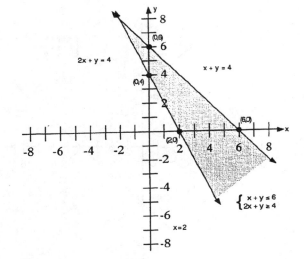

31. $2x + 3y < 12$

boundary: dotted

x-intercept: 6

y-intercept: 4

test point: (0,0)

$2(0) + 3(0) \overset{?}{<} 12$

$0 + 0 \overset{?}{<} 12$

$0 \overset{\checkmark}{<} 12$

$x < y$

boundary: dotted

x-intercept: 0

y-intercept: 0

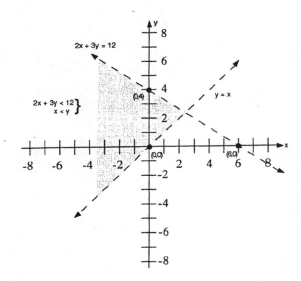

second boundary point: (1,1)

test point: (0,1)

$0 \overset{?}{\leq} 1$

$0 \overset{\checkmark}{\leq} 1$

33. $x + y \leq 4$

boundary : solid

x-intercept: 4

y-intercept: 4

test point: (0,0)]

$0 + 0 \overset{?}{\leq} 4$

$0 \overset{\checkmark}{\leq} 4$

$x - y \leq 4$

boundary: solid

x-intercept: 4

y-intercept: −4

test point: (0,0)

$0 - 0 \overset{?}{\leq} 4$

$0 \overset{\checkmark}{\leq} 4$

$x \geq 0$

boundary: solid

the y-axis

test point: (1,0)

$1 \overset{?}{\geq} 0$

$1 \overset{\checkmark}{\geq} 0$

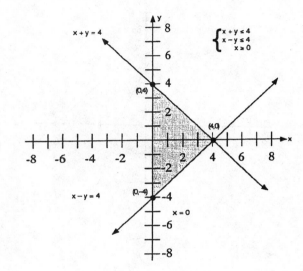

35. domain = {2,4,7}
range = {3,−1,5}

37. domain = {4,3}
range = {2,9,7}

39. domain = {x| −2 ≤ x ≤ 2}
range = {y| −1 ≤ y ≤ 1}

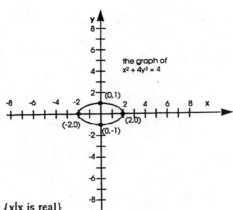

41. domain = {x|x is real}
    range = {y|y is real}

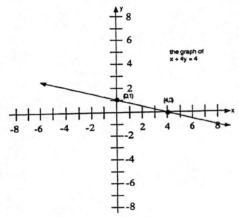

43. domain = {x|x is real}

45. domain = $\left\{x|x^2 - x - 6 \neq 0\right\} = \left\{x|(x-3)(x+2) \neq 0\right\} = \left\{x|x \neq 3 \text{ and } x \neq -2\right\}$

47. function   49. function

51. $f(x) = 2x^2 - 3x - 4$
    $f(-3) = 2(-3)^2 - 3(-3) - 4$
    $\quad\quad = 2(9) + 9 - 4 = 18 + 9 - 4 = 23$

53. $h(x) = 3x + 1$
    $h(x^2) = 3x^2 + 1$

55. $f(x) = 2x^2 - 3x - 4$
    $f(a + 3) = 2(a+3)^2 - 3(a+3) - 4 = 2(a^2 + 6a + 9) - 3(a+3) - 4$
    $\quad\quad = 2a^2 + 12a + 18 - 3a - 9 - 4 = 2a^2 + 9a + 5$

57. $f(x) = 2x^2 - 3x - 4$
    $f(3a) = 2(3a)^2 - 3(3a) - 4 = 2(9a^2) - 9a - 4 = 18a^2 - 9a - 4$

59. $f(x) = 2x^2 - 3x - 4$ $g(x) = \dfrac{1}{x}$

$f(1) = 2(1)^2 - 3(1) - 4$ $g(-1) = \dfrac{1}{-1} = -1$

$\quad = 2(1) - 3(1) - 4 = 2 - 3 - 4 = -5$

So $f(1) + g(-1) = -5 - 1 = -6$

61. $h(x) = 3x + 1$

$h(0) = 3(0) + 1 = 0 + 1 = 1$

$f(x) = 2x^2 - 3x - 4$

$f(h(0)) = f(1) = 2(1)^2 - 3(1) - 4 = 2(1) - 3(1) - 4 = 2 - 3 - 4 = -5$

63. $f(x) = 2x^2 - 3x - 4$ $g(x) = \dfrac{1}{x}$

$f(2) = 2(2)^2 - 3(2) - 4$ $g(2) = \dfrac{1}{2}$

$\quad = 2(4) - 3(2) - 4 = 8 - 6 - 4 = -2$

So $(f + g)(2) = f(2) + g(2) = -2 + \dfrac{1}{2} = -\dfrac{3}{2}$

65. $f(x) = 2x^2 - 3x - 4$

$f(x + 3) = 2(x+3)^2 - 3(x+3) - 4 = 2(x^2 + 6x + 9) - 3(x+3) - 4$

$\quad = 2x^2 + 12x + 18 - 3x - 9 - 4 = 2x^2 + 9x + 5$

So $\dfrac{f(x+3) - f(x)}{3} = \dfrac{2x^2 + 9x + 5 - (2x^2 - 3x - 4)}{3} = \dfrac{2x^2 + 9x + 5 - 2x^2 + 3x + 4}{3}$

$\quad = \dfrac{12x + 9}{3} = 4x + 3$

67. $y = x^3 - 1$

Switch x and y: $x = y^3 - 1$

Solve for y: $x + 1 = y^3$

$\sqrt[3]{x+1} = y$

69. $y = \dfrac{x+1}{x}$

Switch x and y: $x = \dfrac{y+1}{y}$

Solve for y: $xy = y + 1$

$\quad\quad\quad xy - y = 1$

$\quad\quad\quad y(x - 1) = 1$

$\quad\quad\quad\quad y = \dfrac{1}{x-1}$

71.

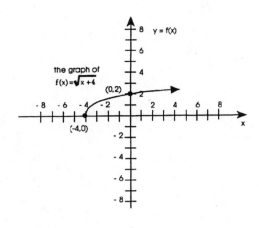

73.

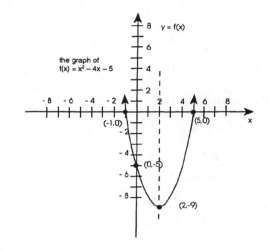

75.

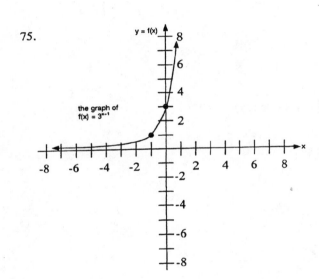

77.

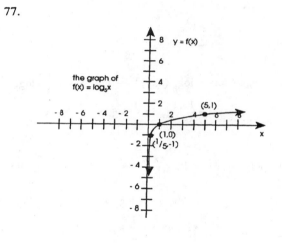

79. $2^6 = 64$

81. $\log_{125} 5 = \dfrac{1}{3}$

83. $27^{\frac{4}{3}} = 81$

85. Let $\log_3 81 = t$

Then $3^t = 81$

$3^t = 3^4$

$t = 4$

87. Let $\log_4 \frac{1}{16} = t$

Then $4^t = \frac{1}{16}$

$4^t = 4^{-2}$

$t = -2$

89. Let $\log_9 \frac{1}{3} = t$

Then $9^t = \frac{1}{3}$

$(3^2)^t = 3^{-1}$

$3^{2t} = 3^{-1}$

$2t = -1$

$t = -\frac{1}{2}$

91. Let $\log_b 1 = t$

Then $b^t = 1$

$b^t = b^0$

$t = 0$

93. $\log_b \sqrt[3]{5xy} = \log_b (5xy)^{\frac{1}{3}} = \frac{1}{3} \log_b 5xy = \frac{1}{3} (\log_b 5 + \log_b x + \log_b y)$

$= \frac{1}{3} \log_b 5 + \frac{1}{3} \log_b x + \frac{1}{3} \log_b y$

95. $\log_3 \frac{x^2 \sqrt{y}}{9wz} = \log_3 \frac{x^2 y^{\frac{1}{2}}}{9wz} = \log_3 x^2 y^{\frac{1}{2}} - \log_3 9wz$

$= \log_3 x^2 + \log_3 y^{\frac{1}{2}} - (\log_3 9 + \log_3 w + \log_3 z)$

$= 2 \log_3 x + \frac{1}{2} \log_3 y - 2 - \log_3 w - \log_3 z$

97. $\log 73,600 = 4.8669$

99. antilog $0.6085 = 4.060$

101. $\log_9 384 = \dfrac{\log 384}{\log 9}$

   $= \dfrac{2.5843}{0.9542}$

   $= 2.7083$

103. $\dfrac{1}{2} \log_8 x = \log_8 5$

   $\log_8 x^{\frac{1}{2}} = \log_8 5$

   $x^{\frac{1}{2}} = 5$

   $x = 25$

105. $5^x = \dfrac{1}{25}$

   $5^x = 5^{-2}$

   $x = -2$

107. $\log_6 x + \log_6 4 = 3$

   $\log_6 4x = 3$

   $6^3 = 4x$

   $216 = 4x$

   $54 = x$

109. $\dfrac{4^{x^2}}{2^x} = 64$

   $\dfrac{(2^2)^{x^2}}{2^x} = 2^6$

   $\dfrac{2^{2x^2}}{2^x} = 2^6$

   $2^{2x^2 - x} = 2^6$

   $2x^2 - x = 6$

   $2x^2 - x - 6 = 0$

   $(2x + 3)(x - 2) = 0$

   $2x + 3 = 0 \ \text{ or } \ x - 2 = 0$

   $x = -\dfrac{3}{2} \ \text{ or } \ x = 2$

111. $9^x = 7^{x+3}$

   $\log 9^x = \log 7^{x+3}$

   $x \log 9 = (x + 3) \log 7$

   $x \log 9 = x \log 7 + 3 \log 7$

   $x \log 9 - x \log 7 = 3 \log 7$

   $x(\log 9 - \log 7) = 3 \log 7$

$$x = \frac{3 \log 7}{\log 9 - \log 7} = \frac{3(0.8451)}{0.9542 - 0.8451} = 23.24$$

113. $y = kx$

    $8 = k(6)$

    $\frac{4}{3} = k$

$y = \frac{4}{3} x$

$y = \frac{4}{3}(20)$

$y = \frac{80}{3}$

115. $x = kyz$

    $10 = k(4)(15)$

    $\frac{1}{6} = k$

$x = \frac{1}{6} yz$

$x = \frac{1}{6}(6)(20)$

$x = 20$

117. $A = P\left(1 + \frac{r}{n}\right)^{nt}$

Substitute $A = 5{,}000$, $P = 3{,}000$, $r = 0.08$, and $n = 4$

$$5{,}000 = 3{,}000\left(1 + \frac{0.08}{4}\right)^{4t}$$

$$5{,}000 = 3{,}000(1 + 0.02)^{4t}$$

$$5{,}000 = 3{,}000(1.02)^{4t}$$

$$\frac{5}{3} = (1.02)^{4t}$$

$$\log \frac{5}{3} = \log (1.02)^{4t}$$

$$\log 5 - \log 3 = 4t \log 1.02$$

$$\frac{\log 5 - \log 3}{4 \log 1.02} = t$$

$$\frac{0.6990 - 0.4771}{4(0.0086)} = t$$

$$6.451 \text{ years} = t$$

119. $A = A_0 e^{-rt}$

Substitute $A = 2$, $A_0 = 20$, and $r = 0.002$

    $2 = 20e^{-0.002t}$

$$\frac{1}{10} = e^{-0.002t}$$

$$\ln\left(\frac{1}{10}\right) = -0.002t$$

$$-\ln 10 = -0.002t$$

$$t = \frac{-\ln 10}{-0.002} = \frac{-2.3026}{-0.002}$$

$$= 1,151 \text{ years}$$

121. $pH = -\log [H_3O^+]$

$8.2 = -\log [H_3O^+]$

$-8.2 = \log [H_3O^+]$

So $[H_3O^+] = $ antilog $(-8.2) = 6.31 \times 10^{-9}$

Cumulative Practice Test: Chapters 10–12

1.
$$\begin{cases} x + y + z = 6 \\ 3x + 2y - z = 11 \\ 2x - 4y - z = 12 \end{cases}$$

Add equations (1) and (2), obtaining equation (4):
$$x + y + z = 6$$
$$\underline{3x + 2y - z = 11}$$
$$4x + 3y = 17$$

Add equations (1) and (3), obtaining equation (5):
$$x + y + z = 16$$
$$\underline{2x - 4y - z = 12}$$
$$3x - 3y = 18$$

Add equations (4) and (5):
$$4x + 3y = 17$$
$$\underline{3x - 3y = 18}$$
$$7x = 35$$
$$x = 5$$

Substitute into equation (4):
$$4x + 3y = 17$$
$$4(5) + 3y = 18$$
$$20 + 3y = 17$$
$$3y = -3$$
$$y = -1$$

Substitute into equation (1):
$$x + y + z = 6$$
$$5 + (-1) + z = 6$$
$$4 + z = 6$$
$$z = 2$$

Solution: $x = 5$, $y = -1$, $z = 2$

CHECK:

$$x + y + z = 6 \qquad 3x + 2y - z = 11 \qquad 2x - 4y - z = 12$$
$$5 + (-1) + 2 \overset{?}{=} 6 \qquad 3(5) + 2(-1) - 2 \overset{?}{=} 11 \qquad 2(5) - 4(-1) - 2 \overset{?}{=} 12$$
$$4 + 2 \overset{?}{=} 6 \qquad 15 - 2 - 2 \overset{?}{=} 11 \qquad 10 + 4 - 2 \overset{?}{=} 12$$
$$6 \overset{\checkmark}{=} 6 \qquad 11 \overset{\checkmark}{=} 11 \qquad 12 \overset{\checkmark}{=} 12$$

3. (a) $\begin{cases} 6x + 5y = 13 \\ 7x + 8y = 26 \end{cases}$

$$D = \begin{vmatrix} 6 & 5 \\ 7 & 8 \end{vmatrix} = 6(8) - 5(7) = 48 - 35 = 13$$

$$D_x = \begin{vmatrix} 13 & 5 \\ 26 & 8 \end{vmatrix} = 13(8) - 5(26) = 104 - 130 = -26$$

$$D_y = \begin{vmatrix} 6 & 13 \\ 7 & 26 \end{vmatrix} = 6(26) - 13(7) = 156 - 91 = 65$$

Then $x = \dfrac{D_x}{D} = \dfrac{-26}{13} = -2$ and $y = \dfrac{D_y}{D} = \dfrac{65}{13} = 5$

CHECK:

$$6x + 5y = 13 \qquad\qquad\qquad 7x + 8y = 26$$
$$6(-2) + 5(5) \overset{?}{=} 13 \qquad\qquad\qquad 7(-2) + 8(5) \overset{?}{=} 26$$
$$-12 + 25 \overset{?}{=} 13 \qquad\qquad\qquad -14 + 40 \overset{?}{=} 26$$
$$13 \overset{\checkmark}{=} 13 \qquad\qquad\qquad 26 \overset{\checkmark}{=} 26$$

(b) $\begin{cases} 4x - 5y + 2z = 17 \\ 3x + 7y - 5z = 2 \\ 5x - 6y + 3z = 21 \end{cases}$

$$D = \begin{vmatrix} 4 & -5 & 2 \\ 3 & 7 & -5 \\ 5 & -6 & 3 \end{vmatrix} = 4 \cdot \begin{vmatrix} 7 & -5 \\ -6 & 3 \end{vmatrix} - (-5) \cdot \begin{vmatrix} 3 & -5 \\ 5 & 3 \end{vmatrix} + 2 \cdot \begin{vmatrix} 3 & 7 \\ 5 & -6 \end{vmatrix}$$
$$= 4(21 - 30) + 5(9 - (-25)) + 2(-18 - 35)$$
$$= 4(-9) + 5(34) + 2(-53)$$
$$= -36 + 170 - 106 = 28$$

$$D_x = \begin{vmatrix} 17 & -5 & 2 \\ 2 & 7 & -5 \\ 21 & -6 & 3 \end{vmatrix} = 17 \cdot \begin{vmatrix} 7 & -5 \\ -6 & 3 \end{vmatrix} - (-5) \cdot \begin{vmatrix} 2 & -5 \\ 21 & 3 \end{vmatrix} + 2 \cdot \begin{vmatrix} 2 & 7 \\ 21 & -6 \end{vmatrix}$$
$$= 17(21 - 30) + 5(6 - (-105)) + 2(-12 - 147)$$
$$= 17(-9) + 5(111) + 2(-159)$$
$$= -153 + 555 - 318 = 84$$

$$D_y = \begin{vmatrix} 4 & 17 & 2 \\ 3 & 2 & -5 \\ 5 & 21 & 3 \end{vmatrix} = 4 \cdot \begin{vmatrix} 2 & -5 \\ 21 & 3 \end{vmatrix} - 17 \cdot \begin{vmatrix} 3 & -5 \\ 5 & 3 \end{vmatrix} + 2 \cdot \begin{vmatrix} 3 & 2 \\ 5 & 21 \end{vmatrix}$$

$$= 4(6-(-105))-17(9-(-25))+2(63-10)$$

$$= 4(111)-17(34)+2(53)$$

$$= 444-578+106 = -28$$

$$D_z = \begin{vmatrix} 4 & -5 & 17 \\ 3 & 7 & 2 \\ 5 & -6 & 21 \end{vmatrix} = 4 \cdot \begin{vmatrix} 7 & 2 \\ -6 & 21 \end{vmatrix} - (-5) \cdot \begin{vmatrix} 3 & 2 \\ 5 & 21 \end{vmatrix} + 17 \cdot \begin{vmatrix} 3 & 7 \\ 5 & -6 \end{vmatrix}$$

$$= 4(147-(-12))+5(63-10)+17(-18-35)$$

$$= 4(159)+5(53)+17(-53)$$

$$= 636+265-901 = 0$$

Then $x = \dfrac{D_x}{D} = \dfrac{84}{28} = 3$, $y = \dfrac{D_y}{D} = \dfrac{-28}{28} = -1$, $z = \dfrac{D_z}{D} = \dfrac{0}{28} = 0$

CHECK:

$4x - 5y + 2z = 17$	$3x + 7y - 5z = 2$	$5x - 6y + 3z = 21$
$4(3) - 5(-1) + 2(0) \overset{?}{=} 17$	$3(3) + 7(-1) - 5(0) \overset{?}{=} 2$	$5(3) - 6(-1) + 3(0) \overset{?}{=} 21$
$12 + 5 + 0 \overset{?}{=} 17$	$9 - 7 - 0 \overset{?}{=} 2$	$15 + 6 + 0 \overset{?}{=} 21$
$17 \overset{\checkmark}{=} 17$	$2 \overset{\checkmark}{=} 2$	$21 \overset{\checkmark}{=} 21$

5. $x - y \le 5$

boundary: solid

x-intercept: 5

y-intercept: -5

test point: $(0,0)$

$0 - 0 \overset{?}{\le} 5$

$0 \overset{\checkmark}{\le} 5$

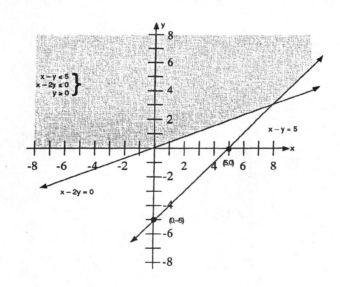

$x - 2y \le 0$

boundary : solid

x-intercept: 0

y-intercept: 0

second boundary point: $(2,1)$

test point: $(0,1)$

$0 - 2(1) \overset{?}{\le} 0$

$0 - 2 \overset{?}{\le} 0$

$-2 \overset{\checkmark}{\le} 0$

$y \ge 0$

boundary: solid

the x-axis

test point: $(0,1)$

$1 \overset{?}{\ge} 0$

$1 \overset{\checkmark}{\ge} 0$

7.   (a) $4x^2 + y^2 = 36$

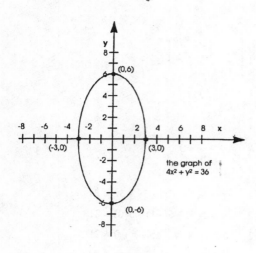

Domain $= \{x \mid -3 \le x \le 3\}$

Range $= \{y \mid -6 \le y \le 6\}$

The relation is not a function since its graph violates the vertical line test.

(b) $y = x^2 - 6x + 9$

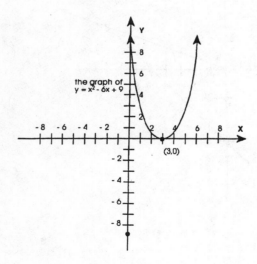

Domain $= \{x \mid x \text{ is real}\}$

Range $= \{y \mid y \ge 0\}$

412

The relation is a function, since its graph satisfies the vertical line test.

9. (a) $f(x) = x^2 - 3x + 2$

$$f(-5) = (-5)^2 - 3(-5) + 2$$
$$= 25 + 15 + 2$$
$$= 42$$

(b) $g(x) = 4x - 1$

$$g(x^2) = 4x^2 - 1$$

(c) $f(x) = x^2 - 3x + 2$

$$f(x + 2) = (x + 2)^2 - 3(x + 2) + 2$$
$$= x^2 + 4x + 4 - 3x - 6 + 2$$
$$= x^2 + x$$

(d) $f(x) = x^2 - 3x + 2$

$$f(x) + 2 = x^2 - 3x + 2 + 2$$
$$= x^2 - 3x + 4$$

(e) $g(x) = 4x - 1$

$$g(5x) = 4(5x) - 1$$
$$= 20x - 1$$

(f) $g(x) = 4x - 1$

$$5g(x) = 5(4x - 1)$$
$$= 20x - 5$$

(g) $g(x) = 4x - 1$ $\qquad f(x) = x^2 - 3x + 2$

$g(0) = 4(0) - 1 = -1$ $\qquad f(g(0)) = f(-1) = (-1)^2 - 3(-1) + 2$

$$= 1 + 3 + 2 = 6$$

(h) $f(x) = x^2 - 3x + 2$ $\qquad g(x) = 4x - 1$

$f(0) = 0^2 - 3(0) + 2 = 2$ $\qquad g(f(0)) = g(2) = 4(2) - 1$

$$= 8 - 1 = 7$$

(i) $f(x) = x^2 - 3x + 2$
$g(x) = 4x - 1$

$$\left(\frac{f}{g}\right)(1) = \frac{f(1)}{g(1)} = \frac{1^2 - 3(1) + 2}{4(1) - 1} = \frac{1 - 3 + 2}{4 - 1} = \frac{0}{3} = 0$$

(j)
$$f(x) = x^2 - 3x + 2$$
$$f(x+3) = (x+3)^2 - 3(x+3) + 2$$
$$= x^2 + 6x + 9 - 3x - 9 + 2$$
$$= x^2 + 3x + 2$$
$$f(x+3) - f(x) = x^2 + 3x + 2 - (x^2 - 3x + 2)$$
$$= x^2 + 3x + 2 - x^2 + 3x - 2$$
$$= 6x$$
$$\frac{f(x+3) - f(x)}{3} = \frac{6x}{3} = 2x$$

11.   (a)   (b)

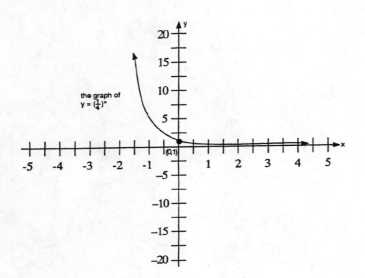

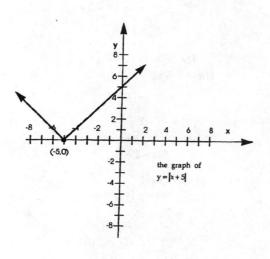

13.   (a)   Let $\log_3 \frac{1}{27} = x$

Then $3^x = \frac{1}{27}$

$$3^x = 3^{-3}$$

$$x = -3$$

So $\log_3 \frac{1}{27} = -3$

(b)   Let $\log_4 8 = x$

        Then $4^x = 8$

          $(2^2)^x = 2^3$

           $2^{2x} = 2^3$

             $2x = 3$

              $x = \dfrac{3}{2}$

      So $\log_4 8 = \dfrac{3}{2}$

15.   (a) $\log 0.00637 = -2.1959$

     (b)   antilog $3.9263 = 8{,}439$

     (c)   ln $94 = 4.5433$

17.   (a) $2x^2 = 8^{3x}$

         $2^{x^2} = (2^3)^{3x}$

         $2^{x^2} = 2^{9x}$

           $x^2 = 9x$

       $x^2 - 9x = 0$

      $x(x - 9) = 0$

            $x = 0$ or $x - 9 = 0$

            $x = 0$ or $x = 9$

     (b) $\log_6 x + \log_6 (x + 5) = 2$

               $\log_6 (x(x + 5)) = 2$

               $\log_6 (x^2 + 5x) = 2$

                   $x^2 + 5x = 6^2$

                   $x^2 + 5x = 36$

             $x^2 + 5x - 36 = 0$

           $(x + 9)(x - 4) = 0$

                  $x + 9 = 0$ or $x - 4 = 0$

                      $x = -9$ or $x = 4$

Reject $x = -9$, since we cannot take logarithms of negative numbers. So the only solution is $x = 4$.

19. $A = P\left(1 + \dfrac{r}{n}\right)^{nt}$

    $A = 10{,}000, \; P = 5{,}000, \; r = 0.1, \; n = 4$

$$10,000 = 5,000\left(1 + \frac{0.1}{4}\right)^{4t}$$

$$10,000 = 5,000(1,025)^{4t}$$

$$2 = (1.025)^{4t}$$

$$\log 2 = \log (1.025)^{4t}$$

$$\log 2 = 4t \log 1.025$$

$$t = \frac{\log 2}{4 \log 1.025} = 7.018 \text{ years}$$

CHAPTER 13
SEQUENCES AND SERIES

Exercises 13.1

1. $-2, 1, 4, 7, \ldots, 25$

3. $7, 11, 15, 19, \ldots, 43$

5. $4, 16, 64, 256, \ldots, 1{,}048{,}576$

7. $0, \dfrac{1}{3}, \dfrac{2}{4}, \dfrac{3}{5}, \ldots, \dfrac{9}{11}$

9. $-\dfrac{1}{3}, \dfrac{1}{4}, -\dfrac{1}{5}, \dfrac{1}{6}, \ldots, \dfrac{1}{12}$

11. $-\dfrac{1}{2}, \dfrac{2}{3}, -\dfrac{3}{4}, \dfrac{4}{5}, \ldots, \dfrac{10}{11}$

13. $6, 7, 10, 17, \ldots, 1{,}029$

15. $4.9, 5.01, 4.999, 5.0001, \ldots, 5.0000000001$

17. $a_n = 4n$

19. $a_n = 2n - 1$

21. $a_n = \dfrac{n}{n+1}$

23. $a_n = \dfrac{(-1)^{n+1}}{n+1}$

25. $a_n = 5(-.1)^n$

27. $\{\$26{,}850, \$27{,}700, \$28{,}550\}$

29. $\{21{,}200, 22{,}472, 23{,}820\}$

31. $\{\$5{,}300, \$5{,}618, \$5{,}955.08, \$6{,}312.38\}$

33. $\{\$31{,}800, \$33{,}708, \$35{,}730.48, \$37{,}874.31\}$

35. $\{30, 15, 7.5, 3.75, 1.875\}$

39. $\{4, 20, 100, 500, 2{,}500\}$

41. $\{6, 28, 138, 688, 3{,}438\}$

43. $\{4, 5, 1, -4, -5\}$

45. $a_{n+1} = 1.06a_n$

Exercises 13.2

1. $S_5 = 2 + 5 + 8 + 11 + 14 = 40$

3. $S_6 = -2 + 0 + 2 + 4 + 6 + 8 = 18$

5. $S_3 = 7 + 4 + 1 = 12$

7. $a_1 = 4, a_2 = 7.$ So $S_2 = 4 + 7 = 11$

9. $b_1 = 4, b_2 = 9, b_3 = 14.$ So $S_3 = 4 + 9 + 14 = 27$

11. $x_1 = 2, x_2 = 4, x_3 = 8, x_4 = 16, x_5 = 32.$ So $S_5 = 2 + 4 + 8 + 16 + 32 = 62$

13. $y_1 = 3, y_2 = 5, y_3 = 9, y_4 = 17, y_5 = 33, y_6 = 65.$ So $S_6 = 3 + 5 + 9 + 17 + 33 + 65 = 132$

15. $a_1 = \dfrac{1}{2}, a_2 = \dfrac{2}{3}, a_3 = \dfrac{3}{4}, a_4 = \dfrac{4}{5}.$

So $S_4 = \dfrac{1}{2} + \dfrac{2}{3} + \dfrac{3}{4} + \dfrac{4}{5} = \dfrac{30}{60} + \dfrac{40}{60} + \dfrac{45}{60} + \dfrac{48}{60} = \dfrac{163}{60}$

17. $b_1 = -\dfrac{1}{2}, b_2 = \dfrac{1}{3}, b_3 = -\dfrac{1}{4}, b_4 = \dfrac{1}{5}, b_5 = -\dfrac{1}{6}.$ So $S_5 = -\dfrac{1}{2} + \dfrac{1}{3} - \dfrac{1}{4} + \dfrac{1}{5} - \dfrac{1}{6} = -\dfrac{30}{60} + \dfrac{20}{60} - \dfrac{15}{60} + \dfrac{12}{60} - \dfrac{10}{60} = -\dfrac{23}{60}$

19. $1 + 2 + 3 + 4 + 5 = 15$

21. $5 + 10 + 15 + 20 + 25 + 30 + 35 = 140$

23. $1 + 4 + 9 + 16 + 25 + 36 = 91$

25. $16 + 54 + 128 + 250 = 448$

27. $11 + 13 + 15 + 17 = 56$

29. $9 + 13 + 17 + 21 + 25 + 29 + 33 = 147$

31. $7 + 17 + 31 + 49 + 71 = 175$

33. $\displaystyle\sum_{k=1}^{5} 2k$

35. $\displaystyle\sum_{k=1}^{7} k^2$

37. (a) $\displaystyle\sum_{i=1}^{n} ca_i = ca_1 + ca_2 + \ldots + ca_n$

$$= c(a_1 + a_2 + \ldots + a_n)$$

$$= c\left(\sum_{i=1}^{n} a_i\right)$$

(b) $\displaystyle\sum_{i=1}^{n} (a_i + b_i) = (a_1 + b_1) + (a_2 + b_2) + \ldots + (a_n + b_n)$

$$= (a_1 + a_2 + \ldots + a_n) + (b_1 + b_2 + \ldots + b_n)$$

$$= \sum_{i=1}^{n} a_i + \sum_{i=1}^{n} b_i$$

38. $\displaystyle\sum_{i=1}^{n} a_i = a_1 + a_2 + \ldots + a_n$

$$\sum_{i=2}^{n+1} a_{i-1} = a_{2-1} + a_{3-1} + \ldots + a_{(n+1)-1}$$

$$= a_1 + a_2 + \ldots + a_n.$$

So $\displaystyle\sum_{i=1}^{n} a_i = \sum_{i=2}^{n+1} a_{i-1}$

Exercises 13.3

1. $a_2 = 5$, $a_3 = 7$; $a_{15} = a_1 + 14d = 3 + 14(2) = 31$

3. $x_2 = -2$, $x_3 = -6$; $x_{15} = x_1 + 14d = 2 + 14(-4) = -54$

5. $z_2 = -\dfrac{3}{2}$, $z_3 = -1$; $z_{15} = z_1 + 14d = -2 + 14\left(\dfrac{1}{2}\right) = 5$

7. $a_5 = 13$, $a_6 = 16$; $a_{15} = a_1 + 14d = 1 + 14(3) = 43$

9. $a_5 = 6$, $a_6 = 8$; $a_{15} = a_1 + 14d = -2 + 14(2) = 26$

11. $a_4 = -13$, $a_5 = -18$; $a_{15} = a_1 + 14d = 2 + 14(-5) = -68$

13. $a_4 = 32$, $a_5 = 40$; $a_{15} = a_1 + 14d = 8 + 14(8) = 120$

15. $a_4 = \dfrac{3}{2},\; a_5 = \dfrac{11}{6};\; a_{15} = a_1 + 14d = \dfrac{1}{2} + 14\left(\dfrac{1}{3}\right) = \dfrac{31}{6}$

17. $a_3 = -\dfrac{4}{3},\; a_4 = -\dfrac{11}{6};\; a_{15} = a_1 + 14d = -\dfrac{1}{3} + 14\left(-\dfrac{1}{2}\right) = -\dfrac{22}{3}$

19. $a_{15} - a_5 = (15 - 5)d$

So $85 - 25 = (15 - 5)d$

$$60 = 10d$$
$$6 = d$$

Then $a_5 = a_1 + (5 - 1)d$

$$25 = a_1 + 4(6)$$
$$25 = a_1 + 24$$
$$1 = a_1$$

21. $a_{12} - a_7 = (12 - 7)d$

So $-41 - (-21) = (12 - 7)d$

$$-20 = 5d$$
$$-4 = d$$

Then $a_7 = a_1 + (7 - 1)d$

$$-21 = a_1 + 6(-4)$$
$$-21 = a_1 - 24$$
$$3 = a_1$$

Finally, $a_5 = a_1 + (5 - 1)d$

$$= 3 + 4(-4)$$
$$= -13$$

23. $a_1 = 26,500,\; d = 800,\; n = 6$
$a_6 = a_1 + (6 - 1)d$

$$= 26,500 + 5(800)$$
$$= 26,500 + 4,000 = 30,500$$

At the end of three years, Harry's yearly salary is $30,500.

25. $a_1 = 65,\; d = 5,\; n = 10$
$a_{10} = a_1 + (10 - 1)d$

$$= 65 + 9(5)$$
$$= 65 + 45 = 110$$

During the twentieth week, Cindy should be bench pressing 110 lbs.

27. $a_1 = 16,\; d = 32,\; n = 6$
$a_6 = a_1 + (6 - 1)d$

$$= 16 + 5(32)$$
$$= 16 + 160 = 176$$

During the sixth second, the object will fall 176 feet.

29. $a_1 = 8$, $d = 2$, $n = 6$

$$S_n = \frac{n}{2}[2a_1 + (n-1)d]$$

$$S_6 = \frac{6}{2}[2(8) + (6-1)2]$$

$$= 3[16 + 5(2)]$$

$$= 3[16 + 10]$$

$$= 3(26) = 78$$

33. $a_1 = 4$, $d = \frac{1}{2}$, $n = 5$

$$S_n = \frac{n}{2}[2a_1 + (n-1)d]$$

$$S_5 = \frac{5}{2}\left[2(4) + (5-1)\frac{1}{2}\right]$$

$$= \frac{5}{2}\left[8 + 4\left(\frac{1}{2}\right)\right]$$

$$= \frac{5}{2}[8 + 2]$$

$$= \frac{5}{2}(10) = 25$$

37. $a_1 = -2$, $d = 2$, $n = 6$

$$S_n = \frac{n}{2}[2a_1 + (n-1)d]$$

$$S_6 = \frac{6}{2}[2(-2) + (6-1)2]$$

$$= 3[-4 + 5(2)]$$

$$= 3[-4 + 10]$$

$$= 3(6) = 18$$

41. $a_1 = 3$, $a_6 = 18$

$$S_n = \frac{n}{2}(a_1 + a_n)$$

$$S_6 = \frac{6}{2}(a_1 + a_n)$$

$$= 3(3 + 18)$$

$$= 3(21) = 63$$

45. $a_1 = 7$, $a_5 = 23$

$$S_n = \frac{n}{2}(a_1 + a_n)$$

$$S_5 = \frac{5}{2}(a_1 + a_5)$$

$$= \frac{5}{2}(7 + 23)$$

$$= \frac{5}{2}(30) = 75$$

49. $S_{50} = \sum\limits_{i=1}^{50} 2i$

$a_1 = 2, \; a_{50} = 100.$

$S_{50} = \dfrac{50}{2}(a_1 + a_{50})$

$\quad\;\; = \dfrac{50}{2}(2 + 100)$

$\quad\;\; = \dfrac{50}{2}(102) = 2,550$

53. $a_{14} - a_4 = (14 - 4)d$

$\quad\; 23 - 3 = (14 - 4)d$

$\quad\quad\; 20 = 10d$

$\quad\quad\quad 2 = d$

$\quad\quad\; a_4 = a_1 + (4 - 1)d$

$\quad\quad\quad 3 = a_1 + 3(2)$

$\quad\quad\quad 3 = a_1 + 6$

$\quad\quad -3 = a_1$

Then $S_{15} = \dfrac{15}{2}[2(-3) + (15 - 1)(2)]$

$\quad\quad\quad\; = \dfrac{15}{2}[-6 + 14(2)]$

$\quad\quad\quad\; = \dfrac{15}{2}[-6 + 28]$

$\quad\quad\quad\; = \dfrac{15}{2}(22) = 165$

57. From exercise 27, we know that $a_6 = 176$.

Then $S_6 = \dfrac{6}{2}(a_1 + a_6)$

$\quad\quad\; = \dfrac{6}{2}(16 + 176)$

$\quad\quad\; = 3(192) = 576$

At the end of 6 seconds, the object will have fallen 576 feet.

61. An important factor that could help Darren decide which offer to accept is how long he plans to stay at the job once he goes to work. As it turns out, the total salary that Darren earns from Firms A and B will be identical if he remains on the job for 9 years.

<u>Firm A:</u> $a_1 = 28,000, \; d = 1,500, \; n = 9$

$S_9 = \dfrac{9}{2}[2(28,000) + (9 - 1)(1,500)]$

$\quad\; = \dfrac{9}{2}[56,000 + 12,000]$

$\quad\; = \dfrac{9}{2}[68,000] = 306,000$

<u>Firm B</u>: $a_1 = 24,000$, $d = 2,500$, $n = 9$

$$S_9 = \frac{9}{2}[2(24,000) + (9-1)(2,500)]$$
$$= \frac{9}{2}[48,000 + 20,000]$$
$$= \frac{9}{2}[68,000] = 306,000$$

If Darren plans to work for less than 9 years, Firm A's offer is better; if he plans to work for more than 9 years, Firm B's offer is better.

Exercises 13.4

1. $a_1 = 2$, $r = 3$; $a_6 = a_1 r^{6-1} = a_1 r^5 = 2(3)^5 = 486$

5. $a_1 = 3$, $r = \frac{1}{3}$; $a_8 = a_1 r^{8-1} = a_1 r^7 = 3\left(\frac{1}{3}\right)^7 = \frac{1}{729}$

9. $a_1 = \frac{1}{2}$, $r = \frac{1}{3}$; $a_5 = a_1 r^{5-1} = a_1 r^4 = \frac{1}{2}\left(\frac{1}{3}\right)^4 = \frac{1}{162}$

11. $a_1 = 3$, $r = 2$; $a_4 = a_1 r^{4-1} = a_1 r^3 = 3(2)^3 = 24$

15. $a_1 = 1$, $a_3 = 16$

So $a_3 = a_1 r^{3-1} = a_1 r^2$

$16 = 1r^2$

$16 = r^2$

$\pm 4 = r$

Then $a_5 = a_1 r^{5-1} = a_1 r^4$

$= 1(\pm 4)^4$

$= 256$

19. $A_4 = a_1 r^4$

$= (16,000)\left(\frac{5}{6}\right)^4$

$= 7,716.05$

In four years, the car will be worth $7,716.05.

23. $A_4 = a_1 r^4$

$= 100,000(1.05)^4$

$= 121,551$

Four years from now, the population of Davisville will be 121,551.

25. $A_5 = a_1 r^5$

      $= 10,000(1.08)^5$

      $= 14,693.30$

At the end of 5 years, Roberta will have \$14,693.30 in the bank.

29. $a_1 = 10, r = \dfrac{1}{2}, n = 5$

      $a_5 = a_1 r^{5-1} = a_1 r^4$

        $= 10\left(\dfrac{1}{2}\right)^4$

        $= \dfrac{10}{16}$

        $= \dfrac{5}{8}$

On the fifth rebound, the ball bounces up $\dfrac{5}{8}$ feet.

31. $a_1 = 3, r = 3, n = 6$

      $S_6 = \dfrac{a_1\left(1 - r^6\right)}{1 - r} = \dfrac{3\left(1 - 3^6\right)}{1 - 3}$

        $= \dfrac{3(-728)}{-2} = 1,092$

35. $a_1 = \dfrac{1}{3}, r = 2, n = 6$

      $S_6 = \dfrac{a_1\left(1 - r^6\right)}{1 - r} = \dfrac{\dfrac{1}{3}\left(1 - 2^6\right)}{1 - 2}$

        $= \dfrac{\dfrac{1}{3}(-63)}{-1} = 21$

39. $\displaystyle\sum_{k=4}^{8} a_k = \sum_{k=1}^{8} a_k - \sum_{k=1}^{3} a_k = S_8 - S_3$

$$a_1 = 1, \ r = \frac{1}{10}, \ n = 8$$

$$S_8 = \frac{a_1\left(1 - r^8\right)}{1 - r} = \frac{1\left(1 - \left(\frac{1}{10}\right)^8\right)}{1 - \frac{1}{10}} = \frac{0.99999999}{0.9} = 1.1111111$$

$$a_1 = 1, \ r = \frac{1}{10}, \ n = 3$$

$$S_3 = \frac{a_1\left(1 - r^3\right)}{1 - r} = \frac{1\left(1 - \left(\frac{1}{10}\right)^3\right)}{1 - \frac{1}{10}} = \frac{0.999}{0.9} = 1.11$$

So $\displaystyle\sum_{k=4}^{8} a_k = 1.1111111 - 1.11 = 0.0011111$

43. Firm A

Firm B

$a_1 = 30,000$

$a_1 = 34,000$

$r = 1.06$

$r = 1.04$

$n = 8$

$n = 8$

(a) $a_8 = a_1 r^7$

(a) $a_8 = a_1 r^7$

$\quad = 30,000(1.06)^7$

$\quad = 34,000(1.04)^7$

$\quad = \$45,108.90$

$\quad = \$44,741.62$

(b) $S_8 = \dfrac{a_1\left(1 - r^8\right)}{1 - r}$

(b) $S_8 = \dfrac{a_1\left(1 - r^8\right)}{1 - r}$

$\quad = \dfrac{30,000\left(1 - (1.06)^8\right)}{1 - 1.06}$

$\quad = \dfrac{34,000\left(1 - (1.04)^8\right)}{1 - 1.04}$

$\quad = \$296,925$

$\quad = \$313,284.50$

45. $a_1 = 2, \ r = \dfrac{1}{2}$. So $S = \dfrac{a_1}{1 - r} = \dfrac{2}{1 - \dfrac{1}{2}} = \dfrac{2}{\dfrac{1}{2}} = 4$

49. $a_1 = 5, \ r = \dfrac{1}{3}$. So $S = \dfrac{a_1}{1 - r} = \dfrac{5}{1 - \dfrac{1}{3}} = \dfrac{5}{\dfrac{2}{3}} = \dfrac{15}{2}$

51. Since $r = 2$, we cannot find the sum S.

53. $.35\overline{35} = .35 + .0035 + .000035 + \dots$

This is a geometric series with $a_1 = .35$ and $r = .01$. So $S = \dfrac{a_1}{1 - r} = \dfrac{.35}{1 - .01} = \dfrac{.35}{.99} = \dfrac{35}{99}$

That is, $.35\overline{35} = \dfrac{35}{99}$.

57. $a_1 = 60$, $r = \dfrac{2}{3}$. So $S = \dfrac{a_1}{1-r} = \dfrac{60}{1-\dfrac{2}{3}} = \dfrac{60}{\dfrac{1}{3}} = 180$

If the ball bounced forever, the total distance that it would fall is 180 feet.

60. $(1-r)\left(1 + r + r^2 + r^3 + \ldots + r^{n-2} + r^{n-1}\right)$

$= (1-r)1 + (1-r)r + (1-r)r^2 + (1-r)r^3 + \ldots + (1-r)r^{n-2} + (1-r)r^{n-1}$

$= (1-r) + \left(r - r^2\right) + \left(r^2 - r^3\right) + \left(r^3 - r^4\right) + \ldots + \left(r^{n-2} - r^{n-1}\right) + \left(r^{n-1} - r^n\right)$

$= 1 + (-r + r) + \left(-r^2 + r^2\right) + \left(-r^3 + r^3\right) + \ldots + \left(-r^{n-1} + r^{n-1}\right) - r^n$

$= 1 + 0 + 0 + 0 + \ldots + 0 - r^n = 1 - r^n$

61.   Consider the ratio of two successive terms of the series: $\dfrac{a_{k+1}}{a_k} = \dfrac{ab^{k+1}}{ab^k} = b$

Since this ratio is constant, the series is geometric, and the common ratio is b.

62.   (a) $a_1 = 3 \cdot 2^1 = 6$, $r = 2$

So $S_{10} = \dfrac{a_1\left(1 - r^{10}\right)}{1-r} = \dfrac{6\left(1 - 2^{10}\right)}{1-2} = \dfrac{6(-1,023)}{-1} = 6,138$

(b) $a_1 = 5 \cdot 3^1 = 15$, $r = 3$

So $S_{10} = \dfrac{a_1\left(1 - r^{10}\right)}{1-r} = \dfrac{15\left(1 - 3^{10}\right)}{1-3}$

$= \dfrac{15(-59,048)}{-2} = 442,860$

Exercises 13.5

1. $6! = 6 \cdot 5 \cdot 4 \cdot 3 \cdot 2 \cdot 1 = 720$

7. $(7-2)! = 5! = 5 \cdot 4 \cdot 3 \cdot 2 \cdot 1 = 120$

17. $\dfrac{8!}{5!3!} = \dfrac{8 \cdot 7 \cdot \cancel{6} \cdot \cancel{5} \cdot \cancel{4} \cdot \cancel{3} \cdot \cancel{2} \cdot \cancel{1}}{\cancel{5} \cdot \cancel{4} \cdot \cancel{3} \cdot \cancel{2} \cdot 1 \cdot \cancel{3} \cdot \cancel{2} \cdot \cancel{1}} = 56$

23. $(a-b)^6 = a^6 + \dfrac{6!}{5!1!}a^5(-b) + \dfrac{6!}{4!2!}a^4(-b)^2 + \dfrac{6!}{3!3!}a^3(-b)^3 + \dfrac{6!}{2!4!}a^2(-b)^4 + \dfrac{6!}{1!5!}a(-b)^5 + (-b)^6$

$= a^6 - 6a^5b + 15a^4b^2 - 20a^3b^3 + 15a^2b^4 - 6ab^5 + b^6$

31. $(a^2 + 2b)^5 = (a^2)^5 + \dfrac{5!}{4!1!}(a^2)^4(2b) + \dfrac{5!}{3!2!}(a^2)^3(2b)^2 + \dfrac{5!}{2!3!}(a^2)^2(2b)^3 + \dfrac{5!}{1!4!}a^2(2b)^4 + (2b)^5$

$= a^{10} + 5a^8(2b) + 10a^6(4b^2) + 10a^4(8b^3) + 5a^2(16b^4) + 32b^5$

$= a^{10} + 10a^8b + 40a^6b^2 + 80a^4b^3 + 80a^2b^4 + 32b^5$

39. The first four terms of the binomial expansion of $(2a + 3b)^8$ are

$$(2a)^8 + \frac{8!}{7!1!}(2a)^7(3b) + \frac{8!}{6!2!}(2a)^6(3b)^2 + \frac{8!}{5!3!}(2a)^5(3b)^3$$

$$= 256a^8 + 3,072a^7b + 16,128a^6b^2 + 48,384a^5b^3$$

43. Substitute $n = 8$ and $r = 3$ into the formula $\dfrac{n!}{(n-r+1)!(r-1)!}x^{n-r+1}y^{r-1}$ to get $\dfrac{8!}{6!2!}x^6y^2 = 28x^6y^2$

49. $\dfrac{7!}{2!5!}(3a^2)^2(-2)^5 = 21(9a^4)(-32) = -6,048a^4$

53. In the expansion of $(x + y)^n$, the literal parts will have the following pattern:

$$x^n, \quad x^{n-1}y, \quad x^{n-2}y^2, \quad \ldots, \quad x^2y^{n-2}, \quad xy^{n-1}, \quad y^n$$

Note that the sum of the exponents in each case is equal to n. $x^{n-k}y^k$ will have a coefficient of

$$\frac{n!}{(n-k)!k!}$$

54. $(1.02)^7 = (1 + 0.02)^7$

$$= (1)^7 + 7(1)^6(0.02) + 21(1)^5(0.02)^2 + 35(1)^4(0.02)^3 + 35(1)^3(0.02)^4$$

$$= +21(1)^2(0.02)^5 + 7(1)^1(0.02)^6 + (0.02)^7$$

$$= 1 + 7(0.02) + 21(0.02)^2 + 35(0.02)^3 + 35(0.02)^4 + 21(0.02)^5 + 7(0.02)^6 + (0.02)^7$$

Since $35(0.02)^4 = 5.6 \times 10^{-6}$ and since the three terms that follow are smaller still, these will not affect a computation to four places. Therefore, $(1.02)^7 \simeq 1 + 7(0.02) + 21(0.02)^2 + 35(0.02)^3 \simeq 1.1487$

55. (a) $\dbinom{6}{2} = \dfrac{6!}{2!4!} = \dfrac{\overset{3}{\cancel{6}} \cdot 5 \cdot \cancel{4} \cdot \cancel{3} \cdot \cancel{2} \cdot \cancel{1}}{\cancel{2} \cdot 1 \cdot \cancel{4} \cdot \cancel{3} \cdot \cancel{2} \cdot \cancel{1}} = 15$

(b) $\dbinom{5}{0} = \dfrac{\cancel{5!}}{0!\cancel{5}!} = \dfrac{1}{0!} = \dfrac{1}{1} = 1$

(c) $\dbinom{9}{3} = \dfrac{9!}{3!6!} = \dfrac{\overset{3}{\cancel{9}} \cdot \overset{4}{\cancel{8}} \cdot 7 \cdot \cancel{6} \cdot \cancel{5} \cdot \cancel{4} \cdot \cancel{3} \cdot \cancel{2} \cdot \cancel{1}}{\cancel{3} \cdot \cancel{2} \cdot 1 \cdot \cancel{6} \cdot \cancel{5} \cdot \cancel{4} \cdot \cancel{3} \cdot \cancel{2} \cdot \cancel{1}} = 84$

(d) $\dbinom{9}{6} = \dfrac{9!}{6!3!} = \dfrac{9!}{3!6!} = 84$, as in (c).

Chapter 13 Review Exercises

1. $-3, -1, 1, 3; a_{12} = 19$

3. $2, 8, 18, 32; x_{12} = 288$

5. $-\dfrac{1}{2}, \dfrac{1}{3}, -\dfrac{1}{4}, \dfrac{1}{5}; a_{12} = \dfrac{1}{13}$

7. $2, 4, 2, 4; x_{12} = 4$

9. $a_n = 5n - 2$

11. $a_n = \dfrac{(-1)^{n+1}}{2n}$

13. $\{23{,}000,\ 23{,}700,\ 24{,}400,\ 25{,}100,\ 25{,}800,\ 26{,}500\}$

15. $A_5 = 500(1.12)^5 = \$881.17$

17. $S_5 = 3 + 6 + 9 + 12 + 15 = 45$

19. $S_6 = 2 + 4 + 8 + 16 + 32 + 64 = 126$

21. $S_6 = 2 + 5 + 8 + 11 + 14 + 17 = 57$

23. $\displaystyle\sum_{i=1}^{6} i = 1 + 2 + 3 + 4 + 5 + 6 = 21$

25. $\displaystyle\sum_{i=1}^{5} 2i = 2 + 4 + 6 + 8 + 10 = 30$

27. $\displaystyle\sum_{n=1}^{6} 3n^2 = 3 + 12 + 27 + 48 + 75 + 108 = 273$

29. $5, 8; \quad a_{10} = a_1 + 9d = 2 + 9(3) = 29$

31. $23, 28; \quad a_{10} = a_1 + 9d = 3 + 9(5) = 48$

33. $-15, -19; \quad a_{10} = a_1 + 9d = 1 + 9(-4) = -35$

35. $x_8 - x_4 = (8 - 4)d$

$22 - 10 = (8 - 4)d$

$12 = 4d$

$3 = d$

Then $x_4 = x_1 + 3d$

$10 = x_1 + 3(3)$

$10 = x_1 + 9$

$1 = x_1$

37. $a_1 = 16, \ d = 32$

$a_8 = a_1 + (8 - 1)d$

$\quad = a_1 + 7d$

$\quad = 16 + 7(32) = 16 + 224 = 240$

The object falls 240 feet during the eighth second.

39. $S_{10} = \dfrac{10}{2}\left[2a_1 + (10-1)d\right]$

$\phantom{S_{10}} = \dfrac{10}{2}[2(5) + 9(3)]$

$\phantom{S_{10}} = 5(10 + 27)$

$\phantom{S_{10}} = 5(37) = 185$

41. $S_8 = \dfrac{8}{2}\left[2a_1 + (8-1)d\right]$

$ = \dfrac{8}{2}[2(3) + 7(7)]$

$ = 4(6 + 49)$

$ = 4(55) = 220$

43. $a_1 = 2,\ a_{30} = 60$.

 So $S_{30} = \dfrac{30}{2}(2 + 60)$

$\phantom{43. So S_{30}} = (15)(62)$

$\phantom{43. So S_{30}} = 930$

45. $a_1 = 1,\ d = 2$

$S_{30} = \dfrac{30}{2}\left[2a_1 + (30-1)d\right]$

$\phantom{S_{30}} = \dfrac{30}{2}[2(1) + 29(2)]$

$\phantom{S_{30}} = 15(2 + 58)$

$\phantom{S_{30}} = 15(60) = 900$

47. $a_8 - a_4 = (8-4)d$

$\ 8 - 5 = (8-4)d$

$\ 3 = 4d$

$\ \dfrac{3}{4} = d$

$a_4 = a_1 + (4-1)d$

$5 = a_1 + 3\left(\dfrac{3}{4}\right)$

$5 = a_1 + \dfrac{9}{4}$

$\dfrac{11}{4} = a_1$

Then $S_{10} = \frac{10}{2}\left[2a_1 + (10-1)d\right]$

$$= \frac{10}{2}\left[2\left(\frac{11}{4}\right) + 9\left(\frac{3}{4}\right)\right]$$

$$= 5\left(\frac{22}{4} + \frac{27}{4}\right)$$

$$= 5\left(\frac{49}{4}\right) = \frac{245}{4}$$

49. $a_5 = a_1 r^4 = 2(3)^4 = 2(81) = 162$

51. $a_6 = a_1 r^5 = 1\left(\frac{1}{3}\right)^5 = 1\left(\frac{1}{243}\right) = \frac{1}{243}$

53. $a_5 = a_1 r^4 = 5(2)^4 = 5(16) = 80$

55. $a_{12} = a_1 r^{11} = 5(2)^{11} = 5(2,048) = 10,240$

Kathy should receive $10,240 in the twelfth month.

57. $a_1 = 3$, $r = \frac{1}{3}$. So $S = \frac{a_1}{1-r} = \frac{3}{1-\frac{1}{3}} = \frac{3}{\frac{2}{3}} = \frac{9}{2}$

59. Since $r = 2$, we cannot find the sum S.

61. $.64\overline{6464} = .64 + .0064 + .000064 + \ldots$

This is a geometric series with $a_1 = .64$ and $r = .01$. So $S = \frac{a_1}{1-r} = \frac{.64}{1-.01} = \frac{.64}{.99} = \frac{64}{99}$. That is,

$.64\overline{6464} = \frac{64}{99}$.

63. $8! = 8 \cdot 7 \cdot 6 \cdot 5 \cdot 4 \cdot 3 \cdot 2 \cdot 1 = 40,320$

65. $9! - 6! = (9 \cdot 8 \cdot 7 \cdot 6 \cdot 5 \cdot 4 \cdot 3 \cdot 2 \cdot 1) - (6 \cdot 5 \cdot 4 \cdot 3 \cdot 2 \cdot 1)$

$\qquad = (6 \cdot 5 \cdot 4 \cdot 3 \cdot 2 \cdot 1)(9 \cdot 8 \cdot 7 - 1)$

$\qquad = 720(504 - 1)$

$\qquad = 720(503)$

$\qquad = 362,160$

67. $x^5 - 5x^4 y + 10x^3 y^2 - 10x^2 y^3 + 5xy^4 - y^5$

69. $(2a)^4 - 4(2a)^3\left(3b^2\right) + 6(2a)^2\left(3b^2\right)^2 - 4(2a)\left(3b^2\right)^3 + \left(3b^2\right)^4$

$= 16a^4 - 96a^3 b^2 + 216a^2 b^4 - 216ab^6 + 81b^8$

71. $(3a)^5 - 5(3a)^4(2b) + 10(3a)^3(2b)^2 - 10(3a)^2(2b)^3$

$= 243a^5 - 810a^4 b + 1,080a^3 b^2 - 720a^2 b^3$

73. $-20\left(2a^2\right)^3(3)^3 = -4,320a^6$

Chapter 13 Practice Test

1.   (a)   4, 7, 10; $x_9 = 28$

      (b)   1, 15, 53; $y_9 = 2(9)^3 - 1 = 1,457$

3. $\left(2(3)^2 + 1\right) + \left(2(4)^2 + 1\right) + \left(2(5)^2 + 1\right) + \left(2(6)^2 + 1\right)$

      $= 19 + 33 + 51 + 73 = 176$

5. $a_5 = a_1 r^4 = 4(2)^4 = 4(16) = 64$

7. $S_4 = \dfrac{a_1\left(1 - r^4\right)}{1 - r} = \dfrac{2\left(1 - 5^4\right)}{1 - 5}$

      $= \dfrac{2(-624)}{-4} = 312$

9. $a_1 = 12$, $r = \dfrac{3}{5}$. So $S = \dfrac{a_1}{1 - r} = \dfrac{12}{1 - \dfrac{3}{5}} = \dfrac{12}{\dfrac{2}{5}} = 30$

    The pendulum would travel 30 feet before coming to rest.

11. $-20\left(3a^2\right)^3(1)^3 = -540a^6$

APPENDIXES

Exercises for Appendix A

3. $y = -4x^2$

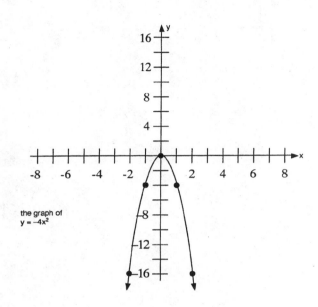

the graph of
$y = -4x^2$

7. $y = (x - 3)^2 + 4$

vertex: (3,4)

axis of symmetry: $x = 3$

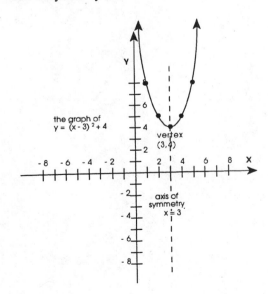

the graph of
$y = (x - 3)^2 + 4$

vertex
(3,4)

axis of
symmetry
x = 3

11. $y = 2(x + 4)^2 - 2$

vertex: $(-4, -2)$

axis of symmetry: $x = -4$

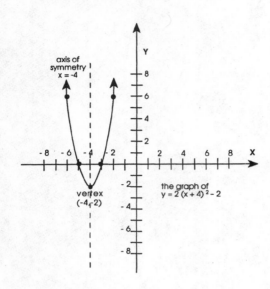

13. $y = 2x^2 - 12x + 19$

$y = (2x^2 - 12x \quad) + 19$

$y = 2(x^2 - 6x \quad) + 19$

$y = 2(x^2 - 6x + 9 - 9) + 19$

$y = 2(x^2 - 6x + 9) + 2(-9) + 19$

$y = 2(x - 3)^2 + 1$

vertex: $(3, 1)$

axis of symmetry: $x = 3$

17. $y = -3x^2 + 30x - 70$

$y = (-3x^2 + 30x \quad) - 70$

$y = -3(x^2 - 10x \quad) - 70$

$y = -3(x^2 - 10x + 25 - 25) - 70$

$y = -3(x^2 - 10x + 25) - 3(-25) - 70$

$y = -3(x - 5)^2 + 5$

vertex: $(5, 5)$

axis of symmetry: $x = 5$

432

21. $y = 2x^2 + 8$

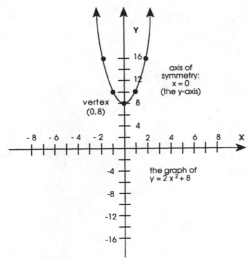

x-intercepts: none

y-intercept: 8

25. $y = x^2 - 2x - 35$

$y = (x^2 - 2x \quad) - 35$

$y = (x^2 - 2x + 1) - 1 - 35$

$y = (x - 1)^2 - 36$

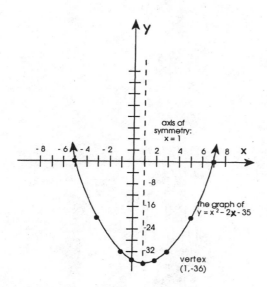

x-intercepts: −5 and 7

y-intercept: −35

29.　$y = -x^2 - 3x - 4$

　　　$y = (-x^2 - 3x\ \ \ \ ) - 4$

　　　$y = -(x^2 + 3x\ \ \ \ ) - 4$

　　　$y = -\left(x^2 + 3x + \dfrac{9}{4} - \dfrac{9}{4}\right) - 4$

　　　$y = -\left(x + \dfrac{3}{2}\right)^2 - \dfrac{7}{4}$

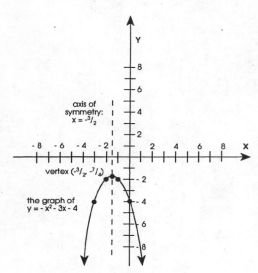

x-intercepts:　none

y-intercept:　−4

33.　$y = 2x^2 + 4x + -1$

　　　$y = (-2x^2 + 4x\ \ \ \ ) - 1$

　　　$y = -2(x^2 - 2x\ \ \ \ ) - 1$

　　　$y = 2(x^2 - 2x + 1 - 1) - 1$

　　　$y = -2(x^2 - 2x + 1) - 2(-1) - 1$

　　　$y = -2(x - 1)^2 + 1$

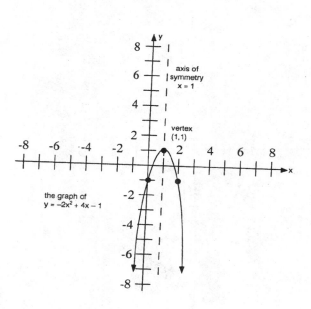

434

x-intercepts: $1 \pm \dfrac{\sqrt{2}}{2}$

y-intercept: -1

35. Since $(x-3)^2$ is a perfect square, it cannot be negative. That is, $(x-3)^2 \geq 0$. Then $2(x-3)^2 \geq 0$ as well, so that $2(x-3)^2 + 5 \geq 0 + 5$. This tells us that $y \geq 5$. The smallest possible value of y is therefore equal to 5. This value occurs when $2(x-3)^2 = 0$, which implies that $x = 3$.

36. Since $(x-h)^2$ is a perfect square, it cannot be negative. That is, $(x-h)^2 \geq 0$. Then $A(x-h)^2 \geq 0$ as well, since

$A > 0$. So $A(x-h)^2 + k \geq 0 + k$, which tells us that $y \geq k$. Conclude that the smallest possible value of y is equal to k. This value occurs when $A(x-h)^2 = 0$, which implies that $x = h$.

37.
$$y = Ax^2 + Bx + C \qquad (A \neq 0)$$
$$y = (Ax^2 + Bx) + C$$
$$y = A\left(x^2 + \frac{B}{A}x\right) + C$$
$$y = A\left(x^2 + \frac{B}{A}x + \frac{B^2}{4A^2} - \frac{B^2}{4A^2}\right) + C$$
$$y = A\left(x^2 + \frac{B}{A}x + \frac{B^2}{4A^2}\right) + A\left(\frac{-B^2}{4A^2}\right) + C$$
$$y = A\left(x + \frac{B}{2A}\right)^2 - \frac{B^2}{4A} + C$$
$$y = A\left(x + \frac{B}{2A}\right)^2 + C - \frac{B^2}{4A}$$
$$y = A\left(x + \frac{B}{2A}\right)^2 + \left(\frac{4AC - B^2}{4A}\right)$$

Comparing this equation with the standard form of the equation of a parabola, we find
$h = -\dfrac{B}{2A}$ and $k = \dfrac{4AC - B^2}{4A}$. So the vertex of the parabola is $\left(-\dfrac{B}{2A}, \ \dfrac{4AC - B^2}{4A}\right)$.

Exercises for Appendix B

5. $\dfrac{(x-2)^2}{25} + \dfrac{(y-5)^2}{81} = 1$

Comparing with standard form, we find that

$h = 2$, $k = 5$, $a = 5$, and $b = 9$. Then the

center of the ellipse is (2,5) and

its vertices are (7,5), (−3,5), (2,14)

and (2,−4). Equations of the axes are $x = 2$

and $y = 5$.

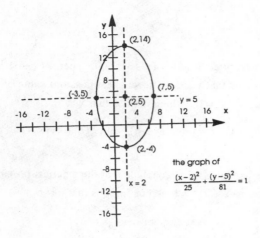

the graph of

$$\frac{(x-2)^2}{25} + \frac{(y-5)^2}{81} = 1$$

11. $$\frac{(x+3)^2}{9} + \frac{(y+1)^2}{4} = 1$$

Comparing with standard form, we find that

h = −3, k = −1, a = 3, and b = 2. Then the

center of the ellipse is (−3,−1), and its vertices

are (0,−1), (−6,−1), (−3,1) and (−3,−3). Equations

of the axes are x = −3 and y = −1.

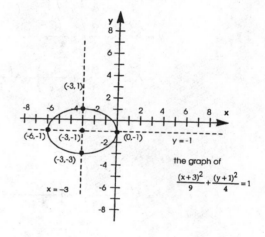

the graph of

$$\frac{(x+3)^2}{9} + \frac{(y+1)^2}{4} = 1$$

15. $$16(x + 1)^2 + 64(y - 2)^2 = 64$$

$$\frac{16(x+1)^2}{64} + \frac{64(y-2)^2}{64} = \frac{64}{64}$$

$$\frac{(x+1)^2}{4} + \frac{(y-2)^2}{1} = 1$$

Comparing with standard form, we find that

h = −1, k = 2, a = 2, and b = 1. Then the

center of the ellipse is (−1,2), and its vertices are

(1,2), (−3,2), (−1,3) and (−1,1). Equations of the

axes are x = −1 and y = 2.

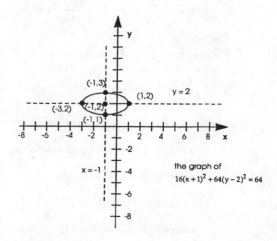

the graph of

$$16(x+1)^2 + 64(y-2)^2 = 64$$

23. $\dfrac{(x-3)^2}{25} - \dfrac{(y-4)^2}{16} = 1$

Comparing with standard form, we find that $h = 3$, $k = 4$,

$a = 5$, and $b = 4$. Then the center of the hyperbola is $(3,4)$,

and its vertices are $(8,4)$ and $(-2,4)$. Equations of the axes

are $x = 3$ and $y = 4$.

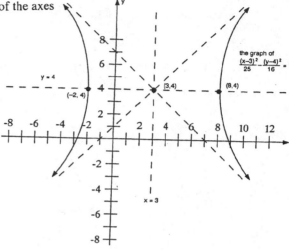

27. $\dfrac{(y+2)^2}{16} - \dfrac{(x-1)^2}{25} = 1$

Comparing with standard form, we find that

$h = 1$, $k = -2$, $a = 5$, and $b = 4$. Then the

center of the hyperbola is $(1,-2)$, and its vertices

are $(1,2)$ and $(1,-6)$. Equations of the axes are

$x = 1$ and $y = -2$.

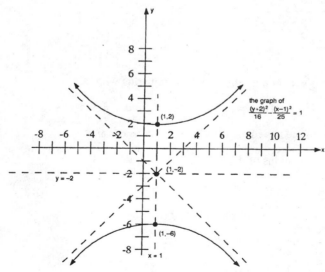

33. $36(x + 1)^2 - 9(y - 2)^2 = 36$

$$\frac{36(x+1)^2}{36} - \frac{9(y-2)^2}{36} = \frac{36}{36}$$

$$\frac{(x+1)^2}{1} - \frac{(y-2)^2}{4} = 1$$

Comparing with standard form, we find that $h = -1$, $k = 2$, $a = 1$, and $b = 2$. Then the center of the hyperbola is $(-1,2)$, and its vertices are $(0,2)$ and $(-2,2)$. Equations of the axes are $x = -1$ and $y = 2$.

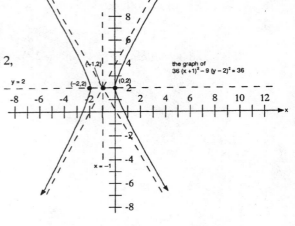

45. $16y^2 - 9x^2 - 96y + 36x = 36$

$16y^2 - 96y - 9x^2 + 36x = 36$

$16(y^2 - 6y) - 9(x^2 - 4x) = 36$

$16(y^2 - 6y + 9) - 9(x^2 - 4x + 4) = 36 + 144 - 36$

$16(y - 3)^2 - 9(x - 2)^2 = 144$

$$\frac{16(y-3)^2}{144} - \frac{9(x-2)^2}{144} = \frac{144}{144}$$

$$\frac{(y-3)^2}{9} - \frac{(x-2)^2}{16} = 1$$

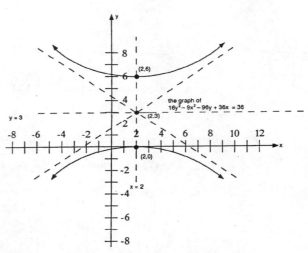

The graph of this equation is a hyperbola. Comparing with standard form, we find that $h = 2$, $k = 3$, $a = 4$, and $b = 3$. Then the center of the hyperbola is $(2,3)$, and its vertices are $(2,6)$ and $(2,0)$. Equations of the axes are $x = 2$ and $y = 3$.

49. $36x^2 + 25y^2 + 216x + 100y = 476$

$36x^2 + 216x + 25y^2 + 100y = 476$

$36(x^2 + 6x) + 25(y^2 + 4y) = 476$

$$36(x^2 + 6x + 9) + 25(y^2 + 4y + 4) = 476 + 324 + 100$$

$$36(x+3)^2 + 25(y+2)^2 = 900$$

$$\frac{36(x+3)^2}{900} + \frac{25(y+2)^2}{900} = \frac{900}{900}$$

$$\frac{(x+3)^2}{25} + \frac{(y+2)^2}{36} = 1$$

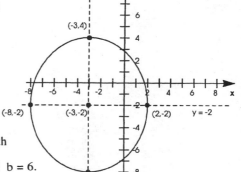

The graph of this equation is an ellipse. Comparing with standard form, we find that h = –3, k = –2, a = 5, and b = 6. Then the center of the ellipse is (–3,–2), and its vertices are (2,–2), (–8,–2), (–3,4) and (–3,–8). Equations of the axes are x = –3 and y = –2.

53. Substitute the given information into the point–slope formula to find $y - k = \pm \dfrac{b}{a}(x - h)$ as the equations of the asymptotes.

Exercises for Appendix C

1. $\log 739 = \log(7.39 \times 10^2) = \log 7.39 + \log 10^2 = \log 7.39 + 2 = 0.8686 + 2 = 2.8686$

5. $\log 280{,}000 = \log(2.80 \times 10^5) = \log 2.80 + \log 10^5 = \log 2.80 + 5 = 0.4472 + 5 = 5.4472$

7. $\log 0.0000553 = \log(5.53 \times 10^{-5}) = \log 5.53 + \log 10^{-5} = \log 5.53 - 5$

$$= 0.7427 - 5 \quad \text{or} \quad -4.2573$$

11. $\log 0.837 = \log(8.37 \times 10^{-1}) = \log 8.37 + \log 10^{-1} = \log 8.37 - 1$

$$= 0.9227 - 1 \quad \text{or} \quad -0.0773$$

15. $\log N = 2.8420$

$N = 6.95 \times 10^2 = 695$

17. $\log N = 0.7308 - 3$

$N = 5.38 \times 10^{-3} = 0.00538$

23. $\log N = 5.8733$

$N = 7.47 \times 10^5 = 747{,}000$

25. $\log N = 0.7803$

$N = 6.03$

29.

$$\begin{matrix} \underline{N} \\ 10\begin{bmatrix} 7\begin{bmatrix} 2{,}840 \\ 2{,}847 \\ 2{,}850 \end{bmatrix} \end{bmatrix} \end{matrix} \qquad \begin{matrix} \underline{\log N} \\ \begin{bmatrix} 3.4533 \\ \\ 3.4548 \end{bmatrix} x \, 0.0015 \end{matrix}$$

$$\frac{7}{10} = \frac{x}{0.0015}$$

$$x = \frac{7}{10}(0.0015) = 0.0011 \text{ (rounded off to 4 decimal places)}$$

Then log 2,847 = 3.4533 + 0.0011 = 3.4544

35.

$$\begin{matrix} \underline{N} \\ 0.01\begin{bmatrix} x\begin{bmatrix} 8.810 \\ \\ 8.820 \end{bmatrix} \end{bmatrix} \end{matrix} \qquad \begin{matrix} \underline{\log N} \\ \begin{bmatrix} 0.9450 \\ 0.9453 \\ 0.9455 \end{bmatrix} 0.0003 \, \Big] 0.0005 \end{matrix}$$

$$\frac{x}{0.01} = \frac{0.0003}{0.0005}$$

$$x = \frac{(0.0003)(0.01)}{0.0005} = 0.006$$

Then antilog 0.9453 = 8.810 + 0.006 = 8.816

39.   Let log N = $(513)^{\frac{1}{2}}$

Then log N = $\log(513)^{\frac{1}{2}} = \frac{1}{2} \log 513 = \frac{1}{2}(\log 5.13 \times 10^2) = \frac{1}{2}(\log 5.13 + 2)$

$$= \frac{1}{2}(0.7101 + 2) = \frac{1}{2}(2.7101) = 1.3551$$

So N = antilog 1.3551 = 2.265 x 10^1 = 22.65

45.   Let N = $(1.31)^{40}$

Then log N = $\log(1.31)^{40}$ = 40 log 1.31 = 40(0.1173) = 4.692

So N = antilog 4.692 = 4.9204 $\times 10^4$ = 49,204 (actual value is approximately 49,074)

49.   log 3.53 = 0.5478

log 3.54 = 0.5490

log 3.55 = 0.5502

log 3.56 = 0.5515

The points (3.53, 0.5478), (3.54, 0.5490), (3.55, 0.5502) and (3.56, 0.5515) appear to be on a straight line with slope $\frac{0.5490 - 0.5478}{3.54 - 3.53} = \frac{0.0012}{0.01} = 0.12$. The straight line is actually an

440

approximation to the graph. The closeness of this approximation for small changes in x justifies the process of linear interpolation.

CHAPTER TESTS, CUMULATIVE REVIEWS, AND CUMULATIVE PRACTICE TESTS

Chapter 1—Test A

1. Let A = $\{x|x$ is an odd factor of $30\}$

    B = $\{y|y$ is a prime between 4 and $24\}$

    (a) List $A \cap B$

    (b) List $A \cup B$

2. Indicate whether each of the following is true or false:

    (a) $0 \in N$

    (b) $5 \in W$

    (c) $-\frac{1}{3} \in Q$

3. Graph the following sets on the real number line:

    (a) $\{x|x \le 2\}$

    (b) $\{t|-5 < t \le 5\}$

4. If the statement is true, state the property that it illustrates. If the statement is not true, write false.

    (a) $x + yz = yz + x$

    (b) $3(mn) = (3m)(3n)$

5. Evaluate the following:

    (a) $-1 - (+2) - (-3) + (-4)$

    (b) $(-3)^3 + (-3)(-5)$

    (c) $|-10 + |4 - 7||$

    (d) $8 - 7[6 - (5 + 4)]$

6. Evaluate the following, given $x = -1$ and $y = 4$:

    (a) $\dfrac{x^2 - y^2}{xy - 1}$

    (b) $(y - x^2)^2$

7. Simplify the following:

    (a) $(-3x^2y)(-4xy^4)(-x^3y^2)$

    (b) $8a^3b^2 - 5a^2b^2 + a^2b^3 - 4a^3b^2$

    (c) $3p - 5(1 - p) + 4(p + 2)$

(d) $2u - \left\{ v - 2\left[u - (v + 3u) \right] \right\}$

8. Two sides of a triangle are equal in length and the third side is 4 less than five times one of the equal sides. Express the perimeter of the triangle in terms of one variable.

9. Dwayne has 16 stamps altogether. Some are blue stamps worth 8¢ each and the rest are green stamps worth 15¢ each. If x represents the number of green stamps, express the number of blue stamps in terms of x. What is the total value of Dwayne's stamps in terms of x?

Chapter 1—Test B

1. Let $A = \left\{ x \mid x \text{ is an even factor of } 24 \right\}$

   $B = \left\{ y \mid y \text{ is a positive multiple of 4 which is less than } 18 \right\}$

   (a) List $A \cap B$

   (b) List $A \cup B$

2. Indicate whether each of the following is true or false:

   (a) $\sqrt{36} \in Q$

   (b) $0 \in W$

   (c) $-\dfrac{2}{3} \in Z$

3. Graph the following sets on the real number line:

   (a) $\left\{ y \mid y < -1 \right\}$

   (b) $\left\{ s \mid -3 \le s \le 4 \right\}$

4. If the statement is true, state the property that it illustrates. If the statement is not true, write false.

   (a) $(x + y)z = z(x + y)$

   (b) $(5c)(5d) = 5(cd)$

5. Evaluate the following:

   (a) $-5 - (+5) + -5 - (-5)$

   (b) $(-2)^4 - (-2)(-4)$

   (c) $\left\| 1 - 6 \right| - 6 \right|$

   (d) $3 - 3[3 - (3 + 3)]$

6. Evaluate the following, given $x = 3$ and $y = -2$:

   (a) $x^2 - y^3$

   (b) $\dfrac{(x - y)^2}{y - xy - x^2}$

443

7. Simplify the following:

   (a) $(-6x^3y)(2x^2y^2)(-x^4y^3)$

   (b) $5p^4q^5 + 4p^5q^4 - 5p^4q^4 - 4p^4q^5$

   (c) $4m + 3(2 - m) + 2(m - 5)$

   (d) $3s - \{t - 3[2s - (t + s)]\}$

8. The middle side of a triangle is 2 more than the shortest side, and the longest side is twice the shortest side. Express the perimeter of the triangle in terms of one variable.

9. A small glass of lemonade sells for 20¢ and a large glass sells for 30¢. Lucy sells 25 glasses of lemonade altogether. If x represents the number of large glasses sold, express the number of small glasses sold in terms of x. How much money does Lucy make selling lemonade?

Chapter 2—Test A

Solve the following equations:

1. $2x + 5 = 8x - 1$

2. $2(x - 1) + 3(x - 2) = 4(x - 3) + 5(x - 4)$

3. $3b - [2 - 3(b - 2)] = 4(b - 3)$

4. $|3x - 1| - 1 = 7$

Solve the following inequalities and graph the solutions on the real number line:

5. $5x - 4 > 4x - 5$

6. $4x + 4[1 - 4(x - 1)] \le -4$

7. $-2 < 3 - x \le 5$

8. $|2x + 3| \le 1$

9. $|4x - 5| > 7$

10. A coin collection contains 45 coins. The silver coins weigh 2 oz. each and the gold coins weigh 3 oz. each. If the total weight of the coins is 117 oz., how many silver coins and how many gold coins are in the collection?

11. Two trains leave from the same station and travel in opposite directions. If one travels 30 kph faster than the other, and they are 660 km apart after two hours, find the speed of each train.

12. Olga has 40 dimes and quarters in her change purse. If the total value of her coins is at most $8.50, what is the maximum number of quarters that Olga can have?

Chapter 2—Test B

Solve the following equations:

1. $6x + 1 = 3x - 5$

2. $3(x - 1) - 4(x - 2) = 5(x - 3) - 2x$

3. $4d - [1 - 4(d - 1)] = 4 - d$

4. $|4x + 3| + 1 = 2$

Solve the following inequalities and graph the solutions on the real number line:

5. $2x - 3 \ge 4x - 5$

6. $3x + 3[x - 3(x + 3)] < -18$

7. $-3 \le 2 - x < 5$

8. $|2x - 1| < 3$

9. $|3x + 2| \geq 8$

10. A bouquet contains 24 silk flowers. Irises cost $3.00 each and lilies cost $4.00 each. If the total bouquet costs $80.00, how many irises and how many lilies are in the bouquet?

11. A bicyclist leaves from a park and travels at the rate of 3 mph. Two hours later, another bicyclist leaves from the same point and travels in the opposite direction at the rate of 5 mph. How long after the first bicyclist leaves will the two be 38 miles apart?

12. Ivan has 15¢ stamps and 25¢ stamps in his desk drawer. If the total value of his stamps is $6.10 and he has 30 stamps altogether, what is the maximum number of 25¢ stamps that Ivan can have?

Chapter 3—Test A

Perform the operations and simplify the following:

1. $(4x^2 - 2xy + y^2)(2x + y)$

2. $(t + 4)^2 - (t - 4)^2$

3. $(a + 3b + 1)(a + 3b - 1)$

4. $(2x^2 + y^3)^2$

Factor the following completely:

5. $12u^4v^2 - 18u^2v^4 + 6u^3v$

6. $8cx - 2c + 12dx - 3d$

7. $a^2 - ab - 72b^2$

8. $8x^2 + 14x - 15$

9. $3x^2 + 4x + 5$

10. $4m^4n^2 - 4m^2n^4$

11. $s^3t^4 + 12s^3t^3 + 36s^3t^2$

12. $b^3 - 27$

13. $(3x - 2y)^2 - 16$

14. $x^3 + 3x^2 + 2x + 6$

15. Find the quotient and remainder using long division: $(x^3 - 5x^2 + 4) \div (x - 2)$

Chapter 3—Test B

Perform the operations and simplify the following:

1. $(x^2 + 3xy + 9y^2)(x - 3y)$

2. $(3 - r)^2 + (3 + r)^2$

3. $(2a - b + 3)(2a - b - 3)$

4. $(c^2 - 3d^3)^2$

Factor the following completely:

5. $15s^3t^2 + 20s^2t^3 - 30st^2$

6. $6mx - 4nx + 9m - 6n$

7. $p^2 - 2pq - 63q^2$

8. $8x^2 + 19x - 15$

9. $4x^2 - 5x + 7$

10. $a^3b^3 - 4ab^5$

11. $w^4z^4 - 8w^3z^4 + 16w^2z^4$

12. $h^3 + 64$

13. $(x + 4y)^2 - 9$

14. $k^3 - 2k^2 + 4k - 8$

15. Find the quotient and remainder using long division: $(x^3 + 4x^2 + 6) \div (x + 3)$

Cumulative Review: Chapters 1–3

In exercises 1–4, list the following sets, given that:

$A = \{x | x \text{ is an odd number, } 4 < x < 18\}$

$B = \{y | y \text{ is a multiple of 5, } 5 \leq y \leq 25\}$

$C = \{z | z \text{ is a prime, } 2 < z \leq 17\}$

1. C

2. $B \cup C$

3. $A \cap C$

4. $B \cap C$

In exercises 5–6, answer true or false.

5. $\sqrt{9} \in W$

6. $-\dfrac{3}{4} \in Q$

In exercises 7–8, graph the given set.

7. $\{x | 3 < x \leq 7\}$

8. $\{x | 3 < x \leq 7, \ x \in I\}$

In exercises 9–10, list the property illustrated by the statement. If the statement is false, state so.

9. The product of two real numbers is a real number.

10. $\dfrac{x + 5}{5} = x$

In exercises 11–14, perform the operations.

11. $1 - 2 + 3 - 4 + 5$

12. $(-3)(-3)(-3) - 3$

13. $(8 - (-5))(3 + (-2))$

14. $8 - (-5) \cdot 3 + (-2)$

In exercises 15–16, evaluate the expressions, given $x = -3$, $y = 3$, and $z = -2$.

15. $|x - y| + |y - z|$

16. $\dfrac{2x^2 - 4y}{3z}$

In exercises 17–18, perform the operations and express your answer in simplest form.

17. $2x^2(x^3 - 3x + 5)$

18. $(-3ab)^2(-2ab^2)$

In exercises 19–22, translate the statements algebraically.

19. Four times the sum of a number and 7.

20. Five less than the sum of three consecutive odd integers.

21. The length of a rectangle is 10 more than twice its width. Express the area and perimeter of the rectangle in terms of its width.

22. Mrs. O'Leary has 14 bills in her wallet. Some are one dollar bills and the rest are five dollar bills. Express the value of her bills in terms of the number of one dollar bills.

In exercises 23–30, solve the equation.

23. $2x - 2 = 5x + 7$

24. $3(a + 2) = 1 - 3(1 - a)$

25. $3x - 2 - (x - 1) = 3 + 2(2 - x)$

26. $2x - 3(2x - 3) = 3x - 2(3x - 2)$

27. $5(x - 1) - 3(2x + 1) = 3x - 4(x + 2)$

28. $x(4x + 1) = 2x(2x + 5)$

29. $|2x + 7| = 3$

30. $3x + 1 = |2x - 11|$

In exercises 31–38, solve the inequalities and graph the solutions on the number line.

31. $5x - 1 > 2x + 1$

32. $3 - 4x \le 4 - 3x$

33. $2(x - 1) + 4(2x + 1) < 5(2x + 3)$

34. $-1 \le 2 - 3x < 8$

35. $|x + 3| > 2$

36. $|3x + 5| < 11$

37. $|2 - 6x| \le 4$

38. $|3 - 4x| \ge 3$

In exercises 39–46, solve the problem algebraically.

39. If 3 times two more than a number is 3 more than two times the number, find the number.

40. A man worked for 52 hours and earned $296. If he was paid at the rate of $5/hour and received $8/hour for working overtime, how many overtime hours did the man work?

41. An electrician bought 22 boxes of light bulbs for $58. Some boxes contained 4 bulbs and cost $3 per box, while the rest contained 3 bulbs and cost $2 per box. How many bulbs did the electrician buy altogether?

42. A gardener charges $16 an hour for trimming hedges and $20 an hour for turning soil. If he works for 9 hours and earns $160, how many hours did he spend on each task?

43. Jane can type at the rate of 50 words per minute while Joan can type at the rate of 80 words per minute. If Jane needs an hour to type a report, how much time would Joan need to type the same report?

44. A secretary can hand-stamp 500 envelopes per hour and must send out 6,000 letters. After working for a while, he decides to use a postal machine that allows him to stamp 1,000 envelopes per hour. If the total job takes him 7 hours, how many envelopes were stamped in each way?

45. The middle side of a triangle is 5" more than the shortest side, and the longest side is twice as long as the shortest side. If the shortest side varies from 6" to 10", what is the range of values for the perimeter of the triangle?

46. Oranges cost 60¢ per pound and apples cost 75¢ per pound. If a customer wishes to buy 20 pounds of apples and oranges and cannot spend more than $13.80, what is the maximum number of pounds of apples that can be bought?

In exercises 47–48, identify the degree of the polynomial.

47. $3 - 5x^3 + 2x - x^5$

48. $6x^4y - x^3y^3 + 8y^5$

In exercises 49–60, perform the operations and express your answer in simplest form.

49. $(4x^3 - x^2 + 8x - 5) - (2x^3 - x + 3)$

50. $-2x^2(xy^2 + 3xy - 4y^2)$

51. $(a + 2b)(2a - b)$

52. $(5u + 4v)(3u - 7v)$

53. $(c - d - 3)(c - d)$

54. $(a - 3b)(a^2 + 3ab + 9b^2)$

55. $(4m + 5n)^2$

56. $(7y - 2z)(7y + 2z)$

57. $(4m + 5n)(4m - 5n)$

58. $(7y - 2z)^2$

59. $(s + 3t - 1)(s + 3t + 1)$

60. $(3p - 2q + 4)^2$

In exercises 61–82, factor as completely as possible.

61. $x^2 - 7x - 30$

62. $c^2 - 16d^2$

63. $r^2 + 14rs + 48s^2$

64. $2a^2 - 7ab + 6b^2$

65. $9m^2 - 18mn + 8n^2$

66. $49u^2 - 64z^2$

67. $100w^2 - 81$

68. $16x^2 + 72xy + 81y^2$

69. $9a^2 - 48ab + 64b^2$

70. $25c^2 + 36d^2$

71. $9a^2 - 48ab - 64b^2$

72. $6h^3 - 27h^2 - 33h$

73. $5v^4 - 3v^3 - 2v^2$

74. $10p^4 + 13p^2q^2 - 3q^4$

75. $9s^2 + 35st^2 - 4t^4$

76. $v^4w + v^3w^4 - 2v^2w^7$

77. $(2x + y)^2 - 9$

78. $(x^2 + y^2)^2 - 25$

79. $r^4 - 81$

80. $(a - b)^2 - 4(a - b) + 3$

81. $27k^3 - 8$

82. $u^3 + u^2 - 4u - 4$

In exercises 83–86, find the quotient and remainder.

83. $(x^2 - 4x + 2) \div (x - 3)$

84. $(3x^2 + 4x + 2) \div (3x - 2)$

85. $(6x^3 + x^2 - 1) \div (2x - 1)$

86. $(x^5 + 1) \div (x^2 + 1)$

Cumulative Test: Chapters 1–3

1. Given the sets

   $A = \{a | a \text{ is a multiple of } 3, \ 5 < a < 25\}$
   $B = \{b | b \text{ is a multiple of } 5, \ 4 < b < 24\}$

   find:

   (a) $A \cap B$

   (b) $A \cup B$

2. What real number property is illustrated by the statement $x(y + z) = x(z + y)$?

3. Evaluate the following:

   (a) $\dfrac{-3-(-3)^2}{3-(-3)}$

   (b) $1-2\{3-4(5-6)\}$

4. Given $x = 2$ and $y = -3$, evaluate $\dfrac{x^2 + 4xy + 3y^2}{-x - y}$

5. Solve the following equations:

   (a) $2x + 1 = 3x - 4$

   (b) $2(x + 2) + 3(x + 3) = 2 - 4(x + 4)$

   (c) $|2x + 1| = 7$

6. Solve the following inequalities and graph the solution set on the number line:

   (a) $2x - 3(1 - x) > 10 - 4(1 + x)$

   (b) $|3 - 2x| \le 7$

   (c) $|5x + 2| > 3$

7. Regular gasoline costs \$1 per gallon, while premium unleaded gasoline costs \$1.20 per gallon. If a gas station sells 72 gallons of gasoline in one hour and collects \$80, how many gallons of each type of gasoline were sold?

8. A shipping crate contains 40 boxes, some weighing 5 lbs. and the others weighing 8 lbs. If the crate cannot hold more than 275 lbs., what is the maximum number of 8 lb. boxes that can be shipped?

9. Perform the operations and simplify:

   (a) $(3a^4 + 4a^2b^2 - 2b^2) - (2a^4 - 3a^2b^2 - b^2)$

   (b) $(2s - 3t)(3s + 4t)$

   (c) $(y^3 - 2z)(y^3 + 2z)$

   (d) $(4m + 5n)^2$

   (e) $(c - 2d - 1)^2$

10. Factor the following completely:

   (a) $a^2 + 3a - 40$

   (b) $12s^2 - 4st - 5t^2$

   (c) $u^6 + 4u^3v + 4v^2$

   (d) $25x^2 - 30xy + 9y^2$

   (e) $(b - 6c)^2 - 16$

   (f) $27r^3 + 8$

11. Find the quotient and remainder: $(3x^3 - 4x^2 + 1) \div (x - 3)$

Chapter 4—Test A

1. Express the following in simplest form.

   (a) $\dfrac{28ab^5}{35a^3b^2}$

   (b) $\dfrac{x^2 - 4x + 4}{x^2 - 4}$

   (c) $\dfrac{4x^4 + 10x^3 - 6x^2}{2x^4 - x^3}$

2. Perform the indicated operations and express your answer in simplest form.

   (a) $\dfrac{3p^2q}{4rs^3} \cdot \dfrac{16s}{9p^4q^4}$

   (b) $\dfrac{2}{5uv} + \dfrac{3u}{2v^2}$

   (c) $\dfrac{m^2 + 3mn - 10n^2}{m^3 + 2m^2n} \div \dfrac{mn + 5n^2}{m^3 - 4mn^2}$

   (d) $\dfrac{9x - 7}{2x - 5} + \dfrac{3x + 8}{5 - 2x}$

   (e) $\dfrac{2z}{z^2 - 1} + \dfrac{z}{z^2 - 3z + 2} - \dfrac{1}{z^2 - z - 2}$

   (f) $\dfrac{6}{x + 2} \div \left(\dfrac{3}{x + 2} - \dfrac{1}{x} \right)$

3. Express as a simple fraction reduced to lowest terms: $\dfrac{\dfrac{4}{x^2} - 1}{3 - \dfrac{6}{x}}$

4. Solve the following:

   (a) $\dfrac{x + 2}{4} - \dfrac{2x - 1}{3} \le 5$

   (b) $\dfrac{2x + 1}{x + 1} = 4 - \dfrac{x + 2}{x + 1}$

   (c) $\dfrac{3}{3x - 2} - \dfrac{1}{3x} = \dfrac{14}{9x^2 - 6x}$

5. Solve explicitly for x: $y = \dfrac{3x - 2}{x + 1}$

6. A 15-lb. cookie assortment is made by mixing cookies that sell for $2/lb. with cookies that sell for $3/lb. How many pounds of each type should be in the assortment if it is to sell for $2.40/lb?

7. Ralph can install an air conditioner in $1\frac{1}{2}$ hours. Ed can do the same job in $1\frac{1}{3}$ hours. How long will it take them if they work together?

Chapter 4—Test B

1. Express the following in simplest form.

   (a) $\dfrac{40s^3t^4}{12s^4t}$

   (b) $\dfrac{x^2 + 4x}{x^2 + 8x + 16}$

   (c) $\dfrac{6x^4 + 15x^3 + 6x^2}{4x^2 + 2x}$

2. Perform the indicated operations and express your answer in simplest form.

   (a) $\dfrac{2wx^3}{5y^2z^2} \div \dfrac{6w^2x}{25y^3}$

   (b) $\dfrac{4}{3kl^2} - \dfrac{3}{4k^2}$

   (c) $\dfrac{p^2 + 4pq + 3q^2}{3p^2 - 3q^2} \cdot \dfrac{2p^2 + 4pq - 6q^2}{pq}$

   (d) $\dfrac{5x + 1}{3x - 1} - \dfrac{x - 3}{1 - 3x}$

   (e) $\dfrac{4k}{k^2 + 2k - 3} - \dfrac{2}{k^2 + 5k + 6} - \dfrac{3k}{k^2 + k - 2}$

   (f) $\left(\dfrac{2}{x - 3} + \dfrac{1}{x} \right) \div \dfrac{6}{x - 3}$

3. Express as a simple fraction reduced to lowest terms: $\dfrac{1 - \dfrac{9}{x^2}}{\dfrac{6}{x} - 2}$

4. Solve the following:

   (a) $\dfrac{3x - 2}{2} + \dfrac{2x - 3}{5} > 6$

   (b) $\dfrac{3x - 10}{x - 4} = 3 - \dfrac{x - 6}{x - 4}$

   (c) $\dfrac{1}{2x} - \dfrac{2}{3x + 1} = \dfrac{3}{6x^2 + 2x}$

5. Solve explicitly for x: $y = \dfrac{2x - 3}{x + 2}$

6. How many pounds of light chocolate that sells for $5 /lb. must be mixed with 8 lbs. of dark chocolate that sells for $8/lb. in order to form a mixture that sells for $6/lb.?

7. It takes Ann 5 hours to complete a filing assignment. If she works together with Fran, she can complete the assignment in 3 hours. How long would it take Fran to complete the assignment if she worked alone?

Chapter 5—Test A

1. Graph the following using the intercept method.

   (a) $4x + 3y = 24$

   (b) $y + 2 = 0$

2. Find the slope of the line satisfying the given conditions:

   (a) passing through $(5,-1)$ and $(2,5)$

   (b) an equation of the line is $-2x + 3y = 7$

3. Find the value of k if a line with slope = 3 passes through the points $(k,-1)$ and $(7,k)$.

4. Write an equation of the line satisfying the given conditions:

   (a) the line passes through the points $(1,3)$ and $(-1,7)$

   (b) the line passes through $(3,3)$ with slope = 4

   (c) the line has an x-intercept of 2 and a y-intercept of 7

   (d) the line passes through $(-1,5)$ and is parallel to $x + 2y = 2$

   (e) the line passes through $(-1,5)$ and is perpendicular to $x + 2y = 2$

   (f) the vertical line passes through $(-5,4)$

5. Graph the inequality $2x - 5y \le 15$.

6. The relationship between a man's age and his annual insurance premium is perfectly linear. If a 30-year old man pays $90 per year and a 50-year old man pays $230 per year, what is the annual premium of a 35-year old man?

7. Solve the following systems of simultaneous equations.

   (a) $\begin{cases} 3x - 2y = 4 \\ 2x - 3y = -9 \end{cases}$

   (b) $\begin{cases} \dfrac{s}{4} = \dfrac{t}{3} + 1 \\ \dfrac{t}{2} = -\dfrac{s}{3} + 7 \end{cases}$

8. Carrie invests $2400 in two bonds. One yields 7% annual interest, while the other yields 10% annual interest. If Carrie receives $213 as her total annual interest, how much did she invest at each rate?

Chapter 5—Test B

1. Graph the following using the intercept method.

   (a) $5x + 2y = -10$

   (b) $x + 3 = 0$

2. Find the slope of the line satisfying the given conditions:

   (a) passing through $(-3,1)$ and $(1,7)$

   (b) an equation of the line is $-4x - 5y = 1$

3. Find the value of k if a line with slope $= -4$ passes through the points $(6,k)$ and $(k,3)$.

4. Write an equation of the line satisfying the given conditions:

   (a) the line passes through the points $(2,-1)$ and $(5,5)$

   (b) the line passes through $(-2,-2)$ with slope -3

   (c) the line has an x-intercept of -3 and a y-intercept of 4

   (d) the line passes through $(4,-1)$ and is parallel to $3x - y = 5$

   (e) the line passes through $(4,-1)$ and is perpendicular to $3x - y = 5$

   (f) the vertical line passes through $(-3,8)$

5. Graph the inequality $4x + 5y \geq 10$

6. The relationship between a man's age and his annual insurance premium is perfectly linear. If a 30-year old man pays $90 per year and a 50-year old man pays $230 per year, how old is a man whose annual premium is $195?

7. Solve the following systems of simultaneous equations.

   (a) $\begin{cases} 3x + 5y = -1 \\ 2x + 3y = -2 \end{cases}$

   (b) $\begin{cases} \dfrac{s}{3} = -\dfrac{t}{5} + 3 \\ \dfrac{t}{2} = \dfrac{s}{5} - 8 \end{cases}$

8. George invests $2700 in two stocks. One pays an annual dividend of 5%, while the other pays an annual dividend of 8%. If George receives $201 as his total annual dividend, how much did he invest at each rate?

Chapter 6—Test A

Perform the indicated operations and express your answer in simplest form with positive exponents only. Assume all variables represent positive real numbers.

1. $(x^3y)(x^2y^3)^3$

2. $(-3x^3y^2)^2(-2y)^3$

3. $(-2a^{-1}b^{-2})^{-1}(3cd^{-2})^2$

4. $\dfrac{(4xy^3)^2}{(-2x^3y)^3}$

5. $\left(\dfrac{2u^{-2}v^{-4}}{3uv^{-1}}\right)^{-3}$

6. $\left(\dfrac{16c^{-5}d^6}{121c^{-6}d^8}\right)^0$

7. $\dfrac{16^{\frac{3}{4}}4^{-2}}{8^{-\frac{2}{3}}}$

8. $\left(\dfrac{a^{-\frac{1}{2}}a^{\frac{2}{3}}}{a^{-2}}\right)^6$

9. $\dfrac{x^{-2}-2x^{-1}}{x^{-3}}$

10. $x^{\frac{3}{2}}\left(x^{\frac{3}{2}}+x^{-\frac{1}{2}}\right)$

11. $\left(2x^{-\frac{1}{2}}-x^{\frac{1}{2}}\right)^2$

12. Evaluate the following:

(a) $(-64)^{\frac{2}{3}}$

(b) $(64)^{-\frac{3}{2}}$

13. Express $4b^{\frac{4}{3}}$ in radical form.

14. If there are 27,500,000 dings in 1 dang and 19,400 dangs in 1 dong, how many dings are in 1 dong?

Chapter 6—Test B

Perform the indicated operations and express your answer in simplest form with positive exponents only. Assume all variables represent positive real numbers.

1. $\left(x^2y\right)^3\left(x^3y^2\right)$

2. $\left(-2x^2y^2\right)^3\left(-3x^2\right)^2$

3. $\left(4p^{-2}q^{-1}\right)^2\left(6r^{-3}s\right)^{-1}$

4. $\dfrac{\left(-3xy^3\right)^3}{\left(6x^2y^3\right)^2}$

5. $\left(\dfrac{2mn^{-3}}{5m^{-2}n}\right)^{-2}$

6. $\left(\dfrac{81u^{-8}v^7}{49u^{-3}v^3}\right)^0$

7. $\dfrac{25^{\frac{3}{2}}5^{-3}}{125^{-\frac{2}{3}}}$

8. $\left(\dfrac{c^{\frac{3}{2}}c^{-\frac{2}{3}}}{c^{-1}}\right)^6$

9. $\dfrac{y^{-2}+3y}{y^{-4}}$

10. $z^{\frac{3}{2}}\left(z^{\frac{1}{2}}-z^{-\frac{1}{2}}\right)$

11. $\left(4x^{-\frac{1}{2}} + x^{\frac{1}{2}}\right)^2$

12. Evaluate the following:

    (a) $(64)^{\frac{3}{2}}$

    (b) $(-64)^{-\frac{2}{3}}$

13. Express $5d^{\frac{2}{5}}$ in radical form.

14. If there are 3,190,000 binks in 1 bonk and 277,000,000 bonks in 1 bunk, how many binks are in 1 bunk?

Cumulative Review: Chapters 4–6

In exercises 1–4, reduce to lowest terms.

1. $\dfrac{30s^4t^2}{35st^3}$

2. $\dfrac{a^2 + 8ab + 16b^2}{a^2 - 16b^2}$

3. $\dfrac{4y^3 - 5y^2z + yz^2}{y^5 - 4y^4z + 3y^3z^2}$

4. $\dfrac{m^2 - 2mn + 4n^2}{m^3 + 8n^3}$

In exercises 5–14, perform the operations. Express your answer in simplest form.

5. $\dfrac{4p^2q^3}{9rs^4} \div \dfrac{2p^3q}{15r^3s^2}$

6. $\dfrac{4x + 12y}{x - 2y} \cdot \dfrac{x^2 - 4y^2}{x^2 + 5xy + 6y^2}$

7. $\dfrac{a}{3b} - \dfrac{3b}{a}$

8. $\dfrac{5x - 1}{9 - 15x} + \dfrac{2}{15x - 9}$

9. $\dfrac{3x^3 + 6x^2 - 9x}{x + 1} \cdot \dfrac{2x + 2}{x^3 - x^2}$

10. $\dfrac{4a^2 + 4ab - 3b^2}{4a^2 - 8ab + 3b^2} \div \dfrac{4a^2 + 9b^2}{4a^2 - 9b^2}$

11. $\dfrac{3}{x + 4} - \dfrac{1}{x - 4}$

12. $\dfrac{5}{x^2 - 3xy - 4y^2} - \dfrac{3}{x^2 - 5xy + 4y^2}$

13. $\dfrac{x^2y - 4y^3}{2xy^2 - y^3} \div \left(\dfrac{4x - 8y}{2xy + y^2} \div \dfrac{2x - y}{3x + 6y}\right)$

14. $\dfrac{9x}{3x + y} - \dfrac{7x}{3x - y} - \dfrac{6y^2}{9x^2 - y^2}$

In exercises 15–16, write as a simple fraction reduced to lowest terms.

15. $\dfrac{1 + \dfrac{4}{x} + \dfrac{3}{x^2}}{1 + \dfrac{3}{x}}$

16. $\dfrac{\dfrac{x}{x + 2y} + \dfrac{2y}{x + 3y}}{\dfrac{2}{x + 2y} - \dfrac{1}{x + 3y}}$

In exercises 17–22, solve the equation or inequality.

17. $\dfrac{3}{x} + \dfrac{1}{3} = 3 + \dfrac{1}{x}$

18. $4 - \dfrac{12}{2x - 3} = 0$

19. $\dfrac{x}{3} + \dfrac{x+2}{4} \le \dfrac{x+8}{6}$

20. $\dfrac{x+7}{x+5} + 2 = \dfrac{2}{x+5}$

21. $\dfrac{2}{x+3} - \dfrac{1}{x-2} - \dfrac{6}{x^2 + x - 6}$

22. $\dfrac{3x-2}{3} - \dfrac{3-2x}{2} > \dfrac{5}{6}$

In exercises 23–26, solve for the given variable.

23. $5s - 4t = 3s + t$ for s

24. $cd + c = 2d - cd$ for c

25. $\dfrac{p + 3q}{p} = 2q$ for p

26. $\dfrac{p + 3q}{p} = 2q$ for q

In exercises 27–30, solve algebraically.

27. The ratio of registered male voters to registered female voters in Monroe County is 9 to 8. If there are 34,000 registered voters in Monroe County, how many of them are males?

28. A 15% solution of calcium chloride is mixed with a 35% solution of calcium chloride in order to form 20 gallons of a 27% solution. How many gallons of each of the original solutions are in the mixture?

29. Alonzo can wash the windows in his house in 3 hours. His son needs 5 hours to do the same job. How long would it take Alonzo and his son to wash the windows if they worked together?

30. Onyx stones are worth $30 each and garnet stones are worth $50 each. A shipment of 70 stones has a total value of $2,940. How many of each type of stone are in the shipment?

In exercises 31–36, sketch a graph of the equation using the intercept method.

31. $2x - 3y = 12$

32. $4x + y = 6$

33. $2y = 5x + 10$

34. $2x = 5y + 10$

35. $6y - 1 = 2$

36. $3 - x = 4$

In exercises 37–42, find the slope of the line.

37. passing through the points $(-5,2)$ and $(-6,6)$

38. passing through the points $(0,k)$ and $(k,0)$, where $k \ne 0$

39. whose equation is $y = 3 + 6x$

40. whose equation is $7x - 3y = 15$

41. parallel to the line passing through the point $(8,1)$ and $(5,-8)$

42. perpendicular to the line having an x-intercept of 3 and a y-intercept of 4

In exercises 43–44, find the values of a satisfying the given condition(s).

43. The line through the points $(a,2)$ and $(5,a)$ has slope 2.

44. The line through the points $(2,1)$ and $(4,a)$ is parallel to the line through the points $(1,-2)$ and $(a,6)$.

In exercises 45–52, find an equation of the line satisfying the given conditions.

45. The line passes through the point (–4,3) and has slope –5.

46. The line passes through the point (8,–6) and has slope 0.

47. The line passes through the points (1,–4) and (3,2).

48. The line passes through the points (–5,11) and (–5,–11).

49. The line has y-intercept 3 and slope –2.

50. The line has x-intercept 3 and slope –2.

51. The line passes through the point (1,–6) and is parallel to the line $7x + 2y = 2$.

52. The line passes through the origin and is perpendicular to the line $5x + 3y = 4$.

In exercises 53–57, graph the inequality on the rectangular coordinate system.

53. $4x - 3y \leq 24$

54. $5x + y > 10$

55. $2y \geq x + 8$

56. $x + 7 < 3$

57. $5 - 3y \leq 11$

58. A survey shows that the relationship between the temperature in a home and the homeowner's utility bill is a linear one. If the bill is $150 when the temperature is 68°F and if it is $250 when the temperature is 73°F, predict the bill that the homeowner would receive if he kept his home temperature at 75°F.

In exercises 59–66, solve the given system of equations.

59. $\begin{cases} 2x + y = -10 \\ -7x + 5y = 1 \end{cases}$

60. $\begin{cases} 3x - 4y = 8 \\ 2x + 5y = 13 \end{cases}$

61. $\begin{cases} 6a + 7b = -2 \\ 7a + 8b = -2 \end{cases}$

62. $\begin{cases} -5u + 8v = 48 \\ 8u + 5v = 30 \end{cases}$

63. $\begin{cases} m = n + 6 \\ n = m - 4 \end{cases}$

64. $\begin{cases} 3s = t + 1 \\ t = 3s - 1 \end{cases}$

65. $\begin{cases} \dfrac{1}{2}p - \dfrac{1}{3}q = -1 \\ \dfrac{1}{4}p + \dfrac{1}{5}q = 5 \end{cases}$

66. $\begin{cases} w + \dfrac{z}{4} = \dfrac{2}{3} \\ \dfrac{w}{3} - z = -\dfrac{1}{2} \end{cases}$

In exercises 67–78, perform the operations and simplify. Express your answer with positive exponents only.

67. $\left(-4xy^2\right)^3 \left(2x^2y\right)$

68. $\dfrac{9x^3y^8}{6x^6y^6}$

69. $\left(\dfrac{5a^3b^2}{15a^2b^4}\right)^2$

70. $\dfrac{\left(-c^2d\right)^3 \left(-cd^2\right)^2}{-2c^3d^2}$

71. $\left(m^{-3}n^{-4}\right)\left(m^{-2}n^3\right)$

72. $\left(5k^{-2}\ell^3\right)^{-2}$

73. $\dfrac{s^{-5}t^{-4}}{s^0t^{-4}}$

74. $\left(-\dfrac{1}{2}\right)^{-5}$

75. $\left(-3u^2v^{-3}\right)^{-1}\left(-2u^{-3}v^{-1}\right)^{-2}$

76. $\left(\dfrac{2pq^{-2}}{\left(2p^{-3}q\right)^2}\right)^{-1}$

77. $\left(w^{-2}+z^{-2}\right)\left(w^2-z^2\right)$

78. $\dfrac{4-k^{-2}}{2-k^{-1}}$

In exercises 79–80, express the number using scientific notation.

79. 4,719.882

80. 0.00001001

81. A pearl will fall through a tube containing molasses at the rate of 0.000000356 inches per minute. How far will the pearl fall between 2 P.M. and 5:30 P.M.?

82. Perform the computations using scientific notation and express your answer in standard notation:

$$\dfrac{(54,000,000)(0.0002)}{(0.009)(40,000)}$$

In exercises 83–84, evaluate.

83. $(-216)^{\frac{2}{3}}$

84. $\left(\dfrac{81}{16}\right)^{-\frac{3}{4}}$

In exercises 85–88, perform the operations and express your answers with positive exponents only.

85. $a^{\frac{4}{5}}a^{-\frac{1}{3}}$

86. $\dfrac{b^{\frac{7}{8}}}{b^{\frac{2}{3}}}$

87. $\left((32)^{-\frac{2}{5}}4^{\frac{3}{2}}\right)^{-2}$

88. $\left(\dfrac{z^{\frac{3}{2}}z^{-\frac{1}{4}}}{z^{\frac{2}{3}}}\right)^{-3}$

In exercises 89–90, change to radical notation.

89. $x^{\frac{2}{5}}$

90. $(3x)^{\frac{4}{3}}$

Cumulative Test: Chapters 4–6

1. Perform the operations and express your answer in simplest form.

   (a) $\dfrac{8x^2+2xy-y^2}{2x^2+5xy+2y^2} \div \dfrac{16x^2-y^2}{x^2+2xy}$

   (b) $\dfrac{3a}{6a-10b}+\dfrac{5b}{10b-6a}$

   (c) $\dfrac{x+3}{x^2-1}-\dfrac{2}{x^2-x}$

2. Express as a simple fraction reduced to lowest terms: $\dfrac{\dfrac{1}{x-2} - 2}{2 + \dfrac{1}{x+2}}$

3. Solve each of the following:

   (a) $\dfrac{1}{x-2} - \dfrac{5}{2x} = \dfrac{8}{x^2 - 2x}$

   (b) $\dfrac{x+3}{5} + \dfrac{x+5}{3} > 6$

   (c) $\dfrac{2x-11}{x-4} = 1 - \dfrac{3}{x-4}$

4. Solve for z if $w = \dfrac{z-5}{z+5}$

5. Conrad can paint a room in $1\frac{1}{2}$ hours, while his assistant can paint the same room in $2\frac{1}{2}$ hours. How long would it take them to paint the room if they work together?

6. Sketch a graph of the equation $4x + 5y = 30$ using the intercept method.

7. Find the slope of the line passing through the points (2,5) and (–1,–1).

8. Find an equation of the line passing through the point (–5,7) and:

   (a) parallel to the line $4x - 8y = 1$

   (b) perpendicular to the line $4x - 8y = 1$

9. Sketch a graph of $8x - 9y < 72$ on a rectangular coordinate system.

10. The relationship between the age of a child and his achievement on a standardized test is exactly linear. If a 5-year old scores 48 on the test and an 8-year old scores 72 on the test, what is the age of a child who scores 104 on the test?

11. Solve the following system of equations:

$$\begin{cases} 4x - 3y = 17 \\ 3x + 4y = -6 \end{cases}$$

12. Perform the operations and simplify. Express your answer with positive exponents.

   (a) $\left(2x^4 y^3\right)^2 \left(-3x^2 y\right)^3$

   (b) $\left(2x^{-4} y^3\right)^{-2} \left(-3x^{-2} y^{-1}\right)^3$

   (c) $\left(\dfrac{2a^{-1} b^{-3}}{8a^{-3} b^{-2}}\right)^{-2}$

   (d) $\dfrac{m^{-1} - n^{-1}}{m^{-2} - n^{-2}}$

13. Express 0.0006001 in scientific notation.

14. Perform the operations using scientific notation. Express your answer in standard form:

$$\dfrac{(1,440,000)(0.000018)}{(30)(9,000)}$$

15. Evaluate $(-32)^{-\frac{3}{5}}$

16. Perform the operations and simplify. Express your answers using positive exponents only. Assume all variables represent positive real numbers.

(a) $\left(x^{\frac{3}{4}}x^{-\frac{2}{3}}\right)^{6}$

(b) $\dfrac{x^{-\frac{1}{2}}x^{-\frac{1}{5}}}{x^{-\frac{3}{10}}}$

Chapter 7—Test A

Perform the indicated operations and express your answer in simplest form. Assume all variables represent positive real numbers.

1. $\sqrt[4]{16x^{12}y^{8}}$

2. $\sqrt[3]{3x^{4}y}\ \sqrt[3]{9x^{2}y}$

3. $\left(4a\sqrt{a^{3}b^{3}}\right)\left(-5a^{3}\sqrt{ab}\right)$

4. $\dfrac{\sqrt{27}}{\sqrt{75}}$

5. $\sqrt{\dfrac{3}{10}}$

6. $\sqrt[4]{\dfrac{81}{8}}$

7. $\dfrac{\left(2c^{2}d\sqrt{3cd^{3}}\right)\left(4c\sqrt{d}\right)}{\sqrt{6c^{5}}}$

8. $\sqrt[3]{3}\ \sqrt[4]{3}$

9. $3\sqrt{20}+\sqrt{80}-4\sqrt{45}$

10. $\left(2+\sqrt{y}\right)^{2}$

11. $\sqrt{28}-2\sqrt{\dfrac{9}{7}}$

12. $\dfrac{\sqrt{10}}{4+\sqrt{10}}$

Solve the following equations:

13. $\sqrt{x+1}-2=5$

14. $\sqrt{x}+1=5$

15. $\sqrt[3]{x+1}+2=5$

16. $x^{\frac{1}{5}}=3$

Express the following in the form of a + bi.

17. i^{35}

18. $(2i+3)(i-3)$

19. $\dfrac{i+1}{2i+1}$

Chapter 7—Test B

Perform the indicated operations and express your answer in simplest radical form. Assume all variables represent positive real numbers.

1. $\sqrt[5]{32x^{20}y^{10}}$

2. $\sqrt[4]{8x^{2}y^{5}}\ \sqrt[4]{2xy^{3}}$

3. $\left(-3p^2\sqrt{p^5q}\right)\left(6pq\sqrt{pq^3}\right)$

4. $\dfrac{\sqrt{20}}{\sqrt{45}}$

5. $\sqrt{\dfrac{11}{5}}$

6. $\sqrt[3]{\dfrac{27}{2}}$

7. $\dfrac{\left(3uv^3\sqrt{2u^5v}\right)\left(2v\sqrt{u}\right)}{\sqrt{6v^7}}$

8. $\sqrt{7}\sqrt[5]{7}$

9. $2\sqrt{48}-\sqrt{75}-3\sqrt{12}$

10. $\left(4-\sqrt{z}\right)^2$

11. $\sqrt{20}-3\sqrt{\dfrac{4}{5}}$

12. $\dfrac{\sqrt{15}}{5-\sqrt{15}}$

Solve the following equations:

13. $\sqrt{x+2}-1=4$

14. $\sqrt{x}+2=4$

15. $\sqrt[3]{x+2}+1=4$

16. $x^{\frac{1}{7}}=2$

Express the following in the form a + bi.

17. i^{62}

18. $(4i-3)(i-2)$

19. $\dfrac{i+3}{3i+2}$

Chapter 8—Test A

1. Solve the following by either the factoring method or the square root method.

   (a) $(y+2)(4y-3)=y^2+7y+2$

   (b) $5-\dfrac{14}{x^2}=3$

2. Solve the following by any method.

   (a) $b^2+3b-3=0$

   (b) $(x+2)(x-4)=3x^2-11$

   (c) $\dfrac{2}{x}-\dfrac{3x}{x+1}=\dfrac{1}{x^2+x}$

3. Use the discriminant to determine the nature of the roots of $3x^2-4x+3=0$.

4. Solve for s: $p=\dfrac{3rs^2}{t}$

5. Solve the following radical equation: $\sqrt{2g} - \sqrt{3 - g} = 1$

6. Solve the following: $z^4 - 5z^2 - 36 = 0$

7. Solve the following and graph the solution set:

    (a) $x^2 - 3x < 4$

    (b) $\dfrac{2x + 1}{x - 3} \geq 0$

8. Find the length of the diagonal of a rectangle with sides of 2" and 4".

9. A man rides his bicycle for 5 miles against the wind and then rides back for 3 miles with the wind. If the entire trip took 3 hours and if the man can ride at the rate of 4 mph when there is no wind blowing, what is the speed of the wind?

Chapter 8—Test B

1. Solve the following by either the factoring method or the square root method.

    (a) $(5y - 2)(y + 3) = 3y^2 + 10y - 1$

    (b) $\dfrac{8}{x^2} + 1 = 5$

2. Solve the following by any method.

    (a) $b^2 - 5b + 2 = 0$

    (b) $(x + 3)(x - 1) = 4x^2 - 6$

    (c) $\dfrac{3}{x} - \dfrac{2x}{x - 2} = \dfrac{-26}{x(x - 2)}$

3. Use the discriminant to determine the nature of the roots of $4x^2 + 3x + 4 = 0$.

4. Solve for b: $a = \dfrac{b^2 c}{5d}$

5. Solve the following radical equation: $\sqrt{3g} - \sqrt{7 - g} = 1$

6. Solve the following: $z^4 + 24z^2 - 25 = 0$

7. Solve the following and graph the solution set:

    (a) $x^2 + 4x \leq -3$

    (b) $\dfrac{3x - 4}{x + 1} > 0$

8. Find the length of a rectangle if its width is 4" and its diagonal is 6".

9. A woman jogs for 8 miles with the wind at her back and then jogs in the opposite direction for 6 miles against the wind. If the entire trip took 5 hours and if the woman can jog at the rate of 3 mph when no wind is blowing, what is the speed of the wind?

Chapter 9—Test A

1. Find the distance between the points $(-4,3)$ and $(3,-4)$.

2. Find the midpoint between the points $(-1,6)$ and $(5,1)$.

3. Identify the center and radius of the following circles.

   (a) $(x+2)^2 + (y-4)^2 = 18$

   (b) $x^2 + y^2 - 8x + 2y - 8 = 0$

4. Sketch a graph of the following parabolas. Label the x- and y-intercepts, vertex, and axis of symmetry.

   (a) $y = -2x^2 + 5x - 3$

   (b) $y = 3x^2 + 6x + 4$

5. Sketch a graph of the ellipse and indicate the vertices: $4x^2 + 5y^2 = 100$

6. Sketch a graph of the hyperbola. Label the vertices and the asymptotes:

   $$\frac{y^2}{9} - \frac{x^2}{12} = 1$$

7. Sketch a graph of the following figures and label the important aspects of the figure.

   (a) $4x^2 + 9y^2 = 144$

   (b) $x^2 + y^2 + 12x - 6y + 9 = 0$

   (c) $4y - y^2 = x$

Chapter 9—Test B

1. Find the distance between the points $(5,-1)$ and $(-1,5)$.

2. Find the midpoint between the points $(2,-4)$ and $(8,3)$.

3. Identify the center and radius of the following circles.

   (a) $(x+4)^2 + (y-1)^2 = 24$

   (b) $x^2 + y^2 - 6x + 10y + 18 = 0$

4. Sketch a graph of the following parabolas. Label the x- and y-intercepts, vertex, and axis of symmetry.

   (a) $y = -3x^2 - 5x + 2$

   (b) $y = 2x^2 - 4x + 3$

5. Sketch a graph of the ellipse and indicate the vertices: $5x^2 + 3y^2 = 75$

6. Sketch a graph of the hyperbola. Label the vertices and the asymptotes:

   $$\frac{y^2}{16} - \frac{x^2}{12} = 1$$

7. Sketch a graph of the following figures and label the important aspects of the figure.

(a) $9x^2 + 4y^2 = 144$

(b) $x^2 + y^2 + 8x - 4y - 5 = 0$

(c) $2y + y^2 = x$

Cumulative Review: Chapters 7–9

In exercises 1–8, express in simplest radical form. Assume all variables represent positive real numbers only.

1. $\sqrt{36r^6s^{14}}$

2. $\sqrt{45v^5w^9}$

3. $\left(4a\sqrt{27a^3b^2}\right)\left(-a\sqrt{3ab^3}\right)$

4. $\sqrt[4]{64p^6q^9}$

5. $\dfrac{\sqrt{125}}{\sqrt{5}}$

6. $\sqrt{\dfrac{7}{c^2d^3}}$

7. $\sqrt[9]{x^3}$

8. $\sqrt[5]{\dfrac{3}{2x^2}}$

In exercises 9–18, perform the indicated operations and simplify as completely as possible.

9. $5\sqrt{11} - 2\sqrt{11} + 4\sqrt{11}$

10. $6\sqrt{40} - 3\sqrt{10} - \sqrt{90}$

11. $2h^2 \cdot \sqrt[5]{h^7} - h^3 \sqrt[5]{h^2}$

12. $\sqrt{\dfrac{5}{8}} + \sqrt{\dfrac{8}{5}}$

13. $\sqrt{7}\left(\sqrt{5} + \sqrt{3}\right) - \sqrt{3}\left(\sqrt{7} - \sqrt{5}\right)$

14. $(\sqrt{14} - \sqrt{7})^2$

15. $(\sqrt{13} + \sqrt{5})(\sqrt{13} - \sqrt{5})$

16. $\left(\sqrt{x} + 1\right)^2 - \left(\sqrt{x+1}\right)^2$

17. $\dfrac{6}{\sqrt{6} - 2} - \dfrac{18}{\sqrt{6}}$

18. $\dfrac{7\sqrt{2}}{4 + \sqrt{2}}$

In exercises 19–22, solve the equation.

19. $\sqrt{2x+1} = 5$

20. $\sqrt{2x} + 1 = 5$

21. $\sqrt[5]{3x+5} = -1$

22. $x^{-\frac{1}{3}} + 1 = -1$

In exercises 23–24, express as i, –1, –i, or 1.

23. i^{97}

24. i^{-10}

In exercises 25–26, express in the form a + bi.

25. $(4 + 7i)(3 - 2i)$

26. $\dfrac{5 - 2i}{2 + 5i}$

In exercises 27–30, solve by the factoring or square root method.

27. $x^2 + 3x - 18 = 0$

28. $x^2 - 4x + 4 = x + 10$

29. $4a^2 - 2a + 1 = 3(a + 1) - 5(a - 2)$

30. $(a + 3)^2 = 32$

In exercises 31–32, solve by completing the square.

31. $z^2 + 4z - 2 = 0$

32. $2z^2 - 12z = 7$

In exercises 33–40, solve by any method.

33. $2y^2 - 2y - 5 = 0$

34. $(c + 1)(c - 4) = 4c^2 - 10$

35. $x - 4 = \dfrac{4}{x - 4}$

36. $(p + 3)(p - 5) = 1$

37. $\dfrac{1}{x - 2} + \dfrac{2}{x} = 1$

38. $(d - 1)(d + 6) = (d + 2)(d + 4)$

39. $\dfrac{6}{h + 1} + \dfrac{3h}{h - 2} = \dfrac{5}{h + 1}$

40. $\dfrac{2}{x + 2} + \dfrac{3}{x + 3} = \dfrac{4}{x + 4}$

In exercises 41–42, solve for the given variable.

41. $a = \dfrac{bc^2}{d}$ (for c)

42. $x^2 - xy - 2y^2 = 0$ (for y)

In exercises 43–44, solve the equation.

43. $\sqrt{5x + 4} = x + 2$

44. $\sqrt{2x} + \sqrt{x - 1} = 3$

In exercises 45–46, solve for the given variable.

45. $\sqrt{4x + 1} = 3y$ (for x)

46. $\sqrt{4x} + 1 = 3y$ (for x)

In exercises 47–48, solve the equation.

47. $16 - x^4 = 0$

48. $s^{\frac{1}{2}} + 2s^{\frac{1}{4}} - 8 = 0$

In exercises 49–52, solve the inequality and graph the solution set.

49. $(x - 3)(x + 6) \geq 0$

50. $3x^2 < x + 2$

51. $\dfrac{x + 1}{x + 4} \leq 0$

52. $\dfrac{3x - 1}{x} > 0$

In exercises 53–56, solve algebraically.

53. The sum of a number and two times its reciprocal is $\dfrac{11}{3}$. Find the number.

54. The length of a rectangle is 1" less than three times its width. If the area of the rectangle is 24 square inches, find the dimensions of the rectangle.

55. The longer leg of a right triangle is 7 cm. more than the shorter leg. If the hypotenuse of the triangle is 3 cm. more than twice the shorter leg, find the lengths of the sides of the triangle.

56. An 8" x 10" photograph is surrounded by a matting of uniform width and then enclosed in a glass frame. If the area of the entire frame is 143 square inches, find the width of the matting.

In exercises 57–60, find the distance and midpoint between P and Q.

57. P(–1,2), Q(2,–2)   58. P(–3,5), Q(3,–5)

59. P(0,1), Q(–4,5)   60. P(1,6), Q(0,7)

In exercises 61–64, graph the circle and indicate its center and radius.

61. $(x+1)^2 + (y-3)^2 = 16$ 62. $x^2 + (y+2)^2 = 12$

63. $x^2 + y^2 - 8x + 8y - 4 = 0$ 64. $x^2 + y^2 + 4x + 6y + 5 = 0$

In exercises 65–70, graph the parabola. Indicate its vertex, axis of symmetry, and intercepts (if they exist).

65. $y = 2x^2$ 66. $y = \frac{1}{2}x^2 - 8$

67. $y = x^2 - 6x$ 68. $y = -x^2 + 4x - 5$

69. $y = -x^2 + 2x + 2$ 70. $x = 4y^2 - 4y + 1$

In exercises 71–73, graph the ellipse. Indicate its vertices.

71. $\frac{x^2}{9} + \frac{y^2}{6} = 1$ 72. $25x^2 + y^2 = 25$

73. $9x^2 + 4y^2 = 1$

In exercises 74–76, graph the hyperbola. Indicate its vertices.

74. $\frac{x^2}{16} - \frac{y^2}{8} = 1$ 75. $25y^2 - 36x^2 = 900$

76. $x^2 - 4y^2 = 1$

In exercises 77–82, identify the figure and, if possible, sketch a graph of its equation. Label the important aspects of the figure.

77. $\frac{y^2}{4} - \frac{x^2}{9} = 4$ 78. $x = 1 - y^2$

79. $\frac{y}{2} - \frac{x}{3} = 2$ 80. $4x^2 + 9y^2 = 9$

81. $x^2 + y^2 - 8y - 9 = 0$ 82. $y = 2x^2 - 6x - 8$

83. The area, A, of a rectangular field that can be enclosed with 30 yards of fencing is related to the length, x, of the field in the following way: $A = 30x - x^2$. What length will produce the maximum area? What is the maximum area?

Cumulative Test: Chapters 7–9

1. Express the following in simplest form. Assume all variables represent positive real numbers.

   (a) $\sqrt{98s^3t^6}$

   (b) $\left(3c\sqrt{8c^2d}\right)\left(-2d\sqrt{6cd^2}\right)$

   (c) $\sqrt[4]{\dfrac{2}{x}}$

2. Perform the operations and simplify as completely as possible.

   (a) $4\sqrt{28} - \sqrt{63} - 5\sqrt{7}$

   (b) $\sqrt{\dfrac{x}{y}} + \sqrt{\dfrac{y}{x}}$

   (c) $\left(\sqrt{7} - \sqrt{5}\right)\left(\sqrt{5} - \sqrt{7}\right)$

   (d) $\dfrac{3\sqrt{3}}{3 + \sqrt{3}}$

3. Solve for x if $\sqrt{3x-1} - 2 = 1$

4. Express the following in the form a + bi: $\dfrac{3 + 4i}{4 + 3i}$

5. Solve the following equations by any method.

   (a) $5x^2 - 3 = 2x$

   (b) $\dfrac{x-2}{3} = \dfrac{2}{x+2}$

   (c) $2x^2 - 10x + 5 = 0$

   (d) $\dfrac{x-1}{x+1} - \dfrac{x+1}{x-1} = 4$

6. Solve for s: $\sqrt{1 - \dfrac{s^2}{2}} = t$

7. Solve for x: $\sqrt{3x-2} = 1 + \sqrt{x-1}$

8. Solve for x and graph the solution on the real number line: $3x^2 - 8x - 3 \le 0$

9. A boy started on a trip across a lake by motorboat. After traveling for 15 miles, his motor failed and he had to row an additional 6 miles in order to reach the opposite shore. If his average speed by motor was 4 mph more than his average rowing speed, and if the entire trip took $5\dfrac{1}{2}$ hours, what was his average rowing speed?

10. Find the distance between the points (–7,–4) and (–4,0) and find the midpoint between the two points.

11. Graph the circle $x^2 + y^2 + 10x = 11$.

12. Graph the parabola $y = 3x^2 - 6x - 9$. Identify intercepts (if they exist) and its vertex.

13. Identify and graph the following figures. Label the important aspects of the figure.

    (a) $49y^2 - 9x^2 = 441$

    (b) $4x^2 + \dfrac{y^2}{4} = 1$

14. A rock is thrown upward from the top of a 96 foot tower. The equation $s = -16t^2 + 48t + 96$ gives the distance (in feet) that the rock is above the ground t seconds after it is thrown. What is the maximum height of the rock?

Chapter 10—Test A

1. Solve the following system of simultaneous equations.

$$\begin{cases} x + 2y + 4z = -6 \\ 3x - y + 3z = 7 \\ 2x + 3y - z = 7 \end{cases}$$

2. Solve the following using augmented matrices:

    (a) $\begin{cases} 2x - 5y = -7 \\ x + y = 7 \end{cases}$

    (b) $\begin{cases} x + 2y = -3 \\ x - 3z = 4 \\ 3x - y - 4z = 9 \end{cases}$

3. Evaluate the following determinants.

    (a) $\begin{vmatrix} 5 & 1 \\ -6 & 2 \end{vmatrix}$

    (b) $\begin{vmatrix} 1 & -1 & 3 \\ 2 & 2 & 0 \\ 4 & -3 & 5 \end{vmatrix}$

4. Solve the following using Cramer's rule.

    (a) $\begin{cases} 4x - 3y = 3 \\ 6x + 12y = -1 \end{cases}$

    (b) $\begin{cases} x - 2y = 1 \\ y - 2z = 5 \\ x - 2z = 7 \end{cases}$

5. A change purse contains nickels, dimes, and quarters. If there are 13 coins in the purse with 3 times as many nickels as quarters, and if the total value of the coins is $1.30, how many of each type of coin are in the purse?

6. Solve the following non-linear system of equations:

$$\begin{cases} x^2 - 3y^2 = 1 \\ x - y = 3 \end{cases}$$

7. Sketch the solution set of the following system of inequalities.

$$\begin{cases} x - 3y \geq 6 \\ y \geq -4 \\ x < 3 \end{cases}$$

Chapter 10—Test B

1. Solve the following system of simultaneous equations.

$$\begin{cases} 2x - y - 2z = -1 \\ x + 3y + 4z = 4 \\ -3x - 2y + 3z = 12 \end{cases}$$

2. Solve the following using augmented matrices:

(a) $$\begin{cases} 4x - 3y = 10 \\ x + 4y = -7 \end{cases}$$

(b) $$\begin{cases} x - 3y = -10 \\ x + 4z = 7 \\ 2x + 2y - 3z = -2 \end{cases}$$

3. Evaluate the following determinants.

(a) $$\begin{vmatrix} -4 & 3 \\ -2 & 5 \end{vmatrix}$$

(b) $$\begin{vmatrix} -1 & 2 & -1 \\ 0 & 4 & 3 \\ 2 & -3 & -2 \end{vmatrix}$$

4. Solve the following using Cramer's rule.

(a) $$\begin{cases} 6x + 2y = 3 \\ 3x - 4y = 4 \end{cases}$$

(b) $$\begin{cases} 2x + y = 1 \\ 2y + z = 10 \\ 2x + z = 2 \end{cases}$$

5. A wallet contains $10 bills, $20 bills, and $50 bills. If there are 15 bills in the wallet with one more $20 bill than $50 bill, and if the total value of the bills is $310, how many of each type of bill are in the wallet?

6. Solve the following non-linear system of equations:

$$\begin{cases} y^2 - 2x^2 = 1 \\ y + 2x = 1 \end{cases}$$

469

7. Sketch the solution of the following system of inequalities.

$$\begin{cases} x + 2y > 4 \\ \quad x \le 1 \\ \quad y < 5 \end{cases}$$

Chapter 11—Test A

1. Identify the domains and ranges of the following relations. (Sketch a graph if necessary.)

   (a) $\{(1,4), (2,-4), (1,0), (-2,4)\}$

   (b) $4x^2 + 9y^2 = 36$

2. Identify the domain of the following relations.

   (a) $y = \sqrt{2 - x}$

   (b) $y = \dfrac{x + 1}{2x - 1}$

3. Identify which of the following relations are functions.

   (a) $\{(3,1), (-3,1)\}$

   (b) $\{(7,5), (-7,5), (7,-5)\}$

   (c) $x = y^2 - 1$

   (d)

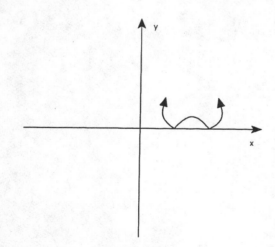

4. Given $f(x) = 2x^2 + x$ and $g(x) = \sqrt{3x - 2}$, find:

   (a) $g(6)$

(b)   f(–1)

(c)   f(x – 1)

(d)   f(g(2))

5.   Sketch a graph of $f(x) = |3 + 2x|$.

6.   Find the inverse of the relation $\{(-4,1), (3,6), (7,1)\}$.

7.   Determine if the following function is a one-to-one function.

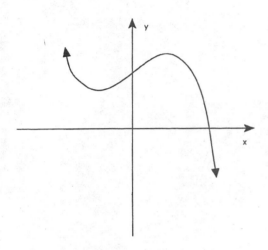

8.   Find the inverse of the following functions.

(a) $f(x) = -\dfrac{1}{2}x + 3$

(b) $f(x) = \dfrac{3x + 1}{x}$

9.   If y varies directly as the cube of x and y is 32 when x is 2, find y when x is 3.

10.   If z varies directly as x and inversely as the square of y and z is 12 when x is 3 and y is 2, find z when x is 18 and y is 6.

Chapter 11—Test B

1.   Identify the domains and ranges of the following relations. (Sketch a graph if necessary.)

(a) $\{(3,-1), (-3,1), (3,1), (1,3)\}$

(b) $9x^2 + y^2 = 81$

2. Identify the domains of the following relations.

   (a) $y = \sqrt{x - 3}$

   (b) $y = \dfrac{x - 2}{2x + 1}$

3. Identify which of the following relations are functions.

   (a) $\{(5,0), (-5,0)\}$

   (b) $\{(8,-4), (-8,4), (-8,-4)\}$

   (c) $x = y^2 + 4$

   (d)

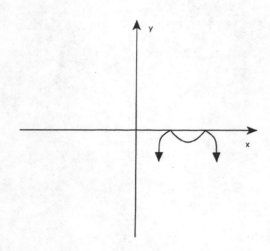

4. Given $f(x) = 3x^2 - x$ and $g(x) = \sqrt{4x + 1}$, find:

   (a) $g(12)$

   (b) $f(-1)$

   (c) $f(x + 2)$

   (d) $f(g(2))$

5. Sketch a graph of $f(x) = |1 - 3x|$.

6. Find the inverse of the relation $\{(5,-2), (0,3), (-6,-2)\}$.

7. Determine if the following function is a one-to-one function.

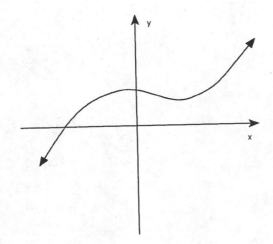

8. Find the inverse of the following functions.

    (a) $f(x) = \dfrac{1}{3}x + 2$

    (b) $f(x) = \dfrac{2x - 1}{x}$

9. If y varies directly as the square of x and y is 45 when x is 3, find y when x is 4.

10. If z varies directly as x and inversely as the cube of y and z is 4 when x is 2 and y is 3, find z when x is 32 and y is 4.

Chapter 12—Test A

1. Sketch the graph of $y = \left(\dfrac{2}{3}\right)^{x}$.

2. Sketch the graph of $y = \log_4 x$.

3. Write in exponential form:

    (a) $\log_4 2 = \dfrac{1}{2}$

    (b) $\log_3 \dfrac{1}{27} = -3$

4. Write in logarithmic form:

    (a) $64^{\frac{2}{3}} = 16$

    (b) $5^{-3} = 0.008$

473

5. Evaluate each of the following logarithms:

(a) $\log_6 \dfrac{1}{36}$

(b) $\log_3 \sqrt{3}$

(c) $\log_{16} 128$

6. Write as a sum of simpler logarithms.

(a) $\log_b \left(\dfrac{x^3 \sqrt{y}}{z} \right)$

(b) $\log_3 \sqrt[4]{27x^3 y^5}$

7. Use a calculator to compute each of the following:

(a) $\log 384{,}000$

(b) antilog (-2.0794)

(c) $\ln 1.234$

(d) $e^{-0.47}$

8. Use the change of base formula to compute the following to three places:

(a) $\log_3 23$

(b) $\log_9 0.941$

9. Solve the following equations.

(a) $4^x = \dfrac{1}{256}$

(b) $\log_3 x + \log_3 (x + 2) = 1$

(c) $9^{1-x} = 27^{x-2}$

(d) $\log 40x - \log(x - 6) = 2$

(e) $4^{x+1} = 6$

10. If $8,000 is invested in an account paying 9.6% interest compounded semi-annually, how much money will be in the account after 4 years?

11. Assuming the population growth model $P = P_0\, e^{0.09t}$, how long will it take for a town's population to increase from 11,000 to 33,000?

Chapter 12—Test B

1. Sketch the graph of $y = \left(\dfrac{3}{2} \right)^x$.

2. Sketch the graph of $y = \log_3 x$.

3. Write in exponential form:

(a) $\log_{27} 3 = \frac{1}{3}$

(b) $\log_2 \frac{1}{16} = -4$

4. Write in logarithmic form:

(a) $16^{\frac{3}{4}} = 8$

(b) $20^{-2} = 0.0025$

5. Evaluate each of the following logarithms:

(a) $\log_4 \frac{1}{64}$

(b) $\log_5 \sqrt{5}$

(c) $\log_8 256$

6. Write as a sum of simpler logarithms.

(a) $\log_b \left(\frac{x^2 \sqrt[3]{y}}{z} \right)$

(b) $\log_4 \sqrt[5]{16x^2y^4}$

7. Use a calculator to compute each of the following:

(a) $\log 918$

(b) antilog (-1.2628)

(c) $\ln 2.468$

(d) $e^{-1.05}$

8. Use the change of base formula to compute the following to three places:

(a) $\log_4 47$

(b) $\log_7 0.092$

9. Solve each of the following equations:

(a) $2^x = \frac{1}{128}$

(b) $\log_4 (x-1) + \log_4 (x+2) = 1$

(c) $9^{2-x} = 27^{x-1}$

(d) $\log 20x - \log(x-8) = 2$

(e) $5^{x-1} = 7$

10. If $7,000 is invested in an account paying 8.8% interest compounded semi-annually, how much money will be in the account after 5 years?

11. Assuming the population growth model $P = P_0 e^{0.12t}$, how long will it take for a town's population to increase from 9,000 to 36,000?

Cumulative Review: Chapters 10–12

In exercises 1–6, solve the given system of equations.

1. $\begin{cases} x - y - z = 1 \\ 3x + y + 2z = 5 \\ 2x - y + 4z = 12 \end{cases}$

2. $\begin{cases} 5x - 3y + 2z = -5 \\ 4x + 2y - 5z = -4 \\ 3x + y - 4z = -5 \end{cases}$

3. $\begin{cases} 2x + 5y - z = 3 \\ 3x - 4y + 2z = -2 \\ 5x + y + z = 2 \end{cases}$

4. $\begin{cases} x + \dfrac{y}{2} - \dfrac{z}{3} = -5 \\ \dfrac{x}{2} - \dfrac{y}{3} + z = 1 \\ \dfrac{x}{3} - y + \dfrac{z}{2} = -5 \end{cases}$

5. $\begin{cases} 2x - 3y - 6z = 9 \\ 4x - 6y - 12z = 18 \\ \dfrac{2}{3}x - y - 2z = 3 \end{cases}$

6. $\begin{cases} 2x + y = 4 \\ 3y + 2z = -8 \\ 3x + 4z = 5 \end{cases}$

In exercises 7–14, evaluate the given determinant.

7. $\begin{vmatrix} 4 & -1 \\ 2 & 7 \end{vmatrix}$

8. $\begin{vmatrix} 2 & 3 \\ -6 & -9 \end{vmatrix}$

9. $\begin{vmatrix} -8 & 2 \\ 2 & 8 \end{vmatrix}$

10. $\begin{vmatrix} 7 & 0 \\ 7 & 7 \end{vmatrix}$

11. $\begin{vmatrix} 1 & 1 & -1 \\ 1 & -1 & 1 \\ -1 & 1 & 1 \end{vmatrix}$

12. $\begin{vmatrix} 2 & -3 & 1 \\ -8 & 12 & -4 \\ 4 & 1 & -5 \end{vmatrix}$

13. $\begin{vmatrix} -3 & 2 & 5 \\ 4 & -7 & 0 \\ -1 & 6 & 0 \end{vmatrix}$

14. $\begin{vmatrix} 5 & 3 & -2 \\ -2 & 3 & 5 \\ 5 & 3 & -2 \end{vmatrix}$

In exercises 15–16, solve using matrices.

15. $\begin{cases} x - 3y = 1 \\ 3x + 2y = 25 \end{cases}$

16. $\begin{cases} x + y + z = -1 \\ x - 3y - 2z = -2 \\ 4x + 2y + z = 3 \end{cases}$

In exercises 17–20, solve the given system of equations using Cramer's rule.

17. $\begin{cases} 4x - 9y = -30 \\ 7x + 5y = -11 \end{cases}$

18. $\begin{cases} 2x + 7y = -20 \\ 5x + 3y = 8 \end{cases}$

19. $\begin{cases} 7x + y - 3z = 3 \\ 2x - 4y + z = 7 \\ 4x + 5y + 6z = 5 \end{cases}$

20. $\begin{cases} x - 2y = -5 \\ y + 4z = 22 \\ 3x - z = 17 \end{cases}$

In exercises 21–28, solve the given system of equations.

21. $\begin{cases} x + 2y = 5 \\ x^2 + y^2 = 5 \end{cases}$

22. $\begin{cases} 2x + y = 5 \\ x^2 + y^2 = 10 \end{cases}$

23. $\begin{cases} y = x^2 - 2x \\ y = x^2 - 4 \end{cases}$

24. $\begin{cases} y = 2x^2 - 3x + 1 \\ y = 4 - 3x - x^2 \end{cases}$

25. $\begin{cases} x^2 + 2y^2 = 11 \\ x^2 - 4y^2 = 5 \end{cases}$

26. $\begin{cases} x^2 + 4y^2 = 4 \\ 4y^2 + 9x^2 = 36 \end{cases}$

27. $\begin{cases} x = y^2 - 1 \\ x - 2y = -1 \end{cases}$

28. $\begin{cases} x^2 - 4y^2 = 4 \\ y^2 - x = 2 \end{cases}$

In exercises 29–34, sketch the solution set to each system of inequalities.

29. $\begin{cases} x + 3y > 3 \\ x - y < -5 \end{cases}$

30. $\begin{cases} 2x - 5y \geq 10 \\ x - 2y \leq 6 \end{cases}$

31. $\begin{cases} 2x - y \geq 3 \\ x + y < 0 \end{cases}$

32. $\begin{cases} 4x + 3y < 12 \\ y \geq 2 \end{cases}$

33. $\begin{cases} x + 2y \geq 6 \\ -x + 2y \leq 6 \\ x < 6 \end{cases}$

34. $\begin{cases} 3x + y > 3 \\ 3x + y < 6 \\ x \geq 0 \\ y \geq 0 \end{cases}$

In exercises 35–38, find the domain and range of each relation and determine whether it is a function.

35. $\{(1,6), (6,1), (-1,-6), (-6,-1)\}$

36. $\{(2,-5), (-1,-5), (4,5)\}$

37. $\{(3,-8), (2,-1), (3,-1)\}$

38. $\{(7,8), (8,9), (-7,9), (-8,8)\}$

In exercises 39–42, sketch the graph of the given relation or function and determine its domain and range.

39. $9x^2 + y^2 = 9$

40. $9x^2 - y^2 = 9$

41. $9x + y = 9$

42. $9x^2 - y = 9$

In exercises 43–46, determine the domain of the given function.

43. $y = 2 + x - x^2$

44. $y = \sqrt{2x + 4}$

45. $y = \dfrac{x^2}{x^2 - 8x - 9}$

46. $y = \dfrac{x}{\sqrt{2 - x}}$

In exercises 47–50, determine which are graphs of functions.

47.

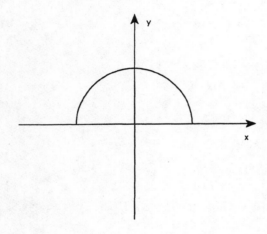

48.

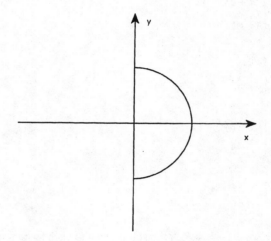

49.

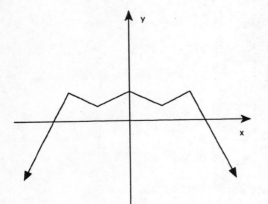

50.

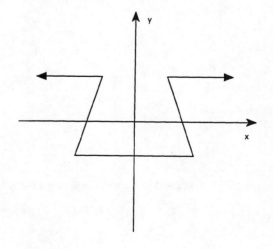

In exercises 51–66, let $f(x) = x^2 - 4x + 1$, $g(x) = \dfrac{2}{x^2}$, and $h(x) = 2 - 3x$. Find:

51. $f(2)$

52. $g(-2)$

53. $(h(x))^2$

54. $h(x^2)$

55. $f(b+2)$

56. $f(b) + f(2)$

57. $f(2b)$

58. $2f(b)$

59. $f(-1) - g(-1)$

60. $g(g(2))$

61. $f(h(-1))$

62. $h(f(-1))$

63. $(h \cdot f)(0)$

64. $h(f(0))$

65. $\dfrac{f(x+4) - f(4)}{4}$

66. $\dfrac{h(x+a) - h(x)}{a}$

In exercises 67–70, find the inverse of the given function if it exists.

67. $y = 2x^3 + 1$

68. $y = \dfrac{5 - 3x}{2}$

69. $y = \dfrac{x+2}{x+3}$

70. $y = x^2 + 4$

In exercises 71–78, sketch the graph of the given function.

71. $f(x) = \sqrt{1+x}$

72. $f(x) = 1 + \sqrt{x}$

73. $f(x) = x^2 + 8x + 12$

74. $f(x) = |2x - 1|$

75. $f(x) = 2^{2x-1}$

76. $f(x) = \left(\dfrac{2}{3}\right)^x + 2$

In exercises 79–84, translate logarithmic statements into exponential form and vice versa.

79. $\log_3 81 = 4$

80. $5^{-2} = \dfrac{1}{25}$

81. $\left(\sqrt[5]{32}\right)^2 = 4$

82. $\log_{10} 0.0001 = -4$

83. $\log_{49} 343 = \dfrac{3}{2}$

84. $8^{-\frac{2}{3}} = \dfrac{1}{4}$

In exercises 85–92, find the given logarithm.

85. $\log_4 64$

86. $\log_{27} 3$

87. $\log_5 \dfrac{1}{125}$

88. $\log_8 128$

89. $\log_{11} \dfrac{1}{11}$

90. $\log_9 \sqrt{3}$

91. $\log_{53} 53$

92. $\log_{78} 1$

In exercises 93–96, write each logarithm as a sum of simpler logarithms, if possible.

93. $\log_b \sqrt[4]{7x^2y^3}$

94. $\log_b \left(\dfrac{bx^3\sqrt{y}}{z^2} \right)$

95. $\log_8 \left(\dfrac{4xy^5}{\sqrt{2z^3}} \right)$

96. $\dfrac{\log_3 a^2}{\log_2 a^3}$

In exercises 97–100, use a calculator to find the following.

97. $\log 848{,}000$

98. $\ln 342$

99. antilog 0.0374

100. $e^{0.78}$

In exercises 101–102, use the change of base formula to compute the following.

101. $\log_6 666$

102. $\log_8 0.55$

In exercises 103–112, solve the given equation.

103. $\dfrac{1}{3}\log_2 x = \log_2 3$

104. $2\log_b(x-1) = \log_b(2x-3)$

105. $6^{2x+1} = \dfrac{1}{36}$

106. $8^x = \dfrac{1}{4}$

107. $\log_{\frac{1}{2}} x + \log_{\frac{1}{2}} 6 = -1$

108. $\log_3 x + \log_3(x-6) = 3$

109. $\dfrac{9^{x^2-1}}{3^{x+1}} = 27$

110. $8^{\sqrt{x+2}} = 4^{\frac{1}{2}x-1}$

111. $5^{x+3} = 3^{x+5}$

112. $\log_b(2x+1) - \log_b(x-2) = \log_b 3$

113. If y varies directly with x, and y = 4 when x = 12, find y when x = 27.

114. If y varies inversely with x, and y = 4 when x = 12, find y when x = 27.

115. If z varies jointly with x and y, and z = 20 when x = $\dfrac{5}{4}$ and y = 8, find z when x = 16 and y = $\dfrac{3}{8}$.

116. If z varies directly with the square of x and inversely with y, and z = 16 when x = 6 and y = 9, find z when x = 5 and y = 100.

117. If \$4,000 is invested at an annual interest rate of 12% compounded monthly, how long will it take for the investment to be worth \$6,000?

118. A bacteria culture grows at the rate of 4% per minute. Using the exponential growth model, how long will it take the culture to grow to 10 times its original size?

119. A radioactive sample decays at the rate of 1.5% per day. Using the exponential decay model, how long will it take 48 grams of a radioactive sample to decay to 8 grams?

120. What is the pH of a solution with a hydronium ion concentration of 2.81×10^{-5}?

121. What is the hydronium ion concentration of a solution whose pH is 9.5?

Cumulative Test: Chapters 10–12

1. Solve the following system of equations:

$$\begin{cases} 5x - y + 3z = 18 \\ 2x + 4y - z = -6 \\ 4x - 3y + 5z = 21 \end{cases}$$

2. Solve the following system using matrices:

$$\begin{cases} x + 5y - z = 2 \\ x - 2y + z = 1 \\ x + 3y = 0 \end{cases}$$

3. Solve the following systems of equations using Cramer's rule.

   (a) $\begin{cases} 9x + 4y = 33 \\ 8x - 5y = 55 \end{cases}$

   (b) $\begin{cases} 2x + 3y - 4z = 16 \\ 3x + 4y - 2z = 24 \\ 4x + 2y - 3z = 5 \end{cases}$

4. Solve the following system of equations:

$$\begin{cases} 2x^2 + 3y^2 = 11 \\ x + 2y = 0 \end{cases}$$

5. Sketch the solution set of the following system of inequalities.

$$\begin{cases} y - x < 3 \\ 3y - x > 3 \\ x \le 0 \end{cases}$$

6. Does the relation $\{(7,-1), (4,1), (1,0), (-3,-1)\}$ define a function? Explain your answer. What are its domain and range?

7. Sketch the graph of each of the following relations and determine its domain and range. Is the relation a function? Explain your answer.

   (a) $x^2 + 4y^2 = 16$

(b) $y = 6x - x^2$

8.   Find the domains of the following functions.

(a) $f(x) = \dfrac{x + 1}{4x^2 - 4x - 3}$

(b) $g(x) = \sqrt{4x + 1}$

9.   Given $f(x) = x^2 - x - 2$ and $g(x) = 3x - 2$, find:

(a) $f(3)$

(b) $f(x^3)$

(c) $g(3x - 2)$

(d) $f(x - 1)$

(e) $f(x) - f(1)$

(f) $g(g(2))$

(g) $(g \cdot g)(2)$

(h) $f(g(1))$

(i) $g(f(1))$

(j) $\dfrac{f(x + t) - f(x)}{t}$

10.   Find the inverse of the function $f(x) = \dfrac{x + 1}{x - 1}$

11.   Sketch the graphs of the following equations.

(a) $y = \left(\dfrac{5}{4}\right)^x$

(b) $y = |4 - 2x|$

12.   Write in exponential form: $\log_{32} \dfrac{1}{16} = -\dfrac{4}{5}$

13.   Find each of the following logarithms.

(a) $\log_{\frac{1}{2}} 8$

(b) $\log_{125} 25$

14.   Express as a sum of simpler logarithms, if possible.

(a) $\log_b \left(bx^2 \sqrt{y}\right)$

(b) $\log_5 \sqrt[3]{\dfrac{25x}{y^4}}$

15.   Use a calculator to find:

(a) $\log 4{,}020{,}000$

(b)   antilog (-1.1101)

16.   Use the change of base formula to compute $\log_2 357$

17. Solve the following equations.

(a) $4^{x^2} = 2^{5x-3}$

(b) $\log_3(x+3) + \log_3(x-3) = 3$

18. If y varies directly as the cube of x and y = 64 when x = 2, find y when $x = \dfrac{1}{4}$.

19. Using the exponential growth model, find the rate of growth if an initial population doubles in size in 8.4 years.

20. What is the pH of a solution with a hydronium ion concentration of 7.93×10^{-5} ?

Practice Final Exam

1. Given x = 2, y = 3, and z = –1, evaluate: $\dfrac{(x-y)^3 - |z-x|}{3y + 7z}$

In problems 2–8, perform the operations and express your answer in simplest form. Assume all variables represent positive real numbers.

2. $(3x - y)(2x + y) - (x - 2y)^2$

3. $\dfrac{16a^2 - b^2}{4a^2 + 4ab + b^2} \div \dfrac{b^2 - 3ab - 4a^2}{8a^3 + 12a^2b + 4ab^2}$

4. $\dfrac{8x}{(x-5)(x+3)} + \dfrac{2x}{(x+5)(x+3)}$

5. $\dfrac{x^{-2} - 4y^{-2}}{2y^{-1} + x^{-1}}$

6. $\left(\dfrac{x^{-\frac{2}{3}} y^{\frac{3}{4}}}{x^{-\frac{1}{2}} y^{\frac{1}{2}}} \right)^{-6}$

7. $\sqrt{27x^5 y^7} + xy^2 \sqrt{12x^3 y^3} - \dfrac{12x^3 y^4}{\sqrt{3xy}}$

8. $81^{\frac{3}{4}} - \left(\dfrac{1}{4} \right)^{-\frac{3}{2}} + 5^0$

In problems 9–18, solve the given equation or inequality.

9. $5(x - 1) + 4(x + 2) = 3(x - 3) + 2(x + 4)$

10. $|4x - 1| \geq 7$

11. $a^4 + 24a^2 = 11a^3$

12. $3x^2 + 10x - 8 < 0$

13. $5x^2 - 10x + 4 = 0$

14. $\dfrac{2z}{z-1} + \dfrac{z+1}{z-2} = 1$

15. $\sqrt{2q} + \sqrt{q-1} = 3$

16. $8^{x^2 - 5} = 4^{x+3}$

17. $\log_3(6y + 3) - \log_3(3y - 2) = 2$

18. $2^{2h+1} = 3^{h-2}$

19. Find the quotient and remainder: $(x^3 + x^2 - 11x + 9) \div (x - 3)$

20. Express in the form a + bi: $\dfrac{4 + 2i}{1 + i} - (3 + 2i)(2 - i)$

21. Perform the operations using scientific notation and express your answer in standard form:

$$\frac{(6,400,000)(0.000027)}{(0.006)(45,000)}$$

In problems 22–26, the line L has equation $5x - 4y = 20$.

22. Sketch L using the intercept method.

23. Find the distance between the intercepts of L.

24. Find the midpoint of the segment of L that connects its intercepts.

25. Find an equation of the line through $(3,-3)$ that is parallel to L.

26. Is the line passing through $(-6,1)$ and $(-1,-2)$ perpendicular to L?

In problems 27–32, sketch the graphs of the given equations. Label the important aspects of the figure.

27. $x^2 + y^2 - 20x + 18y + 37 = 0$

28. $16x^2 + 49y^2 = 64$

29. $16x^2 - 49y^2 = 64$

30. $y = 4 + 2x - x^2$

31. $y = |x + 3| - 2$

32. $y = \left(\frac{5}{2}\right)^x$

33. Solve the following system of equations using Cramer's rule.

$$\begin{cases} 6x + 9y - z = 3 \\ 5x - 2y + 4z = 7 \\ 2x + 7y - 5z = 7 \end{cases}$$

34. Solve the following system of equations.

$$\begin{cases} x^2 + 4y^2 = 4 \\ \quad\; y = x^2 - 1 \end{cases}$$

35. Sketch the solution set of the following system of inequalities.

$$\begin{cases} 2x - y \geq 0 \\ 4x + y \leq 18 \\ \quad\; y > 2 \end{cases}$$

In problems 36–39, let $f(x) = \sqrt{1 - 2x}$.

36. Find the domain of f.

37. Compute $f(x - 1)$.

38. Compute $f\left(f\left(\frac{3}{8}\right)\right)$.

39. Find the inverse of f.

In problems 40–45, solve algebraically.

40. Find three consecutive odd integers with the property that 9 times the first plus 10 times the second is equal to 11 times the third.

41. Mr. King can grade a set of test papers in 3 hours. If he works with a teaching assistant, they can complete the job in 2 hours. How long would it take the teaching assistant to grade the papers if he worked alone?

42. A canoeist travels 12 miles at a certain rate of speed. She then returns to her starting point, traveling 2 mph faster that before. If she wishes to complete the trip in no more that 5 hours, what is the minimum rate at which she must originally travel?

43. Find the value(s) of k if the line through the points $(2k, k^2 - 1)$ and $(-1, 3k)$ is perpendicular to the line with equation $x - 3y = 3$.

44. A bank contains only dimes and quarters, and the total value of the coins is $3.80. If the number of dimes and the number of quarters were reversed, the total value of the coins would $5.30. How many dimes and how many quarters are in the bank?

45. If the sum of $3,000 is invested at an annual interest rate of 10% compounded quarterly for 3.75 years, what is the final value of the investment?

ANSWERS

Chapter 1—Test A

1.   (a)   {5}

    (b)   {1,2,3,5,6,7,10,11,13,15,17,19,23,30}

2.   (a) F   (b)   T

    (c) T

3.   (a)   (b)

4.   (a)   commutative property of addition   (b)   false

5.   (a)   −4   (b)   −12

    (c)   7   (d)   29

6.   (a)   3   (b)   9

7.   (a) $-12x^6y^7$ (b) $4a^3b^2 - 5a^2b^2 + a^2b^3$

    (c) $12p + 3$ (d) $-2u - 3v$

8. $7x - 4$, where x = length of one of the equal sides

9. $16 - x$ blue stamps, worth $7x + 128$ cents

Chapter 1—Test B

1.   (a)   {4,8,12}   (b)   {2,4,6,8,12,16,24}

2.   (a) T   (b)   T

    (c) F

3. (a)

(b)

4. (a) commutative property of multiplication   (b) false

5. (a) -10 (b) 8

    (c) 1   (d) 12

6. (a) 17   (b) -5

7. (a) $12x^9y^6$ (b) $p^4q^5 + 4p^5q^4 - 5p^4q^4$

    (c) $3m - 4$ (d) $6s - 4t$

8. $4x + 2$, where x = length of the shortest side

9. $25 - x$ small glasses, worth $10x + 500$ cents

Chapter 2—Test A

1. 1   2. 6

3. -2 4. $3, -\dfrac{7}{3}$

5. $x > -1$ 6. $x \geq 2$

7. $-2 \leq x < 5$ 8. $-2 \leq x \leq -1$

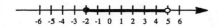

9. $x > 3$ or $x < -\dfrac{1}{2}$

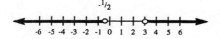

10. 18 silver coins, 27 gold coins

11. slower train: 150 kph, faster train: 180 kph

12. 30 quarters

Chapter 2—Test B

1. -2

2. 5

3. 1

4. $-1, -\dfrac{1}{2}$

5. $x \le 1$

6. $x > -3$

7. $-3 < x \le 5$

8. $-1 < x < 2$

9. $x \ge 2$ or $x \le -\dfrac{10}{3}$

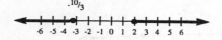

10. 16 irises, 8 lilies

11. 6 hours

12. 16 25¢ stamps

Chapter 3—Test A

1. $8x^3 + y^3$

2. $16t$

3. $a^2 + 6ab + 9b^2 - 1$

4. $4x^4 + 4x^2y^3 + y^6$

5. $6u^2v(2u^2v - 3v^3 + u)$

6. $(4x - 1)(2c + 3d)$

7. $(a - 9b)(a + 8b)$

8. $(4x - 3)(2x + 5)$

9. not factorable

10. $4m^2n^2(m - n)(m + n)$

11. $s^3t^2(t + 6)^2$

12. $(b - 3)(b^2 + 3b + 9)$

13. $(3x - 2y + 4)(3x - 2y - 4)$

14. $(x^2 + 2)(x + 3)$

15. quotient: $x^2 - 3x - 6$, remainder: -8

Chapter 3—Test B

1. $x^3 - 27y^3$

2. $18 + 2r^2$

3. $4a^2 - 4ab + b^2 - 9$

4. $c^4 - 6c^2d^3 + 9d^6$

5. $5st^2(3s^2 + 4st - 6)$

6. $(2x + 3)(3m - 2n)$

7. $(p - 9q)(p + 7q)$

8. $(8x - 5)(x + 3)$

9.   not factorable

10. $ab^3(a + 2b)(a - 2b)$

11. $w^2z^4(w - 4)^2$

12. $(h + 4)(h^2 - 4h + 16)$

13. $(x + 4y - 3)(x + 4y + 3)$

14. $(k^2 + 4)(k - 2)$

15. quotient: $x^2 + x - 3$, remainder: 15

Cumulative Review: Chapters 1–3

1. $\{3,5,7,11,13,17\}$

2. $\{3,5,7,10,11,13,15,17,20,25\}$

3. $\{5,7,11,13,17\}$

4. $\{5\}$

5.   T

6.   T

7.

8.

9.   closure property of multiplication

10. false

11. 3

12. -30

13. 13

14. 21

15. 11

16. -1

17. $2x^5 - 6x^3 + 10x^2$

18. $-18a^3b^4$

19. $4(x + 7)$, where x = the number

20. $3x + 1$, where x = the smallest of the odd integers

21. area = $w(2w + 10)$, perimeter = $6w + 20$, where w = width

22. $70 - 4x$, where x = number of one–dollar bills

23. –3

24. no solution

25. 2

26. 5

27. identity

28. 0

29. –2, –5

30. –12, 2

31. $x > \dfrac{2}{3}$

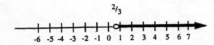

32. $x \geq -1$

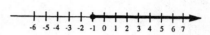

33. identity

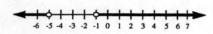

34. $-2 < x \leq 1$

35. $x > -1$ or $x < -5$

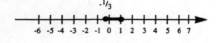

36. $-\dfrac{16}{3} < x < 2$

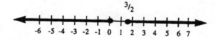

37. $-\dfrac{1}{3} \leq x \leq 1$

38. $x \leq 0$ or $x \geq \dfrac{3}{2}$

39. –3

40. 12 overtime hours

41. 80 bulbs

42. trimming hedges: 5 hours, turning soil: 4 hours

43. 37.5 minutes

44. hand–stamped: 1,000 envelopes, machine–stamped: 5,000 envelopes

45. $29" \leq p \leq 45"$, where p = perimeter

46. 12 pounds of apples

47. 5

48. 6

49. $2x^3 - x^2 + 9x - 8$

50. $-2x^3y^2 - 6x^3y + 8x^2y^2$

51. $2a^2 + 3ab - 2b^2$

52. $15u^2 - 23uv - 28v^2$

53. $c^2 - 2cd + d^2 - 3c + 3d$

54. $a^3 - 27b^3$

55. $16m^2 + 40mn + 25n^2$

56. $49y^2 - 4z^2$

57. $16m^2 - 25n^2$

58. $49y^2 - 28yz + 4z^2$

59. $s^2 + 6st + 9t^2 - 1$

60. $9p^2 - 12pq + 4q^2 + 24p - 16q + 16$

61. $(x + 3)(x - 10)$

62. $(c - 4d)(c + 4d)$

63. $(r + 6s)(r + 8s)$

64. $(2a - 3b)(a - 2b)$

65. $(3m - 2n)(3m - 4n)$

66. $(7u - 8z)(7u + 8z)$

67. $(10w - 9)(10w + 9)$

68. $(4x + 9y)^2$

69. $(3a - 8b)^2$

70. not factorable

71. not factorable

72. $3h(2h - 11)(h + 1)$

73. $v^2(5v + 2)(v - 1)$

74. $\left(5p^2 - q^2\right)\left(2p^2 + 3q^2\right)$

75. $\left(s + 4t^2\right)\left(9s - t^2\right)$

76. $v^2w\left(v - w^3\right)\left(v + 2w^3\right)$

77. $(2x + y - 3)(2x + y + 3)$

78. $\left(x^2 + y^2 - 5\right)\left(x^2 + y^2 + 5\right)$

79. $(r^2 + 9)(r - 3)(r + 3)$

80. $(a - b - 1)(a - b - 3)$

81. $(3k - 2)(9k^2 + 6k + 4)$

82. $(u + 1)(u - 2)(u + 2)$

83. quotient: $x - 1$, remainder: -1

84. quotient: $x + 2$, remainder: 6

85. quotient: $3x^2 + 2x + 1$, remainder: 0

86. quotient: $x^3 - x$, remainder: $x + 1$

Cumulative Test: Chapters 1–3

1. (a) $\{15\}$
   (b) $\{5,6,9,10,12,15,18,20,21,24\}$

2. commutative property of addition

3. (a) -2
   (b) -13

4. 7

5. (a) 5
   (b) -3
   (c) $3, -4$

6. (a) $x > 1$ (b) $-2 \le x \le 5$

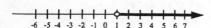   

(c) $x < -1$ or $x > \dfrac{1}{5}$

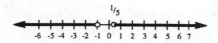

7. regular: 32 gallons, premium unleaded: 40 gallons

8. 25 8–lb. boxes

9. (a) $a^4 + 7a^2b^2 - b^2$ (b) $6s^2 - st - 12t^2$

(c) $y^6 - 4z^2$ (d) $16m^2 + 40mn + 25n^2$

(e) $c^2 - 4cd + 4d^2 - 2c + 4d + 1$

10. (a) $(a - 5)(a + 8)$ (b) $(6s - 5t)(2s + t)$

(c) $\left(u^3 + 2v\right)^2$ (d) $(5x - 3y)^2$

(e) $(b - 6c - 4)(b - 6c + 4)$ (f) $(3r + 2)(9r^2 - 6r + 4)$

11. quotient: $3x^2 + 5x + 15$, remainder: 46

Chapter 4 —Test A

1. (a) $\dfrac{4b^3}{5a^2}$ (b) $\dfrac{x - 2}{x + 2}$

(c) $\dfrac{2(x + 3)}{x}$

2. (a) $\dfrac{4}{3rs^2p^2q^3}$ (b) $\dfrac{4v + 15u^2}{10uv^2}$

(c) $\dfrac{(m - 2n)^2}{mn}$ (d) 3

(e) $\dfrac{3z - 1}{(z + 1)(z - 2)}$ (f) $\dfrac{3x}{x - 1}$

3. $-\dfrac{2 + x}{3x}$

4. (a) $x \ge -10$ (b) no solution

(c) 2

492

5. $x = \dfrac{y + 2}{3 - y}$

6. 9 pounds at \$2/lb., 6 pounds at \$3/lb.

7. $\dfrac{12}{17}$ hours

Chapter 4 —Test B

1. (a) $\dfrac{10t^3}{3s}$ (b) $\dfrac{x}{x + 4}$

   (c) $\dfrac{3x(x + 2)}{2}$

2. (a) $\dfrac{5x^2y}{3wz^2}$ (b) $\dfrac{16k - 9l^2}{12k^2l^2}$

   (c) $\dfrac{2(p + 3q)^2}{3pq}$ (d) 2

   (e) $\dfrac{k - 2}{(k + 2)(k + 3)}$ (f) $\dfrac{x - 1}{2x}$

3. $-\dfrac{x + 3}{2x}$

4. (a) $x > 4$ (b) no solution

   (c) -2

5. $x = \dfrac{2y + 3}{2 - y}$ 6. 16 pounds

7. $7\dfrac{1}{2}$ hours

Chapter 5 —Test A

1. (a) the graph of $4x + 3y = 24$ (b) the graph of $y + 2 = 0$

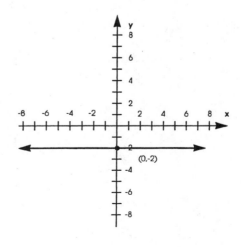

2. (a) -2 (b) $\dfrac{2}{3}$

3. 5

4. (a) $y = -2x + 5$ (b) $y = 4x - 9$

  (c) $y = -\dfrac{7}{2}x + 7$ (d) $y = -\dfrac{1}{2}x + \dfrac{9}{2}$

  (e) $y = 2x + 7$ (f) $x = -5$

5.

6. $125

7. (a) $x = 6$, $y = 7$ (b) $s = 12$, $t = 6$

8. $900 at 7%

  $1,500 at 10%

1.   (a)   the graph of $5x + 2y = -10$ (b)   the graph of $x + 3 = 0$

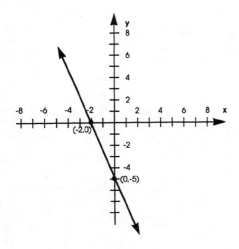 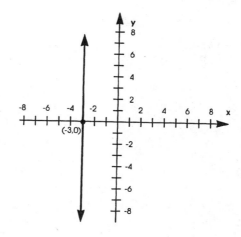

2.   (a) $\dfrac{3}{2}$ (b) $-\dfrac{4}{5}$

3.   7

4.   (a) $y = 2x - 5$ (b) $y = -3x - 8$

      (c) $y = \dfrac{4}{3}x + 4$ (d) $y = 3x - 13$

      (e) $y = -\dfrac{1}{3}x + \dfrac{1}{3}$ (f) $x = -3$

5.

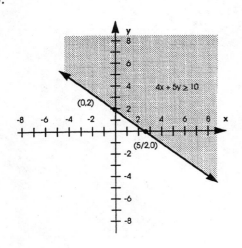

6. 45 years old

7. (a) $x = -7, y = 4$ (b) $s = 15, t = -10$

8. $500 at 5%

   $2,200 at 8%

Chapter 6—Test A

1. $x^9 y^{10}$ 2. $-72x^6 y^7$

3. $\dfrac{-9ab^2 c^2}{2d^4}$ 4. $\dfrac{-2y^3}{x^7}$

5. $\dfrac{27u^9 v^9}{8}$ 6. 1

7. 2 8. a^{13}

9. $x - 2x^2$ 10. $x^3 + x$

11. $\dfrac{4 - 4x + x^2}{x}$

12. (a) 16 (b) $\dfrac{1}{512}$

13. $4\left(\sqrt[3]{b}\right)^4$ or $4\left(\sqrt[3]{b^4}\right)$ 14. 5.335×10^{11}

Chapter 6—Test B

1. $x^9 y^5$ 2. $-72x^{10} y^6$

3. $\dfrac{8r^3}{3p^4 q^2 s}$ 4. $\dfrac{-3y^3}{4x}$

5. $\dfrac{25n^8}{4m^6}$ 6. 1

7. 25 8. c^{11}

9. $y^2 + 3y^5$ 10. $z^2 - z$

11. $\dfrac{16 + 8x + x^2}{x}$

12. (a) 512 (b) $\dfrac{1}{16}$

13. $5\left(\sqrt[5]{d}\right)^2$ or $5\left(\sqrt[5]{d^2}\right)$ 14. 8.836×10^{14}

Cumulative Review: Chapters 4–6

1. $\dfrac{6s^3}{7t}$

2. $\dfrac{a+4}{a-4}$

3. $\dfrac{4y-z}{y^2(y-3z)}$

4. $\dfrac{1}{m+2n}$

5. $\dfrac{10q^2r^2}{3ps^2}$

6. 4

7. $\dfrac{a^2-9b^2}{3ab}$

8. $-\dfrac{1}{3}$

9. $\dfrac{6(x+3)}{x}$

10. $\dfrac{(2a+3b)^2}{4a^2+9b^2}$

11. $\dfrac{2(x-8)}{(x-4)(x+4)}$

12. $\dfrac{2}{(x+y)(x-y)}$

13. $\dfrac{2x+y}{12}$

14. $\dfrac{2(x-3y)}{3x-y}$

15. $\dfrac{x+1}{x}$

16. $x+y$

17. $\dfrac{3}{4}$

18. 3

19. $x \le 2$

20. no solution

21. 1

22. $x > \dfrac{3}{2}$

23. $s = \dfrac{5}{2}t$

24. $c = \dfrac{2d}{2d+1}$

25. $p = \dfrac{3q}{2q-1}$

26. $q = \dfrac{p}{2p-3}$

27. 18,000 male voters

28. 15% solution: 8 gallons, 35% solution: 12 gallons

29. $\dfrac{15}{8}$ hours

30. 28 onyx stones, 42 garnet stones

31. the graph of $2x - 3y = 12$

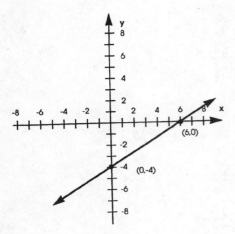

32. the graph of $4x + y = 6$

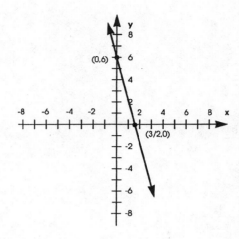

33. the graph of $2y = 5x + 10$

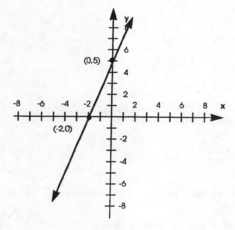

34. the graph of $2x = 5y + 10$

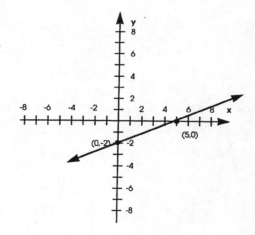

35. the graph of $6y - 1 = 2$

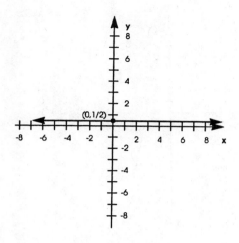

36. the graph of $3 - x = 4$

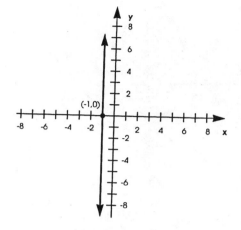

37. −4

38. −1

39. 6

40. $\dfrac{7}{3}$

41. 3

42. $\dfrac{3}{4}$

43. 4

44. 5, −3

45. $y = -5x - 17$

46. $y = -6$

47. $y = 3x - 7$

48. $x = -5$

49. $y = -2x + 3$

50. $y = -2x + 6$

51. $y = -\dfrac{7}{2}x - \dfrac{5}{2}$

52. $y = \dfrac{3}{5}x$

53.

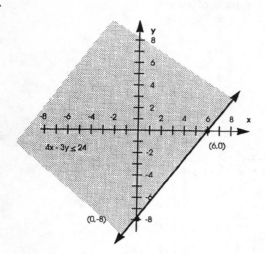

$4x - 3y \leq 24$

(6,0)

(0,-8)

54.

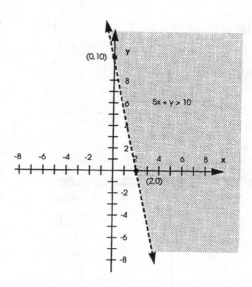

(0,10)

$5x + y > 10$

(2,0)

55.

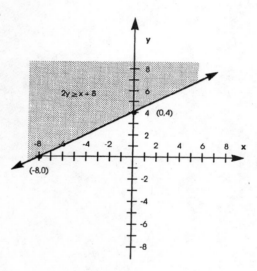

$2y \geq x + 8$

(0,4)

(-8,0)

56.

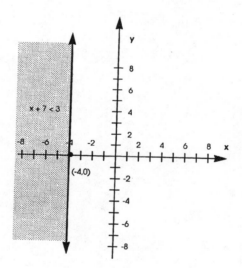

$x + 7 < 3$

(-4,0)

57.

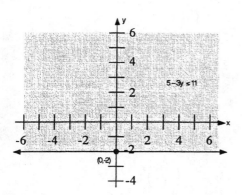

$5 - 3y \leq 11$

(0,-2)

58. $290

59. $x = -3, y = -4$

60. $x = 4, y = 1$

61. $a = 2, b = -2$

62. $u = 0, v = 6$

63. no solution

64. infinitely many solutions, each with $t = 3s - 1$

65. $p = 8, q = 15$

66. $w = \dfrac{1}{2}, z = \dfrac{2}{3}$

67. $-128x^5y^7$

68. $\dfrac{3y^2}{2x^3}$

69. $\dfrac{a^2}{9b^4}$

70. $\dfrac{c^5d^5}{2}$

71. $\dfrac{1}{m^5n}$

72. $\dfrac{k^4}{25\ell^6}$

73. $\dfrac{1}{s^5}$

74. -32

75. $\dfrac{-u^4v^5}{12}$

76. $\dfrac{2q^4}{p^7}$

77. $\dfrac{w^4 - z^4}{w^2z^2}$

78. $\dfrac{2k + 1}{k}$

79. 4.719882×10^3

80. 1.001×10^{-5}

81. 7.476×10^{-5} inches

82. 30

83. 36

84. $\dfrac{8}{27}$

85. $a^{\frac{7}{15}}$

86. $b^{\frac{5}{24}}$

87. $\dfrac{1}{4}$

88. $\dfrac{1}{z^{\frac{7}{4}}}$

89. $\sqrt[5]{x^2}$ or $\left(\sqrt[5]{x}\right)^2$

90. $\sqrt[3]{(3x)^4}$ or $\left(\sqrt[3]{3x}\right)^4$

Cumulative Test: Chapters 4–6

1. (a) $\dfrac{x}{4x + y}$ (b) $\dfrac{1}{2}$

   (c) $\dfrac{x + 2}{x(x + 1)}$

2. $\dfrac{(x + 2)(5 - 2x)}{(x - 2)(2x + 5)}$

3. (a) -2 (b) $x > 7$

   (c) no solution

4. $z = \dfrac{5w + 5}{1 - w}$

5. $\dfrac{15}{16}$ hours

6. the graph of $4x + 5y = 30$

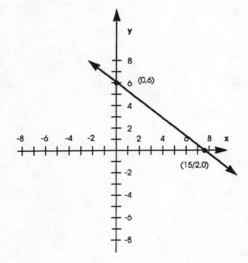

7. 2

8. (a) $y = \frac{1}{2}x + \frac{19}{2}$ (b) $y = -2x - 3$

9.

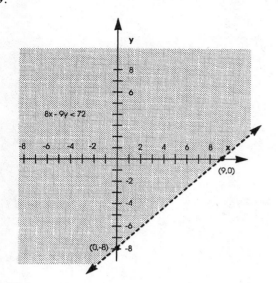

10. 12 years old

11. $x = 2, y = -3$

12. (a) $-108x^{14}y^9$ (b) $-\frac{27x^2}{4y^9}$

    (c) $\frac{16b^2}{a^4}$ (d) $\frac{mn}{n + m}$

13. 6.001×10^{-4}

14. 0.000096

15. $-\frac{1}{8}$

16. (a) $x^{\frac{1}{2}}$ (b) $\frac{1}{x^{\frac{2}{5}}}$

Chapter 7—Test A

1. $2x^3y^2$

2. $3x^2 \sqrt[3]{y^2}$

3. $-20a^6b^2$

4. $\frac{3}{5}$

5. $\frac{\sqrt{30}}{10}$

6. $\frac{3\sqrt[4]{2}}{2}$

7. $4cd^3\sqrt{2}$

8. $\sqrt[12]{3}$

9. $-2\sqrt{5}$

10. $4 + 4\sqrt{y} + y$

11. $\dfrac{8\sqrt{7}}{7}$

12. $\dfrac{2\sqrt{10} - 5}{3}$

13. 48

14. 16

15. 26

16. 243

17. $-i$

18. $-11 - 3i$

19. $\dfrac{3}{5} - \dfrac{1}{5}i$

Chapter 7—Test B

1. $2x^4y^2$

2. $2y^2 \sqrt[4]{x^3}$

3. $-18p^6q^3$

4. $\dfrac{2}{3}$

5. $\dfrac{\sqrt{55}}{5}$

6. $\dfrac{3\sqrt[3]{4}}{2}$

7. $2u^4v\sqrt{3}$

8. $\sqrt[10]{7}$

9. $-3\sqrt{3}$

10. $16 - 8\sqrt{z} + z$

11. $\dfrac{4\sqrt{5}}{5}$

12. $\dfrac{\sqrt{15} + 3}{2}$

13. 23

14. 4

15. 25

16. 128

17. -1

18. $2 - 11i$

19. $\dfrac{9}{13} - \dfrac{7}{13}i$

Chapter 8—Test A

1. (a) $2, -\dfrac{4}{3}$

(b) $\pm\sqrt{7}$

2. (a) $\dfrac{-3 \pm \sqrt{21}}{2}$

(b) $\dfrac{-1 \pm \sqrt{7}}{2}$

(c) $1, -\dfrac{1}{3}$

3. two imaginary roots

4. $s = \pm \dfrac{\sqrt{3prt}}{3r}$

5. 2

6. $\pm3, \pm 2i$

7. (a) $-1 < x < 4$

(b) $x \le -\dfrac{1}{2}$ or $x > 3$

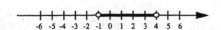

8. $2\sqrt{5}$ inches

9. 2 mph

Chapter 8—Test B

1. (a) $1, -\dfrac{5}{2}$

(b) $\pm\sqrt{2}$

2. (a) $\dfrac{5 \pm \sqrt{17}}{2}$

(b) $\dfrac{1 \pm \sqrt{10}}{3}$

(c) $4, -\dfrac{5}{2}$

3. two imaginary roots

4. $b = \pm \dfrac{\sqrt{5acd}}{c}$

5. 3

6. $\pm1, \pm 5i$

7. (a) $-3 \le x \le -1$

(b) $x < -1$ or $x > \dfrac{4}{3}$

8. $2\sqrt{5}$

9. 1 mph

Chapter 9—Test A

1. $7\sqrt{2}$

2. $\left(2, \dfrac{7}{2}\right)$

3.   (a)   center: $(-2,4)$, radius: $3\sqrt{2}$ (b)   center: $(4,-1)$, radius: 5

4.   (a)   x-intercepts: $1, \dfrac{3}{2}$, y-intercept: -3

           vertex: $\left(\dfrac{5}{4}, \dfrac{1}{8}\right)$, axis of symmetry: $x = \dfrac{5}{4}$

   (b)   x-intercept: none, y-intercept: 4

           vertex: $(-1,1)$, axis of symmetry: $x = -1$

5.   vertices: $(\pm 5, 0)$, $(0, \pm 2\sqrt{5})$ 6.   vertices: $(0, \pm 3)$, asymptotes: $y = \pm \dfrac{\sqrt{3}}{2} x$

7.   (a)   ellipse; vertices: $(\pm 6, 0)$, $(0, \pm 4)$

   (b)   circle; center: $(-6,3)$, radius: 6

   (c)   parabola; intercepts: $(0,0)$, $(0,4)$

           vertex: $(4,2)$, axis of symmetry: $y = 2$

Chapter 9—Test B

1. $6\sqrt{2}$ 2. $\left(5, -\dfrac{1}{2}\right)$

3.   (a)   center: $(-4,1)$, radius: $2\sqrt{6}$ (b)   center: $(3,-5)$, radius: 4

4.   (a)   x-intercepts: $-2, \dfrac{1}{3}$, y-intercept: 2

           vertex: $\left(-\dfrac{5}{6}, \dfrac{49}{12}\right)$, axis of symmetry: $x = -\dfrac{5}{6}$

   (b)   x-intercept: none, y-intercept: 3

           vertex: $(1,1)$, axis of symmetry: $x = 1$

5.   vertices: $(\pm\sqrt{15}, 0)$, $(0, \pm 5)$

6.   vertices: $(0, \pm 4)$, asymptotes: $y = \pm \dfrac{2\sqrt{3}}{3} x$

7.   (a)   ellipse; vertices: $(\pm 4, 0)$, $(0, \pm 6)$

   (b)   circle; center: $(-4,2)$, radius: 5

   (c)   parabola; intercepts: $(0,0)$, $(0,-2)$

           vertex: $(-1,-1)$, axis of symmetry: $y = -1$

Cumulative Review: Chapters 7–9

1. $6r^3s^7$

2. $3v^2w^4\sqrt{5vw}$

3. $-36a^4b^2\sqrt{b}$

4. $2pq^2\sqrt[4]{4p^2q}$

5. 5

6. $\dfrac{\sqrt{7d}}{cd^2}$

7. $\sqrt[3]{x}$

8. $\dfrac{\sqrt[5]{48x^3}}{2x}$

9. $7\sqrt{11}$

10. $6\sqrt{10}$

11. $h^3\sqrt[5]{h^2}$

12. $\dfrac{13\sqrt{10}}{20}$

13. $\sqrt{35}+\sqrt{15}$

14. $21-14\sqrt{2}$

15. 8

16. $2\sqrt{x}$

17. 6

18. $2\sqrt{2}-1$

19. 12

20. 8

21. -2

22. $-\dfrac{1}{8}$

23. i

24. -1

25. $26+13i$

26. $-i$

27. $-6, 3$

28. $6, -1$

29. $\pm\sqrt{3}$

30. $-3\pm4\sqrt{2}$

31. $-2\pm\sqrt{6}$

32. $\dfrac{6\pm5\sqrt{2}}{2}$

33. $\dfrac{1\pm\sqrt{11}}{2}$

34. $-2, 1$

35. $6, 2$

36. $1\pm\sqrt{17}$

37. $1, 4$

38. -14

39. $\dfrac{-2 \pm \sqrt{10}}{3}$

40. $-6 \pm 2\sqrt{3}$

41. $c = \pm \dfrac{\sqrt{abd}}{b}$

42. $y = -x,\ y = \dfrac{x}{2}$

43. 0, 1

44. 2

45. $x = \dfrac{9y^2 - 1}{4}$

46. $x = \dfrac{9y^2 - 6y + 1}{4}$

47. $\pm 2,\ \pm 2i$

48. 16

49. $x \leq -6$ or $x \geq 3$

50. $-\dfrac{2}{3} < x < 1$

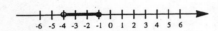

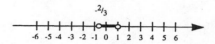

51. $-4 < x \leq -1$

52. $x < 0$ or $x > \dfrac{1}{3}$

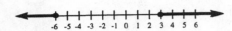

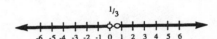

53. 3 or $\dfrac{2}{3}$

54. 3" x 8"

55. 5 cm., 12 cm., 13 cm.

56. $1\dfrac{1}{2}$"

57. distance: 5, midpoint: $\left(\dfrac{1}{2}, 0\right)$

58. distance: $2\sqrt{34}$, midpoint: (0,0)

59. distance: $4\sqrt{2}$, midpoint: (–2,3)

60. distance: $\sqrt{2}$, midpoint: $\left(\dfrac{1}{2}, \dfrac{13}{2}\right)$

61. center: (–1,3), radius: 4

62. center: (0,–2), radius: $2\sqrt{3}$

63. center: (4,–4), radius: 6

64. center: (–2,–3), radius: $2\sqrt{2}$

65. intercept: (0,0), vertex: (0,0), axis of symmetry: $x = 0$

66. intercepts: $(\pm 4,0)$, (0, –8), vertex: (0,–8), axis of symmetry: $x = 0$

67. intercepts: (0,0), (6,0), vertex: (3,–9), axis of symmetry: $x = 3$

68. intercept: (0,–5), vertex: (2,–1), axis of symmetry: $x = 2$

69. intercepts: $(1 \pm \sqrt{3}, 0)$, $(0,2)$, vertex: $(1,3)$, axis of symmetry: $x = 1$

70. intercepts: $\left(0, \frac{1}{2}\right)$, $(1,0)$, vertex: $\left(0, \frac{1}{2}\right)$, axis of symmetry: $y = \frac{1}{2}$

71. vertices: $(\pm 3, 0)$, $(0, \pm\sqrt{6})$

72. vertices: $(\pm 1, 0)$, $(0, \pm 5)$

73. vertices: $\left(\pm \frac{1}{3}, 0\right)$, $\left(0, \pm \frac{1}{2}\right)$

74. vertices: $(\pm 4, 0)$

75. vertices: $(0, \pm 6)$

76. vertices: $(\pm 1, 0)$

77. hyperbola; vertices: $(0, \pm 4)$, asymptotes: $y = \pm \frac{4}{9} x$

78. parabola; intercepts: $(1, 0)$, $(0, \pm 1)$, vertex: $(1,0)$, axis of symmetry: $y = 0$

79. straight line; slope: $\frac{2}{3}$, y-intercept: 4

80. ellipse; intercepts: $\left(\pm \frac{3}{2}, 0\right)$, $(0, \pm 1)$

81. circle; center: $(0,4)$, radius: 5

82. parabola; intercepts: $(-1,0)$, $(4,0)$, $(0,-8)$

vertex: $\left(\frac{3}{2}, -\frac{25}{2}\right)$, axis of symmetry: $x = \frac{3}{2}$

83. 15 yards; 225 square yards

Cumulative Test: Chapters 7–9

1. (a) $7st^3 \sqrt{2s}$ (b) $-24c^2d^2 \sqrt{3cd}$

   (c) $\dfrac{\sqrt[4]{2x^3}}{x}$

2. (a) 0   (b) $\dfrac{(x+y)\sqrt{xy}}{xy}$

   (c) $2\sqrt{35} - 12$ (d) $\dfrac{3\sqrt{3} - 3}{2}$

3. $\dfrac{10}{3}$ 4. $\dfrac{24}{25} + \dfrac{7}{25} i$

5. (a) $1, -\dfrac{3}{5}$ (b) $\pm \sqrt{10}$

(c) $\dfrac{5 \pm \sqrt{15}}{2}$

(d) $\dfrac{-1 \pm \sqrt{5}}{2}$

6. $s = \pm \sqrt{2\left(1 - t^2\right)}$

7. 1, 2

8. $-\dfrac{1}{3} \le x \le 3$

9. 2 mph

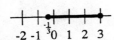

10. distance: 5, midpoint: $\left(-\dfrac{11}{2}, -2\right)$

11. center: $(-5,0)$, radius: 6

12. intercepts: $(-1,0)$, $(3,0)$, $(0,-9)$, vertex: $(1,-12)$

13. (a) hyperbola; vertices: $(0, \pm 3)$, asymptotes: $y = \pm \dfrac{3}{7}\, x$

(b) ellipse; vertices: $\left(\pm \dfrac{1}{2}, 0\right)$, $(0, \pm 2)$

14. 132 feet

Chapter 10—Test A

1. $x = 4$, $y = -1$, $z = -2$

2. (a) $x = 4$, $y = 3$

(b) $x = 1$, $y = -2$, $z = -1$

3. (a) 16

(b) -22

4. (a) $x = \dfrac{1}{2}$, $y = -\dfrac{1}{3}$

(b) $x = 3$, $y = 1$, $z = -2$

5. 6 nickels, 5 dimes, 2 quarters

6. $(7,4)$, $(2,-1)$

7.

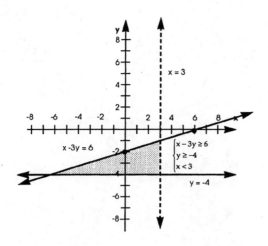

Chapter 10—Test B

1. $x = 1, y = -3, z = 3$

2. (a) $x = 1, y = -2$ (b) $x = -1, y = 3, z = 2$

3. (a) -14 (b) 19

4. (a) $x = \dfrac{2}{3}, y = -\dfrac{1}{2}$ (b) $x = -1, y = 3, z = 4$

5. 8 $10 bills, 4 $20 bills, 3 $50 bills   6. $(2,-3), (0,1)$

7.

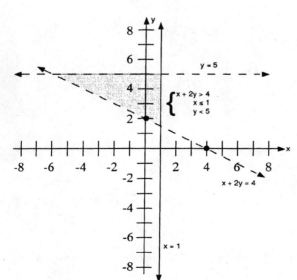

1.   (a)   domain: $\{1, 2, -2\}$

            range: $\{0, 4, -4\}$

   (b)   domain: $\{x | -3 \le x \le 3\}$

            range: $\{y | -2 \le y \le 2\}$

2.   (a)   domain: $\{x | x \le 2\}$

   (b)   domain: $\left\{x | x \ne \dfrac{1}{2}\right\}$

3.   (a) yes   (b)   no

   (c) no   (d)   yes

4.   (a) 4   (b)   1

   (c) $2x^2 - 3x + 1$ (d)   10

5.   the graph of
$$f(x) = |3 + 2x|$$

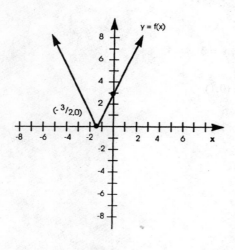

6. $\{(1, -4), (6, 3), (1, 7)\}$ 7.   no

8.   (a) $f^{-1}(x) = -2x + 6$ (b) $f^{-1}(x) = \dfrac{1}{x - 3}$

9.   108   10.   8

Chapter 11—Test B

1. (a) domain: $\{3,-3,1\}$

   range: $\{-1,1,3\}$

   (b) domain: $\{x|-3 \le x \le 3\}$

   range: $\{y|-9 \le y \le 9\}$

2. (a) domain: $\{x|x \ge 3\}$

   (b) domain: $\left\{x|x \ne -\dfrac{1}{2}\right\}$

3. (a) yes   (b) no

   (c) no   (d) yes

4. (a) 7   (b) 4

   (c) $3x^2 + 11x + 10$ (d) 24

5. the graph of

   $f(x) = |1 - 3x|$

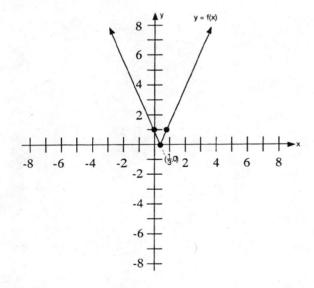

6. $\{(-2,5),(3,0),(-2,-6)\}$ 7. no

8. (a) $f^{-1}(x) = 3x - 6$ (b) $f^{-1}(x) = -\dfrac{1}{x-2}$

9. 80   10. 27

1.   the graph of

        $y = \left(\dfrac{2}{3}\right)^x$

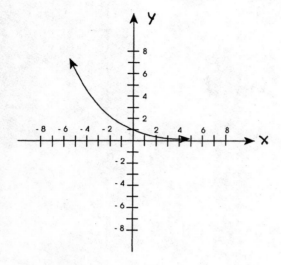

2.   the graph of

        $y = \log_4 x$

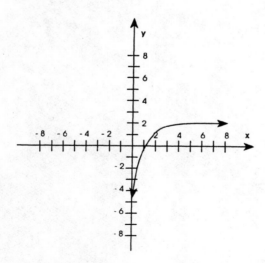

3.   (a) $4^{\frac{1}{2}} = 2$ (b) $3^{-3} = \dfrac{1}{27}$

4. (a) $\log_{64} 16 = \frac{2}{3}$ (b) $\log_5 0.008 = -3$

5. (a) -2 (b) $\frac{1}{2}$

   (c) $\frac{7}{4}$

6. (a) $3 \log_b x + \frac{1}{2} \log_b y - \log_b z$ (b) $\frac{3}{4} + \frac{3}{4} \log_3 x + \frac{5}{4} \log_3 y$

7. (a) 5.5843   (b) 0.00833

   (c) 0.2103   (d) 0.6250

8. (a) 2.854   (b) 0.028

9. (a) -4 (b) 1

   (c) $\frac{8}{5}$ (d) 10

   (e) $\frac{\log 6 - \log 4}{\log 4} \cong 0.2925$

10. $\cong \$11,641$ 11. $\cong 12.20$ years

Chapter 12—Test B

1. the graph of

$$y = \left(\frac{3}{2}\right)^x$$

2.   the graph of

$$y = \log_3 x$$

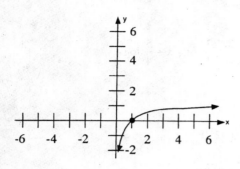

3.   (a) $27^{\frac{1}{3}} = 3$ (b) $2^{-4} = \dfrac{1}{16}$

4.   (a) $\log_{16} 8 = \dfrac{3}{4}$ (b) $\log_{20} 0.0025 = -2$

5.   (a) -3 (b) $\dfrac{1}{2}$

     (c) $\dfrac{8}{3}$

6.   (a) $2 \log_b x + \dfrac{1}{3} \log_b y - \log_b z$ (b) $\dfrac{2}{5} + \dfrac{2}{5} \log_4 x + \dfrac{4}{5} \log_4 y$

7.   (a)   2.9628   (b)   0.0546

     (c)   0.9034   (d)   0.3499

8.   (a)   2.777   (b) -1.226

9.   (a) -7 (b)   2

     (c) $\dfrac{7}{5}$ (d)   10

     (e) $\dfrac{\log 7 + \log 5}{\log 5} \cong 2.2090$

10. $\cong \$10,767$ 11. $\cong 11.56$ years

1. $x = 1, y = -2, z = 2$

2. $x = 0, y = 3, z = 2$

3. no solution

4. $x = -6, y = 6, z = 6$

5. infinitely many solutions, each with $2x - 3y - 6z = 9$

6. $x = 3, y = -2, z = -1$

7. 30

8. 0

9. -68

10. 49

11. -4

12. 0

13. 85

14. 0

15. $x = 7, y = 2$

16. $x = 0, y = 4, z = -5$

17. $x = -3, y = 2$

18. $x = 4, y = -4$

19. $x = 1, y = -1, z = 1$

20. $x = 7, y = 6, z = 4$

21. $(1,2)$

22. $(1,3), (3,-1)$

23. $(2,0)$

24. $(1,0), (-1,6)$

25. $(3,1), (3,-1), (-3,1)$ and $(-3,-1)$

26. $(2,0), (-2,0)$

27. $(-1,0), (3,2)$

28. $(6, 2\sqrt{2}),(6, -2\sqrt{2}), (-2,0)$

29.

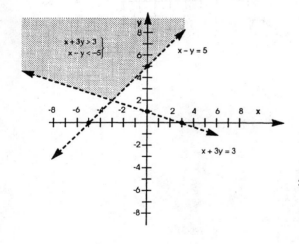

30.

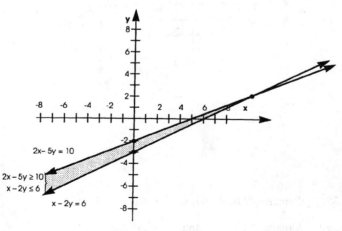

31.

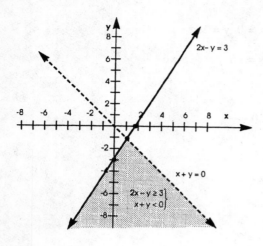

32.

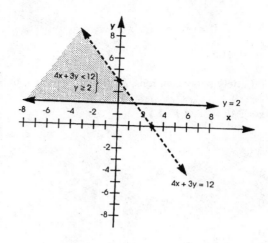

33.

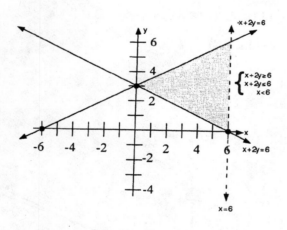

34.

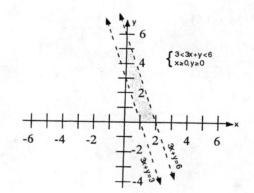

35.   domain: {1,6,–1,–6}, range: {6,1,–6,–1}; yes

36.   domain: {2,–1,4}, range: {–5,5}; yes

37.   domain: {3,2}, range: {–8,–1}; no

38.   domain: {7,8,–7,–8}, range: {8,9}; yes

39. ellipse; vertices: $(\pm 1, 0)$, $(0, \pm 3)$

domain: $\{x \mid -1 \le x \le 1\}$, range: $\{y \mid -3 \le y \le 3\}$

40. hyperbola; vertices: $(\pm 1, 0)$, asymptotes: $y = \pm 3x$

domain: $\{x \mid x \ge 1 \text{ or } x \le -1\}$, range: $\mathbf{R}$

41. straight line; slope: -9, y-intercept: 9, domain: $\mathbf{R}$, range: $\mathbf{R}$

42. parabola; intercepts: $(1,0)$, $(-1,0)$, $(0,-9)$, vertex: $(0,-9)$

axis of symmetry: $x = 0$, domain: $\mathbf{R}$, range: $\{y \mid y \ge -9\}$

43. $\mathbf{R}$

44. $\{x \mid x \ge -2\}$

45. $\{x \mid x \neq 9, x \neq -1\}$

46. $\{x \mid x < 2\}$

47. yes

48. no

49. yes

50. no

51. -3

52. $\dfrac{1}{2}$

53. $4 - 12x + 9x^2$

54. $2 - 3x^2$

55. $b^2 - 3$

56. $b^2 - 4b - 2$

57. $4b^2 - 8b + 1$

58. $2b^2 - 8b + 2$

59. 4

60. 8

61. 6

62. -16

63. 2

64. -1

65. $\dfrac{x^2 + 4x}{4}$

66. -3

67. $y = \sqrt[3]{\dfrac{x - 1}{2}}$

68. $y = \dfrac{5 - 2x}{3}$

69. $y = \dfrac{2 - 3x}{x - 1}$

70. no inverse

71.   the graph of

$$f(x) = \sqrt{1 + x}$$

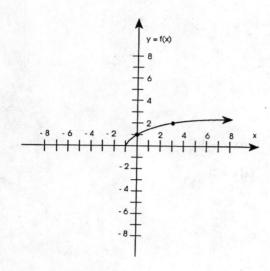

72.   the graph of

$$f(x) = 1 + \sqrt{x}$$

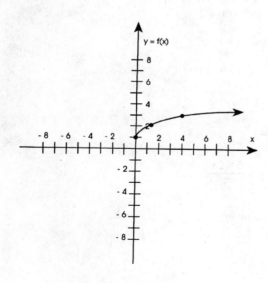

73.   the graph of

$$f(x) = x^2 + 8x + 12$$

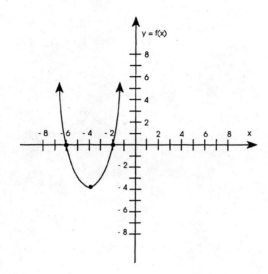

74.   the graph of

$$f(x) = |2x - 1|$$

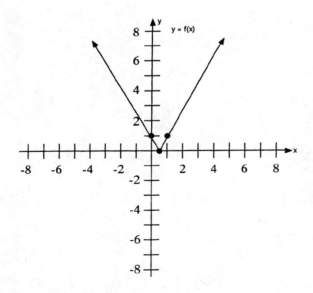

521

75. the graph of

$$f(x) = 2^{2x-1}$$

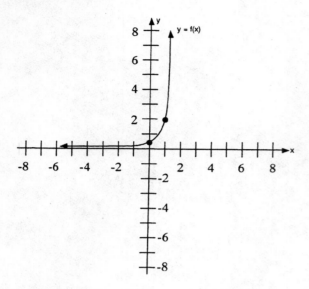

76. the graph of

$$f(x) = \left(\frac{2}{3}\right)^x + 2$$

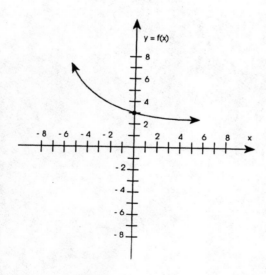

77. the graph of

   $f(x) = \log_3 (x - 2)$

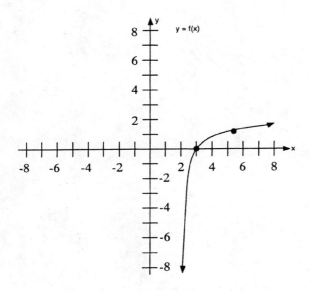

78. the graph of

   $f(x) = \log_3 x - 2$

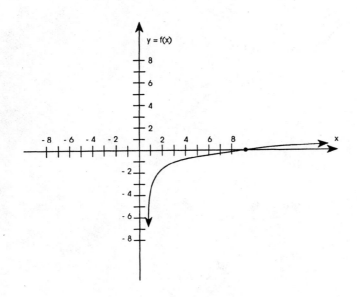

523

79. $3^4 = 81$

80. $\log_5 \frac{1}{25} = -2$

81. $\log_{32} 4 = \frac{2}{5}$

82. $10^{-4} = 0.0001$

83. $49^{\frac{3}{2}} = 343$

84. $\log_8 \frac{1}{4} = -\frac{2}{3}$

85. 3

86. $\frac{1}{3}$

87. -3

88. $\frac{7}{3}$

89. -1

90. $\frac{1}{4}$

91. 1

92. 0

93. $\frac{1}{4} \log_b 7 + \frac{1}{2} \log_b x + \frac{3}{4} \log_b y$

94. $1 + 3 \log_b x + \frac{1}{2} \log_b y - 2 \log_b z$

95. $\frac{1}{2} + \log_8 x + 5 \log_8 y - \frac{3}{2} \log_8 z$

96. $\frac{2 \log_3 a}{3 \log_2 a}$

97. 5.9284

98. 5.8348

99. 1.09

100. 2.1815

101. 3.628

102. -0.287

103. 27

104. 2

105. $-\frac{3}{2}$

106. $-\frac{2}{3}$

107. $\frac{1}{3}$

108. 9

109. $-\frac{3}{2}, 2$

110. 14

111. $\frac{5 \log 3 - 3 \log 5}{\log 5 - \log 3} \cong 1.30$

112. 7

113. 9

114. $\dfrac{16}{9}$

115. 12

116. 1

117. $\cong 3.34$ years

118. $\cong 57.56$ minutes

119. $\cong 119.45$ days

120. 4.5513

121. 3.16×10^{-10}

Cumulative Test: Chapters 10–12

1. $x = 3, y = -3, z = 0$

2. $x = 3, y = -1, z = -4$

3. (a) $x = 5, y = -3$

(b) $x = -2, y = 8, z = 1$

4. $(-2,1), (2,-1)$

5.

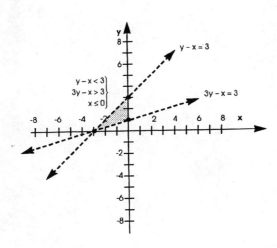

6. Yes, since no two distinct ordered pairs have the same first entry.

domain: $\{7,4,1,-3\}$, range: $\{-1,1,0\}$

7. (a) ellipse; vertices: $(\pm4, 0), (0, \pm2)$

domain: $\{x \mid -4 \le x \le 4\}$, range: $\{y \mid -2 \le y \le 2\}$; not a function

(b) parabola; intercepts: $(0,0), (6,0)$

vertex: $(3,9)$, axis of symmetry: $x = 3$

domain: $\mathbf{R}$, range: $\{y \mid y \le 9\}$; function

8. (a) $\left\{x \mid x \neq \dfrac{3}{2} \text{ and } x \neq -\dfrac{1}{2}\right\}$

   (b) $\left\{x \mid x \geq -\dfrac{1}{4}\right\}$

9. (a) 4   (b) $x^6 - x^3 - 2$

   (c) $9x - 8$ (d) $x^2 - 3x$

   (e) $x^2 - x$ (f) 10

   (g) 16   (h) -2

   (i) -8 (j) $2x + t - 1$

10. $f^{-1}(x) = \dfrac{x+1}{x-1}$

11. (a) the graph of

   $y = \left(\dfrac{5}{4}\right)^x$

(b)   the graph of

$$y = |4 - 2x|$$

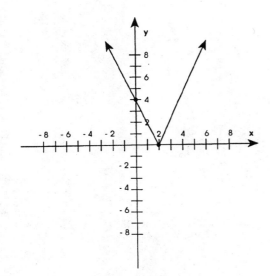

12. $32^{-\frac{4}{5}} = \frac{1}{16}$

13.   (a) −3   (b) $\frac{2}{3}$

14.   (a) $1 + 2 \log_b x + \frac{1}{2} \log_b y$ (b) $\frac{2}{3} + \frac{1}{3} \log_5 x - \frac{4}{3} \log_5 y$

15.   (a) 6.6042   (b) 0.0776

16.   8.480

17.   (a) $1, \frac{3}{2}$ (b) 6

18. $\frac{1}{8}$ 19. $\cong 8.25\%$

20. $\cong 4.1$

Practice Final Exam

1.   −2   2. $5x^2 + 5xy - 5y^2$

3. $\dfrac{-4a(4a + b)}{2a + b}$ 4. $\dfrac{10x}{(x - 5)(x + 5)}$

5. $\dfrac{y - 2x}{xy}$

6. $\dfrac{x}{y^{\frac{3}{2}}}$

7. $x^2y^3\sqrt{3xy}$

8. 20

9. -1

10. $x \geq 2$ or $x \leq -\dfrac{3}{2}$

11. $0, 3, 8$

12. $-4 < x < \dfrac{2}{3}$

13. $\dfrac{5 \pm 2\sqrt{5}}{5}$

14. $\dfrac{3}{2}, -1$

15. 2

16. $3, -\dfrac{7}{3}$

17. 1

18. $\dfrac{\log 2 + 2 \log 3}{\log 3 - 2 \log 2} \cong -10.05$

19. quotient: $x^2 + 4x + 1$, remainder: 12

20. $-5 - 2i$

21. 0.64

22. the graph of

$5x - 4y = 20$

23. $\sqrt{41}$

24. $\left(2, -\dfrac{5}{2}\right)$

25. $y = \dfrac{5}{4} x - \dfrac{27}{4}$

26. no

27. circle; center: $(10,-9)$, radius: 12

28. ellipse; vertices: $(\pm 2, 0)$, $\left(0, \pm \dfrac{8}{7}\right)$

29. hyperbola; vertices: $(\pm 2, 0)$, asymptotes: $y = \pm \dfrac{4}{7} x$

30. parabola; intercepts: $(1 \pm \sqrt{5}, 0)$, $(0,4)$, vertex: $(1,5)$

     axis of symmetry: $x = 1$

31. the graph of

     $y = |x + 3| - 2$

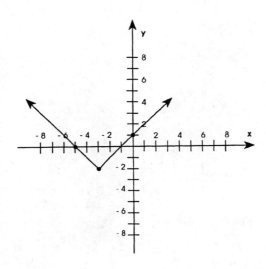

32. the graph of

     $y = \left(\dfrac{5}{2}\right)^{x}$

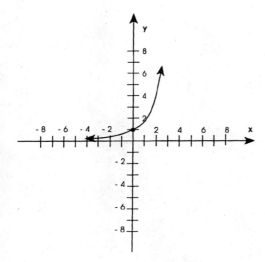

33. $x = 3, y = -2, z = -3$ 34. $(0, -1), \left(\dfrac{\sqrt{7}}{2}, \dfrac{3}{4}\right), \left(-\dfrac{\sqrt{7}}{2}, \dfrac{3}{4}\right)$

35.

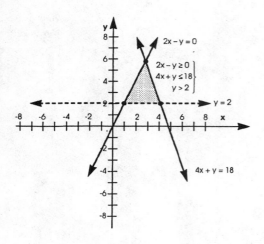

36. $\left\{x \mid x \le \dfrac{1}{2}\right\}$ 37. $\sqrt{3 - 2x}$

38. 0 39. $f^{-1}(x) = \dfrac{1 - x^2}{2}$

40. $3, 5, 7$ 41. 6 hours

42. 4 mph   43. $-1, -2$

44. 18 dimes, 8 quarters   45. $\cong \$4,345$